# Decision Support Systems and Knowledge Management for Sustainable Engineering

# Decision Support Systems and Knowledge Management for Sustainable Engineering

Editors

**Athanasios P. Vavatsikos**
**Dimitrios E. Koulouriotis**
**Ioannis Nikolaou**
**Konstantinos P. Tsagarakis**


MDPI • Basel • Beijing • Wuhan • Barcelona • Belgrade • Manchester • Tokyo • Cluj • Tianjin

*Editors*
Athanasios P. Vavatsikos    Dimitrios E. Koulouriotis    Ioannis Nikolaou
Democritus University of Thrace Democritus University of Thrace Democritus University of Thrace
Greece    Greece    Greece

Konstantinos P. Tsagarakis
Democritus University of Thrace
Greec

*Editorial Office*
MDPI
St. Alban-Anlage 66
4052 Basel, Switzerland

This is a reprint of articles from the Special Issue published online in the open access journal *Sustainability* (ISSN 2071-1050) (available at: https://www.mdpi.com/journal/sustainability/special_issues/decision_support_systems_knowledge_management_sustainable_engineering).

For citation purposes, cite each article independently as indicated on the article page online and as indicated below:

LastName, A.A.; LastName, B.B.; LastName, C.C. Article Title. *Journal Name* **Year**, *Article Number*, Page Range.

**ISBN 978-3-03943-096-3 (Hbk)**
**ISBN 978-3-03943-097-0 (PDF)**

# Contents

# About the Editors

**Athanasios P. Vavatsikos** is an Assistant Professor of Spatial Decision Support Systems in the Department of Production & Management Engineering at the Democritus University of Thrace. He has a degree from the department of Civil Engineering of the Democritus University of Thrace and a Ph.D. in Spatial Decision Support Systems from the School of Engineering from the Democritus University of Thrace, GR. His research interests include: geographical information systems; decision making theory and models; spatial multicriteria and suitability analysis; technical–economic project evaluation; environmental and energy economics; quantitative methods. His research work has been published in more than 20 papers in refereed journals and proceedings of scientific conferences.

**Dimitrios E. Koulouriotis** is a professor of Intelligent Systems in Industrial Production and Management in the Department of Production and Management Engineering at the Democritus University of Thrace. He has a degree from the Department of Electrical and Computer Engineering of the Democritus University of Thrace, an MSc degree from the Department of Electronics and Computer Engineering and a Ph.D. in Intelligent Systems from the Dep. of Production Engineering and Management, both from the Technical University of Crete, GR. His research interests include: intelligent systems, computational intelligence and cognitive engineering, data and knowledge engineering, decision support systems, production and logistics engineering and management, engineering and project management, safety and reliability, TQM and BE, innovation and technology management. His research work has been published in more than 80 papers in major journals.

**Ioannis Nikolaou** is an associate professor of Corporate Environmental Management and Environmental Management Systems" in the Department of Environmental Engineering at the Democritus University of Thrace. His research interests include: corporate environmental management, corporate sustainability, corporate social responsibility, business circular economy models and environmental economics. His research work has been published in more than 60 papers in peer-reviewed journals. He has served as a Guest Editor for many journals.

**Konstantinos P. Tsagarakis** is a professor of Economics of Environmental Science and Technology at the Department of Environmental Engineering at the Democritus University of Thrace. He has a degree from the department of Civil Engineering of the Democritus University of Thrace, a BA degree from the Department of Economics of the University of Crete and a Ph.D. in Public Health from the School of Civil Engineering at the University of Leeds, UK. His research interests include: technical–economic project evaluation; environmental and energy economics; public health economics; environmental and energy behavior; big data; online behavior; environmental performance of firms; quantitative methods. His research work has been published in more than 80 papers in refereed journals. He is an Associate Editor of Water Policy Journal and has served as a Guest Editor for several others.

*Article*

# Risk-Informed Performance Assessment of Construction Projects

**Merkourios Papanikolaou and Yiannis Xenidis ***

Department of Civil Engineering, Aristotle University of Thessaloniki, 541 24 Thessaloniki, Greece;
mpapanik@civil.auth.gr
* Correspondence: ioxen@civil.auth.gr

Received: 25 May 2020; Accepted: 27 June 2020; Published: 1 July 2020

**Abstract:** Construction projects are struggling to reach satisfactory performance in a context full of various and many risks. Despite the long-lasting study of performance in construction no universal approaches have emerged from numerous proposals. A mainstream in dealing with this issue is benchmarking methods with the use of appropriate indicators. This research work adopts such a group of indicators that address more aspects than the usual cost, schedule, and quality considerations. Following suggestions for proper adjustments and clarifications for the appropriate use of this group of indicators, this research proceeds by introducing the risk aspect to the calculation of project performance. Identified risks are modeled through fuzzy numbers to deal with data incompleteness and then they are appropriately added to compose risk coefficients for the performance dimensions expressed through the respective indicators. Then, the overall project's performance is calculated addressing a risk-informed performance assessment of construction projects. The whole process is tested for verification with a theoretical example. The development of a simple yet inclusive and adjustable methodological framework for risk-informed performance assessment is expected to provide practitioners with a relatively easy approach for integrating both the risk and performance management in construction.

**Keywords:** risk management; risk assessment; performance assessment; performance management; performance evaluation; earned value analysis; construction projects; fuzzy numbers

---

## 1. Introduction

A long-standing ascertainment in the construction industry is that a large number of projects does not achieve its objectives and goals, while only 40% of them are carried out in alignment with the objectives and goals set by the management [1]. Complexity in delivering, in combination with the large number of stakeholders that pursue their own—often competing—interests [2], along with the pressures of a highly competitive economic environment constitute a framework where efficient construction project management becomes essential for success.

Numerous academic and professional research efforts have focused on the key factors for successful delivery of construction projects [2–15]; however, it is evident that this is a hard task as it requires the integration of various stakeholders' perspectives [16] that, furthermore, are volatile in terms of their significance in different projects and different environments. Therefore, the efforts to compile lists of commonly accepted key success factors or criteria or indicators, etc., are continuous without achieving, though, the creation of a robust, shared platform for appraising construction projects' success. Table 1 supports this argument by summarizing the identified key success factors for construction projects based on a review of 14 recent studies. As it is shown there, researchers clearly agree on the consideration of time, cost, and quality as key success factors for construction projects; however, the recorded degree of agreement concerning the rest of the identified factors is considerably less.

The findings of this review are consistent with similar findings of other research works that have been even more inclusive in terms of the reviewed literature [2].

**Table 1.** Literature findings on key success factors for construction projects.

| Factors | Studies on Key Success Factors | | | | | | | | | | | | |
|---|---|---|---|---|---|---|---|---|---|---|---|---|---|
| | [3] | [4] | [5] | [6] | [7] | [8] | [9] | [10] | [11] | [12] | [13] | [14] | [15] |
| Time | X | X | X | X | X | X | X | X | X | X | X | X | X |
| Cost | X | X | X | X | X | X | X | X | X | X | X | X | X |
| Quality | X | X | X | X | X | X | X | | | X | X | X | X | X |
| Safety | | X | | | | X | | X | X | X | | | X |
| Client Satisfaction | | X | | | | X | | | | | | | X |
| Team Satisfaction | | | | | | | | | | | | | X |
| Profitability | | | | | | | | | | X | X | | X |
| Risk | | | | | | | | | | X | | | |
| Environmental Impact | | | X | | | | | | | | | | |
| Effectiveness | | | | | | X | | X | | X | | | |
| Billing | | | | | | | | | | | | | X |
| Society Impact | | | X | | | | | | | | | | |
| Staffing | | | | | | | | X | | | | | |
| Procurement | | | | | | | | X | | | | | |

A probing observation of the listed factors in Table 1 leads to the conclusion that all of them with the exception of one are actually constituents of the performance notion, as the latter is perceived in the modern holistic context of performance management [17] that extends beyond the classical approaches [18,19]. The factor that, although relative, is not a constituent of performance is risk.

The notion of risk, which is excessively studied in numerous cases, is generally considered as an effect of the uncertainty and, consequently, as a measure of deviation from an anticipated outcome [20–22]. In this work, it is defined as the outcome of the uncertainty about several aspects of a problem in hand combined with the anticipated consequence of a hazard's occurrence. This approach is generally accepted among those that have been proposed in the long period of the field's development [23].

Managing performance and risks are, so far, two distinguished fields in the context of project management in terms both of content and orientation of the applied processes. Traditionally, risk management in the construction industry focuses rather on alleviating the adverse effect of potential risks, than exploiting the opportunities that may be associated with them [17,24,25]. This attitude ensures a safe operational environment that supports the existence of a minimum level of performance; however, it does not promote value creation through performance optimization [17]. On the other hand, performance management is largely perceived in practice as a sterile monitoring of the achievement level of objectives [10] through the application of measuring systems that quantify, ex post, the efficiency on achieving predetermined goals [17,26]. This practice does not support well-informed predictions for future performance as previous experience and knowledge is solely dependent on past conditions. At the same time, project development mechanisms and their inherent complexities remain largely unexplored, while the collected data during a project's development are not adequately processed or even assessed for their accuracy and reliability [17,18]. These observations clearly indicate the potential of a full integration between performance and risk management not only aiming at aligning processes' carrying out and objectives, but also at exploiting their complementarities to achieve a dynamic performance management approach that shall utilize risk management results for increasing performance and not only securing against performance reduction.

This paper aims at contributing to this course of research by building on previous works to propose a methodology for a risk-informed performance assessment of construction projects. The significance of this work lies mainly in the introduction of a structured mathematical approach for risk-informed performance assessment, which allows the revision of construction projects' performance goals based on the success of managing risks in these projects. This integration is considered to be important as it systematizes project delivery processes, which although they run concurrently during a project's development, they are not integrated to allow full exploitation in setting realistic performance goals.

## 2. Risk and Performance Management Integration Approaches

The integration of risk and performance management frameworks has been proposed in a rather limited number of previous research efforts. Starting from tracing the potential for such an integration, a comparative analysis of strengths and weaknesses between the two frameworks has clearly shown the parallels among them [17]. Furthermore, it was shown that the two frameworks are complementary in serving the main purpose, i.e., reaching performance goals, yet starting from two different start lines. From a performance management point of view, future performance is a pre-determined, standard benchmark, which needs to be fulfilled based on the assumptions and conditions considered at the design phase and their amendments due to real conditions occurring during the materialization phase. This is a rather deterministic approach where performance is pursued up to a certain level. From a risk management point of view, future performance is again a pre-determined, standard benchmark, which, this time, needs to be safeguarded from known or unknown fluctuations and disturbances to design requirements that occur during the project's materialization phase. Therefore, risk management is about ensuring the environment where performance management deploys [17] and when that is achieved, performance is increased [22].

While the potential is identified, the methodological integration between the two frameworks has not significantly progressed. The attempts made so far can be broadly classified in three groups:

- The contextual, which mainly propose a framing of risk and performance in the context of broader frameworks such as strategic or enterprise risk management (ERM). For example, in [18], performance management is identified as an autonomous step for strategy monitoring, while risk management is identified as a component in the strategy realization step. In this case, there is rather a co-operation than an integration of the two frameworks. Other efforts have concluded that ERM and performance are positively related, nevertheless they have not investigated the mechanisms or even the existence of a direct relation between the two frameworks [27–31].

- The controlling, which mainly focus on project control, in the light of uncertainty [32–34]. In these cases, deviations of performance expressed as time and cost are investigated through Monte Carlo simulations addressing this way the modeling of uncertainties in performance assessment. These attempts aim at detecting the potential of performance deviations through a stochastic analysis and support project control rather than utilize risk management for setting performance goals.

- The performative, which are actually combining risk and performance management to set new performance goals during a project's development. In this group of attempts, earned value analysis (EVA) plays a central role by constituting the methodological background for performance measurement that is further enhanced with risk management practices. For example, the introduction of control charts [35] validates the idea of stochastically predicting deviations due to potential risks, but also addresses the idea of moving from benchmark values to statistically determined ones for setting performance goals. This idea is further developed in other research efforts where setting performance goals is more and more enriched with risk management tools and techniques. Based on a combination of EVA, statistical analysis, and Monte Carlo simulations, tolerance limits are set for project's performance control [36], or performance goals are directly estimated with a level of confidence [37]. Even in cases when the required total work is either completely unknown or uncertain, a proper modification of EVA to address fuzzy numbers is proposed for estimating both past and future performance values [14].

The methodological integration that is proposed in this paper falls into the third group and it expands on previous related research that combines EVA and benchmarking. The analytical presentation of this integration is provided in Section 3.

### 3. Proposed Methodological Integration of Risk and Performance Assessment

The proposed methodological integration expands the methodological framework that was initially presented by Nassar in 2005 [15] and has been subjected to enrichments by other researchers. The selection of this framework for proposing an integration between risk and performance assessment was made due to the analytical mathematical structure it presented, the combination between the benchmarking and the analytical calculation approaches, and its performative nature. Nassar's framework is also consistent with the requirements for structuring performance systems concerning clarity of the measured performance [38] and appropriate selection of performance indicators [39].

The proposed framework is presented in two parts. The first part presents the methodological background that this research has used concerning performance assessment and the second part presents the way to integrate this framework with the risk management framework as proposed in this research.

*3.1. Methodological Background for Performance Assessment*

According to the adopted methodological background for performance assessment in this research, construction project performance is successful when the project team, the company, or organization carrying out the project and the client evaluate it so. The evaluation system used is common for these project stakeholders and comprises eight so-called performance dimensions [15], which are represented by respective indicators as shown in Sections 3.1.1–3.1.8

3.1.1. The Cost Dimension

The cost dimension is measured through the cost performance index (*CPI*) taken from EVA. The index measures the achieved efficiency in the project, in terms of value earned for a specific amount of actual costs. The known formula for calculating this index is presented in Equation (1):

$$CPI = EV/AC \tag{1}$$

where *EV* is the earned value, i.e., the obtained value as a percentage of the budgeted work and *AC* is the actual cost spent for this obtained value. Both measures reflect the overall progress until the period where the project's monitoring takes place.

3.1.2. The Time Dimension

The time dimension is measured through the schedule performance index (*SPI*) taken, again, from EVA. The index measures the achieved efficiency in the project, in terms of value earned against the anticipated progress based on the design. The known formula for calculating this index is presented in Equation (2):

$$SPI = EV/PV \tag{2}$$

where *PV* is the planned value, i.e., the budgeted work for a specific period of time based on the project's design. Again, both measures reflect the overall progress until the period where the project's monitoring takes place.

3.1.3. The Billing Dimension

The billing dimension is measured through the billing performance index (*BPI*), which measures the billed against the earned revenue at the time of the project's monitoring. The idea behind this index is related to the efficiency of cash flows for a project under development. While construction projects managers recognize the necessity of controlling costs and schedules, they pay significantly less attention in securing cash flows for the whole duration of the project to avoid periods of shortage of funds that in turn may have adverse impacts for the project's development. Efficient cash flows are those with a positive sign (or marginally neutral) for the project and, moreover, those that are timely processed; therefore, *BPI* is calculated with the formula presented in Equation (3):

$$BPI = BRWP/ERWP \tag{3}$$

where *BRWP* is the billed revenue and *ERWP* is the earned revenue for the delivered worked until the period where the project's monitoring takes place.

### 3.1.4. The Safety Dimension

The safety dimension is measured through the safety performance index (*SFI*), which measures the performance in retaining a safe environment during the development of the construction project. This dimension is very significant as it introduces the safety factor, which has multiple impacts on the project's successful delivery ranging from productivity and cost to reputation and competitiveness issues. The formula for calculating this index is presented in Equation (4):

$$SFI = LTI \times C/M \tag{4}$$

where *LTI* measures the lost time due to safety incidents, *C* is a constant that represents 100 employees with full time annual occupation, and *M* measures the total man-hours required for the project until the period where the project's monitoring takes place.

### 3.1.5. The Profitability Dimension

The profitability dimension is measured through the profitability performance index (*PPI*), which measures the earned revenue against the actual costs incurred until the time of the project's monitoring. The index evaluates the actual profits earned only after subtracting from the revenue all direct and indirect costs and the calculation formula is presented in Equation (5):

$$PPI = ERWP/AC \tag{5}$$

where *ERWP* and *AC* are measures that have been defined in Equations (1) and (3), respectively.

### 3.1.6. The Quality Dimension

The quality dimension is measured through the quality performance index (*QPI*), which addresses the conformance of the realized work with the quality standards and specifications set both by the valid regulatory framework wherein the project is delivered and the customer's requirements. Given that non-compliance requires reworks to reach the desired standards, the index measures the quality of the worked performed by calculating the cost of reworks required as a percentage of the actual costs incurred for the project's construction. The calculation formula is presented in Equation (6):

$$QPI = \text{Total Direct and Indirect Costs of Reworks/Total Field Costs} \tag{6}$$

### 3.1.7. The Team Satisfaction Dimension

The team satisfaction dimension is measured through the team satisfaction index (*TSI*), which addresses the impact of human factor to the project's performance. The effect of the human factor in almost every other aspect of project performance is evident, especially for the construction industry where there is still a large portion of works that are not automated. It could be argued that the impact of the human factor is measured indirectly through the Equations (1)–(6), since the indicators in these equations are affected from the human staff in a construction project. This argument, although reasonable, does not eliminate the need for an index like *TSI*, since the indirect measurement of the human factor's impact is both vague (there is no particular indication as of the exact impact of the human factor in a calculated index value) and partial (not all aspects of the human effect on performance are calculated). *TSI* as defined in [15] addresses 12 specific parameters of team satisfaction with an impact to the project's performance, namely:

1. Involvement with the project;
2. Client/Suppliers response to team's needs;
3. Project Manager response to team's needs;
4. Adequate equipment;
5. Adequate training;
6. Appropriate financial compensation;
7. Clarity of responsibilities;
8. Quality of supervision;
9. Interest in the nature of work;
10. Cooperative environment;
11. Conformity with internal procedures during work;
12. Access to project baselines and progress reports.

For each one of these parameters, the project management team sets the priority and level of satisfaction, and the overall team satisfaction is calculated through the formula presented in Equation (7):

$$TSI = \sum_{1}^{12} Wi \times Ri \tag{7}$$

where $W_i$ is the relative weight of priority with $\sum_{1}^{12} Wi = 1$, and $R_i$ is the level of satisfaction in the range [1, 10] with 10 signifying the maximum level of satisfaction.

3.1.8. The Client Dimension

The client dimension is measured through the client satisfaction index (*CSI*), which addresses the impact of the correspondence to a client's demands as a factor of project's performance. Satisfying a client's demands can be a hard task with multiple impacts. It requires a thorough identification and understanding of the client's expectations and even the mechanisms that lead to their change, in order to: (a) Retain a proper balance between reasonable and exaggerated demands, (b) minimize change requests, and (c) manage expectations successfully to achieve smooth cooperation with the client and good reputation as a supplier, which is a significant business asset in competition. As in the case of *TSI*, *CSI* as defined in [15] addresses 12 specific parameters of client satisfaction with an impact to the project's performance, namely:

1. Understanding of the project requirements;
2. Understanding of client system and procedures;
3. Response to client requests/needs;
4. Flexibility and adjustment to change;
5. Overall capability of contractor's project team;
6. Effective communication;
7. Innovation in problem solving;
8. Performance with respect to cost;
9. Performance with respect to schedule;
10. Performance with respect to service quality;
11. Performance with respect to product quality;
12. Performance with respect to safety procedures.

For each one of these parameters, the project management team sets the priority and level of satisfaction and the overall client satisfaction is calculated through the formula presented in Equation (8):

$$CSI = \sum_{1}^{12} Wi \times Ri \tag{8}$$

where $W_i$ is the relative weight of priority with $\sum_{1}^{12} Wi = 1$, and $R_i$ is the level of achievement in the range [1, 10] with 10 signifying the maximum level of satisfaction.

### 3.1.9. Calculation of the Overall Project Performance

The indicators of the performance system are calculated based on measured data or evaluations according to the nature of each index. The different nature of the indicators results in different ranges of values for them; however, in all cases, it is easy to define the same performance classes for the indices. Table 2 presents the classes and the corresponding values ranges according to the suggestions in [15].

**Table 2.** Performance classes for system's indicators.

| Class | Condition | Index Range |
| --- | --- | --- |
| A | Outstanding Performance | $I > 1.15$ |
| B | Exceeding Target | $1.05 < I \leq 1.15$ |
| C | Within Target | $0.95 < I \leq 1.05$ |
| D | Below Target | $0.85 < I \leq 0.95$ |
| E | Poor Performance | $I \leq 0.85$ |

The different nature of the indicators can also justify a differentiation of their contribution to the overall project's performance. Weighting of the indices can be done through expert judgments or with the application of more systematic methods and techniques such as the analytical hierarchy process (*AHP*). Finally, Nassar [15] has suggested Equation (9) for the calculation of the total project's performance:

$$PI = w_1 \times CPI + w_2 \times SPI + w_3 \times BPI + w_4 \times PPI + w_5 \times SFI + w_6 \times QPI + w_7 \times TSI + w_8 \times CSI \tag{9}$$

where *PI* is the project performance index, and $w_i$ is the relative weight of significance of each performance indicator with $\sum_{1}^{8} wi = 1$.

### 3.2. Improvements of the Performance Assessment Methodological Background

The methodological background for performance assessment was further improved in this research, both in content and processes. In particular, two indicators, *BPI* and *QPI*, were redefined and a particular process of normalization for the calculation of *PI* was proposed. These novel interventions are described in detail in Sections 3.2.1–3.2.3.

### 3.2.1. The Modified Billing Performance Index

*BPI* as defined in [15] focuses on the billing of the performed work, which is important as it shows the readiness of the contractor to claim rewards; however, efficient cash flows are not secured only by proper billing, but also and more importantly by timely payment of the rewards. Although, in terms of logistics, cash flows can be recorded once billing is performed, payments are essential to retain liquidity that is of utmost importance in production, especially for construction projects that present volatilities in expenses. Another aspect is that payments are subjected to factors external to the project, and they are controlled by the client or the funder and not by the project's manager. Given that no payments are feasible without the respective billing of the work performed, the maximum value of *BPI* could be equal to 1, thus signifying full payment for the billed work. *BPI's* adjustment to reflect the above and be consistent with the structural approach of other indices such as *CPI* and *SPI*, is achieved through a new definition of the index, which is presented in Equation (10):

$$BPI = ERWP/BRWP \tag{10}$$

Equation (10) is the reversed Equation (3) and the measures of the two equations have identical interpretations.

### 3.2.2. The Modified Quality Performance Index

*QPI* as defined in [15] considers quality deviations only with respect to the construction phase. Although reworks are mostly expected in that phase, quality deviations may be present in the design phase as well. Moreover, any costs of reworks, i.e., any failures in quality have an impact on the overall project's performance, therefore the baseline to evaluate them should be the project's actual (direct and indirect) costs, instead of the costs in the construction phase. The adjustment to Equation (6) in order to reflect the above, results to Equation (11):

$$QPI = CoR/(AC - CoR) \tag{11}$$

where *CoR* is the total cost of reworks and *AC* is the actual cost of the whole project until the period where the project's monitoring takes place.

### 3.2.3. The Normalization Process

The eight weighted indices that are added to calculate the project's performance according to Equation (9) are not taking values in the range presented in Table 2. Although it might be feasible to consider similar grading for the conditions of those indices that are measured as percentages (e.g., *CPI, SPI*, etc.), this is absolutely not the case for indices as *TSI* and *CSI*, which are measured in 1–10 scales. A normalization is then required for the final calculation, which while is suggested in [15], it is not described or shown there. In the context of this research, a standard normalization process is proposed as following:

1.  A two-dimensional system is created where the $X$ axis represents the ranges that correspond to each performance class as defined in Table 2 in the set of values for the specific indicator and the $Y$ axis represents the same ranges as defined in Table 2.
2.  The marginal values of the respective classes in the two linear grading systems (i.e., the index's rating system and the normalized system) are forming pairs, which are plotted in the $XY$ plane. Then, a graphical presentation is created by drawing the trend line of the set of paired values.
3.  The equation of the trend line is derived with a desired $R^2$ approximating the value of "1" as the best fitting equation is required to transpose from an indicator's rating system to the normalized respective one. For this reason, third degree polynomials are selected, although even linear equations had an $R^2 > 0.98$. It should be mentioned that polynomials of a greater degree could be inappropriate due to overfitting.
4.  The normalization equation can be used for transposing any value given to the indicator's rating system to the normalized rating system for the calculation of the project's performance index.

The described process is presented in Figure 1 for *PPI*, while Table 3 summarizes the normalization equations for all indices that according to [15] had rating systems different than the normalized rating system, which is shown in Table 2.

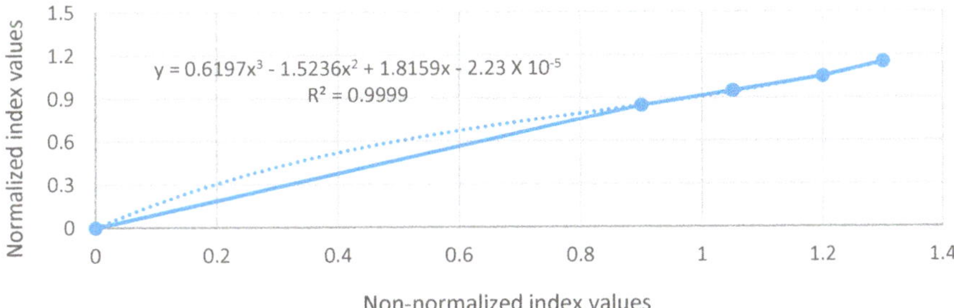

**Figure 1.** An example of the normalization process for profitability performance index (*PPI*).

**Table 3.** Normalization equations for the indices rating systems defined according to [15].

| Indicator | Equation | $R^2$ |
|---|---|---|
| BPI | $y = 4.9657x^3 - 7.8118x^2 + 4.0553x - 1.97 \times 10^{-6}$ | 0.998 |
| PPI | $y = 0.6197x^3 - 1.5236x^2 + 1.8159x - 2.23 \times 10^{-5}$ | 0.999 |
| SFI | $y = 3.3333x^3 - 1x^2 - 0.9333x + 1.15$ | 1.0 |
| QPI | $y = -0.0143x^3 + 0.1167x^2 - 0.35x + 1.2976$ | 1.0 |
| TSI | $y = 0.005x^3 - 0.0923x^2 + 0.6021x - 0.5148$ | 0.999 |
| CSI | $y = 0.005x^3 - 0.0923x^2 + 0.6021x - 0.5148$ | 0.999 |

The normalization equations in Table 3 are respective to the rating systems proposed in Nassar [15]. Nevertheless, the proposed normalization process is simple, accurate, and adjustable to any rating system that has to be defined by the project management team according to the project's conditions and the project's stakeholders' requirements.

### 3.3. Introduction of the Risk Factor in the Performance Assessment Methodology

The purpose of this paper as stated in the introduction is to integrate risk and performance assessment. For that reason, the risk aspect has to be introduced to the selected performance assessment methodology as improved after the modifications in the context of this research. This introduction can be in the form of a risk factor that will be composed according to the reasoning that follows.

A proper risk management entails risk identification, assessment, and response. Given that in the framework of this research, decisions are made based on the project's performance index, the final stage of risk management is excluded, therefore only risk identification and risk assessment are required.

Risk identification can be applied by the project management team through any of the acknowledged methods and techniques available in the vast related literature. The identified risks are noted as $R_{ij}$, where $i$ represents the affected performance index and therefore takes values in the set of [1, 8], while $j$ represents the code number of the risk and takes values in the set of [1, $n$], with $n$ being the total number of risks identified for the index $i$. With this notation, all identified risks are directly related to the performance indices.

Risk assessment follows then with the use of the most generic formula for quantifying risks, which is presented in Equation (12):

$$R = P \times C \tag{12}$$

where $P$ is the probability of the risk to occur and $C$ is the consequence upon occurrence of the risk.

The factors in Equation (12), ideally, should take real values estimated through the application of appropriate models and quantification techniques; however, performing such quantifications in the construction industry can be an extremely hard task due to the following two reasons: (a) The available data are very often incomplete, inaccurate, and generally inappropriate for systematic assessments, and (b) the approaches in practice are more qualitative and experience-based than mathematical and model-based, therefore leading to rough estimations, which are often expressed in linguistic terms (e.g., very high probability, low consequence, etc.). To adjust to this reality and appropriately model it for assessing construction projects risks, fuzzy theory provides an adequate framework that can be easily applied [40,41]. Probability and consequence in Equation (12) can be represented as fuzzy numbers taking values in a range of qualitative linguistic grades similar to those used in everyday practice. The mathematical equivalent of these ratings is selected to be trapezoidal numbers, which are broadly used in the respective literature for the easiness of their use [42]. Table 4 presents the proposed trapezoidal fuzzy numbers for each grade of probability and consequence in the context of this research. Scaling may differ based on the user's risk attitude and experience.

**Table 4.** Performance classes for system's indicators.

| Probability (P) Class | Fuzzy Number | Consequence (C) Class | Fuzzy Number |
|---|---|---|---|
| Absolute certain | (0.8, 0.9, 1.0, 1.0; 1.0) | Very high | (0.7, 0.8, 0.9, 1.0; 1.0) |
| Very frequent | (0.7, 0.8, 0.8, 0.9; 1.0) | High | (0.5, 0.6, 0.7, 0.8; 1.0) |
| Frequent | (0.5, 0.6, 0.7, 0.8; 1.0) | Moderate | (0.3, 0.4, 0.5, 0.6; 1.0) |
| Probable | (0.4, 0.5, 0.5, 0.6; 1.0) | Low | (0.1, 0.2, 0.3, 0.4; 1.0) |
| Occasional | (0.2, 0.3, 0.4, 0.5; 1.0) | Very low | (0, 0.1, 0.2, 0.3; 1.0) |
| Rare | (0.1, 0.2, 0.2, 0.3; 1.0) | | |
| Very rare | (0, 0, 0.1, 0.2; 1.0) | | |

Once the probabilities and consequences are associated with a specific class, the calculation of risk is a simple product of fuzzy numbers that equals another fuzzy number, which is defuzzified through the circumcenter of centroids rule used in [42,43], according to Equations (13) and (14):

$$S_A(X_0, Y_0) = (\frac{a + 2b + 2c + d}{6}, \frac{(2a + b - 3c) \times (2d + c - 3b) + 5w^2}{12w})$$
(13)

$$CRR_{ij} = \sqrt{X_0^2 + Y_0^2}$$
(14)

In Equation (13), *a-d* are the trapezoid's apices and $w$ is the trapezoid's height, while in Equation (14), $CRR_{ij}$ is the final crisp number produced after the application of the defuzzification process, which represents the risk rating of a certain risk $j$ with an impact on an indicator $i$. Once all $R_{ij} = CRR_{ij}$ are calculated, the risk-informed performance index is calculated with the following step-by-step process:

- Step 1: Calculation of the risk associated with each performance index. Several formulae have been proposed in the literature for calculating the total risk from individual risks [40,42,44,45]. To use an equation that addresses crisp numbers and considers the uneven contribution of individual risks to the overall risk, based on their significance as stated by their value, Equation (15) suggested in [44] is adopted in the context of this research:

$$CRR_i = \sqrt{\frac{\sum_1^N (CRR_{ij})^2}{N}}$$
(15)

where $N$ is the number of total risks for a single dimension $i$.

- Step 2: Calculation of the risk level of each performance index. The risk level (*RL*) of a performance index is actually the normalized value of the respective dimension's total risk against the sum of all total risks of all performance dimensions. It is calculated through Equation (16) and it allows the weighting of all performance indices in the range [0, 1]:

$$RL_i = \frac{CRR_i}{\sum_1^8 CRR_i}$$
(16)

- Step 3: Calculation of the risk factor of each performance index. The risk factor ($RF$) is the risk coefficient in the calculation of risk-informed performance. It is calculated for each performance index through Equation (17):

$$RF_i = 1 - RL_i \tag{17}$$

As it is evident from the equation, the more increased is the risk level of the performance dimension, the lower is the value of the risk factor.

- Step 4: Calculation of risk-informed project's performance index. The last step for calculating risk-informed performance is the introduction of the risk factor to Equation (9). The new form of the equation that calculates $PI$ is that of Equation (18):

$$RiPI = w_1 \times RF_1 \times CPI + w_2 \times RF_2 \times SPI + w_3 \times RF_3 \times BPI + w_4 \times RF_4 \times PPI + w_5 \times RF_5 \\ \times SFI + w_6 \times RF_6 \times QPI + w_7 \times RF_7 \times TSI + w_8 \times RF_8 \times CSI \tag{18}$$

As shown in this equation, the project's performance index has been replaced by the risk-informed performance index ($RiPI$), while all performance indices are weighted both in terms of their significance and the risk that is associated to them.

The introduction of the risk coefficient in the assessment of project performance is rationalizing the index's value as it addresses the adverse effect that risks have in maximizing performance. This is useful for decision-making as it indicates the potential reduction of the recorded performance at some given point in time. Such indication along with the respective values of the indices of the various performance dimensions may trigger specific interventions towards dimensional and total project's performance improvement.

*3.4. Extension of the Risk-informed Performance Assessment Methodology*

The introduction of the risk factor to the performance assessment methodology is a first significant step towards the integration of risk and performance dimensions in performance management; however, as addressed, it covers only the necessary calibration of performance assessments due to potential risk occurrences, thus retaining a partial approach regarding both the control capacity and the predictive power that such a methodology should present. For this reason, further extensions are required to the proposed risk-informed performance assessment methodology, for it to become more valuable towards performance control and prediction of future performance.

These extensions are inspired and following the example of previous work [32,33] and especially the introduction of the project risk baseline ($PRB$), which is defined as *"the evolution of the value of project remaining risk over time: the remaining variability of project cost/duration during the project life cycle"* [32]. Nevertheless, while the notion of $PRB$ is very useful, the facts that it is limited to include only the factors of project's cost and duration and, more important, that these factors are not merged but modeled as discrete performance control indicators leads to the need for developing a novel approach that aims at modeling $PRB$ as a comprehensive measure of all factors included in the performance assessment methodology, described in Sections 3.1–3.3. Towards this direction, two new indicators of planned performance ($PP$) and actual performance ($AP$) are defined as following:

- Planned performance ($PP$) is the value of performance that must be achieved at the construction project's completion according to the project's planning and design. The maximum value of $PP$ is taken through Equation (18) when no deviations from planning occur (i.e., $CPI = SPI = BPI = 1$) and is dependent on the values that the construction project manager sets for the factors affecting performance (i.e., indicators calculated from Equations (4), (5), (7), (8) and (11)).
- Actual performance ($AP$) is the value of performance that is actually achieved at the time when the project's control is taking place. At this point, all variables are taking the values recorded according to the project's progress and $AP$ is calculated through Equation (18) (i.e., $AP = RiPI$).

To link *PP* and *AP* with the construction project's completion phase, the two indicators are multiplied with the project's planned percentage of completion (*PPC*) and the project's percentage of completion (*PC*), respectively. The result is the planned performance for a planned percentage of the project's completion (*PP$_{PPC}$*) and the actual performance for the actual percentage of the project's completion (*AP$_{PC}$*). Once the planned and actual performance are linked with the actual project's progress, the performance risk baseline (*PRB*) is defined as the evolution of the value of the project's risks for achieving performance throughout the construction project's life cycle. This definition expands the one given in [32] by further integrating the notions of risk and performance, which is the aim of this research. *PRB* is assessed according to the project's control plan by using Equation (19):

$$PRB_t = PP - PP_{PPC} \tag{19}$$

where *PRB$_t$* is the construction project's performance risk baseline at a specific time *t*. Equation (19) reflects that project's risks for achieving performance are increased at the project's beginning, and they are reducing throughout the project's life cycle and towards project's completion.

The variation of *PRB* between two consecutive project controls is calculated from Equation (20):

$$w_{Pt} = PRB_{t-1} - PRB_t \tag{20}$$

At this point, the construction project's performance is already risk-informed and linked with the project's progress, while it can also be allotted in time. What is still required is to consider the potential tolerance of performance variations that ensure keeping performance under predefined thresholds through the construction project's lifecycle. This tolerance is the last factor that should be considered in project's performance control and estimations of risk-informed future performance and it is introduced to the methodology through the following process:

- Step 1: The project performance buffer (*PPB*) is defined according to Equation (21):

$$PPB = AP_{mean} - AP_p \tag{21}$$

where *AP$_{mean}$* is the project's estimated mean actual performance and *AP$_p$* is the estimated performance for a given level of confidence *p*. Both estimations are based on Monte Carlo simulation, where the simulated variables are the performance indicators that determine *RiPI* (*AP*), while the risk factors and weights in Equation (18) are constants determined by the analyst.

- Step 2: *PPB* is allotted in time according to the project's control plan, based on Equation (22):

$$PB_t = (w_{Pt} \times PPB/\sigma^2{}_{Pp}) \tag{22}$$

where $\sigma^2{}_{Pp}$ is the total performance variation, i.e., $\sigma^2{}_{Pp} = \Sigma w_{Pt}$. The accumulated performance buffer at any given time (*APB$_t$*) indicates the total use of the buffer until the period of the project's control and is calculated according to Equation (23):

$$APB_t = PB_t + APB_{t-1} \tag{23}$$

- Step 3: Project control entails the comparison of the accumulated performance buffer (*APB$_t$*) added to the actual performance at a given time with the planned performance at the same time. This comparison is mathematically expressed by Equation (24):

$$PCI_t = APB_t + AP_{PC} - PP_{PPC} \tag{24}$$

where *PCI$_t$* is the performance control index, which can be either: (a) Positive (*PCI$_t$* > 0), thus indicating an improved performance compared to the planned one or (b) negative (*PCI$_t$* < 0), thus indicating a reduced performance compared to the originally planned.

- Step 4: Having knowledge of the project's status regarding performance achievement, the methodology concludes with the estimation of performance at project's completion. The respective index ($EPAC_t$) is calculated according to Equation (25):

$$EPAC_t = (PP + PCI_t) \tag{25}$$

$EPAC_t$ is the estimated performance at project's completion and a greater value than $PP$ ($EPAC_t > PP$) indicates the existence of a quantified performance surplus that can be managed accordingly, while a lower value than $PP$ ($EPAC_t < PP$) indicates the existence of a quantified performance shortage that should be covered through appropriate interventions.

The full application of the extended risk-informed performance assessment methodology introduces construction project's performance control and construction project's performance estimation at completion, thus providing a full review of the project's performance during its life cycle.

### 3.5. Verification of Risk-Informed Performance Assessment

The verification of the methodological framework is attempted by applying it to a theoretical example that is presented in [32]. This particular example was selected intentionally as it allows comparisons between the two approaches (i.e., this research's and the research presented in [32]) and their results; however, the application of the proposed methodology requires more data that were selected arbitrarily, yet close to the original values to constitute all together a realistic theoretical example.

As shown in Figure 2, the example presents a construction project's segment that comprises four activities, A1–A4. Table 7 presents the planned and actual durations and costs, while the following assumptions are also considered for the theoretical example:

- Fees include 10% profit. Payments are expected to be fully consistent with billing at the design phase; however, the actual cash flows are recorded as presented in Table 5.
- Each activity is carried out from a team of two persons. The total man-hours for the completion of the project are 380. An initial estimation at the design phase for a loss of 10 manhours is not confirmed as the actual manhours lost are as presented in Table 5.
- At the design phase, the cost of reworks is expected to be at 750 €; however, the actual cost of reworks is the one presented in Table 5.
- The evaluation, at the design phase, of all parameters for both Team and Client satisfaction considers a level of satisfaction set at 8; however, the actual values of the levels of satisfaction for each parameter of each factor are those presented in Tables 6 and 8.
- Eleven risks have been identified through the risk identification process.

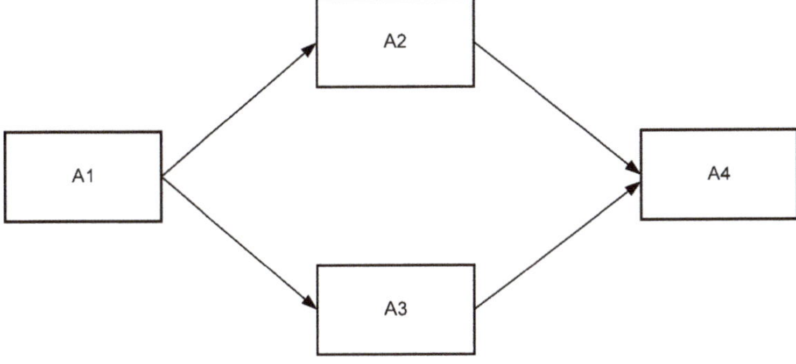


**Figure 2.** Theoretical example: Construction project's activities.

**Table 5.** Performance data upon construction project's completion of the theoretical example.

| Week | Running Activity | Work Completed (%) | AC (€) | Payments (€) | Lost Man-Hours (hrs) | Cost of Reworks (€) |
|---|---|---|---|---|---|---|
| 1 | A1 | 4 | 240 | 0 | 6 | 0 |
| 2 | A1 | 8 | 240 | 0 | 9 | 0 |
| 3 | A1 | 12 | 240 | 0 | 2 | 0 |
| 4 | A2,A3 | 24 | 770 | 0 | 5 | 270 |
| 5 | A2,A3 | 36 | 770 | 0 | 8 | 230 |
| 6 | A2,A3 | 48 | 770 | 3333 | 0 | 0 |
| 7 | A2,A3 | 60 | 770 | 0 | 1 | 500 |
| 8 | A3 | 70 | 270 | 0 | 5 | 0 |
| 9 | A3 | 80 | 270 | 0 | 2 | 0 |
| 10 | A4 | 90 | 375 | 0 | 0 | 0 |
| 11 | A4 | 100 | 375 | 0 | 0 | 0 |
| 12 | - | | 0 | 0 | - | - |
| 13 | - | | 0 | 0 | - | - |
| 14 | - | | 0 | 2266 | - | - |
| **Total** | | | **5090** | **5599** | **38** | **1000** |

**Table 6.** Calculation of team satisfaction index (*TSI*) for indicative weightings and levels of satisfaction.

| Satisfaction Parameter | Weights | Level of Satisfaction | Product |
|---|---|---|---|
| 1 | 0.0795 | 6 | 0.477 |
| 2 | 0.0756 | 7 | 0.530 |
| 3 | 0.0622 | 6 | 0.373 |
| 4 | 0.1009 | 8 | 0.807 |
| 5 | 0.1035 | 8 | 0.828 |
| 6 | 0.1016 | 9 | 0.914 |
| 7 | 0.0795 | 5 | 0.397 |
| 8 | 0.0795 | 5 | 0.397 |
| 9 | 0.0795 | 8 | 0.636 |
| 10 | 0.0795 | 8 | 0.636 |
| 11 | 0.0795 | 6 | 0.477 |
| 12 | 0.0795 | 7 | 0.556 |
| *TSI* | | | **7.028** |

**Table 7.** Cost and time data of the theoretical example.

| Activities | Planned Duration (Weeks) | | | Planned Cost (€) | Actual Duration (Weeks) | Actual Cost (€) |
|---|---|---|---|---|---|---|
| | Minimum | Maximum | Mean | | | |
| A1 | 1 | 3 | 2 | 700 | 3 | 720 |
| A2 | 2 | 6 | 4 | 1900 | 4 | 2000 |
| A3 | 3 | 7 | 5 | 1500 | 6 | 1620 |
| A4 | 1 | 3 | 2 | 700 | 2 | 750 |
| **Total** | | | **9** | **4800** | **11** | **5090** |

The application of Equations (1), (2), (4), (5), (7), (8), (10) and (11) results to the non-normalized values of the indices. Then following the normalization process with the equations shown in Table 3, the final normalized values of the indices are shown in Table 9. Table 9 also presents the weights of each performance dimension for the calculation of the project's performance index.

**Table 8.** Calculation of client satisfaction index (*CSI*) for indicative weightings and levels of satisfaction.

| Satisfaction Parameter | Weights | Level of Satisfaction | Product |
|---|---|---|---|
| 1 | 0.0801 | 6 | 0.481 |
| 2 | 0.0769 | 7 | 0.539 |
| 3 | 0.0657 | 6 | 0.394 |
| 4 | 0.0980 | 8 | 0.784 |
| 5 | 0.1002 | 6 | 0.601 |
| 6 | 0.0985 | 6 | 0.591 |
| 7 | 0.0801 | 7 | 0.561 |
| 8 | 0.0801 | 8 | 0.641 |
| 9 | 0.0801 | 8 | 0.641 |
| 10 | 0.0801 | 7 | 0.561 |
| 11 | 0.0801 | 8 | 0.641 |
| 12 | 0.0801 | 9 | 0.721 |
| *CSI* | | | **7.154** |

**Table 9.** Non-normalized and normalized values of performance indices.

| Performance Index | Non-Normalized Value | Normalized Value | Weights |
|---|---|---|---|
| *CPI* | 0.943 | 0.943 | 0.15 |
| *SPI* | 1.000 | 1.000 | 0.10 |
| *BPI* | 0.595 | 0.693 | 0.10 |
| *SFI* | 0.100 | 1.050 | 0.15 |
| *PPI* | 0.655 | 0.710 | 0.15 |
| *QPI* | 0.244 | 1.218 | 0.15 |
| *TSI* | 7.028 | 0.893 | 0.10 |
| *CSI* | 7.154 | 0.899 | 0.10 |

The introduction of the risk aspect in the calculation of construction project's performance is achieved through the implementation of the risk assessment process (Equations (12)–(17)) that is described in Section 3.2. The calculation results are presented in Tables 10–12.

**Table 10.** Risk quantification with the use of fuzzy numbers.

| Risk | Probability | Consequence | $R_{ij}$ |
|---|---|---|---|
| $R_{11}$ | (0.4, 0.5, 0.5, 0.6; 1.0) | (0.5, 0.6, 0.7, 0.8; 1.0) | (0.20, 0.30, 0.35, 0.48; 1.0) |
| $R_{12}$ | (0.5, 0.6, 0.7, 0.8; 1.0) | (0.5, 0.6, 0.7, 0.8; 1.0) | (0.25, 0.36, 0.49, 0.64; 1.0) |
| $R_{21}$ | (0.5, 0.6, 0.7, 0.8; 1.0) | (0.5, 0.6, 0.7, 0.8; 1.0) | (0.25, 0.36, 0.49, 0.64; 1.0) |
| $R_{22}$ | (0.4, 0.5, 0.5, 0.6; 1.0) | (0.1, 0.2, 0.3, 0.4; 1.0) | (0.04, 0.10, 0.15, 0.24; 1.0) |
| $R_{31}$ | (0.2, 0.3, 0.4, 0.5; 1.0) | (0.1, 0.2, 0.3, 0.4; 1.0) | (0.02, 0.06, 0.12, 0.20; 1.0) |
| $R_{41}$ | (0.7, 0.8, 0.8, 0.9; 1.0) | (0.5, 0.6, 0.7, 0.8; 1.0) | (0.35, 0.48, 0.56, 0.72; 1.0) |
| $R_{42}$ | (0.2, 0.3, 0.4, 0.5; 1.0) | (0.3, 0.4, 0.5, 0.6; 1.0) | (0.06, 0.12, 0.20, 0.30; 1.0) |
| $R_{51}$ | (0.4, 0.5, 0.5, 0.6; 1.0) | (0.1, 0.2, 0.3, 0.4; 1.0) | (0.04, 0.10, 0.15, 0.24; 1.0) |
| $R_{61}$ | (0.7, 0.8, 0.8, 0.9; 1.0) | (0, 0.1, 0.2, 0.3; 1.0) | (0, 0.08, 0.16, 0.27; 1.0) |
| $R_{71}$ | (0.1, 0.2, 0.2, 0.3; 1.0) | (0.3, 0.4, 0.5, 0.6; 1.0) | (0.03, 0.08, 0.10, 0.18; 1.0) |
| $R_{81}$ | (0, 0, 0.1, 0.2; 1.0) | (0, 0.1, 0.2, 0.3; 1.0) | (0, 0, 0.02, 0.06; 1.0) |

Having calculated all the factors of Equation (18), the risk-informed performance index of the construction project is assessed as shown below:

$$RiPI = 0.15 \times 0.852 \times 0.943 + 0.10 \times 0.864 \times 1.000 + 0.10 \times 0.887 \times 0.693 + 0.15 \times 0.850 \times 1.050 + 0.15$$
$$\times 0.885 \times 0.710 + 0.15 \times 0.887 \times 1.218 + 0.10 \times 0.886 \times 0.893 + 0.10 \times 0.888 \times 0.899 = \mathbf{0.818}$$

Based on the *RiPI* calculation, the project's performance falls into class E ($RiPI \leq 0.85$), which is considered as poor performance.

**Table 11.** Crisp risk rating calculation after defuzzification.

| Risk | $X_0$ | $Y_0$ | $CRR_{ij}$ |
|---|---|---|---|
| $R_{11}$ | 0.330 | 0.405 | 0.522 |
| $R_{12}$ | 0.432 | 0.382 | 0.576 |
| $R_{21}$ | 0.432 | 0.382 | 0.576 |
| $R_{22}$ | 0.130 | 0.409 | 0.429 |
| $R_{31}$ | 0.097 | 0.409 | 0.421 |
| $R_{41}$ | 0.525 | 0.393 | 0.656 |
| $R_{42}$ | 0.167 | 0.403 | 0.437 |
| $R_{51}$ | 0.130 | 0.409 | 0.429 |
| $R_{61}$ | 0.125 | 0.401 | 0.420 |
| $R_{71}$ | 0.095 | 0.414 | 0.425 |
| $R_{81}$ | 0.017 | 0.416 | 0.416 |

**Table 12.** Risk levels and risk factors calculations.

| Performance Index | $CRR_i$ | $RL$ | $RF$ |
|---|---|---|---|
| *CPI* | 0.550 | 0.148 | 0.852 |
| *SPI* | 0.508 | 0.136 | 0.864 |
| *BPI* | 0.421 | 0.113 | 0.887 |
| *SFI* | 0.557 | 0.150 | 0.850 |
| *PPI* | 0.429 | 0.115 | 0.885 |
| *QPI* | 0.420 | 0.113 | 0.887 |
| *TSI* | 0.425 | 0.114 | 0.886 |
| *CSI* | 0.416 | 0.112 | 0.888 |

Based on the numerical data for the theoretical example according to planning and design, the application of Equation (18) yields $PP = 0.918$. Then, considering a weekly performance control period for specific scheduled and actual project's progress and applying Equations (19) and (20), Table 13 and Figure 3 summarize the project's actual and planned performance during the project's lifecycle.

**Table 13.** Planned and actual project's progress and performance elements.

| Week | Work Completion as Scheduled (%) | $PP_{PPC}$ | $PRB_t$ | $w_{Pt}$ | Actual Work Completion (%) | Actual Performance (AP) | $AP_{PC}$ |
|---|---|---|---|---|---|---|---|
| 0 | 0 | 0 | 0.918 | 0 | 0 | 0 | 0 |
| 1 | 7.29 | 0.067 | 0.851 | 0.067 | 4 | 0.646 | 0.026 |
| 2 | 14.58 | 0.134 | 0.784 | 0.067 | 8 | 0.642 | 0.051 |
| 3 | 30.73 | 0.282 | 0.636 | 0.148 | 12 | 0.647 | 0.078 |
| 4 | 46.88 | 0.430 | 0.488 | 0.148 | 24 | 0.635 | 0.152 |
| 5 | 63.02 | 0.579 | 0.339 | 0.148 | 36 | 0.630 | 0.227 |
| 6 | 79.17 | 0.727 | 0.191 | 0.148 | 48 | 0.871 | 0.418 |
| 7 | 85.42 | 0.784 | 0.134 | 0.057 | 60 | 0.813 | 0.488 |
| 8 | 92.71 | 0.851 | 0.067 | 0.067 | 70 | 0.812 | 0.568 |
| 9 | 100 | 0.918 | 0 | 0.067 | 80 | 0.814 | 0.651 |
| 10 | - | - | - | - | 90 | 0.813 | 0.732 |
| 11 | - | - | - | - | 100 | 0.818 | 0.818 |

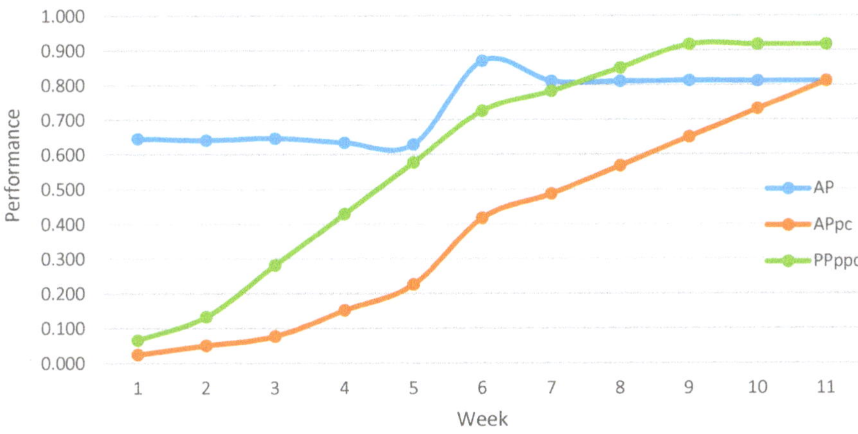

**Figure 3.** Actual and planned performance's evolution during the project's life cycle.

The project performance buffer (*PPB*) for the theoretical example was calculated by performing a Monte Carlo simulation of 5000 iterations with the use of specific software (SPSS v.18). The level of confidence was set to 90% and the expected value was set to 0.918, i.e., the calculated value of *PP*. Then, the simulation returned the result shown in Figure 4, where the mean performance ($AP_{mean}$) is estimated at 0.90, the performance at the confidence level ($AP_{0.90}$) is estimated at 0.824, and *PPB* is estimated at 0.076, according to Equation (21).

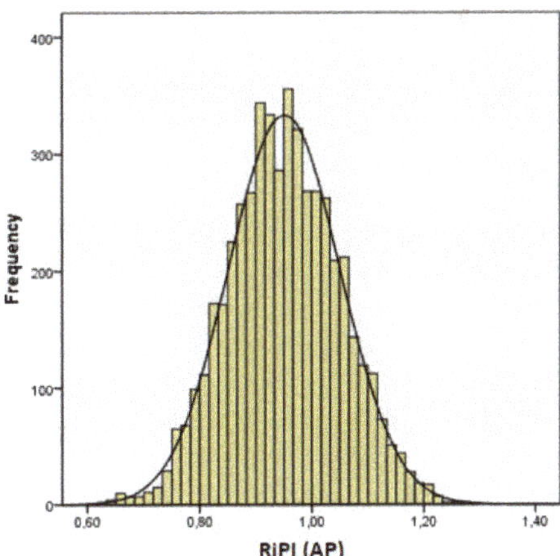

**Figure 4.** Monte Carlo simulation results for actual performance at 90% confidence level.

Finally, the application of Equations (22)–(25) results in the outcome presented in Table 14 regarding the performance control index and the estimation of performance at project's completion. As shown there (column $PCI_t$), the project is constantly underperforming at all control points. Performance is increasingly reduced until the sixth week where the trend is reversed, however, not enough to return to acceptable performance levels until the project's completion. The improvement after the sixth week is due to the effect of the *BPI* index to the project's performance, which is changed after the payment made then (see Table 5). The final outcome is consistent with the analysis made in [32] for the same example, concerning the failure in meeting schedule performance requirements.

As already mentioned, the proposed methodology integrates considerably more constituents of a project's parameters, thus providing a more holistic view and estimation of performance.

**Table 14.** Planned and actual project's progress and performance elements.

| Week | Actual Work Completion (%) | $AP$ | $AP_{PC}$ | $APB_t$ | $PP_{PPC}$ | $PCI_t$ | $EPAC_t$ |
|---|---|---|---|---|---|---|---|
| 1 | 4 | 0.646 | 0.026 | 0.006 | 0.067 | −0.036 | 0.882 |
| 2 | 8 | 0.642 | 0.051 | 0.011 | 0.134 | −0.071 | 0.847 |
| 3 | 12 | 0.647 | 0.078 | 0.023 | 0.282 | −0.181 | 0.737 |
| 4 | 24 | 0.635 | 0.152 | 0.036 | 0.430 | −0.242 | 0.676 |
| 5 | 36 | 0.630 | 0.227 | 0.048 | 0.579 | −0.304 | 0.614 |
| 6 | 48 | 0.871 | 0.418 | 0.060 | 0.727 | −0.249 | 0.669 |
| 7 | 60 | 0.813 | 0.488 | 0.065 | 0.784 | −0.231 | 0.687 |
| 8 | 70 | 0.812 | 0.568 | 0.070 | 0.851 | −0.212 | 0.706 |
| 9 | 80 | 0.814 | 0.651 | 0.076 | 0.918 | −0.191 | 0.727 |
| 10 | 90 | 0.813 | 0.732 | 0.076 | 0.918 | −0.110 | 0.808 |
| 11 | 100 | 0.818 | 0.818 | 0.076 | 0.918 | −0.024 | 0.884 |

Table 14 also presents the estimation of performance at project's completion ($EPAC_t$), which, since the first week, is constantly estimated below the estimated value of $PP$. This estimation does not only confirm the project's underperformance, which is quantified with $PCI_t$, but also provides a quantified value of the distance that has to be covered towards reaching the performance requirements set at the project's design. Covering this distance is a decision-making problem for the project's manager and management team and the proposed methodology provides quantified results of a risk-informed performance assessment to facilitate the decision-making process.

## 4. Discussion and Conclusions

This work contributes to the course of research of assessing construction projects performance through the incorporation of risk management. Although very relevant, the two frameworks remain heavily segmented despite the similarities in the processing and the common output, which is a decision-making factor. The proposed methodological framework for risk-informed construction project performance assessment integrates the benchmarking approach through indicators with a robust yet simple mathematical structure that is fully deployed in this paper. This approach does not take, directly, into consideration the potential interdependencies between performance variables. It rather settles to consider them only at the stage of assigning values to the model's variables, where these values are provided based on several determining factors, including the interdependencies between performance elements. The reason for such an approach is that with respect to this issue, this research focused entirely on addressing the integration between risk and performance; therefore the performance model was selected to be an appropriate adaptation, evolution and, eventually, enrichment of previous work to improve the knowledge background and develop a new framework that incorporates the risk aspect in the form of coefficients to performance dimensions.

The proposed methodological framework is balancing between the required robustness and completeness of a sound approach and the adjustability and subjectivity that decision-making processes require and inherently possess. There are many weightings and ratings that introduce the human aspect of stakeholders and project management teams both concerning performance and risk aspects. At the same time, there are indicators assessed purely based on recorded data, while the adoption of fuzzy numbers for the modeling of risk provides a sound way for dealing with incompleteness of data when this is occurring. The research's most important contribution though is that it performs risk-informed performance control and future performance estimations and provides quantified outputs that can be used from the construction project's management team for deciding appropriate interventions with respect to the achieved and desired project's performance. A comprehensive and easy to apply methodology such as the proposed one may be very useful for project managers in their everyday decision-making tasks and facilitate successful project's performance management.

This research revisits the field of integrating risk and performance in a way that raises more questions and trigger further research. For example, what is the trade-off between costs and benefits from establishing processes and assigning resources to collect all the required data and then analyze them for an integrated risk and performance management? What should be the level of accuracy for modeling risks with fuzzy numbers or for determining the marginal values for the performance classes? What is the amount of the effort required when the several weightings and ratings by expert judgments involved are not considered as constants, but are subjected to changes during the project's lifecycle? What is the amount of the effort required for applying this methodological framework in case of large and complicated projects?

Investigating for the responses of these questions and a constant revision process of the proposed framework are expected to be topics of related future research. Achieving better modeling of the performance variables by including their interdependencies and better calibration of the several weighting factors addressed in the methodology shall advance even more risk-informed performance assessment of construction projects.

**Author Contributions:** Conceptualization, M.P. and Y.X.; methodology, M.P.; validation, M.P.; formal analysis, M.P.; investigation, M.P.; resources, M.P. and Y.X.; data curation, M.P.; writing—original draft preparation, M.P.; writing—review and editing, Y.X.; visualization, M.P. and Y.X.; supervision, Y.X.; project administration, Y.X. All authors have read and agreed to the published version of the manuscript.

**Funding:** This research received no external funding.

**Conflicts of Interest:** The authors declare no conflict of interest.

## References

1. Musawir, A.; Serra, C.E.M.; Zwikael, O.; Ali, I. Project governance, benefit management, and project success: Towards a framework for supporting organizational strategy implementation. *Int. J. Proj. Manag.* **2017**, *35*, 1658–1672. [CrossRef]
2. Lamprou, A.; Vagiona, D. Success criteria and critical success factors in project success: A literature review. *Rel. Int. J.* **2018**, *1*, 276–284.
3. Bryde, D.; Unterhitzenberger, C.; Joby, R. Conditions of success for earned value analysis in projects. *Int. J. Proj. Manag.* **2017**, *36*, 474–484. [CrossRef]
4. Demirkesen, S.; Ozorhon, B. Impact of integration management on construction project management performance. *Int. J. Proj. Manag.* **2017**, *35*, 1639–1654. [CrossRef]
5. Orihuela, P.; Pacheco, S.; Orihuela, J. Proposal of performance indicators for the design of housing projects. *Procedia Eng.* **2017**, *196*, 498–505. [CrossRef]
6. Sanchez, O.P.; Terlizzi, M.A.; de Moraes, H.R.O.C. Cost and time project management success factors for information systems development projects. *Int. J. Proj. Manag.* **2017**, *35*, 1608–1626. [CrossRef]
7. Chen, H.L.; Chen, W.T.; Lin, Y.L. Earned value project management: Improving the predictive power of planned value. *Int. J. Proj. Manag.* **2016**, *34*, 22–29. [CrossRef]
8. Omar, M.N.; Fayek, A.R. Modeling and evaluating construction project competencies and their relationship to project performance. *Autom. Constr.* **2016**, *69*, 115–130. [CrossRef]
9. Warburton, R.D.H.; Cioffi, D.F. Estimating a project's earned and final duration. *Int. J. Proj. Manag.* **2016**, *34*, 1493–1504. [CrossRef]
10. Yun, S.; Choi, J.; de Oliveira, D.P.; Mulva, S.P. Development of performance metrics for phase-based capital project benchmarking. *Int. J. Proj. Manag.* **2016**, *34*, 389–402. [CrossRef]
11. Abd El-Karim, M.S.B.A.; Mosa El Nawawy, O.A.; Abdel-Alim, A.M. Identification and assessment of risk factors affecting construction projects. *HBRC J.* **2015**, *13*, 202–216. [CrossRef]
12. Zavadskas, E.K.; Vilutienė, T.; Turskis, Z.; Šaparauskas, J. Multi-criteria analysis of projects' performance in construction. *Arch. Civ. Mech. Eng.* **2014**, *14*, 114–121. [CrossRef]

13. Cheng, E.W.L.; Ryan, N.; Kelly, S. Exploring the perceived influence of safety management practices on project performance in the construction industry. *Saf. Sci.* **2012**, *50*, 363–369. [CrossRef]
14. Naeni, L.M.; Shadrokh, S.; Salehipour, A. A fuzzy approach for the earned value management. *Int. J. Proj. Manag.* **2011**, *29*, 764–772. [CrossRef]
15. Nassar, N.K. An integrated framework for evaluation of performance of construction projects. In *PMI® Global Congress 2009—North America, Orlando, FL, USA, 10–13 October 2009*; Project Management Institute: Newtown Square, PA, USA, 2009.
16. Pheng, L.S.; Chuan, Q.T. Environmental factors and work performance of project managers in the construction industry. *Int. J. Proj. Manag.* **2006**, *24*, 24–37. [CrossRef]
17. Thekdi, S.; Aven, T. An enhanced data-analytic framework for integrating risk management and performance management. *Reliab. Eng. Syst. Saf.* **2016**, *156*, 277–287. [CrossRef]
18. Cokins, G. *Performance Management: Intergrating Strategy Execution Methodologies, Risk and Analytics*, 1st ed.; John Wiley & Sons, Inc: Hoboken, NJ, USA, 2009.
19. Bassioni, H.A.; Price, A.D.F.; Hassan, T.M. Performance measurement in construction. *J. Manag. Eng.* **2004**, *20*, 42–50. [CrossRef]
20. Aven, T. On the new ISO guide on risk management terminology. *Reliab. Eng. Syst. Saf.* **2011**, *96*, 719–726. [CrossRef]
21. Aven, T. On how to define, understand and describe risk. *Reliab. Eng. Syst. Saf.* **2010**, *95*, 623–631. [CrossRef]
22. Olechowski, A.; Oehmen, J.; Seering, W.; Ben-Daya, M. The professionalization of risk management: What role can the ISO 31000 risk management principles play? *Int. J. Proj. Manag.* **2016**, *34*, 1568–1578. [CrossRef]
23. Aven, T. The risk concept-historical and recent development trends. *Reliab. Eng. Syst. Saf.* **2012**, *99*, 33–44. [CrossRef]
24. Mohammed, H.K.; Knapkova, A. The Impact of total risk management on company's performance. *Procedia Soc. Behav. Sci.* **2016**, *220*, 271–277. [CrossRef]
25. Jafari, M.; Chadegani, A.; Biglari, V. Effective risk management and company's performance: Investment in innovations and intellectual capital using behavioural and practical approach. *Int. Res. J. Financ. Econ.* **2011**, *3*, 780–786.
26. Neely, A.; Adams, C.; Kennerley, M. *The Performance Prism: The Scorecard for Measuring and Managing Business Success*; Prentice Hall: Upper Saddle River, NJ, USA, 2002; pp. 159–160.
27. Gordon, L.A.; Loeb, M.P.; Tseng, C.Y. Enterprise risk management and firm performance: A contingency perspective. *J. Account. Public Policy* **2009**, *28*, 301–327. [CrossRef]
28. McShane, M.K.; Nair, A.; Rustambekov, E. Does enterprise risk management increase firm value? *J. Account. Audit. Financ.* **2011**, *26*, 641–658. [CrossRef]
29. Lundqvist, S.A. An exploratory study of enterprise risk management: Pillars of ERM. *J. Account. Audit. Financ.* **2014**, *29*, 393–429. [CrossRef]
30. Callahan, C.; Soileau, J. Does enterprise risk management enhance operating performance? *Adv. Account.* **2017**, *37*, 122–139. [CrossRef]
31. Florio, C.; Leoni, G. Enterprise risk management and firm performance: The Italian case. *Br. Account. Rev.* **2017**, *49*, 56–74. [CrossRef]
32. Pajares, J.; López-Paredes, A. An extension of the EVM analysis for project monitoring: The cost control index and the schedule control index. *Int. J. Proj. Manag.* **2011**, *29*, 615–621. [CrossRef]
33. Acebes, F.; Pajares, J.; Galán, J.M.; López-Paredes, A. Beyond earned value management: A graphical framework for integrated cost, schedule and risk monitoring. *Procedia Soc. Behav. Sci.* **2013**, *74*, 181–189. [CrossRef]
34. Acebes, F.; Pajares, J.; Galán, J.M.; López-Paredes, A. A new approach for project control under uncertainty. Going back to the basics. *Int. J. Proj. Manag.* **2014**, *32*, 423–434. [CrossRef]
35. Leu, S.-S.; Lin, Y.-C. Project performance evaluation based on statistical process control techniques. *J. Constr. Eng. Manag.* **2008**, *134*, 813–819. [CrossRef]
36. Colin, J.; Vanhoucke, M. Setting tolerance limits for statistical project control using earned value management. *Omega* **2014**, *49*, 107–122. [CrossRef]
37. Lipke, W.; Zwikael, O.; Henderson, K.; Anbari, F. Prediction of project outcome. The application of statistical methods to earned value management and earned schedule performance indexes. *Int. J. Proj. Manag.* **2009**, *27*, 400–407. [CrossRef]

38. Zidane, Y.J.-T.; Andersen, B.; Johansen, A.; Ahmad, S. "Need for Speed": Framework for measuring construction project pace—Case of road project. *Procedia Soc. Behav. Sci.* **2016**, *226*, 12–19. [CrossRef]
39. Langston, C. Development of generic key performance indicators for PMBOK® using a 3D project integration model. *Australas J. Constr. Econ. Build.* **2013**, *13*, 78–91. [CrossRef]
40. Majumder, D.; Debnath, J.; Biswas, A. Risk analysis in construction sites using fuzzy reasoning and fuzzy Analytic Hierarchy Process. *Procedia Technol.* **2013**, *10*, 604–614. [CrossRef]
41. Yeung, J.F.Y.; Chan, A.P.C.; Chan, D.W.M. Fuzzy set theory approach for measuring the performance of relationship-based construction projects in Australia. *J. Manag. Eng.* **2012**, *28*, 181–192. [CrossRef]
42. Samantra, C.; Datta, S.; Mahapatra, S.S. Fuzzy based risk assessment module for metropolitan construction project: An empirical study. *Eng. Appl. Artif. Intell.* **2017**, *65*, 449–464. [CrossRef]
43. Shankar, N.R.; Rao, P.P.B. Ranking fuzzy numbers with a distance method using circumcenter of centroids and an index of modality. *Adv. Fuzzy Syst.* **2011**, *2011*, 1–7.
44. Royer, P.S. Risk management: The undiscovered dimension of project management. *Proj Manag. J.* **2000**, *31*, 6–13. [CrossRef]
45. Zolotukhin, A.; Gudmestad, O. Application of fuzzy sets theory in qualitative and quantitative risk assessment. *Int. J. Offshore Polar Eng.* **2002**, *12*, 1–9.

 *sustainability*

*Article*

# A Case Study on Environmental Sustainability Assessment of Spatial Entities with Anthropogenic Activities: The National Park of Eastern Macedonia and Thrace, Greece

**Despoina Aktsoglou * and Georgios Gaidajis**

Laboratory of Environmental Management and Industrial Ecology, Department of Production Engineering and Management, School of Engineering, Democritus University of Thrace, 67100 Xanthi, Greece; geogai@pme.duth.gr
* Correspondence: daktsog@pme.duth.gr; Tel.: +30-25-4107-9356

Received: 22 February 2020; Accepted: 29 May 2020; Published: 1 June 2020

**Abstract:** The current paper presents a methodological framework that is able to evaluate the carrying capacity of protected areas where various human activities, apart from recreation and tourism, take place. The proposed framework converts the energy and product consumption into land required to satisfy those needs (Ecological Footprint) and compares them with the current land uses and available land (Biocapacity), in order to calculate carrying capacity. To facilitate the evaluation, an algorithm that calculates the Ecological Footprint, the Biocapacity, and the Carrying Capacity of the protected area under study by introducing 48 inputs was developed. The inputs were related to the evaluation of individual indicators assessing energy and product consumption of human activities such as households, tertiary sector, municipal buildings, public lighting, private and public transportation, and tourism. A new unit is introduced, the "equivalent person," since the anthropogenic activities within the boundaries of the protected area contribute in a dissimilar way to the total land requirements. The framework is applied, as case study, in the National Park of Eastern Macedonia and Thrace (NPEMT), Greece, with a view to validate and improve its applicability. Within the NPEMT, habitats of significant biodiversity and ecological value are in coexistence with extensive human activities (urban, rural, tourist, light industrial). The study area covers up to approximately 73,000 ha and its population is estimated at about 29,000 people. The Carrying Capacity of the NPEMT according to the current consumption patterns was estimated at 39,193 equivalent residents, which was higher than the current equivalent residents (36,960), indicating a potential for tourism development at the NPEMT. The Ecological Footprint of the NPEMT was estimated at 181,324 Gha or 4.9 Gha/pers$_{eq}$, slightly higher than the European mean (4.69 Gha/pers$_{eq}$). Among activities, households and private transportation (with approximately 79% and 10%, respectively), among land use, agriculture, livestock, and $CO_2$ emissions (with approximately 36%, 30%, and 30%, respectively), and among products, beef, fruits/vegetables, and beverages (with approximately 22%, 15%, and 14%, respectively) were the main contributors of the total Ecological Footprint of the NPEMT. The area of the NPEMT is able to meet the needs of its population provided that the consumption patterns will be stable. The results encourage the expansion of tourism development, as the tourism activity within the NPEMT is limited compared to other adjacent domestic destinations.

**Keywords:** carrying capacity; regional sustainability assessment; ecological footprint; national park

---

## 1. Introduction

Sustainable development requires human systems to function within specific "green" limits to ensure the sufficient supply of goods and services both to current and future generations [1].

The achievement of sustainability in an area depends on whether the impact of anthropogenic activities are within the "green" range, including those activities that take place outside the examined area but whose impacts affect its environmental status [2]. For this reason, the planning and natural resources management in all spatial scales is essential to aim toward sustainability [3].

Focusing on protected areas, the integration of anthropogenic activities within the boundaries of protected areas has led to increasing concerns regarding the appropriate use levels of parks, forests, lakes, and other environmentally sensitive areas [4]. Alongside the significant ecological habitats within protected areas, extensive human activities are developed such as households, tourism, agriculture, light industry, and transportation. The harmonious coexistence of those diverse activities is a basic concern of the management bodies responsible for protected areas [5]. More and more relevant authorities need to quantify and assess the maximum level of human activities that can be developed to satisfy current and future needs, while in parallel sustaining the environmental and ecological health.

The sustainability of protected areas is associated directly with Carrying Capacity for two reasons: (a) the idea of sustainability reflects a limit, similarly with the concept of Carrying Capacity, and (b) both concepts share the same challenges in formulating the objectives, practices, and actions of improvement [6]. From the early 1960s, due to the fact that public visits were the major threat for protected areas [7–10], research on outdoor recreation has utilized the concept of Carrying Capacity to address the resource and social impacts of visitors [7,11,12]. A number of frameworks [5,13–16] have been developed in order to provide management bodies with a basis for decision-making about the Carrying Capacity of national parks and protected areas, defining it as *"the maximum number of visitors an area can sustain without unacceptable deterioration of the physical environment and without considerably diminishing user satisfaction"* [9,15,17]. Therefore, in the current article, the concepts of Environmental Sustainability and Carrying Capacity are identical.

In Greece, there are twenty (20) protected areas [18] known for their great ecological and educational value, when at the same time, apart from tourism, several other human activities are traditionally and/or legally established within their boundaries. Thus, the assessment of the Carrying Capacity becomes more complex, since activities such as agriculture, livestock, aquaculture, households, light industry, and transportation have to be taken into account. Therefore, the Carrying Capacity assessment of an area must take into account the impact of all the activities that take place within its boundaries.

Serving this challenge, the aim of the specific study was to provide an applicable framework that is able to improve the evaluation and monitoring of the environmental sustainability of protected areas. The proposed framework takes advantages from the results of an extensive literature review we have conducted in a previous work [19]. In this work, 13 methods selected from a pool of 61 methods from a literature review were analyzed, categorized, and were finally evaluated based on specific criteria [19,20]. This analysis pointed out key conclusions related to the efficiency and the applicability of environmental sustainability assessment methods of protected areas. More specifically, the "Resource Availability Assessment" category of methods and especially the "Ecological Footprint" method have been indicated as the most appropriate method for the evaluation of environmental sustainability of protected areas. Moreover, the need for the improvement of the ability of methods to incorporate new activities within the environmental boundaries of the protected area, together with the necessity of the methods to provide environmental sustainability thresholds, in order to evaluate quantitatively whether the performance is sustainable, was also pointed out. An effective environmental sustainability assessment method should take into account the spatial characteristics of the examined area and ensure an adequate balance between the level of complexity and the coverage of key sustainability issues [19,20].

This paper consolidates key findings from our previous work on the evaluation of existing environmental sustainability assessment methods [19] and progresses a step forward by integrating all the information into an applicable framework that focuses on protected areas, taking into consideration all anthropogenic activities within its boundaries. The proposed framework is expected to provide a more holistic approach for the assessment of the environmental status and for the development of a sustainable strategy for a protected area.

## 2. Method Description

In this specific section, the theoretical background and the steps that compose the proposed methodological framework are briefly presented.

### 2.1. Theoretical Background

The concept of Carrying Capacity in general expresses an upper limit of the ability to sustain a living system, whereas beyond that limit, instability, degradation, or irreversible damage will subsequently occur [21]. Therefore, Carrying Capacity can be utilized as a supportive tool of policy and decision-making, in order to resolve the aforementioned challenges.

The assessment of the Carrying Capacity of an area is a case-specific procedure and depends on the nature of the problem to be solved and the objectives set by the researcher. As a result, various Carrying Capacity definitions are available in literature. A widely known definition was introduced by Rees [22], according to which the Carrying Capacity of an ecosystem is *"the maximum population of certain species that can be accommodated in an environment without permanent damage to the productivity of the environment"*.

All human populations need natural resources. The use of resources leads to physical outputs that affect global and local areas and has environmental impacts, such as waste generation and impact on climate change. The availability of natural resources and the environmental impacts from their utilization are the two restrictive parameters regulating the size of the population that can be sustained in a given area [23].

Thus, the majority of existing Carrying Capacity methodologies focus on the environmental constraints of resource consumption to determine population limits [24]. The resource-consumption-focused methodologies [25,26] are universally applicable, reasonably comprehensive, and their data and methodology have been made publicly available [27]. The most common current existing examples of environmental modeling [24,28,29] are based on the Ecological Footprint, *"the amount of land and/or water that is necessary to a population or activity, in order to produce, in a sustainable way, all the natural resources it consumes and assimilate the waste it produces, using the available technology"* [30].

Therefore, both Ecological Footprint and Carrying Capacity are based on similar procedures such as defining an area, selecting resources, and defining relevant indicators for qualification [28]. In order to assess the Carrying Capacity of an entity, its Ecological Footprint per person is compared with its Biocapacity, a term that represents the available biologically productive land that absorbs the impact of consumption along with subsequent waste [25]. The Ecological Footprint calculations refer to the estimation of annual consumption needs of anthropogenic activities that take place on the study area and their conversion to biologically productive land. According to Wackernagel et al. [31], the aforementioned consumption needs are classified into six (6) Ecological Footprint land use types, namely, agricultural products, livestock products, fishery and aquaculture, timber products, $CO_2$ emissions, and built-up surfaces. The Ecological Footprint of consumption for each product (EFc) is calculated as:

$$EF_C = EF_P + EF_{im} - EF_{ex} \tag{1}$$

where $EF_P$ is the Ecological Footprint of production, and $EF_{im}$ and $EF_{ex}$ are the Ecological Footprints embodied in imported and exported commodity flows, respectively.

In order to ensure that all the production procedures as well as the necessary materials both for product production or energy generation will be taken into account in the estimation of the Ecological Footprint, the Life Cycle Analysis (LCA) approach was innovatively applied, leading to a more thorough assessment of human activities and ensuring that all required processes and materials were taken into account in the calculations [32]. The Life Cycle Analysis is conventionally characterized as a "cradle-to-grave" or "closed-loop" approach, as it examines the overall environmental impact of a product, process, or system, taking into account every step of its life—from receipt of raw materials to its construction, its sale, usage, and final disposal into the environment [33].

The Biocapacity calculations refer to the estimation of the existing available biologically productive land. The available land is divided into five land uses, namely, cropland area, grazing land area, marine/inland water area, forest area, and infrastructure area [31].

The calculation of Biocapacity is implemented based on the accounting framework proposed by Wackernagel et al. [31]. The available productive areas were firstly converted into land in terms of world average productivity by multiplying the available hectares of each land use with the corresponding "yield factors" [34]. *"The Yield Factors (YFs) account for countries' differing levels of productivity for particular land uses are country-specific and vary by land use type and year. They may reflect natural factors such as differences in precipitation or soil quality, as well as anthropogenic differences such as management practices"* [34]. Subsequently, the abovementioned lands were converted to Biocapacity by multiplying the world hectares of each land use with the corresponding "equivalence factor" [34]. *"The Equivalence Factors (EQFs) convert the areas of different land uses, at their respective world average productivities, into their equivalent areas at global average bioproductivity across all land use and they vary by land use as well as by year. The rationale behind the Equivalent Factor calculation is to weight different land areas in terms of their inherent capacity to produce human useful biological resources. The weighting criterion is not the actual quantity of biomass produced, but what each hectare would be able to inherently deliver"* [34]. The values of the mentioned parameters of yield factor and equivalence factor are retrieved by available references and databases [35,36]. Finally, the Biocapacities of land uses are summed to the total Biocapacity.

The proposed framework is based on the "Ecological Footprint" calculation framework described above, but goes a big step further by simplifying the Ecological Footprint calculations and introducing a new unit, namely, "equivalent person," in order to enable the evaluation of more than one activity.

*2.2. Methodological Framework Description*

The methodological framework has been developed for estimating the Carrying Capacity of an anthropogenic spatial entity, which refers to the spatial scale below a nation and usually includes a province or a municipality or parts of them that are administrated under a specific management scheme or other specific authority (e.g., National Park) and it is summarized in Figure 1.

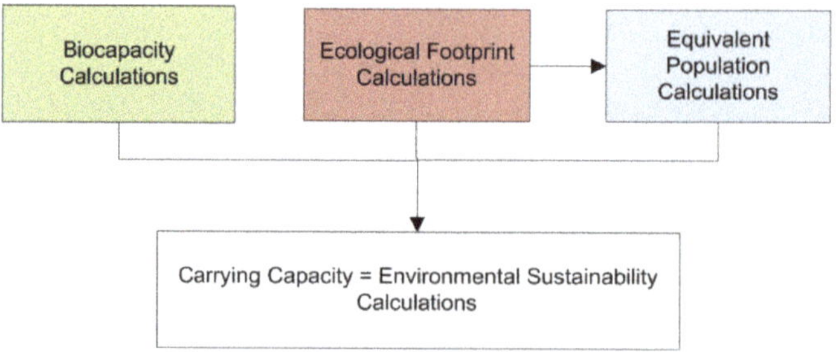


**Figure 1.** The proposed framework.

Based on the above, a comprehensive quantitative definition of Carrying Capacity is given by the authors with the following formula:

$$Carrying\ Capacity\ (max.equivalent\ population) = \frac{Biocapacity\ (available\ land)}{\frac{Ecological\ Footprint\ (required\ land)}{P\ (existing\ equivalent\ population)}} \tag{2}$$

where P is the existing equivalent population of the inhabitants of the spatial entity under study. The unit of "equivalent population" is introduced, since the proposed methodological framework besides the consumption needs of the real human population of the study area estimates the consumption needs of anthropogenic activities, which are not directly dependent on the size of the real population. For example, the annual energy consumption of buildings depends on their size and their use and not on the population size of the study area.

At first, the spatial and time parameters of the framework, namely the geographical boundaries of the study area and the reference year essential for the procedure of data acquiring, are clarified. Following, the calculations are separated in three sectors: (a) Biocapacity calculations, (b) Ecological Footprint calculations, and (c) Calculations of existing Equivalent population.

### 2.2.1. Biocapacity's Accounts

In order to calculate Biocapacity, specific indicators per land use of the CORINE (Coordination of Information on the Environment) land cover methodology [37] were used. These indicators are presented in Figure 2. The available productive land was estimated with the application of a GIS (Geographic Information System) software compatible with the European databases for land uses. Then the available productive lands were converted to Biocapacity per land use by multiplying the available hectares per land use firstly with the yield factors and then with the corresponding equivalent factors, both known from the literature. The total Biocapacity is the sum of all Biocapacities per land use.

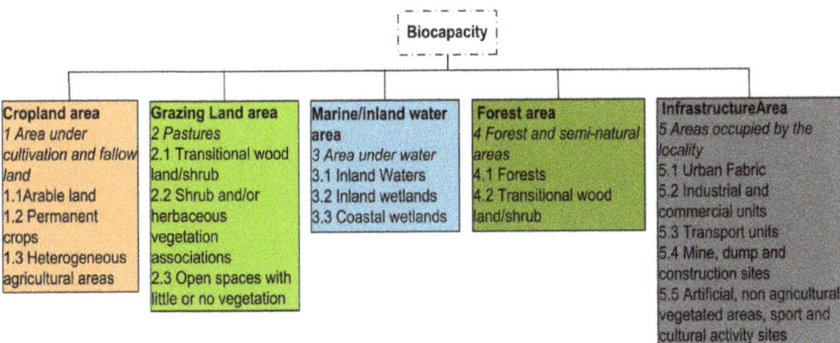


**Figure 2.** The indicators of Biocapacity per land use.

### 2.2.2. Ecological Footprint's Accounts

"The Ecological Footprint theoretically includes all human demands, but in practice, the relevant datasets are not available" [34]. Thus, taking into consideration the availability of datasets from similar studies [35,38] and trying to simplifying the Ecological Footprint calculations framework, the selection of the relevant parameters for energy and product consumption was implemented. The energy and product consumption were related to the needs of households, tertiary sector, municipal buildings, public lighting, public and private transportation, and tourism activities. The selected indicators for the Ecological Footprint assessment are presented in Figure 3, and an indicative sample of them is described in detail in Appendix A.

For the estimation of individual annual consumption, an inventory including indicative quantitative data is implemented. Depending on the availability of data, certain assumptions may be necessary. In order to improve the functionality of the framework, an algorithm that calculates the values of the individual indicators was developed providing that several inputs are known. The algorithm accepts numerous inputs, such as number of residents and tourists, number of buildings per use, area of dwellings, installed power, length of road and railway networks, numbers of vehicles per mean of transportation (cars, motorbikes, etc.), number of passengers and loads for commercial and passenger ships and airplanes, etc. (see Appendix A). The calculations were implemented taking into account assumptions, such as average weight of residents and tourists, average consumption of products per kg of human mass per day, average electrical and thermal energy per m$^2$ per use of building, average surface of buildings per use, average time of public lighting's operation per year, average distance travelled by vehicles on local roads per year, average weight of loads for commercial and passenger ships, etc. (see Appendix A).

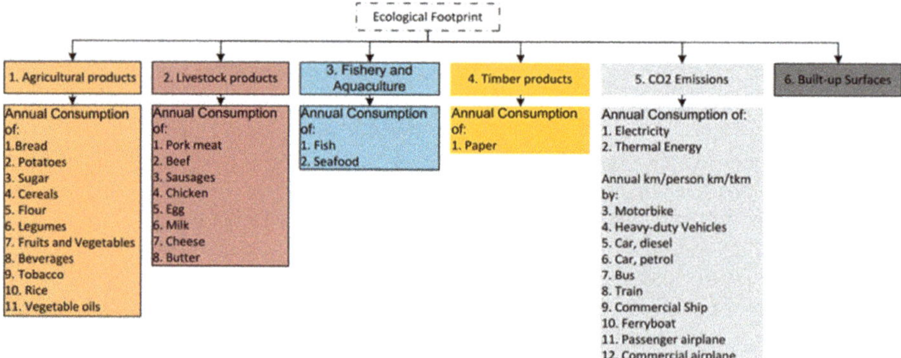


**Figure 3.** The Ecological Footprint's indicators per land use.

For the conversion of the aforementioned annual consumption in required biologically productive land, proper factors were developed with the application of LCA software (SimaPro 7.2) and the methodology Ecological Footprint V 1.02, and were used.

2.2.3. Equivalent Population's Accounts

An innovative part of the proposed framework is the calculation of the existing equivalent population that refers to the estimation of resident population (real and equivalent) of the study area, which contributes to the total annual consumption of resources and energy. The proposed methodological framework for the estimation of the total Ecological Footprint, besides the consumption needs of real residents of the study area, also takes into account the annual consumption of activities that do not depend directly on the resident population, such as the annual fuels consumption for personal transportation, which depends on the annual travelled km by all types of vehicles (cars, motorbikes, trucks, etc.), the annual energy consumption of tertiary sector buildings (hospitals, shops, offices, etc.) that depends on the surface area, etc. Consequently, an equivalence between the consumption needs and the equivalent resident was created, according to the Ecological Footprint that these needs require. By setting the real residents' population of the study area and their needs as a basis, the consumption needs of the remaining anthropogenic activities are matched to equivalent residents. Using this method, every real resident is matched to one equivalent resident, while all the other activities "produce" a corresponding number of equivalent residents. For example, a population of X equivalent residents can be sustained with the Ecological Footprint corresponding to the annual power consumption of tertiary sector buildings, or moreover, a population of Y equivalent residents can be sustained with the Ecological Footprint corresponding to the annual fuel consumption for personal transportation, etc.

The calculation of the existing equivalent residents which contributes to the total Ecological Footprint is implemented by considering the following hypothesis: if $P^*$ real residents need $EF^*$ (area in Gha) to be sustained, then what is the number of residents ($P_{eq}$) able to be sustained in EF (area in Gha)? The existing equivalent resident population is therefore calculated using Equation (3):

$$Peq = \frac{P^*}{EF^*} EF \tag{3}$$

where $P*$ and $EF*$ represent the population and the Ecological Footprint of the real residents, and EF represents the total Ecological Footprint of the spatial entity under study.

## 3. Case Study: National Park of Eastern Macedonia and Thrace, Greece

The proposed framework was applied to the National Park of Eastern Macedonia and Thrace (NPEMT), in Greece, in order to verify and improve its applicability.

### 3.1. Description of the Study Area

The National Park of Eastern Macedonia and Thrace (NPEMT) covers approximately 73,000 ha and includes the wetlands of Delta Nestos and Lakes Vistonida and Ismarida (Figure 4). Within the NPEMT, habitats of significant biodiversity and ecological value are in coexistence with extensive human activities (urban, rural, tourist, industrial). The NPEMT extends to the administrative boundaries of six (6) municipalities, has 43 villages, 10,500 households, and an approximate population of 29,000 people.

**Figure 4.** The geographical boundaries of the National Park of Eastern Macedonia and Thrace (NPEMT).

The main characteristics of the NPEMT area are summarized below:

- The building stock of the NPEMT is constituted by 15,400 dwellings, 820 offices and commercial buildings, 8 healthcare buildings, 100 schools, and 30 hotels.
- The total installed power for public lighting is 700 kW.
- Approximately 3.5 million cars, 4000 motorbikes, and 670,000 heavy-duty vehicles use the national, provincial, and local road networks set within the boundaries of the NPEMT on an annual basis.
- The railway within the boundaries of the NPEMT has a length of 10.5 km and serves 146,000 people annually.
- At the two ports of the area, 9160 arrivals and departures are performed, transferring 140,000 people, 463,000 vehicles, and 194,000 tons of commodities on an annual basis.
- The local airport serves approximately 210,000 people (native and tourists) and 300 tons of commodities annually.
- The total number of visitors that stay at least one night within the NPEMT area is estimated at approximately 25,000.

### 3.2. Implementation of the Methodological Framework

The available land uses and subareas of the NPEMT were measured using the software Mapinfo Professional 12.0 and data from the European program for land cover CORINE (Coordination of Information on the Environment) 2000 [39] (see Table 1). The Biocapacity of the NPEMT was therefore estimated at 192,283 hectares of biologically productive land (area in Gha).

**Table 1.** The calculations of the Biocapacity of the NPEMT.

| Land Uses | Area in ha | Yield Factor * | Equivalent Factor * | Biocapacity (Gha) |
|---|---|---|---|---|
| Cropland Area | 52,011 | 1.5 | 2.2 | 171,472 |
| Grazing Land Area | 4528 | 4.1 | 0.5 | 9283 |
| Marine/ Inland Water Area | 12,284 | 0.8 | 0.4 | 3931 |
| Forest Area | 910 | 1.3 | 1.4 | 1657 |
| Infrastructure Area | 1800 | 1.5 | 2.2 | 5940 |
| Total Biocapacity of the NPEMT | | | | 192,283 |

* Source: [35].

The results from the estimation of Ecological Footprint for the reference year (2013) are presented in Table 2. Relative estimations took place for the Ecological Footprint calculation, deriving by the real resident population of the NPEMT, which was estimated at 143,628 Gha.

The existing equivalent resident population of the NPEMT was calculated using Equation (3) to 36,960 equivalent residents.

$$P_{eq} = \frac{P_1}{EF_1} EF = \frac{29,276}{143,628} 181,324 = 36,960 \text{ equivalent residents}$$

**Table 2.** The calculation of the Ecological Footprint in the totality of the NPEMT.

| Ecological Footprint Subcategory | Ind. Code | Annual Consumptions/Activities | Residents' Ecological Footprint in Gha | Total Ecological Footprint in Gha |
|---|---|---|---|---|
| | Ind. EF 1.1 | Bread | 671 | 676 |
| | Ind. EF 1.2 | Potatoes | 399 | 402 |
| | Ind. EF 1.3 | Sugar | 92 | 92 |
| | Ind. EF 1.4 | Cereals | 728 | 733 |
| | Ind. EF 1.5 | Flour | 260 | 261 |
| 1. Agricultural Products | Ind. EF 1.6 | Legumes | 2509 | 2527 |
| | Ind. EF 1.7 | Fruits and Vegetables | 27,313 | 27,502 |
| | Ind. EF 1.8 | Beverages | 25,759 | 25,938 |
| | Ind. EF 1.9 | Tobacco | 77 | 78 |
| | Ind. EF 1.10 | Rice | 3912 | 3939 |
| | Ind. EF 1.11 | Vegetable oils | 2298 | 2314 |
| | Ind. EF 2.1 | Pork meat | 4141 | 4170 |
| | Ind. EF 2.2 | Beef | 40,965 | 41,249 |
| | Ind. EF 2.3 | Sausages | 1882 | 1895 |
| 2. Livestock Products | Ind. EF 2.4 | Chicken | 1506 | 1516 |
| | Ind. EF 2.5 | Egg | 909 | 916 |
| | Ind. EF 2.6 | Milk | 4111 | 4140 |
| | Ind. EF 2.7 | Cheese | 206 | 207 |
| | Ind. EF 2.8 | Butter | 6 | 6 |
| 3. Fishery/Aquaculture products | Ind. EF 3.1 | Fish | 3928 | 3956 |
| | Ind. EF 3.2 | Seafood | 107 | 107 |

**Table 2.** *Cont.*

| Ecological Footprint Subcategory | Ind. Code | Annual Consumptions/Activities | Residents' Ecological Footprint in Gha | Total Ecological Footprint in Gha |
|---|---|---|---|---|
| 4.Timber products | Ind. EF 4.1 | Paper | 951 | 951 |
| 5. $CO_2$ emissions | Ind. EF 5.1 | Electricity | 11,159 | 21,898 |
| | Ind. EF 5.2 | Thermal Energy | 9739 | 13,717 |
| | Ind. EF 5.3 | Transportation by motorbike | 0 | 830 |
| | Ind. EF 5.4 | Transportation by heavy-duty vehicle | 0 | 9784 |
| | Ind. EF 5.5 | Transportation by car (diesel) | 0 | 1093 |
| | Ind. EF 5.6 | Transportation by car (petrol) | 0 | 5883 |
| | Ind. EF 5.7 | Transportation by bus | 0 | 322 |
| | Ind. EF 5.8 | Transportation by train | 0 | 20 |
| | Ind. EF 5.9 | Transportation by commercial ship | 0 | 3 |
| | Ind. EF 5.10 | Transportation by ferryboat | 0 | 20 |
| | Ind. EF 5.11 | Transportation by passenger aircraft | 0 | 215 |
| | Ind. EF 5.12 | Transportation by commercial aircraft | 0 | 3 |
| 6. Built-up surfaces | Ind. EF 6.1 | Built-up areas * | 0 | 3960 |
| | Total EF of the NPEMT | | 143,628 | 181,324 |

* Built-up areas were estimated for the totality of the NPEMT and not separately for every activity due to lack of available data.

The Carrying Capacity of the NPEMT was calculated using Equation (2) to 39,193 equivalent residents.

$$CC = \frac{Bc}{\frac{EF}{Peq}} = \frac{192,283}{\frac{181,324}{36,960}} = 39,193 \text{ equivalent residents}$$

*3.3. Results and Discussion*

The Ecological Footprint of the NPEMT was estimated at 181,324 Gha or 4.9 Gha/pers$_{eq}$, slightly higher than the European mean (4.69 Gha/pers$_{eq}$) [36], whereas the Biocapacity of the NPEMT was estimated at 192,283 Gha. The Carrying Capacity of the NPEMT was estimated at 39,193 equivalent residents; a number higher than the existing equivalent residents (36,960 equivalent residents). Therefore, the area is able to meet the needs of its population provided that the consumption patterns will be stable or more efficient. Also, there is a considerable margin for the development of new activities within the NPEMT boundaries. More specifically, regarding the tourism activity, although it is a common problem in protected areas worldwide, the NPEMT has a Carrying Capacity surplus of approximately 2000 equivalent residents, indicating that it can support a certain number of extra tourists annually.

The major contributors on the total Ecological Footprint of the NPEMT were the annual consumptions of products and energy related with households (79%) and the annual consumption of energy related with private transportation (10%), as shown in Table 3.

Analyzing the Ecological Footprint of the NPEMT into its land uses as it is presented in Figure 5, it can be observed that the main contributors are "1. Agricultural products" (36%), "2. Livestock products" (30%), and "5. $CO_2$ Emissions" (30%). It is remarkable that the consumption needs of products and energy correspond to approximately 68% and 30%, respectively, of total Ecological Footprint. Among the products, beef annual consumption (22.8%) was identified as the key factor contributing to the Ecological Footprint of the spatial entity under study, while fruits and vegetables (15.2%) and beverages (14.3%) followed. The above results are in agreement with the literature [40], which indicates food consumption as a key contributor to the Ecological Footprint in areas where human activities exist. Potential changes in the nutritional habits of residents and tourists, such as partial replacing of beef, could significantly reduce the total Ecological Footprint.

**Table 3.** The contribution of several human activities on the total Ecological Footprint of the NPEMT.

| Description of Human Activities | Ecological Footprint | % of the Total Ecological Footprint of the NPEMT |
|---|---|---|
| Products and energy consumption for households | 143,628.3 | 79% |
| Energy consumption for buildings of tertiary sector | 10,285.8 | 6% |
| Energy consumption for municipal buildings | 1327.5 | 1% |
| Energy consumption for public lighting | 776.7 | 0% |
| Distance travelled by means of private transportation | 17,304.6 | 10% |
| Distance travelled by means of public transportation | 868.2 | 0% |
| Products and energy consumption for tourism | 3172.7 | 2% |
| Built-up areas | 3960.0 | 2% |
| Total | 181,324.0 | 100 |

The fact that the proposed framework has been applied effectively to the NPEMT is encouraging in relation to its applicability. Also, the development of the algorithm that calculates the Ecological Footprint, the Biocapacity, and the Carrying Capacity of the protected area under study by introducing 48 inputs has reduced the complexity and the expertise needed for the implementation of the proposed framework. The evaluation of the environmental sustainability of the protected area under study has been elaborated in a holistic approach taking into account anthropogenic activities. Moreover, the results of the framework can provide clear information about the progress towards sustainability and the ability to incorporate new activities within the environmental boundaries of the protected area under study.

On the other hand, the authors recognize a certain degree of subjectivity in the evaluation process which is related to the application of certain assumptions required due to the lack of data. All assumptions are supported by statistical data, surveys, literature references [41,42], etc. Moreover, there was no provision for the implementation of an uncertainty and sensitivity analysis method in order to identify and cope with the uncertainties raised by the integration of the LCA method in the proposed framework.

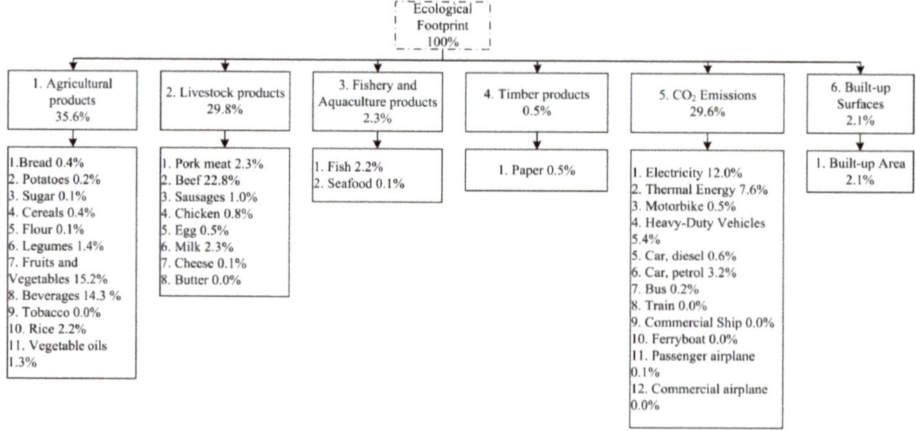


**Figure 5.** The Ecological Footprint results of the NPEMT.

## 4. Conclusions

The current work introduced a methodological framework that is able to evaluate the Environmental Sustainability of a spatial entity with anthropogenic activities and applied it to the National Park of Eastern Macedonia and Thrace. The key advantage of the proposed framework is that it can assess entities that present various human activities, such as households, tertiary sector, tourism, and transportation. In order to achieve this, the proposed methodological framework equates the meaning of Environmental Sustainability to Carrying Capacity, estimating the maximum equivalent population of residents which can live sustainably within a spatial entity. For this estimation, the following are considered: (a) the available biologically productive land (Biocapacity) of the spatial entity, (b) the land needed to satisfy the consumption needs of the anthropogenic activities in products and energy (Ecological Footprint) within the spatial entity, and (c) the real value of the equivalent residents, since all the consumption needs of the anthropogenic activities are not directly connected to the population size. For example, the energy consumption of buildings depends on their surface and use and not on the number of residents. The value of Carrying Capacity comes from the division of Biocapacity to Ecological Footprint per equivalent capita (EF/Peq).

The implementation of the proposed framework contributes to the improvement of the functionality and applicability of the existing methods for evaluating the environmental sustainability of protected areas, while providing clear information about the progress toward sustainability and the ability to incorporate a new activity within the boundaries of the protected area under study. However, particular points, related to the assumptions and the uncertainties, have been identified that need to be improved in order to strengthen the objectivity of the evaluation process and to treat the uncertainties due to the integration of the LCA approach.

Especially, the results from the application of the proposed framework in the NPEMT indicated a positive environmental performance of the region, since the Carrying Capacity is calculated at 39,193 equivalent residents, when the existing equivalent population was estimated at 36,960 equivalent residents. Therefore, the area is able to meet the needs of its equivalent population provided that the consumption patterns will be stable or improved. The results encourage a sensible expansion of touristic activities. Household activities are the main contributor (79%) to the total Ecological Footprint, whereas product consumption is responsible for 68% of the total Ecological Footprint, which is in agreement with other studies.

The authors are planning to continuously improve the effectiveness of the proposed framework by developing supporting software that will enable its fast and reliable implementation and by applying an uncertainty analysis for different scenarios in order to strengthen the reliability and credibility of the integration of the LCA approach in the proposed framework. Another key issue that needs to be examined in the future is the potential assistance of an experts group in order to determine the values of the assumptions needed in a more objective process.

The management bodies of protected areas have to advise the Public Administration on the sustainability of new projects to be carried out within their boundaries and to develop sustainable strategies ensuring the harmonious coexistence of natural capital with proposed anthropogenic activities. Therefore, both the management bodies of the protected areas and the accountable administrative authorities need to assess and quantify the impact of existing and future human activities that can be developed to satisfy current and future needs, while in parallel sustaining the environmental and ecological health. The proposed Environmental Sustainability Assessment framework can serve as a tool to support decision-making through Carrying Capacity estimation and monitoring.

**Author Contributions:** All authors contributed equally to the text. All authors have read and agreed to the published version of the manuscript.

**Funding:** This research received no external funding.

**Acknowledgments:** The authors would like to thank the NPEMT Management Body for supporting the case study implementation.

**Conflicts of Interest:** The authors declare no conflict of interest.

# Appendix A

Table A1. The Ecological Footprint indicators.

| Indicator | Calculation | Available Assumptions | Available Inputs | Activities to Apply |
|---|---|---|---|---|
| [1] Bread: Annual bread consumption | Average consumption of bread in g per kg of human mass per dayX days of consumption X average weightX number of persons | Average weight in kg of adult residents<br>Average weight in kg of adult tourists<br>Average weight in kg of minor residents<br>Average weight in kg of minor tourists<br>(3) Days per year of consumption for residents<br>(4) Days per year of consumption for residents<br>(5) Average consumption of bread in g per kg of human mass per day | (1) Number of adult residents<br>(2) Number of minor residents<br>(3) Number of adult tourists<br>(4) Number of minor tourists | Households<br>Tourism |
| Paper: Annual paper consumption | Average consumption of paper per capita per year X number of residents | Average consumption in t of paper per capita per year in Greece | (1) Number of adult residents<br>(2) Number of minor residents | Households |
| [2] Annual electricity consumption in households | Surface of single dwellings/apartment buildings X average electrical energy consumption per $m^2$ | (1) Average electrical energy consumption per $m^2$ for single dwellings 1980 in $kWh/m^2$<br>(2) Average electrical energy consumption per $m^2$ for single dwellings 2001 in $kWh/m^2$<br>(3) Average electrical energy consumption per $m^2$ for single dwellings 2010 in $kWh/m^2$<br>(4) Average electrical energy consumption per $m^2$ for apartment buildings 1980 in $kWh/m^2$<br>(5) Average electrical energy consumption per $m^2$ for apartment buildings 2001 in $kWh/m^2$<br>(6) Average electrical energy consumption per $m^2$ for apartment buildings 2010 in $kWh/m^2$ [42] | (1) Surface (in $m^2$) of single dwellings built before 1980<br>(2) Surface (in $m^2$) of single dwellings built between 1981 and 2001<br>(3) Surface (in $m^2$) of single dwellings built after 2002<br>(4) Surface (in $m^2$) of apartment buildings built before 1980<br>(5) Surface (in $m^2$) of apartment buildings built between 1981 and 2001<br>(6) Surface (in $m^2$) of apartment buildings built after 2002 | Households |
| [2] Annual electricity consumption in tertiary sector, municipal buildings, and tourism | Number of each type of building X average electrical energy consumption per $m^2$ for each type of building X average surface ($m^2$) of each type of building | (1) Average electrical energy consumption per $m^2$ for offices/commercial buildings 1980 in $kWh/m^2$<br>(2) Average electrical energy consumption per $m^2$ for offices/commercial buildings 2001 in $kWh/m^2$<br>(3) Average electrical energy consumption per $m^2$ for offices/commercial buildings 2010 in $kWh/m^2$<br>(4) Average electrical energy consumption per $m^2$ for healthcare buildings 1980 in $kWh/m^2$<br>(5) Average electrical energy consumption per $m^2$ for healthcare buildings 2001 in $kWh/m^2$<br>(6) Average electrical energy consumption per $m^2$ for healthcare buildings 2010 in $kWh/m^2$<br>(7) Average surface ($m^2$) of offices/commercial buildings built before 1980<br>(8) Average surface ($m^2$) of offices/commercial buildings built between 1981 and 2001<br>(9) Average surface ($m^2$) of offices/commercial buildings built after 2001 | (1) Number of offices/commercial buildings built before 1980<br>(2) Number of offices/commercial buildings built between 1981 and 2001<br>(3) Number of offices/commercial buildings built after 2001<br>(4) Number of healthcare buildings built before 1980<br>(5) Number of healthcare buildings built between 1981 and 2001<br>(6) Number of healthcare buildings built after 2001<br>(7) Number of schools built before 1980<br>(8) Number of schools built between 1981 and 2001<br>(9) Number of schools built after 2001 | Tertiary sector, Municipal buildings<br>Tourism |

**Table A1.** *Cont.*

| Indicator | Calculation | Available Assumptions | Available Inputs | Activities to Apply |
|---|---|---|---|---|
|  |  | (10) Average surface ($m^2$) of healthcare buildings built before 1980<br>(11) Average surface ($m^2$) of healthcare buildings built between 1981 and 2001<br>(12) Average surface ($m^2$) of healthcare buildings built after 2001<br>(13) Average electrical energy consumption per $m^2$ for schools 1980 in kWh/$m^2$<br>(14) Average electrical energy consumption per $m^2$ for schools 2001 in kWh/$m^2$<br>(15) Average electrical energy consumption per $m^2$ for schools 2010 in kWh/$m^2$<br>(16) Average surface ($m^2$) of schools built before 1980<br>(17) Average surface ($m^2$) of schools built between 1981 and 2001<br>(18) Average surface ($m^2$) of schools built after 2001<br>(19) Average electrical energy consumption per $m^2$ for hotels 1980 in kWh/$m^2$<br>(20) Average electrical energy consumption per $m^2$ for hotels 2001 in kWh/$m^2$<br>(21) Average electrical energy consumption per $m^2$ for hotels 2010 in kWh/$m^2$<br>(22) Average surface ($m^2$) for hotels built before 1980<br>(23) Average surface ($m^2$) for hotels built between 1981 and 2001<br>(24) Average surface ($m^2$) for hotels built after 2001 [41] | (10) Number of hotels built before 1980<br>(11) Number of hotels built between 1981 and 2001<br>(12) Number of hotels built after 2002 |  |
| Annual electricity consumption in public lighting | Average time of lighting operation X Installed power for public lighting | (1) Average time of lighting operation per year in hours | (1) Installed power for public lighting in kW | Public Lighting |

**Table A1.** *Cont.*

| Indicator | Calculation | Available Assumptions | Available Inputs | Activities to Apply |
|---|---|---|---|---|
| Distance travelled by means of private and public transportation (motorbikes, cars, and heavy-duty vehicles) | Average km passing by vehicle X Number of private/public vehicles | Average km travelled by vehicles on local roads per year<br>The national percentage of cars fueled by diesel<br>The national percentage of cars fueled by petrol | Number of private motorbikes moving on local roads<br>Number of public motorbikes moving on local roads<br>Number of private heavy-duty vehicles moving on local roads<br>Number of private heavy-duty vehicles moving on each part of the highway set within the boundaries of the protected area<br>km of each part of the highway set within the boundaries of the protected area<br>Number of public heavy-duty vehicles moving on local roads<br>Number of private cars moving on local roads<br>Number of private cars moving on each part of the highway set within the boundaries of the protected area<br>Number of public car moving on local roads. | Public and Private Transportation |
| Distance travelled by means of public transportation (bus) | km by bus within the boundaries of the protected area | | km by bus within the boundaries of the protected area | Public Transportation |
| Distance travelled by means of public transportation (train) | Number of annual passenger X km of local railway | | (1) Number of annual passengers moving by train on local railway<br>(2) km of local railway set within the boundaries of the protected area | Public Transportation |
| Distance travelled by means of public transportation (commercial ship) | t loaded or/and unloaded from/to commercial ship X km boarding port | | t loaded or/and unloaded from/to commercial ship in each port,<br>km boarding in each port | Public Transportation |

**Table A1.** *Cont.*

| Indicator | Calculation | Available Assumptions | Available Inputs | Activities to Apply |
|---|---|---|---|---|
| Distance travelled by means of public transportation (ferryboat) | t loaded or/and unloaded from/to commercial ship X km boarding port | (1) Average weight of a passenger in t<br>(2) Average weight of a heavy-duty vehicle in t<br>(3) Average weight of a bus in t<br>(4) Average weight of a car in t<br>(5) Average weight of a motorbike in t | (1) km boarding at each port<br>(2) Number of passengers loaded to ferryboat at each port<br>(3) Number of heavy-duty vehicles loaded to ferryboat at each port<br>(4) Number of buses loaded to ferryboat at each port<br>(5) Number of cars loaded to ferryboat at each port<br>(6) Number of motorbikes loaded to ferryboat at each port | Public Transportation |
| Distance travelled by means of public transportation (passenger aircraft) | Number of passengers X km traveled by airplane | | (1) Number of passengers arrived by airplane in each airport<br>(2) Number of passengers taking off by airplane from each airport<br>(3) km travelled by airplane during landing at each airport<br>(4) km travelled by airplane during taking off at each airport | Public Transportation |
| Distance travelled by means of public transportation (commercial aircraft) | t loaded/unloaded X km traveled by airplane | | (1) t loaded to airplane at each airport<br>(2) t unloaded from airplane at each airport<br>(3) km travelled by airplane during landing at each airport<br>(4) km travelled by airplane during taking off at each airport | Public Transportation |
| Built-up area | Built-up Areas X Cropland Global Equivalent Factor | | Areas occupied by the locality (buildings, roads, etc.) | |

[1] All the product's consumptions (agricultural, livestock and fishery, and aquaculture products) were calculated similarly to the calculation of "bread consumption," replacing the average consumption of each product in g per kg of human mass per day.[2] The annual thermal energy consumption was calculated similarly to the annual electricity consumption.

## References


1. Daily, G.C. *Nature's Services: Societal Dependence on Natural Ecosystems*, 1st ed.; Island Press: Washington, DC, USA, 1997; pp. 1–412.
2. Graymore, M.L.; Sipe, N.G.; Rickson, R.E. Sustaining Human Carrying Capacity: A tool for regional sustainability assessment. *Ecol. Econ.* **2010**, *69*, 459–468. [CrossRef]
3. Graymore, M.L.M.; Sipe, N.G.; Rickson, R.E. Regional sustainability: How useful are current tools of sustainability assessment at the regional scale? *Ecol. Econ.* **2008**, *67*, 362–372. [CrossRef]
4. Manning, R.E. *Parks and Carrying Capacity: Commons Without Tragedy*; Island Press: Washington, DC, USA, 2007; pp. 1–328.
5. Farrell, T. The Protected Area Visitor Impact Management (PAVIM) framework: A simplified process for making management decisions. *J. Sustain. Tour.* **2002**, *10*, 31–51. [CrossRef]
6. Saarinen, J. Traditions of sustainability in tourism studies. *Ann. Tour. Res.* **2006**, *33*, 1121–1140. [CrossRef]
7. Lawson, S.R.; Manning, R.E.; Valliere, W.A.; Wang, B. Proactive monitoring and adaptive management of social carrying capacity in Arches National Park: An application of computer simulation modeling. *J. Environ. Manag.* **2003**, *68*, 305–313. [CrossRef]
8. Needham, M.; Szuster, B.; Bell, C. Encounter norms, social carrying capacity indicators, and standards of quality at a marine protected area. *Ocean. Coast. Manag.* **2011**, *54*, 633–641. [CrossRef]
9. Prato, T. Modeling carrying capacity for national parks. *Ecol. Econ.* **2001**, *39*, 321–331. [CrossRef]
10. Prato, T. Fuzzy adaptive management of social and ecological carrying capacities for protected areas. *J. Environ. Manag.* **2009**, *90*, 2551–2557. [CrossRef]
11. Wagar, J.A. The Carrying Capacity of Wildlands for Recreation. *Forest Sci.* **1964**, *10*, 1–24.
12. Manning, R. *Studies in Outdoor Recreation. Search and Research for Satisfaction*, 3rd ed.; Cornallis Oregon State University Press: Oregon, OR, USA, 1999; pp. 1–448.
13. Stankey, G.; Cole, D.; Lucas, R.; Petersen, M.; Frissell, S.; Washburne, R. *The Limits of Acceptable Change (LAC) System for Wilderness Planning. General Technical Report*; USDA Forest Service, Intermountain Forest and Range Experiment Station: Ogden, UT, USA, 1985; p. 37.
14. Kuss, F.; Graefe, A.; Vaske, J. *Visitor Impact Management, v.2 The Planning Framework*; National Parks and Conservation Association: Washington, DC, USA, 1990.
15. National Park Service. *Visitor Experience and Resource Protection (VERP) Framework. A Handbook for Planners and Managers*; US Department of the Interior, National Park Service, Denver Service Center: Denver, CO, USA, 1997; pp. 1–108.
16. Payne, R.J.; Nilsen, P. Innovations and Challenges in the Management of Visitor Opportunities in Parks and Protected Area: Commemorating the Work of the Late Robert Graham. In Proceedings of the Workshop held at the University of Waterloo, Heritage Resources Center, Waterloo, ON, Canada, 1 December 1994.
17. Satta, A.; Klaric, Z.; Mangion, M.L.; Travic, A.S. *Guide to Good Practice in Tourism Carrying Capacity Assessment*, 1st ed.; Priority Actions Programme/Regional Activity Centre: Split, Croatia, 2003.
18. Hellenic Parliament. *Law No. 3937/2011: Conservation of Biodiversity and Other Provisions*; Hellenic Parliament: Athens, Greece, 2011.
19. Aktsoglou, D.; Gaidajis, G. Environmental Sustainability Assessment of Spatial Entities with Anthropogenic Activities-Evaluation of Existing Methods. *Sustainability* **2020**, *12*, 2680. [CrossRef]
20. Angelakoglou, K.; Gaidajis, G. A Conceptual Framework to Evaluate the Environmental Sustainability Performance of Mining Industrial Facilities. *Sustainability* **2020**, *12*, 2135. [CrossRef]
21. Liu, H. Comprehensive carrying capacity of the urban agglomeration in the Yangtze River Delta, China. *Habitat Int.* **2012**, *36*, 462–470. [CrossRef]
22. Rees, W.E. Urban ecosystems: The human dimension. *Urb. Ecosyst.* **1997**, *1*, 63–75. [CrossRef]
23. Cohen, J. How Many People Can. The Earth Support? W.W. Norton & Company: New York, NY, USA, 1995.
24. Lane, M.C. The carrying capacity imperative: Assessing regional carrying capacity methodologies for sustainable land-use planning. In Proceedings of the 53rd Annual Meeting of the International Society for the Systems Sciences, Brisbane, Australia, 12–17 July 2009.
25. Peters, C.J.; Wilkins, J.L.; Fick, W.G. Testing a Complete-Diet Model for Estimating the Land Resource Requirements of Food Consumption and Agricultural Carrying Capacity-The New York State Example. *Renew. Agric. Food Syst.* **2007**, *22*, 145–153. [CrossRef]

26. Fairlie, S. Can Britain Feed Itself? *Land* **2007**, *4*, 18–26.

27. Peters, C.J.; Wilkins, J.L.; Fick, W.G. Input and Output Data in Studying the Impact of Meat and Fat on the Land Resource Requirements of the Human Diet and Potential Carrying Capacity-The New York State Example. In *CSS Research Series*; Department of Crop and Soil Sciences: Ithaca, NY, USA, 2005; pp. 1–25.

28. Schroll, H.; Anderson, J.; Kjaergard, B. Carrying Capacity: An approach to local spatial planning in Indonesia. *J. Transdiscipl. Environ. Stud.* **2012**, *11*, 27–39.

29. Azapagic, A.; Perdan, S. *Sustainable Development in Practice: Case Studies for Engineers and Scientists*, 2nd ed.; John Wiley-BlackWeeel Ltd.: West Sussex, England, 2011.

30. Wackernagel, M.; Onisto, L.; Bello, P.; Linares, A.C.; Falfan, I.S.L.; Garcva, J.M.; Guerrero, A.I.S.; Guerrero, M.G.S. *National natural capital accounting with the ecological footprint concept. Ecol. Econ.* **1999**, *29*, 375–390.

31. Wackernagel, M.; Moran, D.; Wermer, P.; Goldfinger, S.; Deumling, D.; Murray, M. *National Footprint and Biocapacity Accounds 2005-The Underlying Calculation Method, 2005 ed.*; Global Footprint Network: Oakland, CA, USA, 2005.

32. Cucek, L.; Klemes, J.J.; Kravanja, Z. A review of Footprint analysis tool for monitoring impacts on sustainability. *J. Clean. Prod.* **2012**, *34*, 9–20. [CrossRef]

33. Haggar, S. *Sustainable Industrial Design and Waste Management: Cradle-to-Grave for Sustainable Development*, 1st ed.; Elsevier Academic Press: San Diego, CA, USA, 2007.

34. Borucke, M.; Moore, D.; Cranston, G.; Gracey, K.; Katsunori, I.; Larsoa, J.; Lazaru, E.; Moralea, J.C.; Wackernagea, M.; Galli, A. Accounting for demand and supply of the biosphere's regenerative capacity: The National Footprint Accounts' underlying methodology and framework. *Ecol. Indic.* **2013**, *24*, 518–533. [CrossRef]

35. Scotti, M.; Bondavalli, C.; Bodini, A. Ecological Footprint as a tool for local sustainability: The municipality of Piacenza (Italy) as a case study. *Environ. Impact Assess. Rev.* **2009**, *29*, 39–50. [CrossRef]

36. Ewing, B.; Goldfinger, S.; Wackernagel, M.; Stechbart, M.; Rizk, S.M.; Reed, A.; Kitzes, J. *The Ecological Footprint Atlas 2008, 2008 ed.*; Global Footprint Network: Oakland, CA, USA, 2008.

37. European Environment Agency. Available online: https://www.eea.europa.eu/publications/COR0-part1 (accessed on 10 November 2019).

38. Castellani, V.; Sala, S. Ecological footprint and life cycle assessment in the sustainability assessment of tourism activities. *Ecol. Indic.* **2012**, *16*, 135–147. [CrossRef]

39. Open Public Data. Land Cover of Greece in 2000, According to the Deliverables of the CORINE Program of the European Union. Available online: http://www.geodata.gov.gr/geodata/index.php?option=com_sobi2&sobi2Task=sobi2Details&catid=16&sobi2Id=54&Itemid (accessed on 10 December 2014).

40. Weber, C.; Matthews, S. Food-Miles and the Relative Climate Impacts of Food Choices in the United States. *Environ. Sci. Technol.* **2008**, *42*, 3508–3513. [CrossRef]

41. Balaras, C.; Gaglia, A.; Georgopoulou, E.; Mirasgedis, S.; Sarafidis, Y.; Lalas, D. European residential buildings and empirical assessment of the Hellenic building stock, energy consumption, emissions and potential energy savings. *Build. Environ.* **2007**, *42*, 1298–1314. [CrossRef]

42. Gaglia, A.; Balaras, C.; Mirasgedis, S.; Georgopoulou, E.; Sarafidis, Y.; Lalas, D. Empirical assessment of the Hellenic non-residential building stock, energy consumption, emissions and potential energy savings. *Energ. Convers. Manag.* **2007**, *48*, 1160–1175. [CrossRef]

 *sustainability*

*Article*

# A Joint Stochastic/Deterministic Process with Multi-Objective Decision Making Risk-Assessment Framework for Sustainable Constructions Engineering Projects—A Case Study

**Panagiotis K. Marhavilas [1,*], Michael G. Tegas [2], Georgios K. Koulinas [1] and Dimitrios E. Koulouriotis [1]**

[1] Department of Production and Management Engineering, Democritus-University of Thrace, Vasil. Sofias 1 St., 67132 Xanthi, Greece; gkoulina@pme.duth.gr (G.K.K.); jimk@pme.duth.gr (D.E.K.)

[2] Department of Engineering Project Management, Faculty of Science & Technology, Hellenic Open University, Parodos Aristotelous 18 St., 26335 Patra, Greece; mtegas@otenet.gr

* Correspondence: marhavil@pme.duth.gr

Received: 29 April 2020; Accepted: 21 May 2020; Published: 23 May 2020

**Abstract:** This study, on the one hand, develops a newfangled risk assessment and analysis (RAA) methodological approach (the MCDM-STO/DET one) for sustainable engineering projects by the amalgamation of a multicriteria decision-making (MCDM) process with the joint-collaboration of a deterministic (DET) and a stochastic (STO) process. On the other hand, proceeds to the application of MCDM-STO/DET at the workplaces of the Greek construction sector and also of the fixed-telecommunications technical projects of OTE SA (that is, the Greek Telecommunications Organization S.A.) by means of real accident data coming from two official State databases, namely of "SEPE" (Labor Inspectorate, Hellenic Ministry of Employment) and of "IKA" (Social Insurance Institution, Hellenic Ministry of Health), all the way through the period of the years2009–2016.Consequently, the article's objectives are the following: (i) The implementation and execution of the joint MCDM-STO/DET framework, and (ii) to make known that the proposed MCDM-STO/DET algorithm can be a precious method for safety managers (and/or decision-makers) to ameliorate occupational safety and health (OSH) and to endorse the sustainable operation of technical or engineering projects as well. Mainly, we mingle two different configurations of the MCDM method, initially the Analytical Hierarchy-Process (the typical-AHP), and afterwards the Fuzzy-Extended AHP (the FEAHP) one, along with the Proportional Risk Assessment Technique (PRAT) and the analysis of Time-Series Processes (TSP), and finally with the Fault-Tree Analysis (FTA).

**Keywords:** MCDM; AHP; fuzzy AHP; PRAT; FTA; time-series; risk assessment; OSH; telecom technical projects; constructions

---

## 1. Introduction

Taking into consideration that one of the intentions of "sustainable engineering" is to create structures that validate evolution without jeopardizing the natural environment's quality, and also the coming generations' capability to meet their own necessities, modern engineering tactics concentrate (among other options) on the systems' design/operation in a manner that admits the sustainable usage of resources. Thus, decision-making processes have to be enhanced by techniques and methods which permit the decision-makers to use a broad variety of sustainable alternatives.

The protection of the employees' health and safety in the workplace is a universally vital aspect, in parallel with the effort of enhancing the productivity and upgrading the sustainable development of a firm. Moreover, the World Health Organization (WHO) defines the sustainable development as a policy *"to meet the needs of today's global population without adversely affecting its health and the environment and without destroying (or endangering) the global resource base, and thus without jeopardizing the ability of the coming generations to meet their needs"* [1–3].

Occupational accidents, on the one hand, have a significant impact upon human body-integrity, and affect the employees, business operation, and overall the sustainability performance of firms, and on the other hand, generate high expenses for the social health and insurance system of any country, and demote the society's sustainability. Each corporation is progressively more concerned with the improvement of sustainability and occupational safety and health (OSH) performance, and this is attained by controlling OSH risks, in accordance with their OSH policy, and in the context of applying risk assessment and analysis (RAA) techniques and the state's legislation. Besides, the Occupational Safety & Health Administration (OSHA) recognizes that actions to ensure occupational safety and health can be incorporated into sustainability efforts and take advantage of the dynamics of sustainability to make companies' workplaces safer and healthier [3–5].

In this sense, the conception of sustainability is used as a frame to categorize existing and/or develop new OSH standards (e.g., ISO 45001, OSHAS 18001, etc.), and also to generate multitudinous efficacious and modern RAA techniques [5].

In general, the variety of RAA assessment methods is such that there are numerous proper approaches for most occasions, and plentiful studies build up hybrid models merging separate techniques. For example, RAA techniques are classified, on the one hand, into quantitative/qualitative/ hybrid techniques, and on the other side, into deterministic and stochastic ones.

However, the examination of the scientific literature with reference to RAA methodologies unveils that every technique presents specific limitations or restrictions in its applicability to miscellaneous accidents, and consequently a separate (or solitary) technique: (i) Can't generate either a realistic risk prediction model or a sufficient RAA process in the worksites, and (ii) can't achieve the optimum risk-assessment result in the workplaces, but on the other hand, future prospects must focus: (i) On the joint estimation/forecasting process, and (ii) on the parallel utilization of a deterministic approach with a stochastic one [6–10].

The above topics constitute the research context of this study, and consequently, its foremost contribution is the under mentioned:(i) The implementation of an alternative RAA methodological framework for upgrading the OSH's situation and sustainability of engineering projects by the combination of a multicriteria decision-making (MCDM) process with the joint collaboration of a stochastic (STO) and a deterministic (DET) process, and (ii) the application of it on the worksites of the Greek construction sector and of fixed-telecommunications technical projects of OTE SA (the Greek Telecommunications Organization SA), by means of real accident data.

The paper is organized into the subsequent sections: (1) Introduction, (2) literature survey, (3) development of the suggested methodological framework, (4) application of the new method on the constructions and telecom sector and analysis of their results, and (5) discussion of the main points and conclusions.

## 2. Literature Survey

Plenty of research studies have been conducted in copious construction industries contributing to the scientific field of OSH by revealing the features (advantages and disadvantages) and also presenting the implementation of numerous RAA techniques. Below, we conduct a survey of the scientific literature concerning RAA techniques, which are used in the OSH field along with the constructions sector, and which are concentrated, on the other hand, on several deterministic (DET), stochastic (STO), and MCDM processes (like PRAT, FTA, TSP, AHP, FEAHP, etc.), that have been incorporated in the newly proposed integrated-method. More explicitly, our study develops a novel method which incorporates three different types of techniques (DET, STO, and MCDM), so the presented literature survey covers these three special classes of techniques.

On the whole, RAA techniques are categorized into three significant groups (e.g., quantitative/ qualitative/hybrid techniques) whereas another methodical categorization comprises the deterministic and the stochastic approaches [8–17].

More specifically, Marhavilas et al. (2011) [8], based on the literature's examination, determined the dominant RAA methods, and categorized them into three principal classes, i.e., the qualitative, the quantitative, and the hybrid (semi-quantitative or qualitative–quantitative) methods, wherein the first ones are grounded mainly on the safety-managers' capability, the second ones express the risk by a mathematical relation in conjunction with real accident data recorded in a workplace, and finally, the third ones are characterized by a high complexity due to their ad-hoc feature. A plethora of other works confirm the results of this study, and also the importance of its classification, like the articles of (i) da Costa et al. (2020) [18], Gul and Celik (2018) [19], AriamuthuVenkidasalapathy et al. (2018) [20], Ozkan and Uluta (2017) [21] *(for quantitative RAA)*, (ii) Athar et al. (2019) [22], Sanmiquel-Pera et al. (2019) [23], Domínguez et al. (2019) [24] *(for qualitative RAA)*, and (iii) Kharzi et al. (2020) [25], Cinar and Cebi (2020) [26], Mutlu et al. (2019) [27], Bora et al., (2019) [28], Kamsu-Foguem and Tiako (2017) [29], Zheng et al. (2017) [30], Mentes and Ozen (2015) [31] (for hybrid RAA).

One of the most practical and skillful numerical formula for evaluating the quantified risk (R) at workplaces is the proportional risk-assessment DET-technique (PRAT) which takes into consideration the severity (S) of a harm (i.e., the potential consequences of an accident), the occurrence likelihood (P) i.e., the probability factor of that harm, and the exposure factor [32–37]. This concept becomes operational, and hence many studies, like the papers of Gul and Celik (2018), Gul et al. (2018), Kokangül et al. (2017), Marhavilas (2015), Marhavilas et al. (2011b), Marhavilas and Koulouriotis (2008) [19,38–42] combined exposure-factors with likelihood-factors to arrive at the frequency (F), and to evaluate, on the one hand, the risk by

$$R = P \cdot S \cdot F \tag{1}$$

and consequently, unsafe OSH situations at worksites by means of exposure, which estimates the hazard's occurrence frequency.

Continuing on, fault-tree analysis (FTA) and event-tree analysis (ETA) are, according to Marhavilas et al. (2014a, 2014b) [43,44], well known DET-methods and valuable analytic tools worldwide, which have been used successfully by reliability experts in failure-analysis and for the reliability and safety of complex technical-systems, and also of OSH-systems. Nowadays, many studies continue incorporating FTAs/ETAs in OSH RAA, like: (i) The work of Mutlu et al. (2019) [27], which depicts a OSH RAA approach in the textile industry by integrating FTA with Failure-Mode and Effect-Analysis (FMEA) and BIFPET, (ii) the articles of Babaei et al. (2018) [45] and Yasli and Bolat (2018) [46], which include FTA in fuzzy OSH RAA methods, (iii) the paper of Gul and Ak (2018) [47], which illustrates the collaboration of MCDM with FTA, (iv) the work of Fuentes-Bargues et al. (2017) [48], which combines Hazard and Operability (HAZOP) and FTA in risk analysis of fuel storage-terminals, and (v) the paper of Marhavilas et al. (2014a) [43], which physically embody FTA in DET OSH-RAA techniques.

Over and above that, the stochastic (STO) processes, like the Time-Series Processes (TSP), Markov-chains, etc., play an important role in the recognition of potential hazardous sources and also in the methodologies of accident analysis, RAA, prediction and modeling, and additionally their contribution in the analysis of accident tendency, and to the forecasting of the safety's situation based on the real accident observations and/or data, is vital [7,17,49–53].

Moreover, there are many studies, in the relative scientific literature, that develop MCDM methods by hybridizing existing, very popular approaches, to improve results in risk assessments. Zheng et al. (2012) [54] employed a variation of fuzzy Analytical Hierarchy-Process (AHP) which uses trapezoidal fuzzy numbers, for risk evaluation under hot and humid environment conditions, and Fattahi and Khalilzadeh (2018) [55] introduced a framework for assessing risks in steel industry using Failure-Mode and Effect-Analysis (FMEA), fuzzy AHP, and Fuzzy Multi-Objective Optimization on the basis of Ratio Analysis (MULTIMOORA) methods. In the study of Mete (2019) [56], a hybrid fuzzy analytical hierarchy process and data envelopment analysis-based framework were proposed for managing uncertainty, while assessing risks in pipeline construction projects. Recently, Oturakci (2019) [57] introduced a novel fuzzy sets and Analytical Hierarchy Process based approach for environmental risks evaluation, while Yucesan and Kahraman (2019) [58] applied the Pythagorean fuzzy analytical hierarchy process method for assisting a team of experts to evaluate risks in hydroelectric power plants. In the study of Boral et al. (2020) [59], a new approach was developed which combines the Fuzzy Analytical Hierarchy Process with the modified FMAIRCA (Fuzzy Multi-Attribute Ideal Real Comparative Analysis) in order to improve the disadvantages of the FMEA technique, used for proactive risk assessment.

It is worth mentioning that there are numerous other published papers, pertaining to OSH management systems and to RAA techniques, and focusing on construction sites. Therefore, Marhavilas (2009, 2015) [40,60], Ardeshir et al. (2014) [61], Anil Kumar et al. (2015) [62], Marhavilas and Vrountas (2018) [63], and Koulinas et al. (2019a; 2019b) [3,64] presented the implementation of various OSH RAA techniques at workplaces of the constructions section.

## 3. The Suggested Methodological Framework

Figure 1 displays the flow sheet of the suggested RAA generic algorithmic framework, which couples a multicriteria decision-making (MCDM) process with both a stochastic (STO) and a deterministic (DET) process. More especially, we join the MCDM-AHP/Fuzzy-Extended AHP (FEAHP) techniques with the DET-PRAT and DET-FTA processes, and also with the STO-TSP method.

Consistent with this diagram, the real sequential data A(t) (for instance accident data) recorded at any workplace can be utilized firstly in the phase of hazard-identification, and subsequently in the DET-PRAT process for quantitatively calculating the risk as a function of time R(t). Moreover, the DET-TSP process can be applied in parallel with DET-PRAT process in order to achieve a joint-evaluation of their results. Particularly, when the time-series data A(t) are sufficient (according to Nyquist-Shannon's sampling-theorem) with reference to a specific time-period, the generated by PRAT time-series R(t) could make-up inputs for TSP, which means they could be analyzed by miscellaneous substantial methods (like the moving-average one, the seasonal-adjustment, the auto-regression, the function-fitting one, etc.), unveiling in many cases significant features for the time-series variation of R(t) and also for the company's RAA situation (such as a long-term trend, a periodicity, etc.).

It is worth noting that the combination of PRAT and TSP techniques can give, by the calculation of risk (R), the essential information to safety-managers in order to (i) identify (taking into consideration the calculated risk level) the most important sources of danger at the companies' worksites, and also (ii) to achieve a preliminary ranking of them. Thus, the resulted ranking (1st ranking) can be precious for evaluating the urgency of investing a restricted budget in specific measures in order to attain the highest OSH-protection with the lowest cost.

To continue, the next segment in the above referred flow sheet is the MCDM one, which represents the application of a multicriteria decision-making technique, like the AHP and/or the FEAHP one, for accomplishing the final ranking (2md ranking) in the sources of danger defined at worksites, in order to achieve the finest (optimum) distribution of an investment budget (i.e., with the minimum cost) to appropriate measures reaching the best OSH-protection.

Ultimately, the proposed RAA generic algorithmic framework is completed by the application of the FTA process in order to: (i) Graphically illustrate possible occurrences, which can result in adverse events (the "base-events"), (ii) relate their sequences, which could lead to a "top-event", and (iii) to calculate the occurrence likelihood of the "top-event". Therefore, the occurrence-probability of the top-event is calculated by using the following equations:

$$P\left(\bigcup_{i=1}^{n} B_i\right) = \sum_{i=1}^{n} P(B_i) - \sum_{i=1}^{n-1}\sum_{j=i+1}^{n} P\left(B_i \cap B_j\right) + \sum_{i=1}^{n-2}\sum_{j=i+1}^{n-1}\sum_{k=j+1}^{n} P\left(B_i \cap B_j \cap B_k\right) - \ldots + (-1)^{n-1} P(B_1 \cap B_2 \cap \ldots \cap B_n) \tag{2}$$

or

$$P\left(\overset{n}{\underset{i=1}{\cup}} B_i\right) = 1 - \prod_{i=1}^{n} [1 - P(B_i)] \tag{3}$$

and

$$P\left(\overset{n}{\underset{i=1}{\cap}} B_i\right) = \prod_{i=1}^{n} [P(B_i)] \tag{4}$$

where $B_i$ ($i = 1, \dots, n$) are the identified (by FTA) "base-events", and taking into account that the first (or equivalently the second) relation is used when the output-event is logically-valid (according to Boolean algebra) if one of the input-events applies (i.e., fulfilling an OR-Gate logical-operation), while the third relation is used when the output-event is logically-valid only if all input-events apply (i.e., through an AND-Gate Boolean-function).

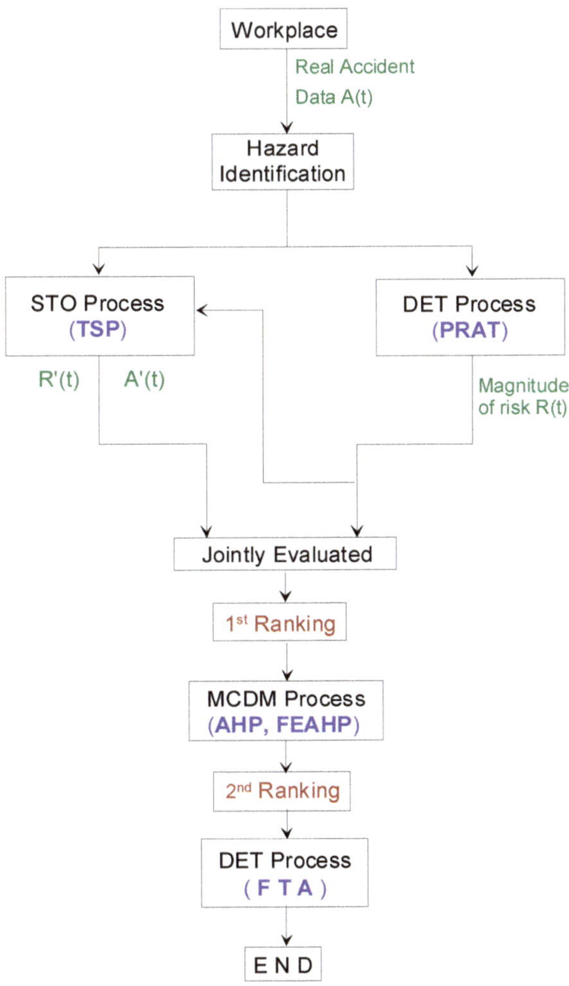


**Figure 1.** The joint stochastic/deterministic with multi criteria decision-making (MCDM) risk-assessment methodological framework.

## 4. Application of the Suggested Methodological Framework on the Constructions Sector

We proceed to the application of the novel methodological framework on the worksites of the Greek construction-sector(C-S)and of fixed-telecommunications technical-projects of OTE SA (the Greek Telecommunications Organization SA), by using real accident data recorded in two official State data sources, namely of "SEPE" (Labor Inspectorate, Hellenic/Greek Ministry of Employment) and of "IKA" (Social Insurance Institution, Hellenic/Greek Ministry of Health) with reference to the years 2009–2016.

The hazard identification (for instance, by HAZOP) and the risk analysis of the sources of danger (or deviations) existing at the worksites of the Greek constructions section are unveiled in Table 1, which depicts the deviations according to the classification of ESAW (European Statistics on Accidents at Work) by EU (2013) [65]. More specifically, the column "c" illustrates the description of the various hazards (or deviations) determined by the usage of "SEPE" and "IKA" statistical databases, while columns "b" and "a" showing the ESAW codes of the deviations, and the general types of the deviations (1st level of categorization), respectively.

**Table 1.** The identified (throughout the period of2009–2016) deviations at the workplaces of the Greek constructions section (C-S) and of fixed-telecommunications technical-projects of OTE SA (the Greek Telecommunications Organization SA), in accordance with the classification of ESAW (European Statistics on Accidents at Work) by EU (2013) [65].

| 1st Level of Hazards<br>Type of Deviation | ESAW-Code<br>(ES-C) | 2nd Level of Hazards<br>Description of Hazards (or Injuries) |
|---|---|---|
| (a) | (b) | (c) |
| Working Environment & Processes | 10 | Deviation due to electrical problems, explosion, fire<br>*[e.g., (i) exposure to or contact with extreme temperature levels, (ii) exposure to or contact with electric current, etc.]* |
| Working Environment & Processes | 20 | Deviation by overflow, overturn, leak, flow, vaporization, emission<br>*(e.g., exposure to or contact with hazardous substances or radiation, etc.)* |
| Working Environment & Processes | 30 | Breakage, bursting, splitting, slipping, fall, collapse of material agent<br>*(e.g., slipping, collapse and being struck by falling objects)* |
| Psychological/Human | 40 | Loss of control (total or partial) of machine, means of transport or handling equipment, hand-held tool, object, animal<br>*[e.g., (i) collision with an immobile object and falling against or being struck by moving objects, (ii) trapping, being crushed-inside or between objects, etc.]* |
| Physical Activity | 50 | Slipping—stumbling and falling—fall of persons<br>*[e.g., (i) falling of person from a height, (ii) falling of person-on the same level]* |
| Physical Activity | 60 | Body movement without any physical stress (generally leading to an external injury) |
| Physical Activity | 70 | Body movement under or with physical stress (generally leading to an internal injury)<br>*[e.g., physical strain—over-exertion]* |
| Psychological/Human | 80 | Shock, fright, violence, aggression, threat, presence |
| Working Environment & Processes | 99 | Other deviations not listed above in this classification |

### 4.1. Application of the PRAT and TSP Processes

The application of PRAT process was achieved by having calculated the risk (with its implementation equation 1), regarding all the hazard sources of Table 1, and using accident data from the above referenced databases ("SEPE" and "IKA") throughout the period of 2009–2016.

In Figure 2, we display time-variations of the magnitude of risk (R) calculated by the PRAT and by means of real data of "SEPE", through the years 2009–2016, that concern the most important hazard-source (i.e., the one with the highest risk) at the workplaces of the construction-sector (C-S) and also of the fixed-telecommunications technical-projects of OTE SA, which has been proven to be the risk-source of *"slipping—stumbling and falling—fall of persons"* (i.e., with ES-C#50).

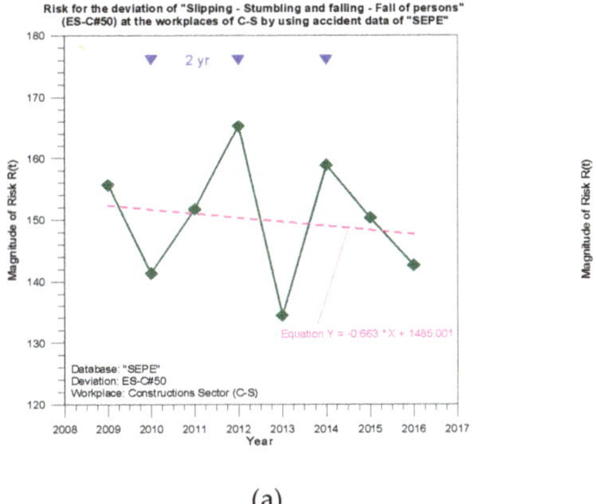
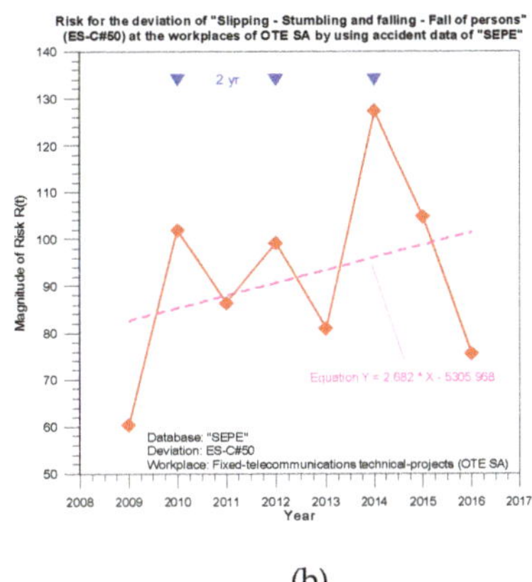

(a)                                    (b)

**Figure 2.** Time-variations of the risk (R) calculated by proportional risk-assessment technique (PRAT) utilizing accident data of "SEPE", during 2009–2016, and concerning the most considerable hazard-source of "slipping—stumbling and falling—fall of persons" (with ES-C#50) at the workplaces of the construction sector (C-S) (**a**) and also of OTE SA (**b**).

In Figure 3, we illustrate time-profiles of risk (R) calculated by the PRAT and via accident data of "IKA", during the same period of 2009–2016, that regard the most significant hazard-source at the worksites of the C-S and of the technical-projects of OTE SA as well, which has been proven to be the risk-source of *"Loss of control (total or partial) of machine, means of transport or handling equipment, hand-held tool, object, animal"*(i.e., with ES-C#40).

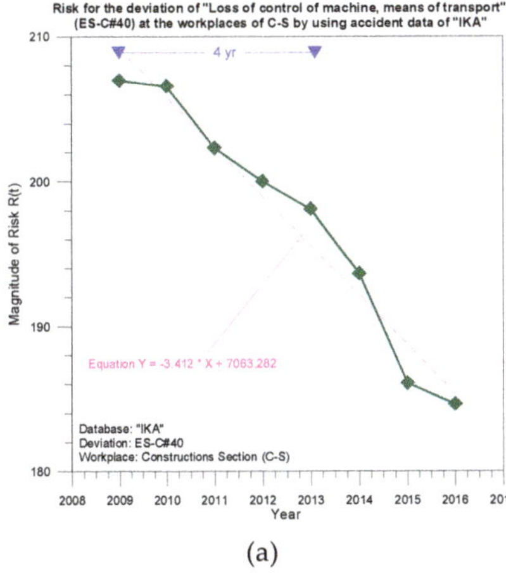
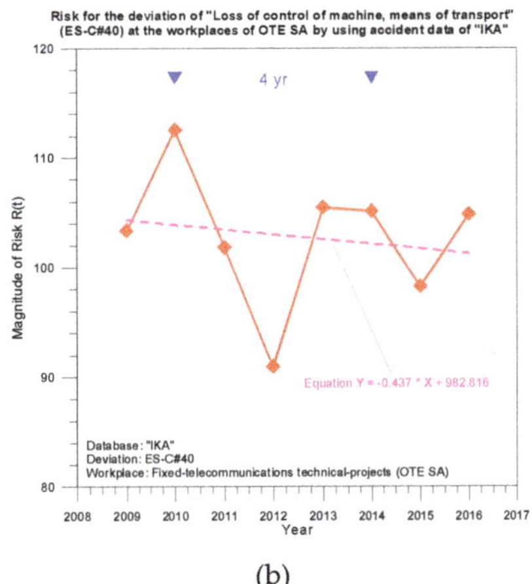

(a)                                    (b)

**Figure 3.** Time-variations of the risk (R) calculated by PRAT utilizing accident data of "IKA", during 2009–2016, and concerning the most considerable hazard-source of "loss of control (total or partial) of machine, means of transport, or handling equipment, hand-held tool, object, animal" (with ES-C#40) at the workplaces of (**a**) the C-S and (**b**) OTE SA.

It is worth noting that the analysis by the TSP process of the time-profiles of the magnitude of R, concerning the deviation with ESAW-code of ES-C#50, shows (by the blue triangular symbols in the graphs 'a' and 'b' of Figure 2) the existence of a periodic fluctuation with a periodicity ($T_1$) that is equal to ~2 years. This seems to be a permanent feature in the "behavior" of the ES-C#50 deviation at the workplaces of the C-S and OTE SA, because apart from the "SEPE" database, this periodicity obviously exists in the graphs 'a' and 'b' of Figure 4, which illustrates the R(t) profiles of ES-C#50 that were recorded at the same workplaces, but grounded, on the other hand, on the data of the "IKA" database.

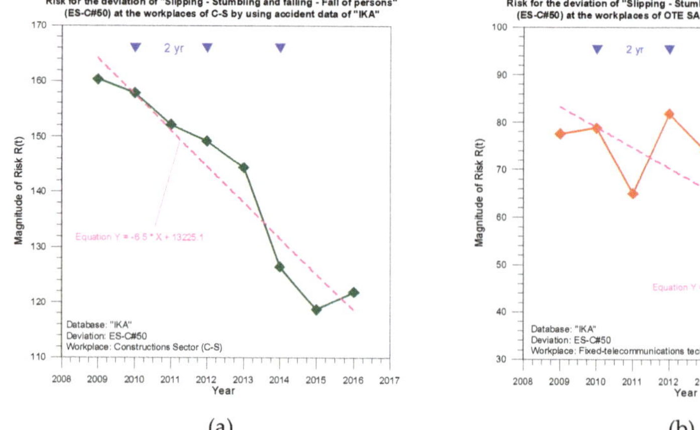


(a)    (b)

**Figure 4.** Time-profiles of the risk (R) calculated by PRAT utilizing accident data of "IKA", during 2009–2016, and concerning the hazard-source of "slipping—stumbling and falling—fall of persons" (with ES-C#50) at the workplaces of the C-S (**a**) and also of OTE SA (**b**).

Moreover, the time-profiles of the risk-value R, concerning the deviation of ES-C#40, illustrate (by the blue triangular symbols in the graphs 'a' and 'b' of Figure 3) the appearance of a periodic fluctuation with a periodicity ($T_2$) equal to ~4 yrs, and accordingly the presence of a harmonic of T1≅2years, i.e., equivalent to T1.

Hence, the presence of a dominant periodic-component (~2yrs) in the most significant hazard source (taking into account the risk-value R) constitutes a steady characteristic for the dynamic behavior of a working environment (or working system), like the workplaces of the C-S and/or of the OTE fixed-telecommunications technical-projects, relatively to OSH safety.

What is more, the time-domain analysis of the time-series of the ES-C#50 deviation shows in its time-profiles (according to Figure 2a, Figure 4a,b) the appearance of a trend factor (or slope) with a negative inclination in the curve of averaging i.e., the average risk-value decreases at the constructions-sector workplaces, during the years 2009–2016, and only the profile of ES-C#50 in drawing 2b, presents a positive inclination or a soft increase in the curve of the risk-average at the workplaces of OTE SA. Besides, the analysis of the time-variations of the ES-C#40 hazard-source displays in its profiles (according to Figure 3a,b and Figure 5b) the appearance of a tendency factor with a negative slope in the curve of averaging, which means the average risk-value decreases at the construction-sector workplaces, during the years 2009–2016, and only the profile of ES-C#40 in drawing Figure 5a indicates a positive slope or an enhancement in the curve of the risk-average.

In addition, the comparison between Figures 2a and 5a, and also between Figures 2b and 5b, reveals that the profile of the deviation ES-C#50 presents all over the period 2009–2016 higher risk-values R in comparison with deviation ES-C#40, confirming that the most significant hazard source at the worksites of the C-S and SA is the ES-C#50 one, according to the "SEPE" database, contributing in this way to the 1st ranking of deviations (as stated by Figure 1). On the other hand, the comparison between Figures 3a and 4a, and also between 3b and 4b, discloses that the profile of deviation ES-C#40 shows in the course of the period 2009–2016 greater risk-values in comparison with deviation ES-C#50, confirming that the most significant hazard-source at the worksites of the C-S and OTE SA is the ES-C#40 one, according to the "IKA" database, contributing to the 1st ranking of Figure 1, as well.

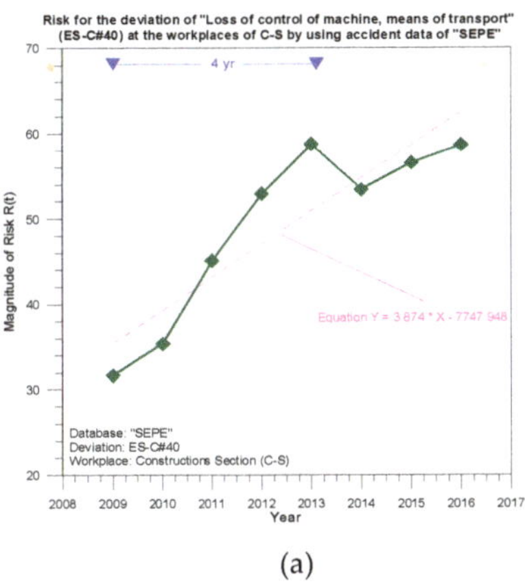
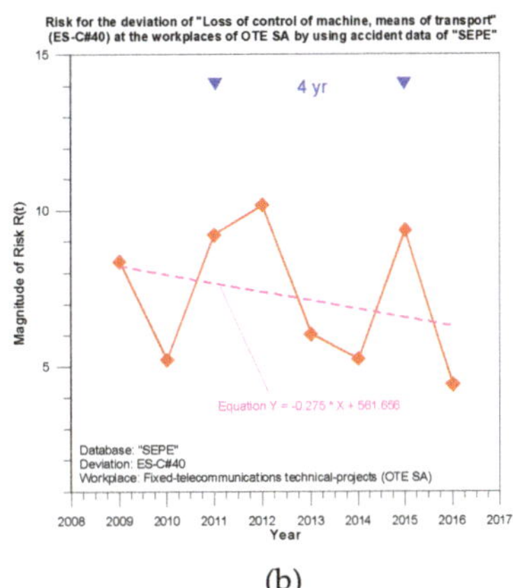





(a)        (b)

**Figure 5.** Time-variations of the risk (R) calculated by PRAT utilizing accident data of "SEPE", during 2009–2016, and concerning the hazard-source of "loss of control (total or partial) of machine, means of transport, or handling equipment, hand-held tool, object, animal" (with ES-C#40) at the workplaces of (**a**) the C-S and (**b**) OTE SA.

We note that the other hazard-sources (i.e., with ES-C #10–#30, and #60–#99) of Table 1 present noticeably lower risk-values R compared to the ones of ES-C#40 and ES-C#50, so the resulted deviation-ranking is: (i) ES-C#40 and ES-C#50, consistent with the "IKA" database, and (ii) ES-C#50 and ES-C#40, according to the "SEPE" database. Consequently, a second and more trustworthy ranking (denoted as 2nd ranking in Figure 1) of the hazard-sources (like the one applied by a MCDM method) is essential, which could be utilized by the safety-responsible (or managers) in the procedure of prioritizing investments in OSH-prevention measures. Inevitably, in the next section, we show the application of the MCDM process.

### 4.2. Application of the Typical-AHP and the Fuzzy-Extended-AHP MCDM-Processes

*The concept of the Typical Analytical Hierarchy Process (AHP):* The Typical-AHP, introduced by Saaty (1990) [66], is one of the most popular multicriteria decision making methods. While applying the AHP, the problem under study is modeled as a hierarchy, and the local and global weights are calculated with respect to each criterion and/or sub-criterion as an alternative of the problem. The method takes as input the judgments of the decision maker via linguistic variables, transforms them to numerical values using a scale, and applies pairwise comparisons for extracting the ranking of factors according to their importance. The major advantage of the AHP method is its integrated function for estimating the decision maker's consistency of judgments, which is a key characteristic for producing reliable results. In that manner, the consistency ratio (CR) is calculated to monitor if the decision maker has a consistent axiom system and is not making judgments "at random". Note that a pairwise comparison matrix is considered as consistent if its CR index is less than or equal to 10%.

*The concept of Fuzzy-Extended-AHP Process (FEAHP):* In the present study, the Fuzzy Extended AHP (FEAHP) proposed by Chang (1996) [67] is used for taking account of the uncertainty of the decision maker's judgments. The FEAHP is a popular extension of AHP because of its "ease of use" feature and of its proven efficiency. In our approach, the only prerequisite from the decision maker are the judgments, which are assigned to quantitative values using the scale proposed by Lamata (2004) [68] and illustrated in Table 2. The corresponding Triangular Fuzzy Number (TFN) value is illustrated in the last column. Note that the value of the standard Saaty's scale [66] is highlighted in bold letters.

**Table 2.** Linguistic variables and corresponding triangular fuzzy numbers (TFNs) by Lamata (2004) [68].

| Importance of Factor *i* over Factor *j* | Fuzzy Number |
|---|---|
| Equal | [1,1,1] |
| Equal to Moderate | [1,2,3] |
| Moderate | [2,3,4] |
| Moderate to Strong | [3,4,5] |
| Strong | [4,5,6] |
| Strong to Very Strong | [5,6,7] |
| Very Strong | [6,7,8] |
| Very Strong to Extremely | [7,8,9] |
| Extremely | [8,9,9] |

The process works as described by Chang (1996) [57], and accordingly, the output triangular fuzzy numbers are transformed into crisp numbers with the average value approach by Zimmermann (2001) [69] as the defuzzification process. Thus, for a triangular fuzzy number M = (l,m,u), the corresponding crisp number C as:

$$C(M) = (l + m + u)/3 \tag{5}$$

*Application of the typical-AHP and FEAHP MCDM processes:* The problem of assessing nine hazard factors with OSH impact at the workplaces of the C-S and OTE SA is modeled as a hierarchy, in which the goal is preventing safety, and the 1st level consists of nine hazard factors (as in Figure 6).

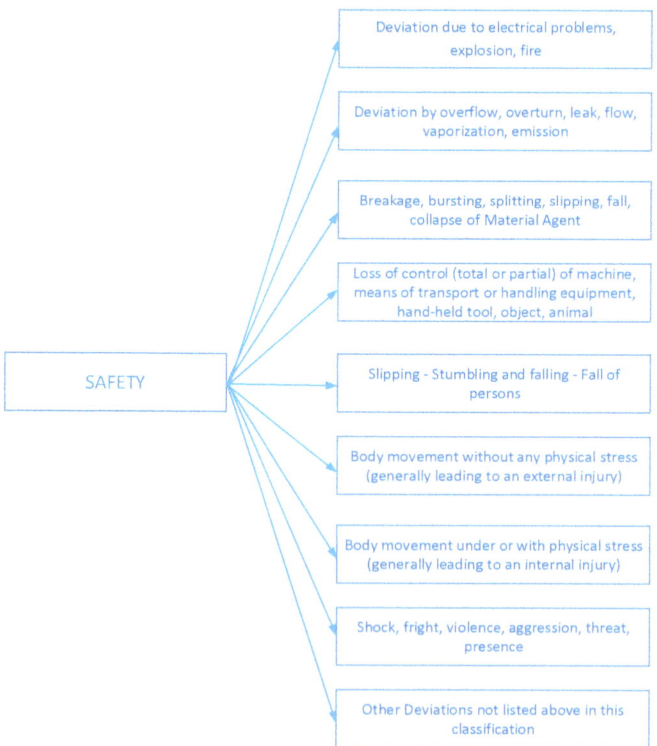


**Figure 6.** The hierarchy used for calculating the weights.

The numbers filled in the comparison matrix (Table 3) by the decision maker correspond to linguistic variables (Table 2) and express his preferences regarding the importance of hazards. As for the consistency of the judgments, the typical-AHP is employed first, and the consistency ratio (CR) is computed. In the present case, CR was found to be less than 0.1, meaning that the resulting weights are reliable, since the judgments of the decision maker are consistent.

**Table 3.** The judgments of the decision maker.

| Safety ESAW Codes (ES-C#) | [10] | [20] | [30] | [40] | [50] | [60] | [70] | [80] | [99] |
|---|---|---|---|---|---|---|---|---|---|
| [10] | 1 | 5 | 1/3 | 1/4 | 1/5 | 3 | 2 | 5 | 5 |
| [20] | | 1 | 1/5 | 1/6 | 1/7 | 1/2 | 1/3 | 2 | 2 |
| [30] | | | 1 | 1/2 | 1/3 | 4 | 3 | 6 | 6 |
| [40] | | | | 1 | 1/2 | 5 | 4 | 7 | 7 |
| [50] | | | | | 1 | 6 | 5 | 8 | 8 |
| [60] | | | | | | 1 | 1/2 | 3 | 3 |
| [70] | | | | | | | 1 | 4 | 4 |
| [80] | | | | | | | | 1 | 2 |
| [99] | | | | | | | | | 1 |

The resulting rankings and weights computed by the typical-AHP and the fuzzy extended FEAHP for every hazard are illustrated in Table 4, with a descending order regarding their importance. Note that these results were extracted using the same decision maker's judgments as shown in the pairwise comparison matrix of the Table 3.

**Table 4.** The results extracted by applying typical and fuzzy extended AHP (FEAHP).

| Hazards' Weights (Typical-AHP) | | Hazards' Ranking (Typical-AHP) | | Hazards' Weights (FEAHP) | | Hazards' Ranking (FEAHP) | |
|---|---|---|---|---|---|---|---|
| [10] | 10.04% | [50] | 31.51% | [10] | 12.59% | [50] | 25.72% |
| [20] | 3.24% | [40] | 22.63% | [20] | 3.95% | [40] | 20.99% |
| [30] | 16.06% | [30] | 16.06% | [30] | 16.67% | [30] | 16.67% |
| [40] | 22.63% | [10] | 10.04% | [40] | 20.99% | [10] | 12.59% |
| [50] | 31.51% | [70] | 7.08% | [50] | 25.72% | [70] | 8.98% |
| [60] | 4.83% | [60] | 4.83% | [60] | 6.21% | [60] | 6.21% |
| [70] | 7.08% | [20] | 3.24% | [70] | 8.98% | [20] | 3.95% |
| [80] | 2.49% | [80] | 2.49% | [80] | 2.86% | [80] | 2.86% |
| [99] | 2.13% | [99] | 2.13% | [99] | 2.03% | [99] | 2.03% |

As for the results of the typical-AHP calculations, the factors "slipping—stumbling and falling—fall of persons" (ES-C#50) and "loss of control (total or partial) of machine, means of transport, or handling equipment, hand-held tool, object, animal" (ES-C#40) are more important, since their weights are 31.51% and 22.63%, respectively. The next more important group includes the factor "breakage, bursting, splitting, slipping, fall, collapse of material agent" (ES-C#30) with a weight of 16.06%, and the (ES-C#10) "deviation due to electrical problems, explosion, fire" with a 10.04% score. Regarding the less important hazards, the sum of their weights is about 19.77% of the total risk, which is considered quite important. More specifically, this group of hazards includes the "body movement under or with physical stress (generally leading to an internal injury)" (ES-C#70), "body movement without any physical stress (generally leading to an external injury)" (ES-C#60), "deviation by overflow, overturn, leak, flow, vaporization, emission" (ES-C#20), and "shock, fright, violence, aggression, threat, presence" (ES-C#80), and "other deviations not listed above in this classification" (ES-C#99).

Applying the Fuzzy Extended AHP (FEAHP) to the judgments of Table 3 results in a similar hazards' ranking, as illustrated in the last column of Table 4. But, notwithstanding this, the relative weights' distance between the first deviation (with ES-C#50) and the second one (with ES-C#40) is very much reduced, meaning that the FEAHP describes the uncertainty of the expert's choices, in a different manner than typical-AHP.

Additionally, it is observed that, except for the first ranked deviation of ES-C#50, which has clearly lower score than before, and the second deviation of ES-C#40, which has almost the same weight, all the other hazards' weights are higher, and they are responsible for 53.29% of the total risk, instead of the total 45.86% calculated with typical-AHP.

Given the fact that these conclusions are extracted from a consistent pairwise comparison matrix and the importance of hazards is close to one another, it is preferable to use FEAHP instead of the typical-AHP under these circumstances, as stated by Chan et al. (2019) [70]. The use of FEAHP ensures that the proposed approach is more efficient since it handles better the judgments' uncertainty, which can affect the health and safety of employees.

### 4.3. Application of the FTA Process

Taking into account the MCDM outcomes extracted (in Table 4) by the utilization of the typical-AHP and FEAHP, along with the result that the hazard-sources' ranking unveiled as the most important deviation at the workplaces of the C-S and of OTE SA, the ES-C#50 one, while as the second-ordered deviation the ES-C#40 one, we complete the proposed RAA generic algorithmic framework by the application of the FTA process, as far as the ES-C#50 at the worksites of the fixed-telecommunications technical-projects of OTE SA is concerned: (i) For graphically illustrating the "base-events" and their sequences to the "top-event", and (ii) for calculating the occurrence probability of the "top-event" (ES-C#50) of the most significant deviation.

Therefore, the resulting FTA drawing is illustrated in Figure 7, where the likelihood of the "top-event" ES-C#50 has been calculated to be P = 51.3% according to the estimated (by the safety manager) likelihoods of the "base-events" E1–E12 depicted in the Table 5 and the utilization of equations (2), (3), and (4) as well. The "intermediate events" G1–G4 (illustrated in Table 6) have also been inserted in this FTA graph. Particularly, the events G1, G2, and G4 constitute the outputs of three transitional OR-gates, while G3 comprises the output of an AND-gate.

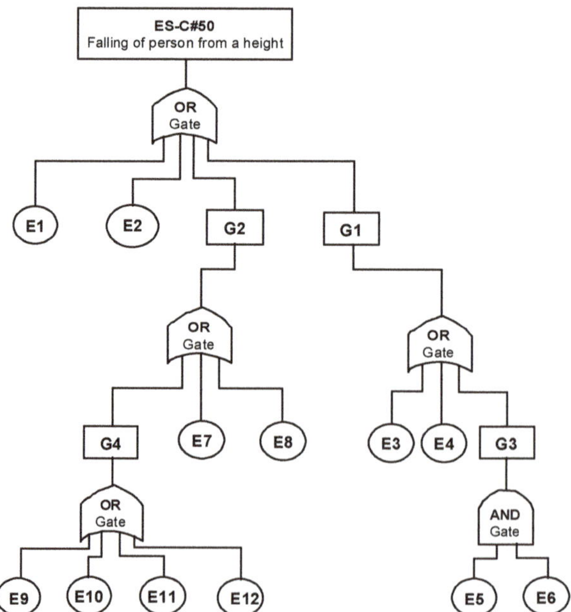


**Figure 7.** The fault-tree analysis (FTA)Drawing, as far as the most significant hazards' source of ES-C#50 at the worksites of the fixed-telecommunications technical-projects of OTE SA is concerned.

**Table 5.** The estimated likelihoods of the "base-events" E1–E2 concerning the FTA graph of the most significant hazards' source of ES-C#50 ("top-event").

| Base-Events | Description of "Base-Events" | Estimated Probability (P) |
|---|---|---|
| E1 | Improper use of tools/equipment | 20.0% |
| E2 | Strong winds | 1.0% |
| E3 | Slippery shoes | 0.1% |
| E4 | Worn seat belts | 5.0% |
| E5 | Lack of helmet | 10.0% |
| E6 | Contact of materials with the head | 10.0% |
| E7 | Pathological causes | 0.1% |
| E8 | Heat/cold | 0.1% |
| E9 | Overtime/repetitive work | 10.0% |
| E10 | Guard | 10.0% |
| E11 | Standing | 10.0% |
| E12 | Intense stress | 10.0% |

**Table 6.** The "intermediate-events" G1–G4 concerning the FTA graph of the most significant hazards' source of ES-C#50 ("top-event").

| Intermediate Events | Description of "Intermediate-Events" | Calculated Probability (P) |
|---|---|---|
| G1 | Improper use of personal protective or damaged equipment | 6.0% |
| G2 | Loss of consciousness/concentration | 34.5% |
| G3 | Negligence | 1.0% |
| G4 | Fatigue | 34.4% |

To boot, the "minimal cut-sets" were determined according to the theory (e.g., [71]), and the fault-tree was first "translated" to its equivalent Boolean equations, and then, by the "top-down" method of replacement, the following Boolean expression (which also constitutes the "minimal cut-set" expression) for the "top-event" was exported:

$$"TOP\_EVENT" = \sum_{i=1}^{4} E_i + (E_5 \cdot E_6) + \sum_{j=7}^{12} E_j \tag{6}$$

Taking into consideration that (i) the probability of the "top-event" is greater than 50% (during the period of 2009–2016), and (ii) the work of Marhavilas and Koulouriotis (2012a) [9], we characterize the ES-C#50 deviation as a high-risky hazard source.

## 5. Discussion and Conclusions

The way that a technical project attains its sustainability goal is implemented by emphasizing, among other aspects, on sustainable living, for instance on the actualization of health, that includes among other things, the buildings' and infrastructure technology [72,73]. What's more, the OSHA accepts that actions to guarantee OSH can be involved in sustainability efforts and take advantage of the dynamics of sustainability to make each workplace safer and healthier [3,4].

Besides, the notion of sustainability is utilized as a frame to sort subsistent (or to grow new) OSH-standards, and also to develop novel and innovative OSH-RAA assessment techniques. Thus, a sizeable part of the scientific literature employs the topic of sustainability as a means for the evolution of OSH-RAA techniques [5]. In general, the diversity of OSH-RAA techniques is such that there are plenty of efficient approaches for most cases, and many studies build up hybrid models merging distinct RAA approaches.

In this sense, this article concentrates on: (i) The implementation of a new RAA methodological framework (the MCDM-STO/DET one) for sustainable engineering projects by the amalgamation of a multicriteria decision-making (MCDM) process with the joint-collaboration of a deterministic (DET) and a stochastic (STO) process, and (ii) the application of MCDM-STO/DET at the workplaces of the Greek construction sector and also of the fixed-telecommunications technical projects of OTE SA (that is the Greek Telecommunications Organization S.A.), by means of real accident data coming from two official State databases, namely of "SEPE" (Labor Inspectorate, Hellenic Ministry of Employment) and of "IKA" (Social Insurance Institution, Hellenic Ministry of Health), all the way through the years of 2009–2016.

Hence, the article's objectives are next: (i) The presentation and execution of the joint MCDM-STO/DET framework at the workplaces of the Greek construction sector and also of the fixed-telecommunications technical projects of OTE SA, and (ii) to make known that the proposed MCDM-STO/DET scheme can be a precious method for safety managers (and/or decision-makers) to ameliorate OSH and also to support the sustainable operation of technical or engineering projects.

Mainly, we mingle two different configurations of the MCDM method, initially the typical AHP technique, and afterwards the fuzzy-extended AHP (FEAHP) one, along with the DET-PRAT and the STO-TSP methods, and finally with the DET-FTA one.

The results of the DET-PRAT utilization could be summarized on the subsequent topics:

1. The most significant hazard-source at the worksites of the C-S and OTE SA is the ES-C#50 one, according to the "SEPE" database (during 2009–2016).
2. The most significant hazard-source at the worksites of the C-S and OTE SA is the ES-C#40 one, according to "IKA" database (during 2009–2016).
3. The resulting hazard-sources ranking based on the calculated risk-value R is, on the one hand: (i) ES-C#50 & ES-C#40, in accordance with the "SEPE" database, and on the other hand, (ii) ES-C#40 & ES-C#50, consistent with the "IKA" database.
4. The maximum value (through 2009–2016) of the magnitude R of the hazard sources, calculated by means of the "SEPE" and "IKA" databases, is about 200.0, that means compulsory measures must be taken earlier than 1.0 year, according to the work of Marhavilas and Koulouriotis (2012b) [10], in order to demote the chance of arising fatal accidents.
5. There are other considerable hazard sources which present a risk value higher than 100.0, and according to the work of Marhavilas and Koulouriotis (2012b) [10], long-term actions are necessary for the extinction of their possible dangers.
6. Besides, other hazard sources present a risk value smaller than 100.0, and according to the previous referenced paper [10], compulsory actions are not essential except for surveillance of the events.

The results of the DET-TSP application by the time-domain analysis of the time-series of accident data and risk values (R) could be summarized as follows:

1. The time-profiles of the ES-C#50 deviation show, in almost all cases, the appearance of a trend factor with a negative inclination in the curve of averaging i.e., the average risk-value decreases at the C-S workplaces (during 2009–2016).
2. Instead, the profile of ES-C#50 presents a positive inclination (or a soft increase) in the curve of the risk-average, only at the workplaces of OTE SA.
3. Besides, the analysis of the time-variations of the ES-C#40 hazard-source displays in its profiles the presence of a tendency factor with a negative slope in the curve of averaging, that means the average risk-value decreases at the workplaces of the C-S and OTE SA (during 2009–2016).
4. Instead, the profile of ES-C#40 indicates a positive slope (or an enhancement) in the curve of the risk-average, only at the workplaces of the C-S.

5.  The analysis of the time-profiles of the magnitude R, concerning the ES-C#50, shows the existence of a periodic fluctuation with a periodicity of ~2 years, which seems to be a permanent feature in the "behavior" of the ES-C#50 deviation at the workplaces of the C-S and OTE SA, according the "SEPE" and "IKA" database.

6.  Likewise, the time-profiles of R, concerning the deviation of ES-C#40, illustrate the appearance of a periodic fluctuation with a periodicity of ~4 years (i.e., a harmonic of ~2years).

In other words, the application of the DET-PRAT technique, along with the DET-TSP one, revealed (by the analysis of the time-profiles of the most significant hazard sources) two significant factors: (i) The presence of a dominant periodic-component (~2years), and (ii) the appearance of a trend factor in the magnitude of risk (R) with a decreasing tension (in almost all the cases), constituting (during 2009–2016) two permanent characteristics for the dynamic behavior of the working environment (or working system) at the workplaces of the C-S and also of the OTE fixed-telecommunications technical-projects, relative to OSH safety, and should be taken into consideration by any responsible (or manager, officer, etc.) for OSH issues, both at the business level and also at the State level.

Taking into consideration that the resulting rankings of the hazard-sources (at the workplaces of the C-S and of OTE SA) by using PRAT and the data of "SEPE" and "IKA" were dissimilar and/or contradictory, a second (final) and more trustworthy ranking of the hazard-sources was achieved by the application of a MCDM method (in its two different configurations, the typical-AHP and the FEAHP one) in order to attain the optimum distribution of an investment budget to appropriate measures reaching the best OSH-protection (i.e., in the procedure of prioritizing investments in OSH-prevention measures). The typical-AHP and the FEAHP were employed in the present study for assessing hazards with an efficient and "easy-to-use" multicriteria method and its fuzzy extension. The resulting rankings show: (i) The necessity of using the FEAHP in OSH applications, since it can better handle the uncertainty of the decision maker's judgments, and (ii) that the most significant hazard-source is ES-C#50 (while as second, the ES-C#40 one).

Finally, we completed the suggested RAA algorithmic framework by the application of the DET-FTA process, as far as the most significant deviation ES-C#50 at the worksites of the fixed-telecommunications technical-projects of OTE SA is concerned, for illustrating the "base-events" and their sequences to the "top-event", and for calculating the probability of the "top-event" (ES-C#50), which was $P = 51.3\%$ (i.e., a high-risk hazard source).

The partnership of PRAT with TSP technique can give, by the computation of risk (R), indispensable information to safety-managers in order to identify the most significant hazard-sources at the corporations' worksites, and also to realize a preliminary ranking (1st ranking) of them. This ranking can be worthy in evaluating the urgency of investing a restricted budget in particular measures for attaining the highest OSH-protection with the minimal cost. Moreover, the next module in the suggested algorithmic framework is the MCDM one, which represents the application of a multicriteria decision-making technique, for accomplishing the final (definitive) ranking (2nd ranking) of the hazardous sources at the worksites, in order to achieve the finest (optimum) distribution of an investment budget (i.e., with the minimal cost) for appropriate measures for reaching the best OSH-protection. In addition, the FTA technique could also efficiently facilitate the aforementioned 1st and 2nd rankings.

It is worth mentioning that to facilitate the calculations, we could use the following integrated software packages: (i) The GRAPHER by Golden Software Inc, and the SPSS one by SPSS Inc, for achieving the TSP analysis, and (ii) the Fault-Tree+ one by ISOGRAPH Inc for the FTA analysis.

Besides, the results of the MCDM calculations, for both Typical-AHP and FEAHP, have been extracted using MS-Excel spreadsheets developed by one of the authors. Our results, calculated by the Typical-AHP, could be confirmed using the software available at http://fuzzymcdm.upol.cz/. This software applies the Typical-AHP (as presented by Saaty; 1990) [66], but it supports a different extension of AHP to fuzzy logic (as presented by Holeček and Talašová; 2016) [74] than the FEAHP performed in this study, i.e., calculations with FEAHP only include the application of the equations as described in the work of Chang (1996) [67]. Because of this, the results with this software might be slightly different.

The main limitations of using this approach are listed below:

- The AHP theory tries to measure the relative importance of alternatives with respect to each criterion by using pairwise comparisons. In order to do pairwise comparisons, experts use a conversion scale with crisp values, for expressing their ideas. So, an important limitation of AHP is the usage of crisp values to reflect human thinking.
- Another limitation is coming from CR, which is a verification of the rational judgment performed by the experts.
- Moreover, an additional limitation concerns the pairwise comparisons of criteria, which are carried out by several experts with a required significant working-experience in occupational safety.
- As a classic multicriteria decision support tool, AHP is a subjective process, and the resulting results, either rankings or weights, are dependent on the way that the judgments are imposed and the criteria are compared to the rest. In other words, another decision maker (or a group of experts) could make "different" judgments regarding the relative importance of each pair of criteria, resulting in different resulting weights for the criteria and/or different rankings. This drawback leads to the need for applying group-decision-making processes for merging different judgments.
- Besides, when there are slight differences between the criteria's weights, it is possible that the AHP could lead to the selection of a suboptimal alternative, instead of the optimal one. This phenomenon could be reduced by considering many criteria to be pairwise compared. A review of the AHP drawbacks can be found in the study of Whitaker (2007) [75].

This multiparty approach can assist safety risk managers by the prioritization of hazards, to reduce the impact of the most important risks. In addition, it could be used by risk managers to allocate more effectively the constrained budget and to reduce mortgage costs, and thus, for reserving extra funds for spending on safety measures, and to further reduce other risk factors' impact. The main contribution of this study is the cooperative use of a proven efficient multicriteria decision-making method with a joint STO/DET one for identifying and prioritizing hazards.

Accordingly, as a future improvement, the FTA technique could be utilized for contributing more capably to the 1st and 2nd ranking of the hazard-sources, according to the configuration of Figure 8, which constitutes an alternative form of the proposed RAA framework (illustrated in Figure 1).

As a general conclusion, every RAA approach has specific limitations and/or restrictions in its applicability to miscellaneous accidents, and consequently, a separate technique can't accomplish the optimum risk-assessment result at the workplaces. On the other hand, collaborative methodologies, like the suggested MCDM-STO/DET one, would be efficient for sustainable engineering projects. Thus, the novelty of this article is fulfilled through the suggested newfangled risk assessment and analysis (RAA) methodological approach by the amalgamation of a multicriteria decision-making (MCDM) process with the joint-collaboration of a deterministic (DET) and a stochastic (STO) process in order to upgrade OSH and to validate the sustainable operation of technical or engineering projects, as well.

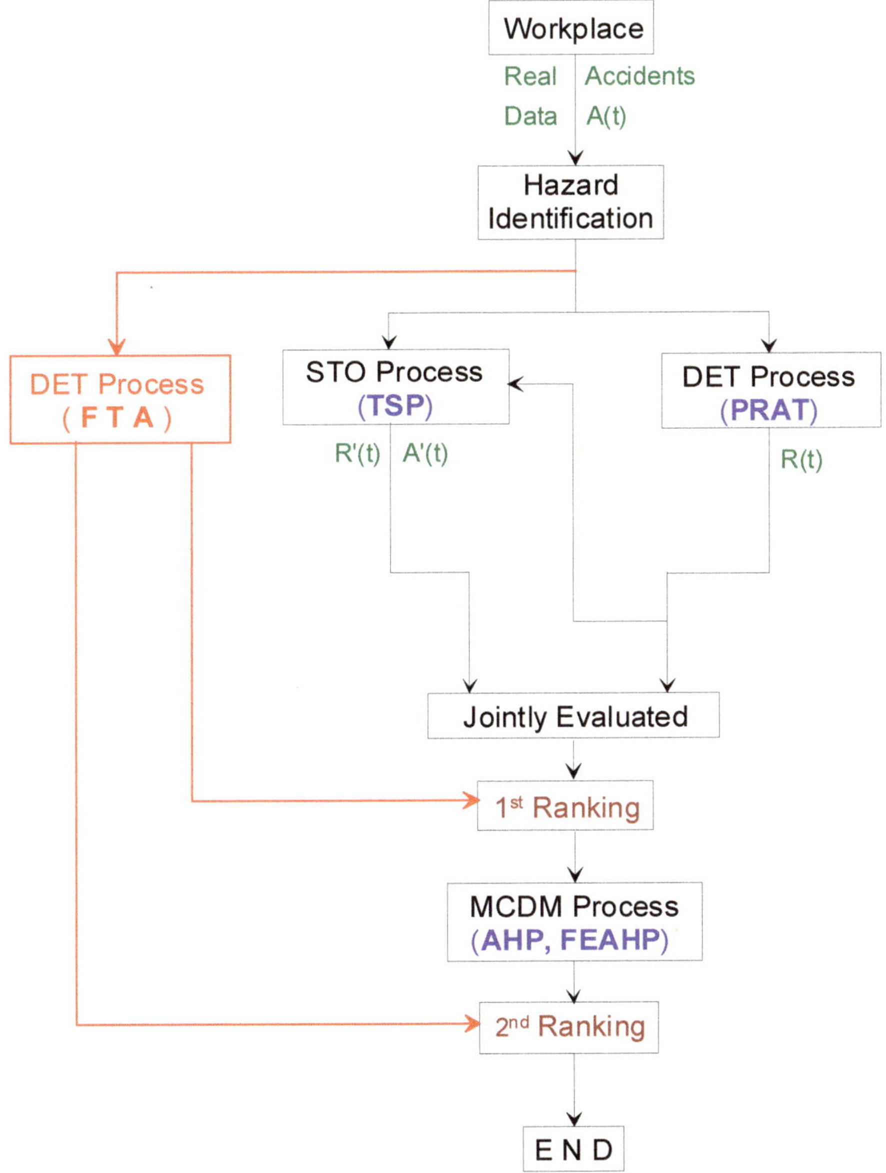


**Figure 8.** An alternative form of the proposed joint stochastic/deterministic/MCDM RAA methodological framework.

**Author Contributions:** Conceptualization, P.K.M.; data curation, P.K.M., M.G.T., G.K.K.; formal analysis, P.K.M.; investigation, P.K.M.; methodology, P.K.M., DEK; writing—original draft, P.K.M., G.K.K.; writing—review and editing, P.K.M., G.K.K., D.E.K. All authors have read and agreed to the published version of the manuscript

**Funding:** This research received no external funding.

**Conflicts of Interest:** We declare no conflict of interest.

## Abbreviations

| | |
|---|---|
| AHP | Analytical Hierarchy-Process |
| C-S | Constructions Sector |
| DET | Deterministic process |
| ETA | Event Tree Analysis |
| ESAW | European Statistics on Accidents at Work |
| ES-C | ESAW-Code |
| FEAHP | Fuzzy Extended Analytical Hierarchy-Process |
| FMEA | Failure Mode and Effect Analysis |
| FTA | Fault Tree Analysis |
| HAZOP | Hazard and Operability |
| IKA | Social Insurance Institution, Hellenic/Greek Ministry of Health |
| MCDM | Multi-Criteria Decision-Making |
| OSHA | Occupational Safety & Health Administration |
| OSH | Occupational Safety and Health |
| OTE | OTE SA—the Greek Telecommunications Organization |
| PRAT | Proportional Risk Assessment Technique |
| RA | Risk Assessment |
| RAA | Risk Analysis and Assessment |
| SEPE | Labor Inspectorate, Hellenic/Greek Ministry of Employment |
| STO | Stochastic process |
| TSP | Time Series Processes |
| WHO | World Health Organization |

## References


1. WHO Global Strategy on Occupational Health for All: The Way to Health at Work. 1994. Available online: https://www.who.int/occupational_health/publications/globstrategy/en/index3.html (accessed on 5 April 2020).
2. Boileau, P.-E. Sustainability and prevention in occupational health and safety. *Ind. Health.* **2016**, *54*, 293–295. [CrossRef]
3. Koulinas, G.K.; Demesouka, O.E.; Marhavilas, P.K.; Vavatsikos, A.P.; Koulouriotis, D.E. Risk Assessment Using Fuzzy TOPSIS and PRAT for Sustainable Engineering Projects. *Sustainability* **2019**, *11*, 615. [CrossRef]
4. OSHA Sustainability in the Workplace: A New Approach for Advancing Worker Safety and Health. 2016. Available online: https://www.osha.gov/sustainability/docs/OSHA_sustainability_paper.pdf (accessed on 5 April 2020).
5. Marhavilas, P.K.; Koulouriotis, D.E.; Nikolaou, I.; Tsotoulidou, S. International Occupational Health and Safety Management-Systems Standards as a Frame for the Sustainability: Mapping the Territory. *Sustainability* **2018**, *10*, 3663. [CrossRef]
6. Soares, C.G.; Teixeira, A.P. Risk assessment in maritime transportation. *Reliab. Eng. Syst. Saf.* **2001**, *74*, 299–309. [CrossRef]
7. Zheng, X.; Liu, M. An overview of accident forecasting methodologies. *J. Loss Prev. Process Ind.* **2009**, *22*, 484–491. [CrossRef]
8. Marhavilas, P.K.; Koulouriotis, D.E.; Gemeni, V. Risk Analysis and Assessment Methodologies in the Work Sites: On a Review, Classification and Comparative Study of the Scientific Literature of the Period 2000-2009. *J. Loss Prev. Process Ind.* **2011**, *24*, 477–523. [CrossRef]
9. Marhavilas, P.K.; Koulouriotis, D.E. Developing a new alternative risk assessment framework in the work sites by including a stochastic and a deterministic process: A case study for the Greek Public Electric Power Provider. *Saf. Sci.* **2012**, *50*, 448–462. [CrossRef]
10. Marhavilas, P.K.; Koulouriotis, D.E. A combined usage of stochastic and quantitative risk assessment methods in the worksites: Application on an electric power provider. *Reliab. Eng. Syst. Saf.* **2012**, *97*, 36–46. [CrossRef]
11. Lassarre, S. Analysis of progress in road safety in ten European countries. *Accid. Anal. Prev.* **2001**, *33*, 743–751. [CrossRef]

12. Reniers, G.L.L.; Dullaert, W.; Ale, B.J.M.; Soudan, K. The use of current risk analysis tools evaluated towards preventing external domino accidents. *J. Loss Prev. Process Ind.* **2005**, *18*, 119–126. [CrossRef]

13. Verlinden, S.; Deconinck, G.; Coupé, B. Hybrid reliability model for nuclear reactor safety system. *Reliab. Eng. Syst. Saf.* **2012**, *101*, 35–47. [CrossRef]

14. Whiteside, M.B.; Pinho, S.T.; Robinson, P. Stochastic failure modeling of unidirectional composite ply failure. *Reliab. Eng. Syst. Saf.* **2012**, *108*, 1–9. [CrossRef]

15. Guikema, S.D.; Quiring, S.M. Hybrid data mining-regression for infrastructure risk assessment based on zero-inflated data. *Reliab. Eng. Syst. Saf.* **2012**, *99*, 178–182. [CrossRef]

16. Stewart, M.G.; O'Connor, A. Probabilistic risk assessment and service life performance management of load bearing biomedical implants. *Reliab. Eng. Syst. Saf.* **2012**, *108*, 49–55. [CrossRef]

17. Marhavilas, P.K.; Koulouriotis, D.E.; Spartalis, S.H. Harmonic Analysis of Occupational-Accident Time-Series as a Part of the Quantified Risk Evaluation in Worksites: Application on Electric Power Industry and Construction Sector. *Reliab. Eng. Syst. Saf. Elsevier* **2013**, *112*, 8–25. [CrossRef]

18. Da Costa, M.A.B.; Brandão, A.L.T.; Santos, J.G.F.; Pinto, J.C.; Nele, M. Quantitative FTA using Monte Carlo analyses in a pharmaceutical plant. *Eur. J. Pharm. Sci.* **2020**, *146*, 105265. [CrossRef]

19. Gul, M.; Celik, E. Fuzzy rule-based Fine–Kinney risk assessment approach for rail transportation systems. *Hum. Ecol. Risk Assess.* **2018**, *24*, 1786–1812. [CrossRef]

20. AriamuthuVenkidasalapathy, J.; Mannan, M.S.; Kravaris, C. A quantitative approach for optimal alarm identification. *J. Loss Prev. Process Ind.* **2018**, *55*, 213–222. [CrossRef]

21. Ozkan, N.F.; Uluta, B.H. A novel to approach to quantify the risk probabilities for a risk analysis methodology. *Adv. Intell. Syst. Comput.* **2017**, *487*, 341–347. [CrossRef]

22. Athar, M.; Shariff, A.M.; Buang, A. A review of inherent assessment for sustainable process design. *J. Clean. Prod.* **2019**, *233*, 242–263. [CrossRef]

23. Sanmiquel-Pera, L.; Bascompta, M.; Anticoi, H.F. Analysis of a historical accident in a Spanish coal mine. *Int. J. Environ. Res. Public Health* **2019**, *16*, 3615. [CrossRef] [PubMed]

24. Domínguez, C.R.; Martínez, I.V.; Piñón Peña, P.M.; Rodríguez Ochoa, A. Analysis and evaluation of risks in underground mining using the decision matrix risk-assessment (DMRA) technique, in Guanajuato, Mexico. *J. Sustain. Min.* **2019**, *18*, 52–59. [CrossRef]

25. Kharzi, R.; Chaib, R.; Verzea, I.; Akni, A. Safe and sustainable development in a hygiene and healthy company: Using the decision matrix risk assessment (DMRA) technique: A Case study. *J. Min. Environ.* **2020**. [CrossRef]

26. Cinar, U.; Cebi, S. A hybrid risk assessment method for mining sector based on QFD, fuzzy logic and AHP. *Adv. Intell. Syst. Comput.* **2020**, *1029*, 1198–1207. [CrossRef]

27. Mutlu, N.G.; Altuntas, S. Risk analysis for occupational safety and health in the textile industry: Integration of FMEA, FTA, and BIFPET methods. *Int. J. Ind. Ergon.* **2019**, *72*, 222–240. [CrossRef]

28. Bora, S.; Bhalerao, Y.; Goyal, A.; Niu, X.; Garg, A. Computation of safety design indexes of industry vehicle operators based on the reach angle, the distance from elbow to ground and the popliteal height. *Int. J. Ind. Ergon.* **2019**, *71*, 155–164. [CrossRef]

29. Kamsu-Foguem, B.; Tiako, P. Risk information formalisation with graphs. *Comput. Ind.* **2017**, *85*, 58–69. [CrossRef]

30. Zheng, X.-Z.; Wang, F.; Zhou, J.-L. A Hybrid Approach for Evaluating Faulty Behavior Risk of High-Risk Operations Using ANP and Evidence Theory. *Math. Probl. Eng.* **2017**, *2017*, 7908737. [CrossRef]

31. Mentes, A.; Ozen, E. A hybrid risk analysis method for a yacht fuel system safety. *Saf. Sci.* **2015**, *79*, 94–104. [CrossRef]

32. Fine, W.T.; Kinney, W.D. Mathematical evaluation for controlling hazards. *J. Saf. Res.* **1971**, *3*, 157–166.

33. Hammer, R.W. *Handbook of System and Products Safety*; Prentice-Hall, Inc.: Englewood Cliffs, NJ, USA, 1972.

34. Kinney, G.F.; Wiruth, A.D. *Practical Risk Analysis for Safety Management*; NTIS report number NWC-TP-5865; Naval Weapons Center: China Lake, CA, USA, 1976.

35. Woodruff, J.M. Consequence and likelihood in risk estimation: A matter of balance in UK health and safety risk assessment practice. *Saf. Sci.* **2005**, *43*, 345–353. [CrossRef]

36. Reniers, G.L.L.; Dullaert, W.; Ale, B.J.M.; Soudan, K. Developing an external domino prevention framework: Hazwim. *J. Loss Prev. Process Ind.* **2005**, *18*, 127–138. [CrossRef]

37. Van der Hoeven, E.T. CE-Marking: Creating a model for applying the EMC, LVD, and Machinery Directive. MSc Thesis, University of Twente, School of Management and Governance, Enschede, The Netherlands, 2012.
38. Gul, M.; Guven, B.; Guneri, A.F. A new Fine-Kinney-based risk assessment framework using FAHP-FVIKOR incorporation. *J. Loss Prev. Process Ind.* **2018**, *53*, 3–16. [CrossRef]
39. Kokangül, A.; Polat, U.; Dağsuyu, C. A new approximation for risk assessment using the AHP and Fine Kinney methodologies. *Saf. Sci.* **2017**, *91*, 24–32. [CrossRef]
40. Marhavilas, P.K. Risk Assessment Techniques in the Worksites of Occupational Health-Safety Systems with Emphasis on Industries and Constructions. Ph.D. Thesis, Department of Production and Management Engineering, Democritus University of Thrace, Xanthi, Greece, 2015. Available online: http://hdl.handle.net/10442/hedi/35612 (accessed on 10 April 2020).
41. Marhavilas, P.K.; Koulouriotis, D.E.; Mitrakas, C. On the development of a new hybrid risk assessment process using occupational accidents data: Application on the Greek Public Electric Power Provider. *J. Loss Prev. Process Ind.* **2011**, *24*, 671–687. [CrossRef]
42. Marhavilas, P.K.; Koulouriotis, D.E. A risk estimation methodological framework using quantitative assessment techniques and real accidents' data: Application in an aluminum extrusion industry. *J. Loss Prev. Process Ind.* **2008**, *21*, 596–603. [CrossRef]
43. Marhavilas, P.K.; Koulouriotis, D.E.; Mitrakas, C. Fault and Event-Tree techniques in occupational health-safety systems-Part I: Integrated risk-evaluation scheme. *Environ. Eng. Manag. J.* **2014**, *13*, 2097–2108. [CrossRef]
44. Marhavilas, P.K.; Koulouriotis, D.E.; Mitrakas, C. Fault and Event-Tree techniques in occupational health-safety systems-Part II: Statistical analysis. *Environ. Eng. Manag. J.* **2014**, *13*, 2371–2386. [CrossRef]
45. Babaei, M.; Roozbahani, A.; Shahdany, S.M.H. Risk Assessment of Agricultural Water Conveyance and Delivery Systems by Fuzzy Fault Tree Analysis Method. *Water Resour. Manag.* **2018**, *32*, 4079–4101. [CrossRef]
46. Yasli, F.; Bolat, B. A risk analysis model for mining accidents using a fuzzy approach based on fault tree analysis. *J. Enterp. Inf. Manag.* **2018**, *31*, 577–594. [CrossRef]
47. Gul, M.; Ak, M.F. A comparative outline for quantifying risk ratings in occupational health and safety risk assessment. *J. Clean. Prod.* **2018**, *196*, 653–664. [CrossRef]
48. Fuentes-Bargues, J.L.; González-Cruz, M.C.; González-Gaya, C.; Baixauli-Pérez, M.P. Risk analysis of a fuel storage terminal using HAZOP and FTA. *Int. J. Environ. Res. Public Health* **2017**, *14*, 705. [CrossRef] [PubMed]
49. Hanea, D.M.; Jagtman, H.M.; Ale, B.J.M. Analysis of the Schiphol Cell Complex fire using a Bayesian belief net based model. *Reliab. Eng. Syst. Saf.* **2012**, *100*, 115–124. [CrossRef]
50. Lisnianski, A.; Elmakias, D.; Laredo, D.; Ben Haim, H. A multi-state Markov model for a short-term reliability analysis of a power generating unit. *Reliab. Eng. Syst. Saf.* **2012**, *98*, 1–6. [CrossRef]
51. Khakzad, N.; Khan, F.; Amyotte, P. Risk-based design of process systems using discrete-time Bayesian networks. *Reliab. Eng. Syst. Saf.* **2013**, *109*, 5–17. [CrossRef]
52. Wang, F.; Ding, L.Y.; Luo, H.B.; Love, P.E.D. Probabilistic risk assessment of tunneling-induced damage to existing properties. *Expert Syst. Appl.* **2014**, *41*, 951–961. [CrossRef]
53. Biyikli, O.; Aydogan, E.K. A new model suggestion to estimate the probability value in occupational health and safety risk assessment. *Appl. Math. Inf. Sci.* **2016**, *10*, 663–671. [CrossRef]
54. Zheng, G.; Zhu, N.; Tian, Z.; Chen, Y.; Sun, B. Application of a trapezoidal fuzzy AHP method for work safety evaluation and early warning rating of hot and humid environments. *Saf. Sci.* **2012**, *50*, 228–239. [CrossRef]
55. Fattahi, R.; Khalilzadeh, M. Risk evaluation using a novel hybrid method based on FMEA, extended MULTIMOORA, and AHP methods under fuzzy environment. *Saf. Sci.* **2018**, *102*, 290–300. [CrossRef]
56. Mete, S. Assessing occupational risks in pipeline construction using FMEA-based AHP-MOORA integrated approach under Pythagorean fuzzy environment. *Hum. Ecol. Risk Assess.* **2019**, *25*, 1645–1660. [CrossRef]
57. Oturakci, M. A new fuzzy-based approach for environmental risk assessment. *Hum. Ecol. Risk Assess.* **2019**, *25*, 1718–1728. [CrossRef]
58. Yucesan, M.; Kahraman, G. Risk evaluation and prevention in hydropower plant operations: A model based on Pythagorean fuzzy AHP. *Energy Policy* **2019**, *126*, 343–351. [CrossRef]
59. Boral, S.; Howard, I.; Chaturvedi, S.K.; McKee, K.; Naikan VN, A. An integrated approach for fuzzy failure modes and effects analysis using fuzzy AHP and fuzzy MAIRCA. *Eng. Fail. Anal.* **2020**, *108*. [CrossRef]

60. Marhavilas, P.K. Risk Estimation in the Greek Constructions' Worksites by using a Quantitative Assessment Technique and Statistical Information of Occupational Accidents. *J. Eng. Sci. Technol. Res. (JESTR)* **2009**, *2*, 51–55. [CrossRef]
61. Ardeshir, A.; Maknoon, R.; Rekab Islamizadeh, M.; Jahantab, Z. Safety risk assessment effective on occupational health in high-rise building construction projects with fuzzy approach. *Iran Occup. Health* **2014**, *11*, 82–95.
62. Anil Kumar, C.N.; Sakthivel, M.; Elangovan, R.K.; Arularasu, M. Analysis of Material Handling Safety in Construction Sites and Countermeasures for Effective Enhancement. *Sci. World J.* **2015**, *742084*. [CrossRef] [PubMed]
63. Marhavilas, P.K.; Vrountas, P.T. Risk Assessment in the Constructions Sector of EU Countries: Application of a Methodological Framework using Quantitative Techniques and Occupational Accidents' Data throughout the period 1996–2011. *J. Eng. Sci. Technol. Res. (JESTR)* **2018**, *11*, 66–73. [CrossRef]
64. Koulinas, G.K.; Marhavilas, P.K.; Demesouka, O.E.; Vavatsikos, A.P.; Koulouriotis, D.E. Risk Analysis and Assessment in the worksites using the Fuzzy-Analytical Hierarchy Process and a Quantitative Technique – A case study for the Greek Construction sector. *Saf. Sci.* **2019**, *112*, 96–104. [CrossRef]
65. European Union (EU). *European Statistics on Accidents at Work (ESAW)-Summary Methodology*; EUROSTAT, European Commission: Luxembourg, 2013; ISBN 978-92-79-28419-9. [CrossRef]
66. Saaty, T.L. How to make a decision: The analytic hierarchy process. *Eur. J. Oper. Res.* **1990**, *48*, 9–26. [CrossRef]
67. Chang, D.-Y. Applications of the extent analysis method on fuzzy AHP. *Eur. J. Oper. Res.* **1996**, *95*, 649–655. [CrossRef]
68. Lamata, M.T. Ranking of alternatives with ordered weighted averaging operators. *Int. J. Intell. Syst.* **2004**, *19*, 473–482. [CrossRef]
69. Zimmermann, H.J. *Fuzzy Set Theory—And Its Applications*; Springer Science & Business Media: New York, NY, USA, 2001.
70. Chan, H.K.; Sun, X.; Chung, S.-H. When should fuzzy analytic hierarchy process be used instead of analytic hierarchy process? *Decis. Support Syst.* **2019**, *125*. [CrossRef]
71. Haasl, D.F.; Roberts, N.H.; Vesely, W.E.; Goldberg, F.F. *Fault Tree Handbook*; NUREG-0492, TRN: 82-003645; U.S. Nuclear Regulatory Commission: Washington, DC, USA, 1981; p. 20555. Available online: https://www.nrc.gov/docs/ML1007/ML100780465.pdf (accessed on 10 April 2020).
72. ICST&D. International Conference on Sustainable Technology and Development. Shenzhen, China, 22–25 August 2020. Available online: https://www.elsevier.com/events/conferences/international-conference-on-sustainable-technology-and-development (accessed on 10 April 2020).
73. Marhavilas, P.K.; Filippidis, M.; Koulinas, G.K.; Koulouriotis, D.E. A HAZOP with MCDM Based Risk-Assessment Approach: Focusing on the Deviations with Economic/Health/Environmental Impacts in a Process Industry. MDPI Sustainability, section: Sustainable Engineering and Science, Special Issue: Decision Support Systems and Knowledge Management for Sustainable Engineering. *Sustainability* **2020**, *12*, 993. [CrossRef]
74. Holeček, P.; Talašová, J. A free software tool implementing the fuzzy AHP method. In Proceedings of the 34th International Conference on Mathematical Methods in Economics 2016, Liberec, Czech Republic, 6–9 September 2016; pp. 266–271, ISBN 978-80-7494-296-9.
75. Whitaker, R. Criticisms of the Analytic Hierarchy Process: Why they often make no sense. *Math. Comput. Modeling* **2007**, *46*, 948–961. [CrossRef]

*Article*

# The Methods and Techniques of Strategic Environmental Assessment. Comparative Evaluation of Greek and International Experience

**Antonios Souloutzoglou * and Anastasia Tasopoulou**

Postgraduate Programme "The Technologies of Environmental Legislation", Department of Environmental Engineering, Democritus University of Thrace, 67100 Xanthi, Greece; ntasopo@yahoo.gr
* Correspondence: anton-soul@hotmail.com

Received: 15 March 2020; Accepted: 16 April 2020; Published: 18 April 2020

**Abstract:** Strategic Environmental Assessment (SEA) is a process used for the evaluation of environmental impacts at a higher level of decision-making than that of each individual project, i.e., during the preparation and implementation of policies, programs, and plans, with the objective of incorporating the aspect of sustainable development in the early stages of planning. However, the "strategic" nature of SEA is the cause of frequent confusion among the responsible competent institutions and professionals regarding the selection of the most appropriate methods and techniques for each individual situation. At the international level, current research indicates a certain ambiguity in the use of methods and techniques in every step of SEA processes. In Greece, despite the implementation of SEA procedures in the preparation of a number of plans and programs after the transposal of the SEA Directive, to date, no attempt has been made to systematize the current experience and practice from the utilization of methods and techniques. The objective of the present study is to analyze and provide a comparative evaluation of the Greek and international experience, on the one hand with the systematic examination of Strategic Environmental Impact Assessments, and on the other with primary research, through questionnaires addressed to Greek practitioners. An important observation is that, both in Greece and abroad, there are inadequacies in the existence and/or the use of guidance manuals for the selection of the most appropriate methods and techniques, and only a limited range of methods and techniques are used in comparison with those catalogued in international literature.

**Keywords:** strategic environmental assessment; strategic environmental impact assessment; methods; techniques; Greece

---

## 1. Introduction

Strategic Environmental Assessment (SEA) is a relatively recent environmental process, having made its debut roughly 30 years ago in countries such as the USA and Canada, and international organizations, such as the World Bank, with the objective of transitioning from the "polluter pays" principle to that of prevention [1]. It is, in essence, a process that aims to incorporate the environmental aspect both prior to and during the designing of policies, plans, and programs (PPPs). This integration produces environmentally sound strategies, with a resultant increase in the certainty that any projects and operations which will subsequently be adopted, as well as any other form of general environmental intervention, will not be harmful [2].

SEA requires an exact process, and a defined arrangement of procedures. This process was gradually integrated in a number of states. In Europe, SEA procedures were comprehensively debated and adopted by the European Parliament and Council in July 2001, with Directive 2001/42/EC (SEA Directive). This directive was transposed significantly later into Hellenic environmental law, in August 2006, with the Joint Ministerial Decision MEPPPW/SES/10717/28.08.2006 (SEA JMD), which was later amended by the Ministerial Decision 40238/2017. Furthermore, the required procedures include the description of the current status of the environment, the examination of alternatives, the identification, description, and evaluation of any significant environmental effects, the integration of the results of the environmental assessment and public consultation into the plan or program under approval. and finally, the monitoring of any future impacts stemming from their implementation.

The aforementioned series of procedures is carried out through the use of various methods and techniques. However, international literature finds confusion between the terms methods and techniques, as many researchers and authors consider them synonymous. Typical examples of this are the Geographic Information Systems (GIS). Both Noble and Storey [3] and Therivel [4] agree that GIS are used for the collection, management, classification, and presentation of data, however, the first authors classify them into methods, while Therivel classifies them into techniques.

Many of these methods and techniques, over 350, are cited in international literature, according to Lee [5]. In particular, certain simple methods and techniques, which do not require specialized equipment and quantitative records, have been documented, an example of such being expert judgment. On the other hand, there are methods and techniques which provide adequate levels of accuracy in their results, through the use of specialized software and quantitative records, such as GIS, while those which utilize matrices for the documentation and evaluation of impacts are also quite widespread. However, the terms "methods" and "techniques" are often conflated, with many practitioners and authors considering them to be synonymous. Their differentiation is down to the fact that techniques provide data, while by contrast, methods relate to the identification and description of possible impacts, as well as the compilation and classification of data [3].

Despite the fact that individual SEA methodologies are provided by the existing legislative frameworks, including a number of activities, there are no universal methods and techniques adhered to by all participating parties. In general terms, SEA methods were drawn from those utilized in both Environmental Impact Assessment (EIA) surveys, and in policy and planning analysis methods. The ambiguity surrounding SEA, according to Noble et al. [6], which is due exactly to its "strategic" nature, is often the cause of confusion among the responsible competent institutions and practitioners regarding the selection of the most appropriate methods and techniques for each individual situation; this results in only a limited number ultimately being utilized. Furthermore, individual methods and techniques are utilized on a greater scale and at more steps of SEA.

In Greece, SEA processes have been implemented in the preparation of a number of plans and programs following the transposal of the SEA Directive. Zagorianakos [7], by conducting an ex-ante appraisal of the implementation of SEA Directive, came to the conclusion that a series of methodological, institutional, political, and legislative problems in Greece make the implementation of the SEA Directive rather problematic. Subsequently, the evaluation of the degree of application of some key features of the SEA during the transport planning process in the Athens Olympics in 2004 demonstrated that the methodological and procedural flexibility of the SEA is the key to its implementation in countries wherein the environmental assessment of strategic actions is either absent or underdeveloped [8]. However, no surveys or studies have been carried out as of yet in order to systematize the existing experience and practice from the development of methods and techniques during SEA processes.

The present study discusses the research and analysis of the methods and techniques applied during the SEA process. This analysis is not carried out independently, however; firstly, a correlation is made between the methods and techniques, and the existing experience and practice, both in Greece and abroad. Furthermore, a comparative assessment of the international and Greek experience is carried out in order to discern similarities and differences regarding the adoption and utilization of the methods and techniques.

## 2. Background

SEA procedures have been widely adopted worldwide, without this necessarily meaning that we are leading to a convergence as regards the legislative framework or the means used between states, due to the variety of legal, cultural, and institutional contexts. Based on the existing literature, the present section seeks to provide insight into SEA identity, institutional framework and its application on a global level, bringing the European experience into focus, and followed by a brief analysis of the position of SEA in the Hellenic Environmental Law. Consequently, an overview of the SEA methodological approach is attempted, and the most frequent methods and techniques used in the relevant procedures are identified and systematized.

### 2.1. Identity and International Institutional SEA Framework

Numerous definitions have been offered in international literature for SEA, by researchers who have studied both its theory and practice. The most comprehensive of these was formulated by Partidario [9] (p. 19), according to whom, "SEA is a systematic, on-going process for evaluating, at the earliest appropriate stage of publicly accountable decision-making, the environmental quality, and consequences, of alternative visions and development intentions incorporated in policy, planning, or programme initiatives, ensuring full integration of relevant biophysical, economic, social and political considerations". Essentially, SEA is utilized at the international level in order to assess environmental impacts during planning, at a higher level than that of the project in question, i.e., during the preparation and enactment of PPPs with the objective of integrating the aspect of sustainable development at an early stage of planning.

The widespread adoption of SEA procedures, which have been included in the legal framework of roughly 60 states, directly correlates with the increased understanding of the relationship between development and the environment [10]. It was initially connected with the expansion of procedures and practices utilized within the context of EIA projects at higher levels of decision-making [11], while, primarily in recent years, gradual efforts have been made for the reinforcement of its "strategic" approach [12].

The primary attributes of SEA, according to Sadler and Verheem [13] are:

1. Assurance that the necessary attention is given to the environmental aspect in the development of PPPs.
2. Reinforcement and improvement of EIA at project level.
3. Enablement of definition and organization of cumulative impacts.
4. Promotion of sustainable development.
5. Promotion of consultation among various governmental bodies and interested parties, and increase of public participation in the configuration of PPPs.

In 1969, the USA was the first state to enact legislation incorporating EIA procedures through the National Environmental Policy Act (NEPA). The NEPA also includes provisions for the assessment of environmental impacts at the strategic or extra-project level, a procedure titled Programmatic Environmental Impact Statement (PEIS) [1]. At the UN level, the United Nations Economic Commission for Europe ratified the SEA Protocol, within the framework of the Fifth Ministerial Conference "Environment for Europe", which was held in Kiev, Ukraine, in May 2003. The protocol came into effect on 21 July 2010, and has been ratified by the EU and 32 other nations [14].

On the other hand, in the European Union (EU), the Habitats Directive (92/43/EEC) was the Directive that required the assessment of potential environmental impacts before the approval of the implementation of the plan or project on Natura 2000 sites [15]. However, there was no comprehensive legislative framework to incorporate the environmental aspect at the PPP level until 2001. The first institutionalized assessments of environmental impacts pertained to the project level were through Directives 85/337/EEC and 97/11/EC (EIA Directives). The European Commission submitted three reports (in 1993, 1997, and 2003) to the European Parliament regarding their implementation and effectiveness. The common conclusion of all three of these reports was that, despite the significant progress which had been achieved by the member states regarding their implementation, high-level environmental protection remained a desideratum.

It thus became apparent that, in order to achieve effective, high-level environmental protection over time, the environmental aspect needed to be integrated into the higher levels of decision making, which established the framework for subsequent permits regarding the execution of projects, i.e., at the level of strategic planning of policies, drafting, and programming [16]. However, the lack of a comprehensive institutional SEA framework provided an incentive for certain member states to adopt such systems in the 1980s and 1990s [4]. Regardless, the utilization of a variety of different procedures demonstrated the need for adoption of a comprehensive institutional SEA framework at the EU level, leading to the submission of a preliminary proposal for discussion by the European Commission in 1990 [17]. The publication of the official proposal from the European Commission was submitted in December 1996, and it became the subject of in-depth debate. It was ultimately adopted in July 2001 as Directive 2001/42/EC. Table 1 presents the SEA steps and outputs, according to Thérivel [4].

**Table 1.** Strategic Environmental Assessment (SEA) steps and outputs [4].

| No | SEA Step | What to Decide | What to Record |
|---|---|---|---|
| 1 | Identify SEA objectives, indicators, and targets | What environmental and sustainability objectives, targets, and/or indicators to test the plan options and statements against | List of SEA objectives, indicators, and targets where relevant |
| 2 | Describe environmental baseline, including future trends; identify environmental issues and problems | What environmental and sustainability issues and constraints to consider during decision-making | Data on the baseline environment; list of relevant environmental and sustainability issues and constraints |
| 3 | Identify links to other relevant strategic actions | What other strategic actions influence the strategic action in question, and how | List of relevant strategic actions, their requirements, and any constraints or conflicts with the strategic action in question |
| 4 | Identify (more) sustainable alternatives for dealing with the problems and implementing the strategic action objective | What alternatives or options to consider | List of alternatives or options |
| 5 | Prepare Scoping Report; consult | What to include in the Scoping Report | Results of steps 1–4; agreed written statement of how to proceed with subsequent SEA steps |
| 6 | Predict and evaluate impact of alternatives/statements; compare alternatives; mitigate impacts of chosen alternative(s)/statements | What are the effects of the alternatives/options and statements on the environmental/sustainability objectives and indicators; what are the preferred alternatives; what mitigation measures to include | Summary of effects of alternatives/options and statements on the environment and sustainability; list of preferred alternatives; explanation of why these are preferred; mitigation measures proposed |
| 7 | Write the SEA report; establish guidelines for implementation | How to present the data from steps 1–6 | Prepare the SEA report |
| 8 | Consult | Whom to consult; how to respond to consultation results | How consultation results were addressed |
| 9 | Monitor the environmental/sustainability impacts of the strategic action | How to deal with any negative impacts of the strategic action | How the strategic action's impacts will be monitored and significant effects dealt with |

Several authors have conducted research on the SEAs legislative framework, its application and effectiveness in areas such as urban planning in Italy and in Spain [18–26]. Specifically, Spain has incorporated the SEA Directive through the Law 9/2006, which was later replaced by the Law 21/2013, and more recently by the Law 9/2018. The latest was a medium to drive significant changes that have been made to the SEA processes in order to ensure the protection of Natura 2000 sites [18]. It is important to highlight the specific differentiation of the authority that is responsible for developing and adopting the plan or program and the relevant authority that is responsible for assessing the environmental impacts of the proposed plan or program [26].

In the case of Italy, the SEA Directive was incorporated into Italian law by national decree 152/2006, and amended in 2008 and 2010. The distinctive characteristic of the Italian system is the regional diversification of SEA procedures due to the legal provisions varying between regions [26]. However, due to the inadequacy of existing legislation, the regional legislative framework has only complied with the minimum requirements of the European directive, along with the Italian national decree [19]. Notwithstanding this shortcoming, SEA has gradually introduced concepts such as carrying capacity and resilience at the planning stage. Simultaneously, it is also leading to progress of public consultation and decision makers' accountability [25].

In summary, the SEA legislative framework constitutes a perpetually developing process worldwide, with the objective of environmental assessment, evaluation, and management of the environmental impacts of human activity. This development does not necessarily result in the convergence and integration of SEA procedures between states. The differing approaches, which have been and continue to be developed, are due to the divergent, among others, legal, political, cultural, and institutional frameworks of each state. Chacker et al. [27], after comparing the implementation of SEA in 12 selected countries, discovered, despite the noted variety in procedures, the existence of significant similarities:

1.  The procedures for determining screening (field of application) as well as scoping (extent and magnitude of application) are legally binding.
2.  In the majority of countries, the content of SEA pertains to the assessment of the environmental impacts of strategic proposals.
3.  The development of an environmental report is a prerequisite in all the states examined by the study.
4.  Almost all of the studied countries require the consideration of alternatives.
5.  The majority of countries require both impact mitigation and monitoring measures.
6.  Public participation is an integral element of the entirety of the SEA systems.
7.  Qualitative assurance of SEA is required in all systems, with the exception of Canada, Denmark, and New Zealand.

### 2.2. SEA in Hellenic Environmental Law

Before the transposal of the SEA Directive, there were no official procedures in place in Hellenic environmental law for the implementation of SEA procedures during the preparation of plans and programs [28]. The evaluation of environmental impacts was limited to the assessment of projects and activities through EIA. The absence of environmental protection provisions at the initial planning stages was characteristic in the case of plans or programs of a strategic nature, such as the construction of the Egnatia Motorway. However, in the relevant Hellenic literature, certain projects of wide scope and extent have been documented, which incorporated procedures similar to SEA in the drafting of the corresponding EIA reports [28,29]. One such notable example was the EIA report on the installations put into place for the Athens Olympics in 2004.

However, despite the existence of this specific legal gap, and the explicit wording of the SEA Directive that "Member states shall bring into force the laws, regulations and administrative provisions necessary to comply with this directive before 21 July 2004" (Article 13, L197/34), it was transposed into Hellenic law two years after the deadline (no. MEPPPW/SES/107017/28.08.2006 JMD and its amendment with MD 40238/2017). The competent authority for the supervision, evaluation, processing, and approval of the preliminary environmental screening and environmental report (SEIA) of the SEA is the Directorate of Environmental Licensing. This authority belongs to the Ministry of the Environment and Energy. The SEA is defined, according to this Joint Ministerial Decision, as the procedure for assessment of the environmental impacts of a plan or program, which includes:

1. The writing of a Strategic Environmental Impact Assessment (SEIA).
2. Public consultation.
3. The consideration of both the SEIA and the relevant public consultation during the decision-making process.
4. Information regarding the relevant decision.

It is worth noting at this point that there is a clear differentiation between the SEA JMD and the corresponding SEA Directive. The SEA JMD namely consists of 11 articles and 4 annexes, whereas the SEA Directive consists of 15 articles and 2 annexes, and it is a transposition of the Directive into the measures and particularities of Greek reality. Moreover, in the SEA JMD, we can highlight that:

1. Scoping (Article 3 and Annex I) is explicitly defined, as opposed to the breadth of the scope of the Directive. It also includes specific types of plans and programs, the most prominent of those can be considered: The urban and spatial planning, Community Operational Programs, as well as those co-funded by the EU.
2. The concept of environmental screening is clearly stipulated, and the enactment regarding the associated procedure is specifically established (Article 5 and Annex IV).
3. A specific definition of the environmental report is given (SEIA—article 2), and its content is clearly defined (Article 6 and Annex III).
4. There is no separate article of public consultation. Nevertheless it is more clearly defined (Article 7). The public consultation process involves the transmission of the SEIA to the competent public authorities while it is published to the public concerned.

*2.3. SEA Methodological Approach*

According to Partidario [30], a model is considered strategic when it includes a visualization of long-term goals, flexibility in the management of complex systems, ability to adapt to dynamic environments and conditions, and focus on the crucial elements. SEA exhibits a strategic approach when the definition of the long-term goals, the adoption of courses of action and the application of the necessary resources have all been assured [31]. Within the framework of its "strategic" nature, SEA must, by necessity, ensure that the strategic actions do not exceed the boundaries, beyond which the impacts will cause irreparable damage to the environment [4]. Therefore, when the SEA decisions and procedures are adhered to at a higher level, their reassessment at lower levels is rendered unnecessary, resulting in conservation of time and resources [4]. The strategic programming framework is demonstrated in Figure 1.

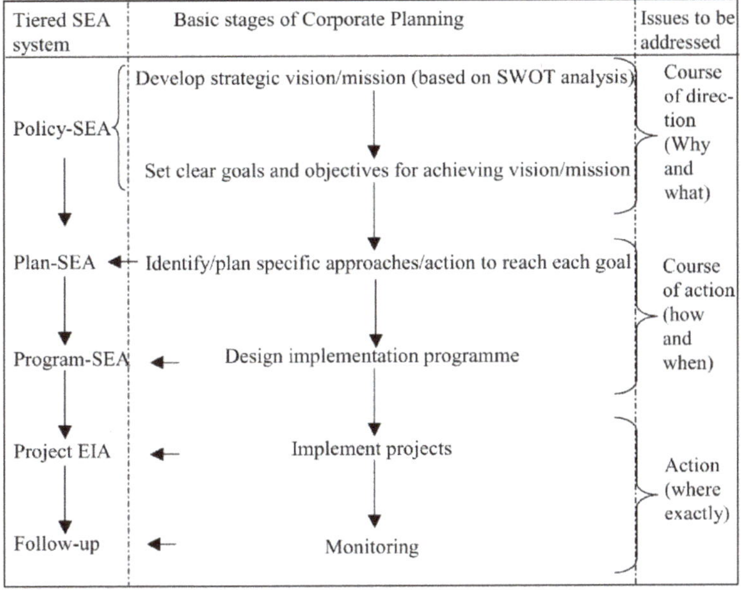

**Figure 1.** Basic strategic planning model ([32], following [33–35]).

An overview of the bibliography reveals that there is no common reception in the SEA approach. Over the years, various researchers and organizations have suggested SEA types, on the basis of the spatial design and the objective of the evaluation (regional and sectoral policy—[36]), the role played by SEA in the protection of environmental matters during the decision-making process (marginal, compliance, constructive—[37]), the means by which SEA is connected to or interacts with policy-making or planning procedures (single opportunity, parallel, integrated, decision centered—[30]) or based on a general SEA typology (formal, near-equivalent, para-SEA, [38]). In Figure 2, we have decided to present the approach formulated by Noble and Nwanekezie [39], due on the one hand to it being less complex, centered on the fundamental principles of SEA, while on the other reflecting the most recent scientific thought on the means by which SEA will improve strategic decision-making.

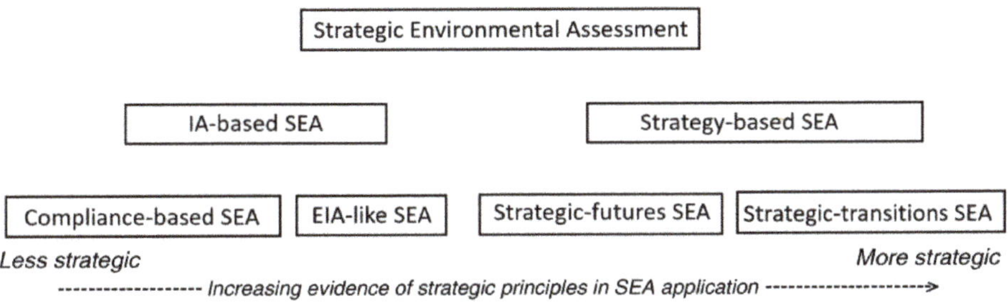


**Figure 2.** Impact assessment-based and strategic-based conceptualization of SEA [39].

According to the authors, SEA can function within a wide strategic scope, to either a lesser or greater extent. In the former case, SEA is based on predetermined procedures, correspondent to those rooted in EIA. Notable examples are the SEA outlined in European Directive 2001/42/EC and The Canadian CEAA Directive [38]. By contrast, in the latter case, during SEA processes, the focus of the evaluation is centered on the issue. The incentive is to locate the appropriate PPP or a suitable alternative strategy, which will provide a resolution to the issue, with the objective of achieving the intended PPP goals [40].

*2.4. SEA Methods and Techniques*

The methods and techniques used in SEA procedures are undefined, in comparison to the exact, practical tools used in other scientific fields [41]. This ambiguity is due to the fact that there are, as of yet, no strictly defined methods and techniques at every step of SEA. Furthermore, the prevailing attitude views these as a collection of tools, of which users are encouraged to select those most suitable for each situation [42].

According to the general prerequisites [43,44], SEA methods and techniques should:

- Be fit for purpose, i.e., they should be able to address key issues and fit into the decision-making process.
- Be integrated into the decision-making timetable.
- Allow the integration of various substantive aspects, i.e., various administrations, sectors, and procedures.
- Allow for the accommodation of uncertainties.
- Be transparent, robust, and relevant to the objective and the practices.
- Be understandable to all parties participating in SEA.
- Be cost-effective.
- Offer alternatives, in addition to measures designed to mitigate any negative impacts.
- Allow the comparison of alternatives.

These prerequisites ought to be considered more as general principles, rather than criteria for the selection, in each individual case, of the most suitable SEA methods and techniques. According to Therivel [4], the factors which determine which of these are most appropriate for each individual case are the scale (international, national, regional, local), the "strategic-ness" (policy, plan, program) of the strategic actions, the type of decision, the audience who would use the outputs, the context in which SEA is conducted, the availability of time, resources, personnel, and equipment, as well as the types of data required by the methods and techniques. Finally, a common defining factor is the tradition and mindset of each individual state, researcher, or practitioner.

A large number of methods and techniques is available for use in SEA processes. According to Lee [5], a study carried out in the early 1980s on behalf of two Dutch ministries compiled descriptions of 350 different methods and techniques in North American and Western European literature. Nevertheless, of this wide array, only a small number were utilized in practice [43–45]. Although an in-depth analysis goes beyond the scope of the present paper (see in greater detail [2,4,43,44,46–49]), Table 2 presents methods and techniques which are extensively referenced in international literature, and categorizes them into two modules. In the first (methods and techniques works for), the methods and techniques are determined according to the level and scope of their implementation (PPP, large–small area, land use plan, sectoral plan). In the second (methods and techniques copes with), the categorization pertains to the quality of utilized data (incomplete or not, uncertain or not), their nature (qualitative or quantitative), and the resources required for their processing.

**Table 2.** Extent of application of SEA methods and techniques ([44] (p. 351), and own processing based on: [2,4,43,46–49]).

| Methods and Techniques | Key Situation in Which SEA Methods and Techniques Could Be Used | | | | | | | | | | |
| | Methods and Techniques Works for | | | | | | Methods and Techniques Copes with | | | | |
| | Policy Level | Plan Level | Program Level | Large Area | Small Area | Land Use Plan | Sectoral Plan | Incomplete Data | Uncertain Data | Qualitative Data | Few Resources |
|---|---|---|---|---|---|---|---|---|---|---|---|
| Carrying capacity analysis | √ | √ | √ | √ | √ | √ | √ | √ | √ | - | ? |
| Checklists | ? | ? | √ | √ | √ | √ | √ | √ | √ | √ | √ |
| Compatibility appraisal | √ | √ | √ | √ | √ | √ | √ | √ | √ | √ | √ |
| Cost-benefit analysis | - | - | √ | √ | √ | √ | √ | - | √ | - | - |
| Ecological footprint analysis | √ | √ | √ | √ | √ | √ | √ | √ | √ | - | ? |
| Expert judgment | √ | √ | √ | √ | √ | √ | √ | √ | √ | √ | √ |
| GIS | ? | √ | √ | √ | √ | √ | √ | ? | ? | ? | ? |
| Guiding questions | √ | √ | √ | √ | √ | √ | √ | √ | √ | √ | √ |
| Indicators | √ | √ | √ | √ | √ | √ | √ | √ | √ | √ | √ |
| Land use partitioning anal. | - | ? | √ | √ | √ | ? | √ | √ | - | - | - |
| Life cycle analysis | - | - | √ | √ | √ | - | √ | - | √ | ? | - |
| Literature/case review | √ | √ | √ | √ | √ | √ | √ | √ | √ | √ | √ |
| Matrices | √ | √ | √ | √ | √ | √ | √ | √ | √ | √ | √ |
| Modeling | - | ? | √ | ? | √ | ? | √ | ? | ? | - | - |
| Multi-criteria analysis | - | - | √ | ? | √ | ? | √ | - | √ | √ | ? |
| Network analysis | √ | √ | √ | ? | √ | √ | √ | √ | √ | √ | √ |
| Overlay maps | - | √ | ? | - | √ | √ | √ | ? | ? | ? | ? |
| Participatory methods | ? | ? | √ | ? | √ | √ | √ | √ | √ | √ | √ |
| Quality of life assessment | √ | √ | √ | √ | √ | √ | √ | √ | √ | √ | ? |
| Risk assessment | - | ? | √ | ? | √ | √ | √ | ? | √ | ? | - |
| Scenario/Sensitivity analysis | ? | √ | √ | √ | √ | ? | √ | - | ? | - | - |
| SWOT analysis | √ | √ | √ | √ | √ | √ | √ | √ | √ | √ | √ |
| Vulnerability analysis | √ | √ | √ | - | √ | √ | √ | ? | √ | - | - |

√ = fully, ? = partly, - = not.

The first conclusion is that the majority of methods and techniques can be utilized, either comprehensively or in part, at all levels and scales, regardless of the quality of the utilized data and required resources. The column titled Qualitative data, in which the methods and techniques are categorized as either qualitative or quantitative, is indicative of both their potential and their limitations. In particular, the qualitative methods and techniques are applicable at all levels of PPP, with the exception of the multi-criteria analysis, in contrast with numerous quantitative methods and techniques, which cannot be applied at the policy level, such as cost-benefit analysis, modeling, etc. The qualitative methods and techniques are applicable even in cases of incomplete or uncertain data and, due to not requiring specialized equipment, can even be utilized when there are limited available resources. Notable examples are expert judgment, guiding questions, and literature/case review. In other words, the qualitative methods and techniques contribute to improve certain SEA steps, making the process more simple and rapid, require little in the way of equipment, and can utilize both qualitative and quantitative data, while simultaneously taking into account existing political sensitivities. By contrast, the quantitative methods and techniques are, in their majority, time-consuming, complicated, and expensive, due to requiring large volumes of data and specialized equipment. However, they are objective, scientifically sound, and produce comprehensible results, while also being useful for application at the planning and programming level, regardless of extent and design. Notable examples are the cost-benefit analysis and the land use partitioning analysis.

Table 3 distinguishes the methods and techniques depending on the extent to which they are applicable at the various steps of SEA. Due to the fact the SEA Directive, and consequently the SEA JMD, both refer to the EIA-like SEA analysis of the steps of SEA, this was chosen as the means of analysis. The differences between this and the figure of Therivel and Wood [44] lie in the addition of the screening column, which aims to pinpoint every element used in the process of preliminary environmental screening, and in the identify alternatives column, with the objective of identifying all those which define alternatives.

Initially, and in contrast with Table 2, it becomes evident that the majority of the methods and techniques cannot be implemented at every step of SEA. Notable exceptions are the participatory methods, which are applicable, either comprehensively or in part, to the entirety of SEA steps. Furthermore, there are methods and techniques which are applicable to the majority of SEA steps, such as expert judgment, guiding questions, indicators, GIS, etc. The majority of these methods and techniques manage primarily qualitative data, exhibit particular advantages, and are noted for their ease of use. On the other hand, the qualitative methods and techniques, due to their various inherent disadvantages, which were previously described, are not suitable for application at the steps of preliminary environmental screening, description of the current status, and monitoring of environmental impacts. Nevertheless, due to their scientific approach and the objectivity of their results, they are notably effective, and are commonly applicable in the identification, description, and evaluation of the environmental impacts of PPP. Indicative examples of such methods and techniques are cost-benefit analysis, modeling, and scenario/sensitivity analysis.

**Table 3.** Methods and techniques applicable at various steps of SEA ([44] (p. 352), and own processing based on: [2,4,43,46–49]).

| Methods and Techniques | SEA Steps during Which SEA Methods and Techniques Could Be Used | | | | | | | | | | |
|---|---|---|---|---|---|---|---|---|---|---|---|
| | Screening | Describe Baseline/Current Status | Identify Impacts | Predict Impacts | Evaluate Impacts | Identify Alternatives | Compare Alternatives | Identify Cumulative/Indirect Impacts | Propose Mitigation | Public Participation | Monitoring |
| Carrying capacity analysis | - | √ | √ | √ | - | √ | - | ? | √ | - | - |
| Checklists | √ | √ | - | √ | √ | - | √ | - | √ | ? | √ |
| Compatibility appraisal | - | - | - | - | - | - | - | - | √ | ? | - |
| Cost-benefit analysis | - | - | √ | √ | √ | √ | √ | - | √ | ? | - |
| Ecological footprint analysis | - | √ | √ | √ | - | √ | - | ? | √ | - | - |
| Expert judgment | √ | √ | √ | √ | √ | √ | √ | √ | √ | - | √ |
| GIS | √ | √ | √ | √ | √ | - | √ | √ | √ | √ | ? |
| Guiding questions | √ | √ | √ | √ | √ | √ | √ | √ | √ | - | √ |
| Indicators | √ | √ | √ | √ | √ | √ | √ | √ | √ | - | √ |
| Land use partitioning anal. | - | - | - | √ | √ | - | ? | ? | √ | - | - |
| Life cycle analysis | - | - | √ | - | √ | √ | √ | √ | √ | √ | - |
| Literature/case review | √ | √ | √ | √ | √ | √ | √ | √ | √ | - | √ |
| Matrices | √ | - | √ | √ | √ | ? | √ | - | √ | √ | √ |
| Modeling | - | - | √ | √ | ? | √ | √ | ? | √ | - | - |
| Multi-criteria analysis | - | - | - | - | √ | √ | √ | - | √ | ? | - |
| Network analysis | - | - | √ | √ | ? | - | ? | √ | √ | ? | - |
| Overlay maps | √ | √ | √ | √ | √ | - | √ | √ | √ | √ | ? |
| Participatory methods | ? | √ | √ | √ | √ | ? | √ | ? | √ | √ | ? |
| Quality of life assessment | ? | ? | √ | √ | √ | - | ? | - | √ | √ | ? |
| Risk assessment | ? | - | - | √ | √ | - | √ | ? | √ | - | - |
| Scenario/Sensitivity analysis | - | - | - | √ | √ | √ | √ | ? | √ | - | - |
| SWOT analysis | - | - | √ | √ | √ | √ | √ | - | - | - | - |
| Vulnerability analysis | - | - | √ | - | √ | √ | √ | - | - | - | - |

√ = fully, ? = partly, - = not.

## 3. Methodology

In order to broaden the scope of the research, both primary and secondary sources were utilized. Bibliographical references, as secondary sources, were the basis for the examination of international and Greek practices in the adoption of SEA methods and techniques. Subsequently, an effort was made to gather information relevant to the selection of SEA methods and techniques in Greece from the primary sources. The survey was carried out through the filling out of questionnaires by SEIA practitioners. The final conclusions of the paper comprise and utilize the findings of both the primary and secondary research.

### 3.1. Secondary Research Methodology

A representative sample of SEIA was selected in order to examine the Greek practices in the adoption of SEA methods and techniques. The following selection criteria were implemented:

- Plans and programs which were carried out under the auspices, in particular, of the National Strategic Reference Framework 2014–2020 Operational Programmes, General Urban Plans, Special Spatial Plans, Specific Spatial Development Plans for Public Property, or for Strategic Investments, the development of tourist ports, agriculture, energy, waste transportation, and management.
- Selection of national, regional, and local plans and programs.
- Selection of the most recent, in terms of both date and sector, SEA.
- An effort was made to select the plans and programs which were submitted for the preliminary environmental screening process.
- The methods and techniques were indicated or explicitly located in the various individual SEA steps. Table 4 displays the results of this selection process.

**Table 4.** Greek SEA plans and programs under consideration.

| No | Strategic Environmental Assessment |
|----|-------------------------------------|
| 1 | National Hazardous Waste Management Plan (March 2016) |
| 2 | Specific Spatial Development Plans for Public Property "Asteras Vouliagmenis" (July 2013) |
| 3 | Specific Spatial Development Plans for Strategic Investments, location "Karapetis" Mykonos Island (June 2019) |
| 4 | Special Spatial Plan for the Thessaloniki Exhibition Centre (January 2019) |
| 5 | Operational Programme Competitiveness Entrepreneurship Innovation 2014–2020 (January 2015) |
| 6 | Regional Operational Programme of Sterea Ellada 2014–2020 (September 2014) |
| 7 | Exploration and exploitation of hydrocarbons in the "Southwest Crete" and Western Crete marine areas (June 2018) |
| 8 | Modification of the General Urban Plan of the Municipality of Halandri—Attica (April 2019) |
| 9 | General Urban Plan of the Municipal Unit of Kallithea of Rhodes island (April 2016) |
| 10 | New master plan of Marina Vouliagmeni (October 2017) |
| 11 | General Urban Plan of the Municipality of Argithea—Regional Unit of Karditsa (March 2016) |
| 12 | Rural Development Programme 2014–2020 (July 2014) |
| 13 | Strategic Transport Investment Framework (November 2014) |

The deeper understanding of the relationship between development and environment led to the worldwide adoption of SEA processes, which resulted in the writing of numerous corresponding environmental surveys over the past several years. In order to better compare Greek and international practices, a representative sample of reports from the EU member states was compiled, due to the common SEA institutional framework, which governs them after the transposal of the SEA Directive. Furthermore, additional focus was placed on the examination of SEA processes in plans and programs at all levels (international, national, regional, and local) and of different objectives. The methods and techniques utilized therein were indicated or explicitly placed in the various distinct SEA steps. Finally, reports from open online sources, primarily governmental websites, were also chosen. The aforementioned criteria were also used in this case. The examined reports are depicted in Table 5.

**Table 5.** European SEA environmental reports under consideration.

| No | Strategic Environmental Assessment |
|----|------------------------------------|
| 1 | UK Offshore Energy SEA (March 2016) |
| 2 | Derby City Council Local Transport Plan 3 SEA (March 2011) |
| 3 | Adriatic-Ionian Operational Programme 2014–2020 SEA (October 2014) |
| 4 | Transport Development Strategy of the Republic of Croatia 2017–2030 SEA (May 2017) |
| 5 | Cooperation Programme of the Baltic Sea Region 2014–2020 SEA (May 2014) |
| 6 | Polish Nuclear Programme SEA (January 2011) |
| 7 | Cross Border Cooperation Programme Hungary—Austria 2014–2020 SEA (April 2014) |
| 8 | Clonburris Strategic Development Zone SEA (September 2017) |
| 9 | Westminster City Council Municipal Waste Manager Strategy 2016–2026 SEA (September 2013) |
| 10 | South Lanarkshire Local Development Plan SEA (March 2017) |

It should be noted that during the critical assessment of the selected samples, the proper or improper utilization of methods and techniques was not examined, nor was the possibility of selection of alternative methods and techniques. Furthermore, we took into account the processes for the determination of the sectors of implementation (screening) and the extent of application (scoping), in the cases where the corresponding reports were available online.

*3.2. Primary Research Methodology*

The objective of the primary research was to document the current situation regarding the selection, adequacy, and implementation of current methods and techniques, as well as the current methodological guidance, through the filling out of questionnaires by practitioners experienced in the composition of SEIA. In total, 14 practitioners filled out the questionnaire, while officials from the competent authority of the Ministry of the Environment and Energy were unable to respond to similar requests, citing their heavy workload. The questions were the following:

- Do you face any difficulties during the composition of SEIA?
- Rate 1 to 10 the severity of each criterion for the selection of methods and techniques during the SEIA composition process: 1. Availability of human resources, 2. availability of financial resources, 3. availability of quantitative or qualitative data, 4. professional experience of practitioner, 5. scale of implementation of a plan or program, 6. sector wherein the plan or program is taking place, 7. type of decision, 8. timeline regarding the completion of the report.
- Do the methods and techniques used in the elaboration of SEIA ensure the following? 1. Be fit for purpose, i.e., they should be able to address key issues and fit into the decision-making process, 2. be integrated into the decision-making timetable, 3. allow the integration of various substantive aspects, i.e., various administrations, sectors, and procedures, 4. allow for the accommodation of uncertainties, 5. be transparent, robust, and relevant to the objective and the practices, 6. be understandable to all parties participating in SEA, 7. be cost-effective, 8. offer alternatives, in addition to measures designed to mitigate any negative impacts, 9. allow the comparison of alternatives.
- Are there any guidance manuals for the selection of methods and techniques during the SEIA composition process?
- If there are such guidance manuals, have you used them?

- Do you consider the existing methodological guidance for conducting the SEIA, which is provided by the Greek authorities, sufficient?
- According to your perspective, is the SEA an effective tool for meeting the goals set by environmental legislation? If not, would the utilization of different methods and techniques mitigate any resultant issues?
- To evaluate both the knowledge and the application of the recorded methods and techniques, as well as their application in the respective steps of the SEA, tables were given for completion. These tables summarize all the methods and techniques recorded during the bibliographic review.

## 4. Results

The results of the research, which was carried out based on the previously described methodology, are presented in three parts: Applied methods and techniques on a global level, applied methods and techniques in Greece, and primary research findings.

### 4.1. Applied Methods and Techniques on a Global Level

Following the comprehensive examination of environmental reports in Table 5, a categorization of the SEA methods and techniques by step, according to the frequency of their use, was carried out, as can be observed in Table 6.

Noteworthy conclusions can be drawn from Table 6. Initially, only a limited number of the methods and techniques referenced in international literature were implemented, ten to be precise. Practitioners therefore tend to utilize methods and techniques which they have already used previously. In other words, their professional experience played a crucial role in the selection of the appropriate methods and techniques.

Out of all the utilized methods and techniques, only the GIS and modelling are capable of processing quantitative data, while the data utilized by the others is primarily qualitative. In other words, practitioners exhibited a clear bias in favor of simple, expedited processes, which can be implemented regardless of the quality and quantity of utilized data, with a minimum of personnel and equipment.

The most frequently used methods and techniques were the expert judgment and the matrices. The former was implemented, for the most part, during the screening and scoping processes, while the second was implemented during the determination, evaluation, and assessment of the immediate, indirect, and cumulative impacts, as well as during the specification of measures designed to mitigate these impacts. Of the remaining methods and techniques, the participatory methods were widely implemented during the determination of the extent and breadth of implementation (scoping), while certain reports, the majority of which were carried out in the United Kingdom, also implemented them in the other SEA steps. The GIS were utilized for the production of maps and composite diagrams, and were widely implemented, in tandem with literature/case review, primarily in the description of the current situation. Finally, the majority of reports included the utilization of indicators in order to track the impacts of the plans and programs.

A point of particular interest is the variation in the number of utilized methods and techniques at each individual SEA step. In particular, only two methods and techniques were utilized during the monitoring of impacts. By contrast, the documented methods and techniques utilized during the steps of determination and evaluation of impacts, and determination and comparison of alternative choices were more numerous; eight and seven, respectively. The notable variation in the utilization of methods and techniques during each individual SEA step indicates that, during the screening process, for example, the almost universal implementation of expert judgments yielded precise, objective results, regardless of the scale, nature, and objective of the plans and programs. This is in contrast with the step of determination and comparison of alternative choices, in which the variety of selected methods and techniques indicates the opposite conclusion.

**Table 6.** SEA methods and techniques by step of European SEAs.

| Methods/Techniques | Screening | Scoping | SEA Steps | | | | | | |
|---|---|---|---|---|---|---|---|---|---|
| | | | Describe Baseline Status | Identify—Predict Impacts | Evaluate Impacts | Identify Cumulative/Indirect Impacts | Identify—Compare Alternatives | Propose Mitigation | Monitoring |
| Checklists | + | | | | | | | + | |
| Expert judgment | +++ | +++ | | ++ | + | ++ | ++ | ++ | |
| GIS | | | +++ | ++ | + | + | + | | |
| Guiding questions | | + | | + | | | + | | |
| Indicators | | | + | + | | | | | +++ |
| Literature/case review | + | + | +++ | + | | + | | | + |
| Matrices | | | ++ | +++ | +++ | +++ | +++ | ++ | |
| Modeling | | | | | | | + | | |
| Multi-criteria analysis | | | | + | ++ | + | + | + | |
| Participatory methods | + | +++ | | + | + | + | + | + | |

+++ used in 50% or more of the examined cases, ++ used in 26–49% of the examined cases, + used in 25% or less of the examined cases.

Finally, during the examination of the methodological guidance for the composition of SEA, it occurred that only half of the examined cases referred explicitly to the use of guidance handbooks. In particular, the reports carried out in the United Kingdom and Ireland presented in detail the guidance documents which were utilized, with the universal application of the manual "A practical guide to the Strategic Environmental Assessment Directive", while the SEA Transport Development Strategy of the Republic of Croatia 2017–2030, which was written in Croatian, contains a general reference to guidance manuals of the European Commission.

*4.2. Applied Methods and Techniques in Greece*

A preliminary point worth noting is that during the examination of selected SEA, it was found that guidance manuals for the selection of the most appropriate methodological approach were directly referred to in only seven cases. The manual "Handbook on SEA for Cohesion Policy 2007–2013" was universally implemented during the processes of particular SEA. The general objective of this manual was the advancement of the utilization of SEA for the incorporation of environmental issues in the plans and programs of the EE "Cohesion Policy" during the period of 2007–2013. Two cases utilized the manual "A practical guide to the Strategic Environmental Assessment Directive", which is a guide for the implementation of SEA in the plans and programs of the United Kingdom that fall within the scope of the SEA Directive. Furthermore, of particular note is the "circular for the observation of Strategic Environmental Analysis of Operational Programmes of Operational Period 2007–2013", issued by the Ministry of Finance and the Economy, which concerns the basic principles, scope, and general organization of the environmental monitoring of Operational Programs. One SEA, that of the Strategic Framework of Transportation Investment, utilized yearly reports which were produced by TERM, established by the European Environmental Service in 2000 with the objective of determining and utilizing the most appropriate indicators for the monitoring of transportation and the environment. However, the aforementioned manuals do not contain a distinct methodological guide for the comprehension and selection of the most appropriate methods and techniques in each separate case. Essentially, their almost exclusive focus is on directions for the compliance with the demands of the SEA Directive. Even though there are references to methods and techniques, these are indicative and refer to only a few of the SEA processes. The remaining SEA exhibited references to an overview of the relevant literature, which includes books and relevant reports prepared for the corresponding plans and programs.

In order to examine and evaluate the utilized SEA methods and techniques of Table 4, they were categorized according to their frequency of use by step, as depicted in Table 7.

The first conclusion which can be drawn from Table 7 is that only 12 of the 23 total documented methods and techniques were implemented in SEA processes in the entirety of the plans and programs. Subsequently, it was observed that the practitioners displayed an evident tendency to choose qualitative methods and techniques, with 10 out of a total of 12 being used for the processing of exclusively qualitative data. The expert judgment occupies a prominent position among them, due to being implemented most frequently and at the majority of SEIA steps. It is, however, of note that in most of the SEIA, this particular method was documented as an assessment provided by members of the team which composed the study who specialized in a particular field (e.g., geology, etc.). There are no references, in any study, to whether a particular technique, such as the Delphi Technique, was utilized in order to produce the assessment, or if this was produced solely based on the training, knowledge, or experience of the individual team members. Accordingly, an equally popular technique was the utilization of risk assessment matrices, primarily in table (or register) form, during the steps of environmental screening, determination, and evaluation of alternative choices, as well as in the evaluation and assessment of environmental impacts. Literature/case review was often utilized, primarily during the steps which required the use of bibliographical data, the most prominent of which was the description of the current situation. Finally, special reference must be made to the participatory methods. Despite the fact that they were not widely implemented in the majority of SEIA, they were applied during both the determination of alternative choices and in the selection of the final proposal in the SEIA of local spatial plans (MMP, OCSRP), in tandem with the completion of their first step. In particular, in the General Urban Plan of the Municipal Unit of Kallithea of Rhodes island and the General Urban Plan of the Municipality of Argithea—Regional Unit of Karditsa, the participatory methods included the transmission of the first step of the report to the competent public authorities. This was followed by a concurrent publication for the interested public, both with a relevant post in a specific website and open presentations.

The quantitative methods and techniques are on the opposite end of the frequency of use spectrum. In particular, out of all the documented methods and techniques which process quantitative data, only the GIS and modeling were utilized. The former was utilized during the majority of steps, albeit primarily in the description of the current state of the environment, for the production of maps and composite diagrams. The utilization of the latter, on the other hand, was notably limited in scope, a characteristic example being the use of a model for the population growth of the municipal unit of Argithea until 2031, with the application of various different scenarios during the implementation of the local spatial plan.

According to Table 7, the variation in the number of methods and techniques utilized during each separate SEA step is particularly noteworthy. Only two methods and techniques were selected in both the processes of prevention, mitigation, and treatment of environmental impacts and for their monitoring. In the first case, the expert judgment, which was essentially an assessment produced by the research team, was universally implemented in the SEIA, while in the second the method of indicators, was preferred, in order to ensure the capability of timely discovery and treatment of likely impacts. By contrast, the step of determination and evaluation of alternative choices featured the utilization of no less than seven different methods and techniques. The implementation of various methods and techniques, as well as combinations thereof, during this particular step, demonstrates that each selection of the appropriate methods and techniques directly correlates with factors such as, for example, the sector, the scope of implementation of the plans and programs, etc. Finally, the plans which underwent the environmental screening process, despite utilizing a total of five different methods and techniques, all utilized the compatibility appraisal in combination with the use of matrices.

This methodological approach, i.e., the preference of the practitioners for particular methods and techniques, is substantiated, for the most part indirectly, by a large number of them. Initially, the breadth of differentiation in both the content and the planning level, which is observable in the plans and programs subject to SEA, in addition to the fact that the guidance books describe a wide variety of methods and techniques, without clear instructions, has led practitioners to select a limited number of methods and techniques. Implicitly, the existing knowledge and experience of the practitioners is a crucial criterion in the selection of the methods and techniques, which often supersedes others, such as the scale of implementation of the plan or program, etc.

Accordingly, the evident bias in favor of qualitative, instead of quantitative, methods and techniques, is due to two crucial factors. On the one hand, the immaturity of the plans and programs, due to the lack of detailed characteristics of the projects and activities which comprise them, often makes the quantitative evaluation of environmental changes impossible. On the other hand, the quantitative methods and techniques constitute, for the most part, composite computational tools which require massive volumes of data, as well as specialized personnel and equipment. Essentially, due to the similarity of the quantitative methods and techniques with the EIA approach, their use is made difficult at the SEA level of plans and programs.

To conclude, the breadth of the methods and techniques utilized in the SEA under examination is relatively limited, and is comprised of the most common qualitative methods and a small number of quantitative techniques. In order to achieve the SEA objectives, practitioners selected, according to their own publications, a composite of methods and techniques, i.e., the combined utilization of two or more methods and techniques. This was observed in the steps of specification and evaluation of alternative choices, and in the assessment and evaluation of environmental impacts.

**Table 7.** Utilized SEA methods and techniques by key SEA steps.

| Methods/Techniques | Screening | SEIA (Strategic Environmental Impact Assessment) Steps | | | | | | | | Public Consultation |
|---|---|---|---|---|---|---|---|---|---|---|
| | | General Data | Scoping | Plan or Program Description | Identify—Compare Alternatives | Describe Baseline Status | Identify—Predict-Evaluate Impacts | Propose Mitigation | Monitoring | |
| Compatibility appraisal | +++ | | | | | | | | | |
| Checklists | + | | | | | | | | | |
| Expert judgment | | | +++ | +++ | + | | + | +++ | + | The process of stakeholder engagement is regulated and involves the transmission of the report to the competent public authorities while it is published to the public concerned. |
| GIS | ++ | This step does not require a defined methodology for its completion | + | ++ | + | +++ | + | | | |
| Guiding questions | | | + | | + | | +++ | | | |
| Literature/case review | | | +++ | ++ | | +++ | + | | + | |
| Matrices | +++ | | | + | +++ | + | +++ | | | |
| Participatory methods | | | + | | + | | | | | |
| Indicators | | | | | | | | | +++ | |
| Modeling | | | | | | + | | | | |
| Multi-criteria analysis | | | | | ++ | | ++ | | | |
| SWOT analysis | +++ | | | + | + | | | | | |

+++ used in 50% or more of the examined cases, ++ used in 26–49% of the examined cases, + used in 25% or less of the examined cases.

### 4.3. Primary Research Findings

The research through questionnaires has the objective of enriching the findings of the study. Half of the participants stated that they encountered difficulties in the composition of SEIA, the most commonly encountered of which were the following:

- Difficulty in the specification and categorization of the means of intervention of the plan or program under evaluation.
- Difficulty in the collection of sufficient and reliable data for the description of the current situation of the regions under examination.
- Inadequate specification of alternative choices from the practitioners in charge of drafting the plan documents, resulting in difficulties in the documentation of their environmental impacts.
- Lack of cooperation and direction from the competent authorities with regards to the determination of evaluation criteria for the impacts, the adequacy of available data, the implemented methodology, and the suggested measures of treating the impacts.
- Incomprehensible and often incomplete legislation.

In the composition of a SEIA, the most common criterion in the selection of methods and techniques is the "sector of the plan or program", followed by the "availability of qualitative or quantitative data". The "availability of financial resources" is an equally crucial factor. By contrast, factors of lesser significance are the "study completion timeframe" and "type of decision", while the professional experience of the involved practitioner was a frequent, albeit not decisive, factor in the selection of methods and techniques. Finally, the criteria "scale of implementation of the plan or program" and "availability of human resources" were subject to the most significant divergence of opinions among the participants.

The majority of participants stated that the methods and techniques utilized during the composition of SEIA were appropriate for their purpose. In other words, they were in a position to treat the basic issues which occurred, and were also suited to the decision-making process. Subsequently, the participants considered that the utilized methods and techniques offered adequate alternative choices and effective measures in order to mitigate any negative environmental impacts, while at the same time allowing for the effective comparison between alternative choices. Simultaneously, according to the majority of the participants, the methods and techniques facilitated the incorporation of various substantive environmental aspects. However, there was a number of observations among the participants regarding the shortcomings in their implementation. The majority of the participants believe that the utilized methods and techniques do not allow the effective treatment of uncertainties, are not transparent and relevant to the stated objective, are not comprehensible to all those involved in the SEA process, and finally, are not cost-effective. Finally, the vast majority of participants believe that the utilized methods and techniques contribute to time-consuming processes, resulting in their exclusion from the decision-making timeframe.

As for the existence and utilization of guidance manuals for the selection of methods and techniques during the composition of SEIA, half of the participants consider that these manuals exist, and use them. The remainder of the participants reported that no such manuals exist, at either the national or the European level, as the only guidance comes from the corresponding legislative documents. The guidance manuals used by the participants are documented in Table 8.

**Table 8.** Guidance manuals for the selection of methods and techniques in the composition of SEIA.

| No | Guidance Manuals |
| --- | --- |
| 1 | Applying Strategic Environmental Assessment. Good practice guidance for development co-operation, Organization for Economic Co-operation and Development, 2006 |
| 2 | A sourcebook on SEA of transport infrastructure plans and programs, European Commission, 2005 |
| 3 | Environmental assessments of Plans, Programmes and Projects. Rulings of the Court of Justice of the European Union, European Union, 2017 |
| 4 | Guidance on Integrating Climate Change and Biodiversity into Strategic Environmental Assessment, European Union, 2013 |
| 5 | Handbook on SEA for Cohesion Policy 2007–2013, Greening Regional Development Programmes Network, 2006 |
| 6 | Implementation of Directive2001/42 on the assessment of the effects of certain plans and programs on the environment, European Union, 2004 |
| 7 | Opinion of the Committee of the Regions on improving the EIA and SEA Directives, European Commission, 2010 |
| 8 | Strategic Environmental Assessment Better Practice Guide—methodological guidance for strategic thinking in SEA, Partidario M, 2012 |
| 9 | Streamlining environmental assessment procedures for energy infrastructure Projects of Common Interest, European Union, 2013 |
| 10 | Study concerning the application and effectiveness of the SEA Directive, European Commission, 2009 |
| 11 | Study concerning the preparation of the report on the application and effectiveness of the SEA Directive, European Union, 2016 |
| 12 | The Relationship between the EIA and SEA Directives. Final Report to the European Commission, Imperial College London, 2005 |
| 13 | Implementation of Directive 2001/42 "On Strategic Environmental Assessment" in the preparation of the Operational Programmes for the 2007–2013 programming period, Ministry of Environment, Spatial Planning and Public Works, 2006 |

However, none of the aforementioned documents propose the use of the most effective methods and techniques during SEA procedures. Certain documents, such as the "Handbook on SEA for Cohesion Policy 2007–2013" and the "Sourcebook on SEA of transport infrastructure plans and programmes", offer a more generalized and adaptable methodological approach for the SEIA guidance of particular types of plans and programs, in order for these to conform to the SEA Directive. Furthermore, documents such as the "Study concerning the preparation of the report on the application and effectiveness of the SEA Directive" and "The Relationship between the EIA and SEA Directives", are meant to offer information to the European Commission regarding the progress of the EU member states in the implementation of the SEA Directive, as well as to emphasize the points at which the SEA and EIA Directives overlap. In other words, the aforementioned documents are exclusively studies for the preparation of reports of the European Commission regarding the implementation and efficacy of the SEA Directive.

The results of the study show the inadequacy of the current methodological guidance for the composition of SEIA by the competent authorities in Greece. The lack of methodological guidance from the competent authorities is, according to the participants of the survey, due to the following reasons:

- There is no specialized institutional framework which the competent authorities can refer to for SEA methodology.
- There is a lack of trained personnel at the competent authorities for SEIA monitoring, evaluation, and approval.

- The observations of the competent authorities are limited to the comprehensiveness of the submitted SEIA file, as well as, occasionally, to the necessary supplemental elements, which usually follow the comments of the relevant competent authorities during public consultation.
- In order to cover the requirements of the JMD 107017/2006, practitioners adapt the methodology of carrying out the SEA according to their own perceptions and scientific and professional background.

Based on these findings, the suggested means of improving the current methodological guidance are the following:

- Incorporation of the conclusions drawn by the European Council reports regarding the implementation of the SEA Directive within the current legislative framework.
- Specialization of the institutional framework with the objective of offering clarifications and suggesting the necessary methods and techniques by step, depending on the sector of the plan or program for which the SEA is being carried out.
- Writing of guidance manuals for the selection of methods and techniques, by either a public body or private entity, according to the updated institutional framework. A particularly useful addition would be the inclusion of representative good practices.
- Reinforcement of the competent authorities with personnel, with concurrent training of current personnel, in order to achieve, on the one hand, substantive monitoring and evaluation of the SEIA being drafted, and on the other, effective monitoring of the environmental impacts from the implementation of the plan or program.

The participants in the survey were given a list (table) with the most frequent methods/techniques referred to in the literature, and were asked to indicate those that they know and those they use during the SEA process. The results are shown in Table 9.

One evident observation is that the majority of the participants were aware of almost the entirety of the SEA methods and techniques. A notable exception was those under the heading "Others", suggested by two of the participants. Asked why they did not utilize all the methods of which they were aware, the participants offered the following answers:

- These methods were not required by the current institutional framework.
- They were deemed unsuitable for implementation in SEA processes.
- There was a lack of necessary data for the implementation of composite analysis methods and techniques.
- There is a lack of financial and human resources for the implementation of specialized methods and techniques.

These reasons correlate with the previous analysis regarding the difficulties faced by practitioners during the composition of SEIA (inadequacy of the current institutional framework, lack of methodological guidance, limited available financial resources).

The analysis of the implemented methods and techniques arrived at equally useful conclusions. Table 9 demonstrates that the majority of the participants only implement 11 of the methods and techniques. The possibility of their utilization, according to the level and scale of implementation, the quality and type of utilized data, as well as the necessary resources for their conclusion, are depicted in Table 10.

According to the above table, the implemented methods and techniques are notably convenient, as, in their majority, they can be applied, either in part or entirely, at all levels and scales of the plans and programs, irrespective of the quality of the utilized data and the necessary resources. Furthermore, most of these methods and techniques, with the exceptions of the GIS and cost-benefit analysis, contribute to simple and expedient processes, do not require specialized equipment, have a limited implementation cost, and can be applied during most of the SEIA steps. A further point of interest is the widespread utilization of the cost-benefit analysis, as it compares the benefits and costs of a program based on its monetary worth. However, the composition of SEIA also includes the assessment and evaluation of environmental and social objectives, which are difficult to evaluate by monetary worth. Finally, this particular technique is a time-consuming and costly process, which requires large volumes of data, and is limited to programs and projects, rather than policies and plans.

**Table 9.** Methods and techniques known and used by the participants.

| Methods/Techniques. | Do You Know Them; | | Do You Use Them; | |
|---|---|---|---|---|
| | Yes | No | Yes | No |
| Carrying capacity analysis | +++ | + | ++ | +++ |
| Checklists | +++ | + | +++ | ++ |
| Compatibility appraisal | ++ | +++ | + | +++ |
| Cost-benefit analysis | +++ | + | +++ | ++ |
| Ecological footprint analysis | +++ | + | ++ | +++ |
| Expert judgment | +++ | ++ | +++ | ++ |
| GIS | +++ | + | +++ | + |
| Guiding questions | +++ | ++ | ++ | +++ |
| Indicators | +++ | + | +++ | + |
| Land use partitioning analysis | +++ | ++ | + | +++ |
| Life cycle analysis | +++ | + | ++ | +++ |
| Literature/case review | +++ | + | +++ | ++ |
| Matrices | +++ | + | +++ | + |
| Modeling | +++ | + | + | +++ |
| Multi-criteria analysis | +++ | + | +++ | + |
| Network analysis | +++ | ++ | + | +++ |
| Overlay maps | +++ | ++ | ++ | +++ |
| Participatory methods | +++ | ++ | +++ | ++ |
| Quality of life assessment | +++ | ++ | ++ | +++ |
| Risk assessment | +++ | + | +++ | ++ |
| Scenario/Sensitivity analysis | +++ | ++ | ++ | +++ |
| SWOT analysis | +++ | + | +++ | ++ |
| Vulnerability analysis | +++ | ++ | + | +++ |
| Other defined by the participants: | | | | |
| Critical factors | + | +++ | + | +++ |
| Ecosystem services | + | +++ | + | +++ |
| GHG emission calculators | + | +++ | + | +++ |
| Spheres of influence and Ecosystem chains | + | +++ | + | +++ |
| CO2MPARE | + | +++ | + | +++ |
| PEST | + | +++ | + | +++ |
| AKIS | + | +++ | + | +++ |

+++ identified by 50% or more of participants, ++ identified by 26–49% of participants, + identified by 25% or less of participants.

**Table 10.** Methods and techniques implemented by the majority of the participants.

| Methods and Techniques | Plan Level | Program Level | Large Area | Small Area | Land Use Plan | Sectoral Plan | Incomplete Data | Uncertain Data | Qualitative Data | Few Resources |
|---|---|---|---|---|---|---|---|---|---|---|
| Checklists | + | ++ | ++ | ++ | ++ | ++ | ++ | ++ | ++ | ++ |
| Cost-benefit analysis | | ++ | ++ | ++ | ++ | ++ | | ++ | | |
| Expert judgment | ++ | ++ | ++ | ++ | ++ | ++ | ++ | ++ | ++ | ++ |
| GIS | ++ | ++ | ++ | ++ | ++ | ++ | + | + | + | + |
| Indicators | ++ | ++ | ++ | ++ | ++ | ++ | ++ | ++ | ++ | ++ |
| Literature/case review | ++ | ++ | ++ | ++ | ++ | ++ | ++ | ++ | ++ | ++ |
| Matrices | ++ | ++ | ++ | ++ | ++ | ++ | ++ | ++ | ++ | ++ |
| Multi-criteria analysis | | ++ | + | ++ | + | ++ | | ++ | ++ | + |
| Participatory methods | + | ++ | + | ++ | ++ | ++ | ++ | ++ | ++ | ++ |
| Risk assessment | | + | ++ | + | ++ | ++ | + | ++ | + | |
| SWOT analysis | ++ | ++ | ++ | ++ | ++ | ++ | ++ | ++ | ++ | ++ |

++ = fully, + = partly, (blank) = not.

Table 11 demonstrates the methods/techniques which the participants implement during the composition of SEIA, by step. *GIS* were the most popular technique, as they were utilized by all of the participants, with the majority thereof utilizing them during the steps of alternative choices, description of the current situation, and assessment and evaluation of impacts. Accordingly, the expert judgment was implemented by the majority of the participants during the steps of prevention, mitigation, and treatment of impacts. The techniques of indicators and literature/case review are particularly noteworthy for their universal application, the former during the monitoring of environmental impacts, and the latter during the description of the current situation. Finally, the matrices and the multi-criteria analysis were equally utilized in the corresponding steps, the most notable of which was the assessment and evaluation of the impacts, where they were utilized by the majority of the participants.

The variation in the number of methods and techniques utilized at each separate SEA step is of particular note. During the monitoring of environmental impacts, almost all the participants utilized indicators in combination with others, such as expert judgment, GIS, etc. Correspondingly, during the description of the current situation, literature/case review, in addition to GIS, was the favored choice of all the participants. The most noteworthy step, however, was that of the assessment and evaluation of impacts, during which the majority of participants utilized GIS, matrices and multi-criteria analysis, in combination with a number of other methods and techniques, such as cost-benefit analysis, checklists, etc.

The next conclusion which can be drawn from Table 11 is that the entirety of the methods and techniques were applied by the participants in the majority of the various SEIA steps. However, it is troubling that certain, particular methods and techniques were utilized by the participants during the step of the general data. We must keep in mind that this is the initial step in the study, which contains broad references to the plan or program under evaluation, the initiation of planning of the plan or program, as well as the practitioner responsible for composing the SEIA. In essence, this step does not require a defined methodology for its completion.

The entirety of the previously presented methods and techniques contribute to composite processes, and require trained, experienced personnel to carry them out; their ineffective implementation may result in erroneous decisions. The aforementioned considerations highlight certain issues regarding the actual, substantive understanding of the utilized methods and techniques by all those involved in SEA processes, and their effective application at each step.

Finally, the study recorded the participants' opinions on whether SEA is an effective process for the achievement of the objectives set in place by environmental legislation. According to the majority of the participants, SEA is indeed an effective process, due to covering the insufficiencies of Greek environmental law regarding the assessment of environmental impacts at the "strategic" level, i.e., at a higher level than that of each individual project. However, in order for SEA to assume a substantive role during the early stage of planning, certain amendments must be made. It is telling that at this point, there is no consensus among the participants. Some of them support that, since the current methods and techniques were drawn from those used in EIA, the utilization of more "strategic" methods and techniques stands to mitigate the problem. The majority of participants, however, believe the current methods and techniques are adequate for the fulfillment of the objectives of SEA, although systematic methodological guidance is required for their more effective utilization, primarily during the processes of assessment, evaluation, treatment, and monitoring of environmental impacts.

**Table 11.** Methods and techniques implemented during key SEA steps by the participants.

| Methods/Techniques | Screening | SEIA Steps | | | | | | | | Public Consultation |
|---|---|---|---|---|---|---|---|---|---|---|
| | | General Data | Scoping | Plan or Program Description | Identify—Compare Alternatives | Describe Baseline status | Identify—Predict-Evaluate Impacts | Propose Mitigation | Monitoring | |
| Carrying capacity analysis | + | | + | + | + | + | + | | + | |
| Checklists | ++ | + | + | + | + | + | ++ | + | ++ | |
| Compatibility appraisal | + | | + | | + | | + | | | |
| Cost-benefit analysis | + | | ++ | + | ++ | | ++ | ++ | + | |
| Ecological footprint analysis | + | | | + | + | + | ++ | + | + | |
| Expert judgment | + | + | ++ | + | + | + | ++ | +++ | ++ | |
| GIS | + | + | + | ++ | +++ | +++ | +++ | ++ | ++ | |
| Guiding questions | + | | + | + | | + | + | | + | |
| Indicators | ++ | + | + | | + | ++ | ++ | ++ | +++ | The process of stakeholder engagement is regulated and involves the transmission of the report to the competent public authorities while it is published to the public concerned |
| Land use partitioning analysis | | + | + | + | + | + | + | + | + | |
| Life cycle analysis | | | | | + | | ++ | + | + | |
| Literature/case review | ++ | + | ++ | + | + | +++ | ++ | ++ | + | |
| Matrices | + | + | + | + | ++ | + | +++ | ++ | + | |
| Modeling | | | + | | + | + | ++ | + | + | |
| Multi-criteria analysis | + | | + | | ++ | + | +++ | ++ | + | |
| Network analysis | + | | + | | + | + | + | + | + | |
| Overlay maps | + | + | + | + | + | ++ | + | + | + | |
| Participatory methods | + | + | ++ | + | ++ | + | + | + | + | |
| Quality of life assessment | + | | + | + | | + | + | + | + | |
| Risk assessment | | | | | + | | ++ | ++ | + | |
| Scenario/Sensitivity analysis | | | | | + | | + | + | + | |
| SWOT analysis | + | + | + | + | ++ | + | ++ | + | + | |
| Vulnerability analysis | | | + | | + | + | ++ | + | + | |
| Other: PEST, AKIS | | | + | + | | | | | | |

+++ filled by 50% or more of participants, ++ filled by 26–49% of participants, + filled by 25% or less of participants.

## 5. Discussion and Conclusions

Based on the findings of the preceding study, the Greek and international practices are comparatively evaluated, taking into account two aspects: Methodological guidance for the composition of SEA, and applied methods and techniques. Subsequently, the conclusions of the primary research regarding both the effectiveness of SEA in Greece and the utilization of methods and techniques will be stated, and certain proposals will be suggested for the elimination of recorded anomalies.

The examination of the methodological guidance for the composition of SEA yielded common results in both international and Greek practice. In particular, the study of representative examples from both Greece and abroad demonstrated that in only half of the examined cases there was a clear reference to the utilized guidance documents, while primary research revealed that only half of the participating practitioners were aware of, and utilized, guidance manuals.

Furthermore, it was found that certain guidance manuals were utilized with greater frequency by the practitioners. At the international level, for which we have detailed information only of those applied in the United Kingdom and Ireland, the "A practical guide to the Strategic Environmental Assessment Directive" is by far the most popular guide. Correspondingly, in Greece, the "Handbook on SEA for Cohesion Policy 2007–2013", is universally applied, while "A practical guide to the Strategic Environmental Assessment Directive" and "Sourcebook on SEA of transport infrastructure plans and programmes" were also regularly consulted. A defining element of the methodological inadequacy in Greece is the lack of any domestic guidance manual, with the guidelines for the composition of SEA being determined by the provisions of JMD 107017/2006 and certain circulars that have been issued.

As was mentioned previously, the aforementioned guidance documents offer a general, flexible methodological approach for the guidance of SEA, with the objective of conforming to the SEA Directive. However, even though some of these documents offer instructions for each individual SEA step, they do not include exact, substantive methodological guidance for the comprehension and selection of the most appropriate methods and techniques for each individual SEA step. Additionally, it appears that certain documents, which were discovered in the international SEA practices (e.g., United Kingdom SEA), have been adapted to the requirements and demands of each state, precluding the possibility of their utilization by other EU member states.

As for the applied SEA methods and techniques, the study demonstrated that similar methods and techniques are utilized in practice, both internationally and in Greece, with their number being from 10 to 12. A number of factors influence the selection of particular methods and techniques by the practitioners of the states in which the SEA Directive has been transposed. First, the common legislative framework which governs the states, which adopts an EIA-like SEA approach, is a crucial factor, due to demanding particular methods and techniques. Similarly, the practitioners utilized specific guidance documents, which recommended a similar approach in the selection of methods and techniques, resulting in the utilization of comparable criteria, as these were established in the course of the primary research, for the selection of the most appropriate methods and techniques during SEA processes.

The majority of applied methods and techniques contribute to simple and expedited processes, process qualitative data, and can be implemented even in cases of inadequate or uncertain information, while, due to not requiring specialized equipment, they can be utilized even when financial resources are limited (with expert judgment, matrices, literature/case review, indicators as notable examples). Of the quantitative methods and techniques, GIS was most widely used in the production of maps and composite diagrams.

There are also evident similarities in the variation of the number of utilized methods and techniques at each individual SEA step. The steps of assessment and evaluation of impacts, and of alternative, saw the use of the widest variety of methods and techniques of those documented in the selected case studies, with the majority of both Greek and European practitioners utilizing risk assessment matrices of primarily table form, in combination with the other methods and techniques. By contrast, the fewest methods and techniques were utilized during the step of monitoring of environmental impacts, with the universal utilization of environmental indicators.

To sum up the findings of the primary study, Greek practitioners encounter significant difficulties during SEA processes, the most important of which have to do with: (a) Incomprehensible, incomplete legislation, (b) difficulties in the determination and categorization of the types of intervention of the plan or program under examination, (c) difficulties in the collection of sufficient and reliable data, (d) insufficient definition of alternative by the practitioners in charge of drafting the plan documents, and (e) lack of cooperation and guidance by the competent authorities.

The current methodological guidance has been deemed inadequate. This particular shortcoming is observable primarily in the lack of a specialized institutional framework for SEA methodology and the lack of trained personnel at the competent authorities. This results in, on the one hand, comments being expressed by the latter being limited to the comprehensiveness of the submitted SEIA file, and on the other the methodology being specialized on the initiative of the practitioners themselves, in order to cover the requirements of the JMD 107017/2006.

Despite the applied methods and techniques being considered appropriate for their objective, the lack of methodological guidance and the selection criteria for the SEA methods and techniques which the practitioners themselves set in place, as they were presented in Section 4.3, result in the selection of methods and techniques which cannot treat uncertainties, are characterized by a lack of transparency, and are not always relevant to the stated objective, while also being incomprehensible to all those involved in the SEA, and lacking cost-effectiveness. The preference for qualitative rather than quantitative methods and techniques is correlated, on the one hand, with the immaturity of the plans and programs, and on the other, with the complexity of the quantitative methods, which, as a rule, contribute to more time-consuming processes, and require the processing of large volumes of data by specialized personnel utilizing specialized equipment.

The more than ten years of experience from the introduction of the SEA Directive into Greek environmental law demonstrate the coverage of a significant deficiency regarding the assessment of environmental impacts during the early planning stage of plans and programs. Despite the fact that SEA has been recognized by the majority of the Greek practitioners who participated in the study as an effective environmental process, the extent to which this particular process has contributed to the goal of sustainable development remains a controversial issue. The ambiguity which, due exactly to its "strategic" nature, characterizes SEA, the current legislation, which, by contrast, requires the adherence to processes based on the corresponding practices of EIA, and the lack of methodological guidance, constitute a tripartite obstacle practitioners have not been able to bypass with the application of the current methods and techniques.

In order to mitigate existing anomalies, crucial changes must be made. These must be immediately applicable, without requiring complex, time-consuming and, primarily, bureaucratic procedures. Within this framework, our suggestion is the determination of technical specifications for the SEA process by sector for plans or programs, in addition to the writing of guidance manuals by sector for plans or programs, as extensions of the technical specifications of the corresponding sector. Furthermore, we propose the writing of a manual of indicators for the recording and monitoring of environmental impacts from the implementation of the plans or programs, by sector and type of plan or program, in addition to training courses for the personnel of the authorities responsible for the planning of plans and programs and the general monitoring and approval of environmental reports and practitioners, in the implementation of the most appropriate methods and techniques during SEA processes. Finally, we propose the broader utilization of participatory methods, such as the citizens' jury, during the step of evaluation of alternative of the plan or program.

**Author Contributions:** Conceptualization, A.T.; methodology, A.T. and A.S.; research conduction, data collection and analysis, visualization, A.S.; research supervision, A.T.; writing, review & editing of the paper, A.S. and A.T. All authors have read and agreed to the published version of the manuscript.

**Funding:** This research received no external funding.

**Conflicts of Interest:** The authors declare no conflict of interest.

## References


1. Sadler, B.; Aschemann, R.; Dusik, J.; Fischer, T.; Partidario, M.; Verheem, R. *Handbook of Strategic Environmental Assessment*; Earthscan: London, UK, 2011.
2. OECD (Organisation for Economic Co-operation and Development). *Applying Strategic Environmental Assessment: Good Practice Guidance for Development Co-operation*; OECD Publishing: Paris, France, 2006; Available online: https://www.oecd.org/environment/environment-development/37353858.pdf (accessed on 10 March 2020).
3. Noble, B.F.; Storey, K. Towards a structured approach to strategic environmental assessment. *J. Environ. Assess. Policy Manag.* **2001**, *3*, 483–508. [CrossRef]
4. Therivel, R. *Strategic Environmental Assessment in Action*; Earthscan: London, UK, 2004.
5. Lee, N. Bridging the gap between theory and practice in integrated assessment. *Environ. Impact Assess. Rev.* **2006**, *26*, 57–78. [CrossRef]
6. Noble, B.F.; Gunn, J.; Martin, J. Survey of current methods and guidance for strategic environmental assessment. *Impact Assess. Proj. Apprais.* **2012**, *30*, 139–147. [CrossRef]
7. Zagorianakos, E. An ex-ante appraisal of the possible implementation of the proposed directive on strategic environmental assessment (SEA) in England and Greece. *Plan. Pract. Res.* **1999**, *14*, 445–465. [CrossRef]
8. Zagorianakos, E. Athens 2004 olympic games' transportation plan: A missed opportunity for strategic environmental assessment (SEA) integration? *J. Transp. Geogr.* **2004**, *12*, 115–125. [CrossRef]
9. Partidário, M.R. (Ed.) *Perspectives on Strategic Environmental Assessment*, 1st ed.; CRC Press: New York, FL, USA, 1999.
10. Fundingsland Tetlow, M.; Hanusch, M. Strategic environmental assessment: The state of the art. *Impact Assess. Proj. Apprais.* **2012**, *30*, 15–24. [CrossRef]
11. Lee, N.; Walsh, F. Strategic environmental assessment: An overview. *Proj. Apprais.* **1992**, *7*, 126–136. [CrossRef]
12. Bina, O. A critical review of the dominant lines of argumentation on the need for strategic environmental assessment. *Environ. Impact Assess. Rev.* **2007**, *27*, 585–606. [CrossRef]
13. Sadler, B.; Verheem, R. Conclusions and Recommendations of the International SEA Effectiveness Study. In Proceedings of the 16th Annual Meeting of the International Association for Impact Assessment (IAIA), Estoril, Portugal, 17–23 June 1996.
14. UNECE (United Nations Economic Commission for Europe). Protocol on Strategic Environmental Assessment to the Convention on Environmental Impact Assessment in Transboundary Context. 2003. Available online: https://www.unece.org/fileadmin/DAM/env/eia/documents/legaltexts/protocolenglish.pdf (accessed on 10 March 2020).
15. Peterson, K.; Kose, M.; Uustal, M. Screening decisions concerning the likely impacts of plans and projects on natura 2000 sites. *J. Environ. Assess. Policy Manag.* **2010**, *12*, 185–214. [CrossRef]
16. Kassios, K.; Lamprou, M. *Distinction Levels of Environmental Impact Assessment Procedures (EIA) and Strategic Environmental Assessment (SEA)*; National Technical University of Athens: Athens, Greece, 2004; Available online: https://www.hydro.ntua.gr/2004-05-21-conference/kassios-lambrou-text.pdf (accessed on 10 March 2020). (In Greek)
17. Kläne, C.; Albrecht, E. Purpose and Background of the European SEA Directive. In *Implementing Strategic Environmental Assessment. Environmental Protection in the European Union*; Schmidt, M., Joao, E., Albrecht, E., Eds.; Springer Science & Business Media: Berlin/Heidelberg, Germany, 2005; Volume 2, pp. 15–29.
18. Arce-Ruiz, R.M.; Soria-Lara, J.A.; González-Del-Campo, A. SEA effectiveness in Spain: Insights from practice. *Impact Assess. Proj. Apprais.* **2019**, *37*, 327–333. [CrossRef]
19. Baresi, U.; Vella, K.J.; Sipe, N.G. Bridging the divide between theory and guidance in strategic environmental assessment: A path for Italian regions. *Environ. Impact Assess. Rev.* **2017**, *62*, 14–24. [CrossRef]
20. Bragagnolo, C.; Geneletti, D. Dealing with land use decisions in uncertain contexts: A method to support Strategic Environmental Assessment of spatial plans. *J. Environ. Plan. Manag.* **2014**, *57*, 50–77. [CrossRef]
21. Campagna, M.; Di Cesare, E.A.; Matta, A.; Serra, M. Bridging the Gap between Strategic Environmental Assessment and Planning: A Geodesign Perspective. *Int. J. E-Plan. Res.* **2018**, *7*, 34–52. [CrossRef]

22. Lai, S.; Leone, F.; Zoppi, C. Strategic Environmental Assessment and Enhancement of Ecosystem Services: A Study concerning Spatial Planning in Sardinia (Italy). In *Urban Planning and Renewal*; Wolfe, M., Ed.; Nova Science Publishers, Inc.: New York, NY, USA, 2018; pp. 61–83.

23. Lamorgese, L.; Geneletti, D. Sustainability principles in strategic environmental assessment: A framework for analysis and examples from Italian urban planning. *Environ. Impact Assess. Rev.* **2013**, *42*, 116–126. [CrossRef]

24. García, J.L.M.I. Strategic environmental assessment for metropolitan plans of coastal areas. The case of Valencia. *Int. J. Sustain. Dev. Plan.* **2017**, *12*, 1272–1281.

25. Rega, C.; Bonifazi, A. Strategic Environmental Assessment and spatial planning in Italy: Sustainability, integration and democracy. *J. Environ. Plan. Manag.* **2014**, *57*, 1333–1358. [CrossRef]

26. Rega, C.; Singer, J.P.; Geneletti, D. Investigating the substantive effectiveness of Strategic Environmental Assessment of urban planning: Evidence from Italy and Spain. *Environ. Impact Assess. Rev.* **2018**, *73*, 60–69. [CrossRef]

27. Chaker, A.; El-Fadl, K.; Chamas, L.; Hatjian, B. A review of strategic environmental assessment in 12 selected countries. *Environ. Impact Assess. Rev.* **2006**, *26*, 15–56. [CrossRef]

28. Psaila, A.; Statha, T. Strategic Environmental Assessment: A New Institution. In Proceedings of the Electronic Proceedings of the International Exhibition and Conference on Environmental Technology, Athens, Greece, 3–6 February 2005; Technical Chamber of Greece: Athens, Greece. Available online: http://library.tee.gr/digital/m2045/m2045_psaila.pdf (accessed on 10 March 2020). (In Greek).

29. Tatsis, L. Strategic Environmental Impact Assessment and Spatial Planning. Νόμος+Φύση 2006. Available online: https://nomosphysis.org.gr/10606/stratigiki-ektimisi-periballontikon-epiptoseon-kai-xorikos-sxediasmos-septembrios-2006/ (accessed on 10 March 2020). (In Greek).

30. Partidário, M.R. Strategic Environmental Assessment Better Practice Guide: Methodological Guidance for Strategic Thinking in SEA. Portuguese Environment Agency and Redas Energeticas Nacionais, Lisbon. Available online: http://www.civil.ist.utl.pt/shrha-gdambiente/SEA_Guidance_Portugal.pdf (accessed on 10 March 2020).

31. Noble, B.F. Strategic environmental assessment: What is it? & what makes it strategic? *J. Environ. Assess. Policy Manag.* **2000**, *2*, 203–224.

32. Marshall, R.; Fischer, T.B. Regional electricity transmission planning and SEA: The case of the electricity company ScottishPower. *J. Environ. Plan. Manag.* **2006**, *49*, 279–299. [CrossRef]

33. McNamara, C. Strategic Planning (in Nonprofit or For-Profit Organizations). 2002. Available online: http://wk.ixueshu.com/file/76c00f2a9720734f.html (accessed on 10 March 2020). (as referenced in [20]).

34. Fischer, T.B. Lifting the fog on SEA—Towards a categorization and identification of some major SEA tasks: Understanding policy-SEA, plan-SEA and programme-SEA. In *Environmental Assessment in the Nordic Countries*; H. Bjarnadóttir, H., Ed.; Nordregion: Stockholm, Sweden, 2000; pp. 39–46.

35. Roberts, P. *Environmentally Sustainable Business—A Local and Regional Perspective*; Paul Chapman Publishing Ltd.: London, UK, 1995.

36. World Bank. *Sectoral Environmental Assessment. Environmental Assessment Sourcebook Update 4*; World Bank: New York, NY, USA, 1993; Available online: http://documents.worldbank.org/curated/en/256691468740431408/pdf/multi0page.pdf (accessed on 10 March 2020).

37. Partidário, M.R. Does SEA change outcomes? Discussion paper No. 2009/31. In *18th International Transport Research Symposium*; Forum, I.T., Ed.; OECD/ITF: Paris, France, 2009; Available online: https://www.oecd-ilibrary.org/docserver/5kmmnc5ln3r0-en.pdf?expires=1583926903&id=id&accname=guest&checksum=61C578BEB043AD7CA345167421B44B9F (accessed on 10 March 2020).

38. Dalal-Clayton, B.; Sadler, B. *Strategic Environmental Assessment: A Sourcebook and Reference Guide to International Experience*; Earthscan: London, UK, 2005.

39. Noble, B.; Nwanekezie, K. Conceptualizing strategic environmental assessment: Principles, approaches and research directions. *Environ. Impact Assess. Rev.* **2017**, *62*, 165–173. [CrossRef]

40. Partidário, M.R. Scales and associated data—What is enough for SEA needs? *Environ. Impact Assess. Rev.* **2007**, *27*, 460–478. [CrossRef]

41. Liou, M.L.; Yeh, S.C.; Yu, Y.H. Reconstruction and systemization of the methodologies for strategic environmental assessment in Taiwan. *Environ. Impact Assess. Rev.* **2006**, *26*, 170–184. [CrossRef]

42. Brown, A.L.; Thérivel, R. Principles to guide the development of strategic environmental assessment methodology. *Impact Assess. Proj. Apprais.* **2000**, *18*, 183–189. [CrossRef]

43. Fischer, T.B. *Theory and Practice of SEA: Towards a More Systematic Approach*; Earthscan: London, UK, 2007.
44. Therivel, R.; Wood, G. Tools for SEA. In *Implementing Strategic Environmental Assessment. Environmental Protection in the European Union*; Schmidt, M., Joao, E., Albrecht, E., Eds.; Springer Science & Business Media: Berlin/Heidelberg, Germany, 2005; Volume 2, pp. 349–362.
45. Fischer, T.B. *Strategic Environmental Assessment in Transport and Land Use Planning*; Earthscan: London, UK, 2002.
46. Bell, S.; Morse, S. *Sustainability Indicators: Measuring the Immeasurable?* 2nd ed.; Routledge: London, UK, 2012.
47. Finnveden, G.; Nilsson, M.; Johansson, J.; Persson, Å.; Moberg, Å.; Carlsson, T. Strategic environmental assessment methodologies—Applications within the energy sector. *Environ. Impact Assess. Rev.* **2003**, *23*, 91–123. [CrossRef]
48. Dusík, J.; Jurkeviciute, A.; McGuinn, J. *Handbook on SEA for Cohesion Policy 2007–2013*; Greening Regional Development Programmes Network: Exeter, UK, 2006.
49. Rauschmayer, F.; Risse, N. A framework for the selection of participatory approaches for SEA. *Environ. Impact Assess. Rev.* **2005**, *25*, 650–666. [CrossRef]

 *sustainability*

*Article*

# An AHP-SWOT-Fuzzy TOPSIS Approach for Achieving a Cross-Border RES Cooperation

**Aikaterini Papapostolou, Charikleia Karakosta *, Georgios Apostolidis and Haris Doukas**

Energy Policy Unit (EPU-NTUA), Decision Support Systems Laboratory, School of Electrical and Computer Engineering, National Technical University of Athens, 15780 Athens, Greece; kpapap@epu.ntua.gr (A.P.); george_apo_8@hotmail.com (G.A.); h_doukas@epu.ntua.gr (H.D.)
* Correspondence: chkara@epu.ntua.gr; Tel.: +30-210-772-2083

Received: 28 January 2020; Accepted: 2 April 2020; Published: 4 April 2020

**Abstract:** The emerging need to tackle climate change and mitigate greenhouse gas emissions has led to the consolidation of interest in renewable energy sources (RES) setting specific targets in the European area. To achieve the ambitious targets set, Member States are given the opportunity to cooperate with one or more of their developing neighboring countries. The aim of this paper is to develop a methodological framework based on the combination of the Strengths, Weaknesses, Opportunities and Threats (SWOT) analysis with the Analytic Hierarchy Process (AHP) and the Fuzzy Technique for Order of Preference by Similarity to Ideal Solution (Fuzzy TOPSIS) methods for adopting the most appropriate strategic plan, in order to establish a successful energy cooperation that will create beneficial conditions for all the involved parties. The results could be important in facilitating decision makers to assess the role and design of this cooperation mechanism. Key insights will also emerge with regards to opportunities for energy strategy cooperation between Europe and its neighboring countries.

**Keywords:** renewable energy sources; energy policy; SWOT analysis; multi-criteria analysis; AHP; fuzzy TOPSIS; Morocco; Egypt

---

## 1. Introduction

### 1.1. Background and Motivation

Nowadays, there is plenty of scientific evidence to prove that climate is changing due to the increasing quantities of greenhouse gas (GHG) emissions, for which human activity is mainly responsible [1]. It is also a fact that as the years go by, global energy demand is rising dramatically, which is the main cause of the bulk of these emissions, as more and more fossil fuels are burnt in order to meet those demands. In order to tackle climate change, the European Commission (EC) has already adopted a series of measures to facilitate the clean energy transition in its energy sector in the future. Recently, the EC Clean Energy for all Europeans package entered into force including policies and legislation regarding renewable energy sources (RES), energy efficiency and GHG emissions reduction, while the recast Renewable Energy Directive (RED) [2,3] envisages an ambitious, binding target of 32% for RES in the European Union (EU) energy mix by 2030.

It is worth mentioning also that the international regulatory framework, as well as the international initiatives, call for increased cooperation, as a crucial factor to fully exploit the vast RES potential worldwide. This will constitute a win-win situation as it will allow us to meet climate change objectives in a cost-effective way and enable countries to develop their economies in a more sustainable way. In addition, it will allow the development of a competitive industry in the field of low-carbon technologies [4].

To meet the ambitious energy targets set, Member States are given the opportunity to cooperate with one or more of the neighboring countries towards this direction. More specifically, one or more Member States may cooperate with one or more developing neighboring countries on joint projects within the territory of the latter, with regard to electricity generation from RES. Any amount of electricity generated by such installations may be taken into account for the purposes of measuring compliance with the Member States' national overall targets, if certain demanding conditions are met.

As Karakosta et al. (2013) [5] noted and compared to the other cooperation mechanisms envisaged by the EU legislation, barriers to the implementation of the cooperation mechanism on joint projects between EU and developing countries include poor grid infrastructure (in order for the energy to be transferred into the community), geopolitical unrest, risks of limited public acceptance, existing legal limitations and complex financing schemes.

However, and despite the potential difficulties that lie in the implementation of the mechanism, especially considering that any such project should be able to attract private funding, there are major benefits as well [6]. In addition, joint projects with developing countries, although quite complex considering the involved parties, different country contexts, regulations, infrastructures etc., could be a crucial instrument striving towards international RES cooperation to foster the social, economic and environmental benefits of RES electricity (RES-E) projects.

### 1.2. Contributions

The core objective of this paper is to assess, through case studies and integrated analysis to what extent cooperation with developing neighboring countries can help Europe achieve its RES targets and beyond this, trigger the deployment of RES-E projects in the host countries and create synergies and mutually beneficial circumstances for all involved parties [7]. In order to support the development of cooperation mechanisms and implement a successful collaboration, the current and future situation of the host country need to be examined, so as to be able to develop the most appropriate energy policies [8]. This could be achieved through the analysis of Strengths, Weaknesses, Opportunities and Threats (SWOT) of the country under examination.

SWOT analysis is used in order to assess a host country's present situation as conducive to implement RES projects under the cross-border cooperation framework. The aim is to identify factors that are favorable (Strengths, Opportunities) and unfavorable (Weaknesses, Threats) to the development of this cross-border cooperation. The SWOT analysis intents to identify win-win actions for both EU Member States and neighboring countries by providing answers to the following questions [9]:

- How can we exploit the most from the regions' strengths to increase the cumulative welfare of both EU and the regional countries?
- How can we circumvent the regions´ weaknesses by choosing the best technology options?
- How can we create opportunities and then capitalize on them?
- How can we manage possible threats, so that the RES targets will not be compromised by unfavorable evolution?

However, SWOT analysis cannot be considered as a sufficient stand-alone tool to solve this energy planning problem. The complex nature of this process requires the use of multi-criteria decision making (MCDM) methods, which seem to be extremely powerful tools and able to deal with the different aspects that these problems include [8,10,11]. MCDM methods have been applied to many energy-related problems, such as energy planning and selection, energy resource allocation, energy policy, management of building energy, transportation systems, and electric utility planning [6,12–15]. Such problems have been discussed either from the perspective of a single criterion decision problem, such as maximizing profit or minimizing cost, or in relation to complex multi-criteria decision problems [16]. According to Wang et al. 2009 [17], the most frequently used criteria are investment cost, $CO_2$ emissions, efficiency, operation and maintenance cost, land use, fuel cost, and job creation.

For this study, Analytic Hierarchy Process (AHP) is utilized in order to determine the weights of the criteria that will be then used to assess the alternative proposed strategies to be followed towards a successful implementation of cooperation mechanisms. After the criteria weights calculation, the Fuzzy Technique for Order of Preference by Similarity to Ideal Solution (Fuzzy TOPSIS) method is used, in order to rank the proposed strategies that emerge towards the promotion of cross border cooperation, since it is widely used to solve decision making problems.

In the current study, the perspectives of a cross-border energy cooperation between the EU and Morocco, which was selected as a potential host country, are examined. The reason why this country was selected has to do with the fact that the development of such a cooperation seems extremely feasible. First of all, Morocco is Spain's southern border and only a relatively narrow body of water separates the two countries. This is a favorable factor for this direction, considering that this kind of cooperation requires electrical interconnection between the countries. It is also worth mentioning that Morocco is the only North African country that has an interconnection with a European country. More specifically, a submarine cable connecting the country with Spain already exists, which is both important for importing energy and also for potential energy exports in the following years. Moreover, the country has strong potential in developing RES, while an upward trend in the energy demand in the past several years can be observed [18].

Finally, the results of the pilot application in Morocco were compared with the results of an additional case study. This paper also provides a comparative analysis between Morocco and Egypt, so as to obtain a clearer picture of the pros and cons of the applied methodology.

*1.3. Decision Support Methods Review*

1.3.1. SWOTAnalysis

SWOT analysis is a well-known analytical tool, which has been widely applied for strategic decision-making processes [19], in regional energy planning and management [20,21], as well as in renewable energy schemes [22,23]. In particular, SWOT analysis has been used so far in order to investigate and assess the current status of RES in different regions [23,24] yielding a good basis for formulating policy recommendations regarding enhanced utilization of RES. The use of SWOT analysis for exploring energy sector conditions and developing an environmental strategic plan could enable a correct comprehension of the current energy situation and serve as a basis for objectives and strategy proposals [25]. Lei et al. 2019 [26] exploit through the SWOT analysis a new opportunity for African countries to develop their solar power resource through mutually beneficial cooperation between Africa and China within the framework of the Belt and Road Initiative (BRI). Agyekum et al. 2020 [27] used the SWOT analysis to assess Ghana's nuclear power program. The research found out that in Ghana there are a lot of strengths and opportunities in investing in nuclear. However, issues such as a porous security system, corruption, porous borders and policy discontinuity are threats to the smooth implementation and operation of a nuclear power plant.

Kamran et al. 2020 [28] performed a SWOT analysis as a reference point that diagnoses the feasibility of current status and future roadmap to nurture the renewable energy sector in Pakistan. Igliński et al. 2016 [29] examined the history, current state and prospects for the development of the wind power sector in Poland including a SWOT analysis of wind power investment. Studies also exist in the literature that combine the SWOT analysis with MCDM methods. Ervural et al. 2018 [30] used a combined Analytic Network Process (ANP) and fuzzy TOPSIS method with SWOT analysis in order to evaluate Turkey's energy planning strategies. The results showed that the most important priority was to turn the country into an energy terminal by effectively using the geo-strategic position within the framework of the regional cooperation. Wang et al. 2020 [31] integrated the fuzzy AHP and SWOT model for choosing and assessing the strategic renewable energy technologies in Pakistan by considering four indicators and 17 sub-indicators. The finding of that study demonstrated that socio-political and economic criteria were the influential indicators for the selection of renewable energy sources. Khan 2018 [32] evaluated the prioritized the strategies for stimulating the growth of the Iranian Compressed Natural Gas (CNG) market through the application of SWOT analysis along with a modified Fuzzy Goal Programming. Finally, Solangi et al. 2019 [33] evaluated strategies for sustainable energy planning in Pakistan through an integrated SWOT-AHP and Fuzzy-TOPSIS approach. The results of the study reveal that providing low-cost and sustainable electricity to residential, commercial, and industrial sectors is a highly prioritized energy strategy.

### 1.3.2. AHP

The AHP framework is a popular tool for formulating and analyzing decisions, which is extremely useful for ranking alternatives, as well as calculating the weights of different criteria through pairwise comparisons [34]. AHP establishes a balance between quantitative and qualitative factors, as it makes it possible to incorporate judgments on intangible qualitative criteria alongside tangible quantitative criteria [35]. The AHP method is based on three basic rules/factors: first, structure of the model; second, comparative judgment of the alternatives and the criteria; third, synthesis of the priorities. Based on the above, it is clear that AHP has two main advantages: mathematical simplicity and flexibility. These two are probably the reasons why AHP is a favorite research tool in many fields, including energy management and renewable energy sources.

Available literature is abundant with examples of AHP method application in various fields, including environment and energy management [36]. Ghimire et al. 2018 [37] identified and ranked through AHP, the barriers to developing renewable energy in Nepal. Twenty-two barriers were identified and categorized into six types of barriers: social, policy and political, technical, economic, administrative, and geographic. Political instability and transportation problems are ranked first and second in overall barriers. In the same year, Ozdemir and Sahin et al. 2018 [38] examined three different locations in Turkey to find the best place for setting up a solar photovoltaic power plant through AHP, which was used to evaluate locations taking into consideration both quantitative and qualitative factors which play an effective role on the electricity production. Recently, Colak et al. 2020 [39] explored the optimal site selection for solar photovoltaic power plants using Geographic Information System (GIS) and AHP having as a case study the Malatya Province in Turkey. Keleey et al. 2018 [40] highlighted the importance of foreign direct investment (FDI) for the development of renewable energy in developing countries by using the AHP method to clarify the relative significance of the determinants in the location decisions of foreign wind and solar energy investors. Finally, Wu et al. (2019) [41] introduced a new approach using the AHP model under an interval type-2 fuzzy weighted averaging set to evaluate the performance of renewable energy projects based on the sustainability view. The results of that proposed method found that the GHG emission reduction had the best rank among other criteria.

### 1.3.3. FuzzyTOPSIS

Fuzzy multi-criteria methods constitute one approach to evaluate alternative decisions, which involve subjective judgments and are made by a group of experts. A pairwise comparison process is used to assist decision makers to make comparative judgments, while absolute judgments are made using a linguistic evaluation method [42].

TOPSIS is one of the known classical and most popular MCDM methods that was developed by Hwang and Yoon in 1981 [43]. TOPSIS is a widely accepted multi-attribute decision-making technique owing to its simultaneous consideration of the ideal and the anti-ideal solutions, and easily programmable computation procedure. Its basic principle has to do with the fact that the chosen alternative should have the shortest distance from the positive ideal solution (PIS) and the farthest from the negative ideal solution (NIS), compared to the others. Having to use crisp values is one of the weak points in the crisp evaluation process. TOPSIS method is not able to deal with decision-makers' ambiguities and uncertainties which cannot be handled by crisp values. The use of fuzzy set theory offers the decision makers the opportunity to incorporate unquantifiable information, incomplete information; non-obtainable information and partially ignorant facts into a decision model [44].

Consequently, fuzzy TOPSIS and its extensions are developed to solve ranking and justification problems [10,42]. It meets specific requirements when uncertain and imprecise knowledge, as well as possibly vague preferences must be considered [45]. This method allows the fuzzy values to be used in the decision problem as it offers a realistic approach by using linguistic assessments instead of numerical values [46,47]. Fuzzy TOPSIS method has been applied in various fields, which shows an excellent performance in the decision making of alternatives selection [48–50].

Moreover, as far as assessments in the field of energy policy are concerned, fuzzy TOPSIS has been applied in many different studies [51]. It has been also used to evaluate the viability of renewable energy projects [15]. Papapostolou et al. 2017 [10] presented a new extension of fuzzy TOPSIS method for prioritization of alternative energy policy scenarios to realize targets of renewable energy in 2030. Rani et al. 2020 [52] ranked and chose the renewable energy sources in MCDM problems based on fuzzy TOPSIS. Çolak and Kaya (2017) [53] developed a new model in order to evaluate renewable energy alternatives with the use of AHP and TOPSIS methods under interval type-2 fuzzy. According to the findings the wind energy was the best source among the available renewable energy sources. Karunathilake et al. (2019) [54] used a combination of the fuzzy TOPSIS method and life cycle thinking to select and assess different renewable energy sources. Ligus and Peternek, 2018 [55] proposed a hybrid MCDM model based on fuzzy AHP and fuzzy TOPSIS in order prioritize low-emission energy technologies development in Poland through criteria relevant to the sustainable development policy goals in Poland. The research results show that renewable energy technologies should be utilized instead of nuclear energy.

### 1.4. Manuscript Organisation

Apart from this introductory section, the rest of the paper is organized as follows. Section 2 gives an overview of the method followed for the assessment of a potential transnational cooperation in the field of RES, as well as the methodological steps of the SWOT-AHP-fuzzy TOPSIS.

Section 3 includes the application of the proposed model for the country of Morocco, as well as a comparison of the obtained results with the respective results obtained from the application in Egypt.

Section 4 includes the discussion of the results and in Section 5, the main conclusions of the paper are summarized and key points for further research are proposed.

## 2. Materials and Methods

### 2.1. Overview of the Proposed Methodology

The establishment of a successful energy cooperation with the host countries, requires the assessment of the host countries current situation, so as Europe to define the appropriate strategic plan towards this direction [7].

Consequently, there is a need to assess, through case studies analysis, the role and design of this cooperation mechanism with regards to:

1. helping Europe achieve (or overfull fill) its RES targets in a cost-effective way, and
2. helping developing countries deploy RES, firstly, to meet their increasing energy needs and, secondly, for exports to the EU.

The following figure (Figure 1) illustrates the methodology applied in order to draw the necessary conclusions considering the effectiveness of the implementation of a cross border cooperation between the EU and developing countries.

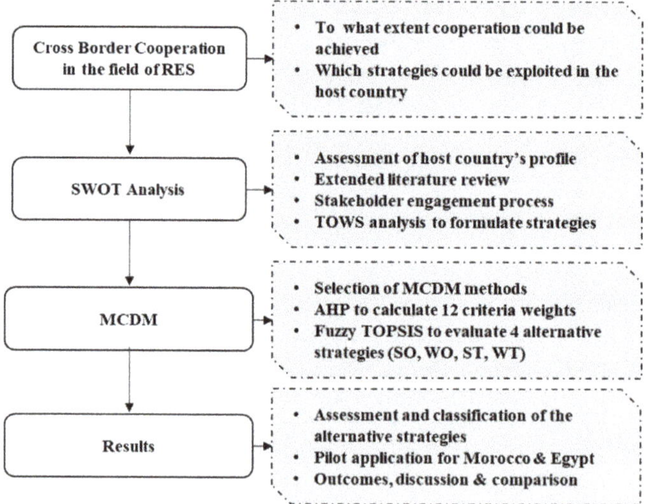

**Figure 1.** Proposed methodology for addressing the problem.

Firstly, the areas to be evaluated, in order to study the extent to which the neighboring countries can participate in such an energy cooperation, were identified. Then, taking these axes into account, 12 criteria were adopted, which refer to the above-mentioned axes and give a clear picture of the current situation in the country under consideration. After setting the criteria, an extensive study in the international literature and in online sources followed to gather information that characterize a host country in each of the criteria. This process was followed by the identification of the strengths, weaknesses, opportunities and threats existing in each of these 12 criteria and a SWOT analysis for the host country was completed. Finally, through Threats, Opportunities, Weaknesses and Strengths (TOWS) analysis, the four alternative strategies (SO, WO, ST, WT) were obtained [25,56,57]. The TOWS analysis has been widely used to define strategies based on a previously conducted SWOT analysis. Thus, according to the specific TOWS matrix, strategies can be developed, on the basis of the identified strengths, weaknesses, opportunities and threats [57]. More specifically the four alternative strategies are defined as:

- SO: Strategy that uses the internal strengths to take advantage of opportunities.
- ST: Strategy that uses the strengths to minimize threats.
- WO: Strategy that improves weaknesses by taking advantage of opportunities.
- WO: Strategy that tries to eliminate weaknesses to avoid threats.

Subsequently, after an extensive literature review, the most appropriate MCDM methods for the specific problem were selected. These MCDM methods were applied to assess and classify the alternative strategies from the most to the least preferable according to decision maker's preferences. Finally, after the strategies' classification, the most suitable strategy towards achieving a successful energy cross border cooperation was identified.

*2.2. Implementation Steps of the Proposed Methodology*

The basic steps of the proposed methodology consist of the following items (Figure 2).

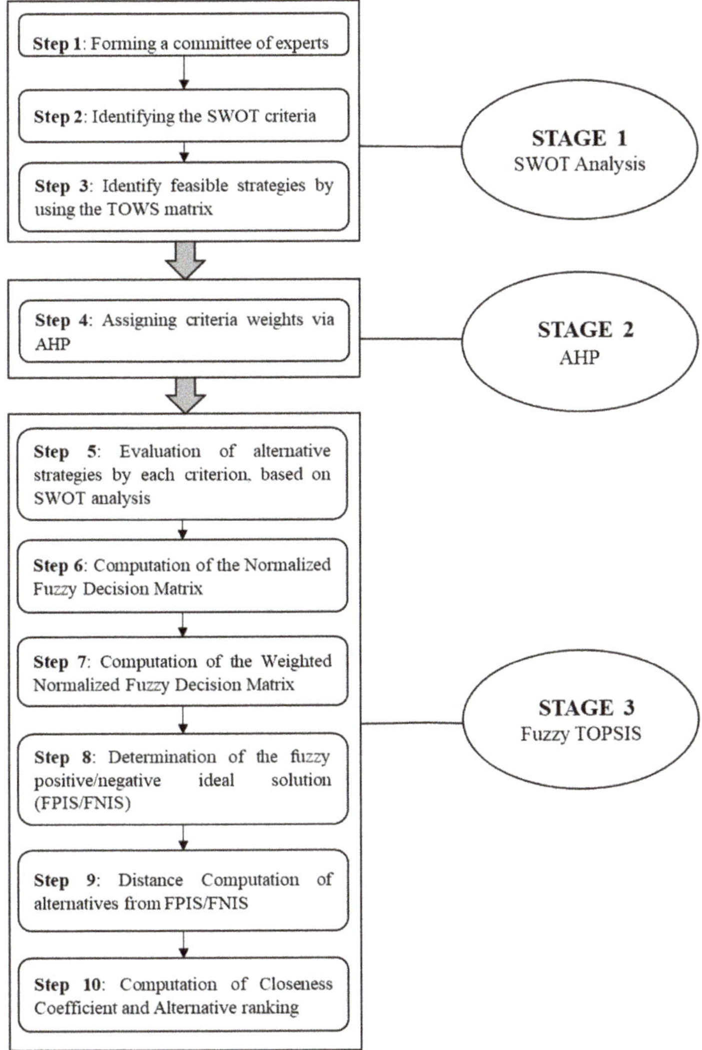


**Figure 2.** The proposed model.

The proposed model for the project selection problem, composed of AHP and fuzzy TOPSIS methods, consists of three basic stages: (1) SWOT Analysis completion, (2) AHP computations, (3) evaluation of alternatives with fuzzy TOPSIS and determination of the final rank.

*Step 1*. Forming a committee of experts

There are a number of "group-based" research techniques available to determine the views or perceptions of individuals, in relation to specific topics [58,59]. By applying the most appropriate to the case technique, at the end of this step a set of K decision makers are defined. Their main target is to define the problem and to examine all the different aspects of it, in detail.

*Step 2*. Identifying the SWOT criteria

SWOT is a powerful strategic analysis tool that combines the study of the strengths and weaknesses of an organization, territory, project or sector with the study of opportunities and threats in its environment with the aim to define a development strategy. It is worth mentioning that a project's strengths and weaknesses demonstrate the project's internal characteristics and are controllable; and opportunities and threats are determined by external factors on which the project has no direct control but can react to its own advantage.

Based on Karakosta et al. 2016 [9], SWOT analysis is approached differently in this paper, compared to how SWOT is typically applied to companies. More specifically, the strengths and weaknesses account for the current host country's situation assessment regarding the implementation of cooperation mechanisms, while the opportunities and threats take into account the possible future (internal and external) developments that may have the ability to change the current situation (Figure 3).

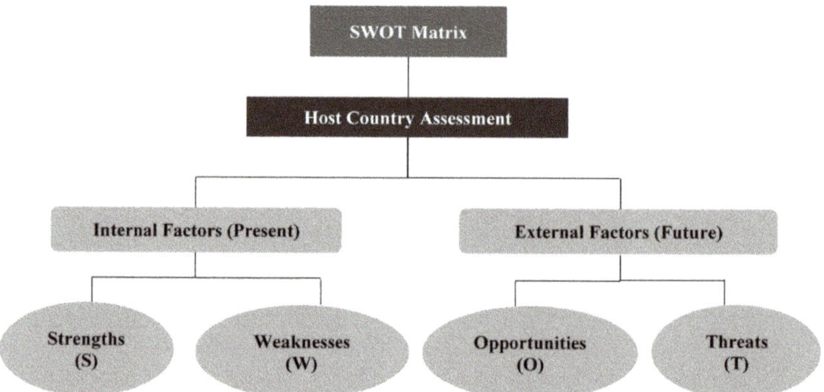


**Figure 3.** Strengths, Weaknesses, Opportunities and Threats (SWOT) analysis approach.

In the above context, a three-level framework is followed in order to cover important dimensions regarding the assessment of a host country for the implementation of cooperation mechanism. These three levels consist of the energy profile, the investment profile as well as the socio-environmental profile of the host country.

Firstly, the energy profile of the host country is examined. This profile includes information about the current energy situation of the host country as well as the targets set, while data on the development of RES are also presented. This profile emphasizes investment and economic factors, which characterize each country, with particular emphasis on the general framework (political, regulatory, institutional) applied by the government of the country in the energy sector. Without any political will and support and, thus, without the regulatory framework to facilitate and encourage the implementation of joint projects, investments in these projects are hardly feasible. Finally, from a socio-environmental point of view, if the public discourse is mainly characterized by oppositional groups arguing against joint projects, policy makers may not be willing to create favorable investment conditions in terms of support mechanism and regulation.

Within this three-level structure, 12 pivotal factors (criteria) were identified reflecting the favorable and non-favorable conditions for joint RES projects, between EU and neighboring countries (Figure 4). The energy profile criteria are: energy system strategy and energy outlook ($C_1$), RES industry development ($C_2$) and energy security ($C_3$). The investment profile criteria are: market structure of energy system ($C_4$), grid and Interconnections ($C_5$), RE regulatory and policy framework ($C_6$), institutional framework for RES-E deployment ($C_7$), financial risks and uncertainty ($C_8$), investment facilitation ($C_9$) and RES-E capacity and potential ($C_{10}$) The socio-environmental criteria are: public acceptance ($C_{11}$) and, finally, environmental and social effects ($C_{12}$). These assessment criteria were presented to experts and stakeholders in order to gather feedback and proceed with the SWOT analysis based on the responses received.

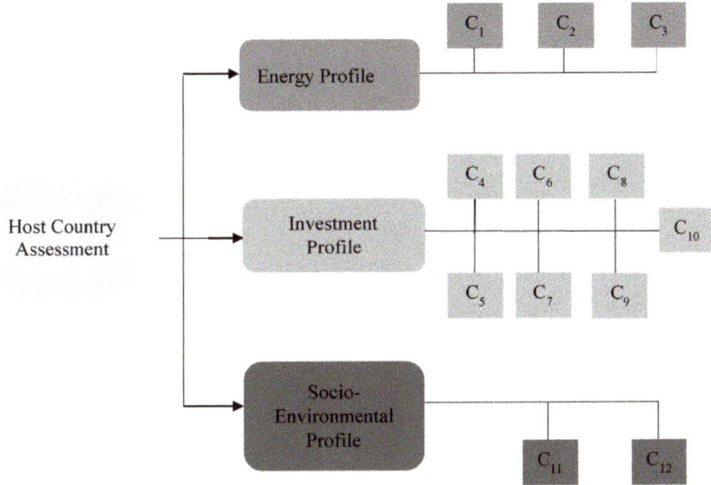



**Figure 4.** SWOT analysis levels and indicators.

***Step 3***. Identify feasible strategies by using the TOWS matrix

After identifying the strength, weakness, opportunity and threat factors, the TOWS matrix is developed based on their combinations - manifesting into four pairwise SO, ST, WO and WT of strategies (Table 1). In the strategies identified as SO, the optimal use of internal strengths and external opportunities are determined. In the strategies identified as WO, the use of external opportunities would reduce or eliminate internal weaknesses. In the strategies identified as ST, the use of internal strengths would reduce or completely eliminate external threats. For the strategies identified as WT, the decrease in external threats would be achieved by considering internal weaknesses [60]. In this paper, these four types will be assessed and ranked using MCDM methods.

**Table 1.** Threats, Opportunities, Weaknesses and Strengths (TOWS) analysis matrix [58].

| | | Internal Factors | |
| --- | --- | --- | --- |
| | | Strengths (S) | Weaknesses (W) |
| **External Factors** | **Opportunities (O)** | SO Strategy | WO Strategy |
| | **Threats (T)** | ST Strategy | WT Strategy |

***Step 4***: Assigning criteria weights via AHP

In this step, the 12 criteria that are used in the problem of the host country assessment are weighted by using AHP method. In this phase, pairwise comparison matrixes are formed to determine the criteria weights. For the purpose of this study, the Saaty's nine-point scale was used [61]. The decision-making team make individual evaluations using the scale provided in Table 2, to determine the values of the elements of pairwise comparison matrixes.

**Table 2.** Nine-point intensity important scale.

| Definition | Intensely of Importance |
| --- | --- |
| Equally important | 1 |
| Moderately more important | 3 |
| Strongly more important | 5 |
| Very strong more important | 7 |
| Extremely more important | 9 |
| Intermediate more important | 2,4,6,8 |

Let C = {C$_j$ |j = 1, 2,...., n} be the set of criteria (in this study n = 12). Considering the criteria, the pairwise comparison of the criterion with the criterion j yields a square matrix $\widetilde{A}$ where, a$_{ij}$ denotes the comparative importance of criterion i with respect to criterion j. In the matrix, a$_{ij}$ = 1, when i = j and a$_{ji}$ = 1/ a$_{ij}$.

$$\widetilde{A} = \begin{bmatrix} a_{11} & a_{12} & \cdots & a_{1n} \\ a_{21} & a_{22} & \cdots & a_{2n} \\ \vdots & \vdots & \ddots & \vdots \\ \alpha_{n1} & a_{n2} & \cdots & \alpha_{nn} \end{bmatrix} \tag{1}$$

The calculation of the normalized matrix $\widetilde{A}_n$ follows. More specifically we calculate the normalized matrix $\widetilde{A}_n$ by summation of each column and then dividing each element by the respective column total, where, element c$_{ij}$ is the normalized element,

$$c_{ij} = \frac{a_{ij}}{\sum_{i=1}^{n} a_{ij}} \tag{2}$$

and

$$\widetilde{A}_n = \begin{bmatrix} c_{11} & c_{12} & \cdots & c_{1n} \\ c_{21} & c_{22} & \cdots & c_{2n} \\ \vdots & \vdots & \ddots & \vdots \\ c_{n1} & c_{n2} & \cdots & c_{nn} \end{bmatrix} \tag{3}$$

Next, we calculate the mean of each row to obtain the normalized principal Eigen vector i.e.,

$$\begin{bmatrix} \frac{1}{n}\sum c_{1j} \\ \cdot \\ \cdot \\ \cdot \\ \cdot \\ \frac{1}{n}\sum c_{nj} \end{bmatrix}, \text{ for } j = 1, 2, \ldots, n. \tag{4}$$

Here, several iterations are done by squaring the normalized matrix till the difference in principal Eigen vectors in previous and iterated matrix becomes almost zero or negative. This final iterated matrix gives the principal eigen vector (or priority vector), which is the weight of each criterion. The principal eigen value of the matrix A has to be calculated, it is called $\lambda_{max}$. It is calculated by multiplying the column sums of matrix A with the principal Eigen vector (e) i.e.,

$$\lambda_{max} = \begin{bmatrix} \sum_{i=1}^{n} a_{i1} & \cdots\cdots & \sum_{i=1}^{n} a_{in} \end{bmatrix} \times \begin{bmatrix} e_1 \\ \vdots \\ \vdots \\ e_n \end{bmatrix}. \tag{5}$$

In the last step of this phase, we calculate the consistency index (CI) = ($\lambda_{max}$ - n) / (n-1) and also obtain the Random Index (RI), for the number of criteria used in decision making, which is four (n = 12) for the case under consideration and the respective value of RI = 1.54. Finally, we calculate the consistency ratio (CR) = CI / RI. Usually, a CR of 0.10 (10%) or less is considered acceptable.

*Step 5*: Evaluation of alternative strategies by each criterion, based on SWOT analysis

Within this step, it is suggested that the decision maker uses the linguistic variables to evaluate the ratings of alternatives strategies, with respect to various criteria. The evaluation is based on the content of the SWOT Analysis, in each of these 12 criteria. For the evaluation of the strategies a seven-point scale has been used as presented in the Table 3.

**Table 3.** Linguistic Variables and Fuzzy Numbers.

| Linguistic Variables for The Ratings | |
|---|---|
| **Linguistic Variables** | **Fuzzy Numbers** |
| Very Poor (VP) | (0, 0, 1) |
| Poor (P) | (0, 1, 3) |
| Medium Poor (MP) | (1, 3, 5) |
| Fair (F) | (3, 5, 7) |
| Medium Good (MG) | (5, 7, 9) |
| Good (G) | (7, 9, 10) |
| Very Good (VG) | (9, 10, 10) |

***Step 6***: Computation of the Fuzzy Decision Matrix and the Normalized Fuzzy Decision Matrix

A fuzzy multi-criteria group decision making problem can be concisely expressed in matrix format as:

$$\widetilde{D} = \begin{bmatrix} \widetilde{x}_{11} & \widetilde{x}_{12} \cdots & \widetilde{x}_{1n} \\ \widetilde{x}_{21} & \widetilde{x}_{22} & \widetilde{x}_{2n} \\ \vdots & \ddots & \vdots \\ \widetilde{x}_{m1} & \widetilde{x}_{m2} \cdots & \widetilde{x}_{mn} \end{bmatrix}, \tag{6}$$

$$\widetilde{W} = \begin{bmatrix} \widetilde{w}_1 & \widetilde{w}_2 \cdots & \widetilde{w}_n \end{bmatrix}, \tag{7}$$

where $\widetilde{x}_{ij} \lor i,j$ are linguistic variables, while $\widetilde{w}_j$, j=1,2, … ,n are already calculated via AHP in step 4. These linguistic variables can be described by triangular fuzzy numbers $\widetilde{x}_{ij} = \left( a_{ij}, b_{ij}, c_{ij} \right)$. To avoid the complicated normalization formula used in classical TOPSIS, the linear scale transformation is used here to transform the various criteria scales into a comparable scale. Therefore, the normalized fuzzy decision matrix denoted by $\widetilde{R}$, can be obtained.

$$\widetilde{R} = \left[ \widetilde{r}_{ij} \right]_{mxn} \tag{8}$$

where:

$$\widetilde{r}_{ij} = \left( \frac{a_{ij}}{c_j^*}, \frac{b_{ij}}{c_j^*}, \frac{c_{ij}}{c_j^*} \right) \tag{9}$$

and

$$c_j^* = \max_i c_{ij} \quad \text{benefit criteria} \tag{10}$$

$$\widetilde{r}_{ij} = \left( \frac{a_j^-}{c_{ij}}, \frac{a_j^-}{b_{ij}}, \frac{a_j^-}{a_{ij}} \right) \tag{11}$$

And

$$a_j^- = \min_i a_{ij} \quad \text{cost criteria} \tag{12}$$

***Step 7***: Computation of the Weighted Normalized Fuzzy Decision Matrix

Considering the different importance of each criterion, the weighted normalized fuzzy decision matrix can be constructed as:

$$\widetilde{V} = \begin{bmatrix} \widetilde{v}_{ij} \end{bmatrix}_{mxn} i = 1, 2, \ldots, m; j = 1, 2, \ldots, n \tag{13}$$

Where

$$\widetilde{u}_{ij} = \widetilde{r}_{ij}(.)\widetilde{w}_j \tag{14}$$

***Step 8:*** Determination of the fuzzy positive/negative ideal solution (FPIS/FNIS)

According to the weighted normalized fuzzy decision matrix, it is evident that the elements $\widetilde{u}_{ij} \forall$ i, j are normalized positive triangular fuzzy numbers and their ranges belong to the closed interval [0, 1]. Then, the fuzzy positive-ideal solution (FPIS, A*) and the fuzzy negative-ideal (FNIS, A⁻) solution can be defined as:

$$A^* = \left(\widetilde{u}_1^*, \widetilde{u}_2^* \ldots, \widetilde{u}_n^*\right) \tag{15}$$

where

$$\widetilde{u}_j^* = \max_i \{u_{ij3}\}\, i = 1,\, 2,\, \ldots,\, m;\, j = 1,\, 2,\, \ldots,\, n \tag{16}$$

$$A^- = \left(\widetilde{u}_1^-, \widetilde{u}_2^- \ldots, \widetilde{u}_n^-\right) \tag{17}$$

Where

$$\widetilde{u}_j^- = \min_i \{u_{ij1}\}\, i = 1,\, 2,\, \ldots,\, m;\, j = 1,\, 2,\, \ldots,\, n \tag{18}$$

***Step 9***: Distance Computation of alternatives from FPIS/FNIS
The distance of each alternative from A* can be currently calculated as:

$$d_i^* = \sum_{j=1}^{n} d\left(\widetilde{u}_{ij}, \widetilde{u}_j^*\right),\, i = 1,\, 2, \ldots,\, m \tag{19}$$

The distance of each alternative from A⁻ can be currently calculated as:

$$d_i^- = \sum_{j=1}^{n} d\left(\widetilde{u}_{ij}, \widetilde{u}_j^-\right),\, i = 1,\, 2, \ldots,\, m \tag{20}$$

where d(.,.) is the distance measurement between two fuzzy numbers.
For triangular fuzzy numbers, the distance between two fuzzy numbers m and n is expressed by:

$$d(\widetilde{m}, \widetilde{n}) = \sqrt{\frac{1}{3}\left[(m_1 - n_1)^2 + (m_2 - n_2)^2 + (m_3 - n_3)^2\right]}, \tag{21}$$

where: $\widetilde{m} = (m_1, m_2, m_3)$ and $\widetilde{n} = (n_1, n_2, n_3)$.

***Step 10***: Computation of closeness coefficient and alternative ranking
The closeness coefficient $CC_i$ enabling the alternatives classification can be calculated as:

$$CC_i = \frac{d_i^-}{d_i^* + d_i^-},\, i = 1, 2, \ldots m \tag{22}$$

The closeness coefficient ranges between (0, 1) and the alternative with the largest closeness coefficient is the ideal solution. Besides, the alternative defining the ideal solution is also the one with the shortest distance from the positive ideal solution FPIS and with the shortest distance from the negative ideal solution FNIS. Thus, the ranking of alternatives in descending order comes as a result [30].

## 3. Application: The Case of Morocco

The proposed methodology was applied to the country of Morocco in order to determine the national strategic plan for the development of a successful cross border cooperation in the field of RES. Initially, an overview of the country's current situation regarding the energy profile, the investment profile and the socio-environmental profile, is presented while the main steps of the proposed methodology and its specific implementation follow. In order to validate the results of the study, the same methodology was also applied for the case of Egypt. The results are presented and compared to those of Morocco.

*3.1. Overview of Morocco's Current Situation*

### 3.1.1. Energy Profile

The energy sector of Morocco is characterized by many imports of fossil fuels, a fact which makes the country completely dependent on others to meet domestic energy needs [62]. It is important to mention that the strategic geographical position offers Morocco the opportunity to be a center of activity with network interconnections. In order to support this a number of actions have taken place, including legislations, the establishment of renewable energy and energy efficiency agencies and the engagement of different domestic and international stakeholders [63,64]. In the meantime, in 2016 the country hosted the United Nations Conference of Parties (COP22) summit in Marrakesh, while its government has set a new national climate policy strengthening its role in international action on climate change by transforming Morocco's energy strategy to significantly improve the country's standards and meet the goals of Paris Agreement [65].

The growing energy demand because of the growing population, combined with the lack of primary energy wealth, have led the country into an energy imbalance [66]. It should be distinctively noted that demand forecasts suggest that Morocco must double its total installed capacity of electricity from 2020 to 2030, to meet the demand, a fact which shows the current energy crisis in the country [67].

As far as the renewable energy is concerned, it offers an excellent opportunity both for the rehabilitation of the country from the imports of fossil fuels, and the prospect of green energy exports to Europe. Conversely, the constant use of conventional resources is a direct consequence of the increase of greenhouse gas emissions, which is a major domestic problem. At the end of 2018 renewable energy share in the national electricity mix reached 35% (it is envisaged an expansion of renewable electricity to a share of 52% of installed capacity by 2030) [68]. However, while the share of renewables in electricity is progressing fast, its share in total final consumption decreased considerably over the past decade, given the expanding energy demand. To meet the challenges, the government of the country adopted an ambitious energy policy relative to the renewable energy sector, which includes targets to increase their participation in the domestic generated electricity, while the operation of both wind farms, as well as solar thermal power plants has increased [69]. The country has also made good use of the best available technologies in its large-scale concentrated solar power (CSP) projects, and pioneers innovative hybrid solutions, with photovoltaic and thermal solar storage [70]. Morocco has increased the hours of storage in its CSP plants and is investing in new interconnections and market integration with European neighbors, ensuring greater security, efficiency and flexibility of the power system [71].

The electricity selling prices remain quite low in relation to the production and transportation costs while, furthermore, stability and balancing issues of the network and energy supply problems must be solved in order to enable the successful penetration of RES in the energy sector of the country [72]. It is also worth mentioning that the power grid needs network reinforcements between south and north where the wind and solar potential are the highest. To this end, ONEE (Office National de l' Électricité et de l' Eau Potable) has, as its main aim, the grid expansion, which will also support interconnections for exports mainly across the Mediterranean [73].

### 3.1.2. Investment Profile

In an increasingly unstable international economic environment, the economy in Morocco remains robust as it has managed to resist the wider economic crisis of the past years [74]. The country's electricity market is almost fully liberalized, which can foster a potential cooperation with the European countries in the field of renewable energy while, at the same time, encouraging and attracting foreign investments is a key component of the strategy of the country. According to the latest edition of the Renewable Energy Country Attractiveness Index (RECAI) Morocco ranks 14th among 40 countries on the attractiveness of their renewable energy investment and deployment opportunities [75].

In recent years significant progress both at the institutional level and at the legislative level has been noted, while reform efforts in the energy sector are made by reorganizing the functioning of

ONEE [76]. Several institutional reforms have been introduced in order to support the implementation of the National Energy Strategy. The national regulatory authority (ANRE) and the Moroccan Agency for Sustainable Energy (MASEN) are dedicated to the development of integrated renewable energy projects, while The Institute for Research in Solar Energy and Renewable Energies (IRESEN) and The Agence Marocaine pour l' Efficacité Énergétique (AMEE) are implementing projects about energy efficiency in collaboration with the private sector [71].

For Morocco, the EU is by far its biggest economic partner, accounting for well over half of its trade and investment [72]. As a result, in December 2018, a plan that accounts for a sustainable energy trading system was signed between Morocco, Germany, France, Spain and Portugal. Furthermore, Morocco launched a vast project of economic modernization program (Industrial Acceleration Plan 2014-2020) to attract more Foreign Direct Investments (FDI) [77]. Finally, at a funding level, some organizations, such as the Energy Development Fund and the Energy Investment Corporation operate with the financial support of the local RES sector [74].

As far as the renewable investments are concerned, the quality of Moroccan renewable energy resources and the attractive investment climate have resulted in lower costs for wind and solar photovoltaics, which are now competitive with the price of fossil fuel-based electricity. However, while Morocco's location is very favorable for the generation of solar and wind electricity, significant investment barriers that hinder RES-E development at scale still prevail [78]. Investors' risk perceptions are particularly high for RES-E technologies, which are characterized by high capital expenditures, and for developing countries. Nowadays, it is a big challenge for Morocco to divert current investments from conventional technologies to low-carbon, and raise additional financial resources for operationalizing the RES-E transition, given the anticipated increases in electricity demand [79].

### 3.1.3. Socio-Environmental Profile

Morocco, as is the case with most North African countries, is particularly vulnerable to climate change and the possible rise in the sea level, while there is a heavy dependence on water resources, mainly in agriculture [80,81]. According to the Intergovernmental Panel on Climate Change (IPCC) methodology, the development of climate scenarios for Morocco reveals a tendency to increase the mean annual temperature (between 0.6 °C and 1.1 °C) and decrease the mean annual volume of precipitation by about 4% in 2020 compared to 2000 levels [82].

In the social sector of the country, although good living conditions prevail in general and the growth rates are stable, some important issues remain unresolved. One of these is the high unemployment levels, which mainly affect young people [74]. However, a new Country Partnership Framework for Morocco (CPF) was designed to support Morocco's efforts at improving social cohesion and reducing social and territorial disparities and was discussed by the Board of Executive Directors on February 19, 2019. Covering the Fiscal years 2019 to 2024, it has the overarching goal of contributing to social cohesion by improving the conditions for growth and job creation and reducing social and territorial disparities [83].

In this direction, the benefits the community of the country will reap on the development of RES are important and for this reason it is considered necessary to design action planning for further information and for the awareness of the public [84].

### 3.2. Application of the Proposed Methodology

The main steps of the proposed methodology and its specific implementation are given below for the case of Morocco:

***Steps 1-2-3:*** Based on the abovementioned methodology, in this research the experts' team, which acted as the decision maker for the problem, used twelve different criteria $C_j$, j = {1,2,3,4,5,6,7,8,9,10,11,12}, in order to assess different types of strategies. The main goal is, on the one hand, to identify which of these strategies is the most suitable for developing energy cooperation between EU and Morocco, and on the other hand, to recognize how strengths, weaknesses, opportunities and threats should be used towards this direction. It is worth mentioning that for each one of these 12 criteria, a SWOT analysis has been conducted considering an extended literature review, examining potential opportunities and barriers that promote or hinder the development and implementation of cooperation mechanisms in the country.

After determining the strengths, weaknesses, opportunities, and threats for each of the abovementioned criteria, four alternative strategies were identified from the SWOT analysis. More specifically, these strategies are SO, WO, ST as well as WT. This paper aims to assess which type, is the most appropriate for developing an energy cooperation with Morocco, after taking under consideration the information provided by the SWOT analysis, in each of the 12 criteria. Morocco's SWOT Analysis is presented in the following figure (Figure 5).

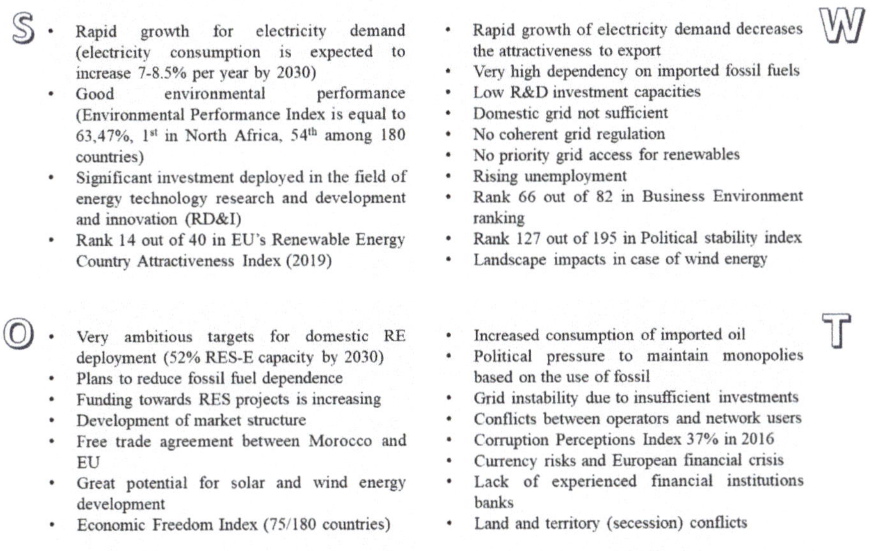

**S**
- Rapid growth for electricity demand (electricity consumption is expected to increase 7-8.5% per year by 2030)
- Good environmental performance (Environmental Performance Index is equal to 63,47%, 1st in North Africa, 54th among 180 countries)
- Significant investment deployed in the field of energy technology research and development and innovation (RD&I)
- Rank 14 out of 40 in EU's Renewable Energy Country Attractiveness Index (2019)

**W**
- Rapid growth of electricity demand decreases the attractiveness to export
- Very high dependency on imported fossil fuels
- Low R&D investment capacities
- Domestic grid not sufficient
- No coherent grid regulation
- No priority grid access for renewables
- Rising unemployment
- Rank 66 out of 82 in Business Environment ranking
- Rank 127 out of 195 in Political stability index
- Landscape impacts in case of wind energy

**O**
- Very ambitious targets for domestic RE deployment (52% RES-E capacity by 2030)
- Plans to reduce fossil fuel dependence
- Funding towards RES projects is increasing
- Development of market structure
- Free trade agreement between Morocco and EU
- Great potential for solar and wind energy development
- Economic Freedom Index (75/180 countries)

**T**
- Increased consumption of imported oil
- Political pressure to maintain monopolies based on the use of fossil
- Grid instability due to insufficient investments
- Conflicts between operators and network users
- Corruption Perceptions Index 37% in 2016
- Currency risks and European financial crisis
- Lack of experienced financial institutions banks
- Land and territory (secession) conflicts

**Figure 5.** SWOT Analysis for Morocco.

EU and Moroccan stakeholders' involvement was the key guarantee that the proper questions were addressed, that assumptions were valid and that results were meaningful and could open opportunities for market development.

It should be mentioned that the experts' team that played the role of the decision maker in our study was identified and selected to participate in the evaluation process within the framework of the project "Bringing Europe and Third countries closer together through renewable Energies (BETTER)", started in July 2012 and ended in 2015 and carried out under the Intelligent Energy - Europe program [84]. BETTER intends to address RES cooperation between the EU and neighboring developing countries in several dimensions. Within the framework of the BETTER project, North Africa countries were examined as far as their potential in developing cooperation mechanisms. To this end, relevant stakeholders were engaged through a stakeholder consultation process. Bilateral meetings with experts in the field, as well as stakeholders' workshops provided feedback to the SWOT analysis and the importance of our evaluation criteria.

*Step 4:* In this step, the weights of the criteria used in the evaluation process are calculated with the use of the AHP method. In this phase, the decision maker is given the task of forming individual pairwise comparison matrix (Table 4), by using the scale given in Table 2.

**Table 4.** Matrix for criteria (pairwise comparison).

| $C_i$ | $C_1$ | $C_2$ | $C_3$ | $C_4$ | $C_5$ | $C_6$ | $C_7$ | $C_8$ | $C_9$ | $C_{10}$ | $C_{11}$ | $C_{12}$ |
|---|---|---|---|---|---|---|---|---|---|---|---|---|
| $C_1$ | 1 | 1/4 | 1/4 | 1/5 | 1/2 | 1/7 | 1/7 | 1/6 | 1/5 | 1/8 | 1/3 | 2 |
| $C_2$ | 4 | 1 | 1 | 1/2 | 3 | 1/4 | 1/4 | 1/3 | 1/2 | 1/5 | 2 | 5 |
| $C_3$ | 4 | 1 | 1 | 1/2 | 3 | 1/4 | 1/4 | 1/3 | 1/2 | 1/5 | 2 | 5 |
| $C_4$ | 5 | 2 | 2 | 1 | 4 | 1/3 | 1/3 | 1/2 | 1 | 1/4 | 3 | 6 |
| $C_5$ | 2 | 1/3 | 1/3 | 1/4 | 1 | 1/6 | 1/6 | 1/5 | 1/4 | 1/7 | 1/2 | 3 |
| $C_6$ | 7 | 4 | 4 | 3 | 6 | 1 | 1 | 2 | 3 | 1/2 | 5 | 8 |
| $C_7$ | 7 | 4 | 4 | 3 | 6 | 1 | 1 | 2 | 3 | 1/2 | 5 | 8 |
| $C_8$ | 6 | 3 | 3 | 2 | 5 | 1/2 | 1/2 | 1 | 2 | 1/3 | 4 | 7 |
| $C_9$ | 5 | 2 | 2 | 1 | 4 | 1/3 | 1/3 | 1/2 | 1 | 1/4 | 3 | 6 |
| $C_{10}$ | 8 | 5 | 5 | 4 | 7 | 2 | 2 | 3 | 4 | 1 | 6 | 9 |
| $C_{11}$ | 3 | 1/2 | 1/2 | 1/3 | 2 | 1/5 | 1/5 | 1/4 | 1/3 | 1/6 | 1 | 4 |
| $C_{12}$ | 1/2 | 1/5 | 1/5 | 1/6 | 1/3 | 1/8 | 1/8 | 1/7 | 1/6 | 1/9 | 1/4 | 1 |

Then, the normalization of the pairwise comparison matrix follows (Table 5).

**Table 5.** Normalized matrix for criteria.

| $C_i$ | $C_1$ | $C_2$ | $C_3$ | $C_4$ | $C_5$ | $C_6$ | $C_7$ | $C_8$ | $C_9$ | $C_{10}$ | $C_{11}$ | $C_{12}$ |
|---|---|---|---|---|---|---|---|---|---|---|---|---|
| $C_1$ | 0.019 | 0.011 | 0.011 | 0.013 | 0.012 | 0.023 | 0.023 | 0.016 | 0.013 | 0.033 | 0.010 | 0.031 |
| $C_2$ | 0.076 | 0.043 | 0.043 | 0.031 | 0.072 | 0.040 | 0.040 | 0.032 | 0.031 | 0.053 | 0.062 | 0.078 |
| $C_3$ | 0.076 | 0.043 | 0.043 | 0.031 | 0.072 | 0.040 | 0.040 | 0.032 | 0.031 | 0.053 | 0.062 | 0.078 |
| $C_4$ | 0.095 | 0.086 | 0.086 | 0.063 | 0.100 | 0.053 | 0.053 | 0.048 | 0.063 | 0.066 | 0.094 | 0.094 |
| $C_5$ | 0.038 | 0.014 | 0.014 | 0.016 | 0.023 | 0.026 | 0.026 | 0.019 | 0.016 | 0.038 | 0.016 | 0.047 |
| $C_6$ | 0.133 | 0.172 | 0.172 | 0.188 | 0.143 | 0.159 | 0.159 | 0.192 | 0.188 | 0,132 | 0.156 | 0.125 |
| $C_7$ | 0.133 | 0.172 | 0.172 | 0.188 | 0.143 | 0.159 | 0.159 | 0.192 | 0,188 | 0.132 | 0.156 | 0.125 |
| $C_8$ | 0.114 | 0.129 | 0.129 | 0.125 | 0.120 | 0.079 | 0.079 | 0.096 | 0.125 | 0.088 | 0.125 | 0.109 |
| $C_9$ | 0.095 | 0.086 | 0.086 | 0.063 | 0.100 | 0.053 | 0.053 | 0.048 | 0.063 | 0.066 | 0.094 | 0.094 |
| $C_{10}$ | 0.152 | 0.215 | 0.215 | 0.251 | 0.167 | 0.317 | 0.317 | 0.288 | 0.251 | 0.265 | 0.187 | 0.141 |
| $C_{11}$ | 0.057 | 0.021 | 0.021 | 0.021 | 0.048 | 0.032 | 0.032 | 0.024 | 0.021 | 0.044 | 0.031 | 0.063 |
| $C_{12}$ | 0.01 | 0.009 | 0.009 | 0.010 | 0.008 | 0.020 | 0.020 | 0.014 | 0.010 | 0.030 | 0.008 | 0.016 |

The results obtained from the computations based on the pairwise comparison matrix provided in Table 5, are presented in Table 6, which includes the final results of AHP.

**Table 6.** Results obtained by Analytic Hierarchy Process (AHP).

| Criteria | $w_i$ | $\lambda_{max}$ | RI | CI | CR |
|---|---|---|---|---|---|
| $C_1$ | 1.8% | | | | |
| $C_2$ | 5% | | | | |
| $C_3$ | 5% | | | | |
| $C_4$ | 7.5% | | | | |
| $C_5$ | 2.4% | | | | |
| $C_6$ | 16% | 12.68 | 1.54 | 0.062 | 0.04 |
| $C_7$ | 16% | | | | |
| $C_8$ | 11% | | | | |
| $C_9$ | 7.5% | | | | |
| $C_{10}$ | 23% | | | | |
| $C_{11}$ | 3.5% | | | | |
| $C_{12}$ | 1.3% | | | | |

According to Table 6, the consistency ratio of the pairwise comparison matrix is calculated as 0.04 < 0.1. Therefore, the weights are shown to be consistent and they are used in the strategy assessment process. Based on the weights presented in the second column of Table 6, the final rank of the criteria is presented (Table 7).

**Table 7.** Final rank of the criteria.

| Criteria | Weight | Final Rank |
|---|---|---|
| RES-E capacity and potential ($C_{10}$) | 0.23 | 1 |
| RE regulatory and policy framework ($C_6$) | 0.16 | 2 |
| Institutional framework for RES-E deployment ($C_7$) | 0.16 | 2 |
| Financial risks and uncertainty ($C_8$) | 0.11 | 3 |
| Market structure of energy system ($C_4$) | 0.075 | 4 |
| Investment facilitation ($C_9$) | 0.075 | 4 |
| RES industry development ($C_2$) | 0.05 | 5 |
| Energy Security ($C_3$) | 0.05 | 5 |
| Social Acceptance ($C_{11}$) | 0.035 | 6 |
| Grid and Interconnections ($C_5$) | 0.024 | 7 |
| Energy System Strategy and Energy Outlook ($C_1$) | 0.018 | 8 |
| Environmental and social effects ($C_{12}$) | 0.013 | 9 |

In the following figure (Figure 6) the importance of the criteria according to decision maker judgment is depicted in a radar chart.

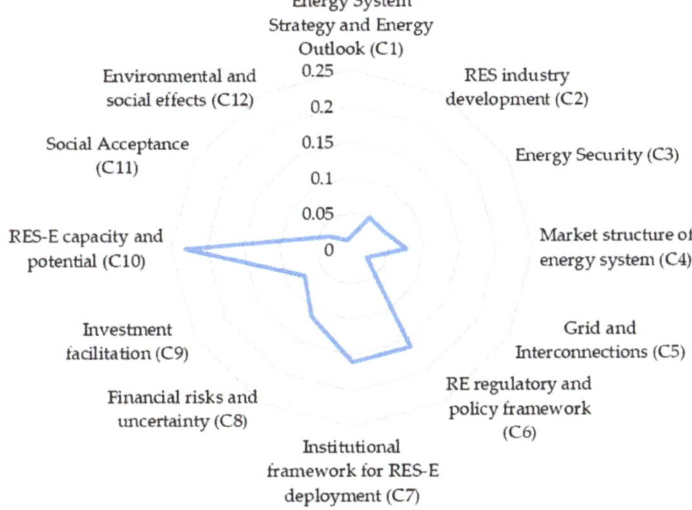


**Figure 6.** Evaluation criteria weights.

***Step 5***: Through the Fuzzy TOPSIS, the linguistic variables are defined to evaluate the ratings of alternatives with respect to these criteria. Using a seven-point scale transformation for the alternatives' ratings S = (VP, P, MP, F, MG, G, VG) the linguistic variables can be converted into fuzzy numbers. Thus, the calibrated conversion scales were constructed through which the four alternative strategies for each of the 12 criteria were evaluated.

Based on the aforementioned scale, the decision maker assessed the alternative policy strategies at EU level to each one of the criteria, taking into account the Morocco's SWOT analysis (Tables 8 and 9).

**Table 8.** Ratings of the alternative strategies by the decision maker

| Criteria | SO | WO | ST | WT |
|---|---|---|---|---|
| $C_1$ | VG | G | F | MG |
| $C_2$ | MG | F | MP | F |
| $C_3$ | MG | MP | MP | F |
| $C_4$ | MG | F | P | VP |
| $C_5$ | VG | F | MG | MP |
| $C_6$ | G | MP | MG | MP |
| $C_7$ | G | MG | F | P |
| $C_8$ | MG | F | F | MP |
| $C_9$ | MG | MP | F | P |
| $C_{10}$ | G | MG | MG | MP |
| $C_{11}$ | F | MP | MP | F |
| $C_{12}$ | MG | MP | F | F |

**Table 9.** Fuzzy decision matrix.

| Criteria | SO | WO | ST | WT |
|---|---|---|---|---|
| $C_1$ | (9, 10, 10) | (7, 9, 10) | (3, 5, 7) | (5, 7, 9) |
| $C_2$ | (5, 7, 9) | (3, 5, 7) | (1, 3, 5) | (3, 5, 7) |
| $C_3$ | (5, 7, 9) | (1, 3, 5) | (1, 3, 5) | (3, 5, 7) |
| $C_4$ | (5, 7, 9) | (3, 5, 7) | (0, 1, 3) | (0, 0, 1) |
| $C_5$ | (9, 10, 10) | (3, 5, 7) | (5, 7, 9) | (1, 3, 5) |
| $C_6$ | (7, 9, 10) | (1, 3, 5) | (5, 7, 9) | (1, 3, 5) |
| $C_7$ | (7, 9, 10) | (5, 7, 9) | (3, 5, 7) | (0, 1, 3) |
| $C_8$ | (5, 7, 9) | (3, 5, 7) | (3, 5, 7) | (1, 3, 5) |
| $C_9$ | (5, 7, 9) | (1, 3, 5) | (3, 5, 7) | (0, 1, 3) |
| $C_{10}$ | (7, 9, 10) | (5, 7, 9) | (5, 7, 9) | (1, 3, 5) |
| $C_{11}$ | (3, 5, 7) | (1, 3, 5) | (1, 3, 5) | (3, 5, 7) |
| $C_{12}$ | (5, 7, 9) | (1, 3, 5) | (3, 5, 7) | (3, 5, 7) |

*Step 6*: In this step the normalized fuzzy decision matrix (Table 10) is calculated.

**Table 10.** Normalized fuzzy decision Matrix

| Criteria | SO | WO | ST | WT |
|---|---|---|---|---|
| $C_1$ | (0.9, 1, 1) | (0.7, 0.9, 1) | (0.3, 0.5, 0.7) | (0.5, 0.7, 0.9) |
| $C_2$ | (0.55, 0.77, 1) | (0.33, 0.55, 0.77) | (0.11, 0.33, 0.55) | (0.33, 0.55, 0.77) |
| $C_3$ | (0.55, 0.77, 1) | (0.11, 0.33, 0.55) | (0.11, 0.33, 0.55) | (0.33, 0.55, 0.77) |
| $C_4$ | (0.55, 0.77, 1) | (0.33, 0.55, 0.77) | (0, 0.11, 0.33) | (0, 0, 0.11) |
| $C_5$ | (0.9, 1, 1) | (0.3, 0.5, 0.7) | (0.5, 0.7, 0.9) | (0.1, 0.3, 0.5) |
| $C_6$ | (0.7, 0.9, 1) | (0.1, 0.3, 0.5) | (0.5, 0.7, 0.9) | (0.1, 0.3, 0.5) |
| $C_7$ | (0.7, 0.9, 1) | (0.5, 0.7, 0.9) | (0.3, 0.5, 0.7) | (0, 0.1, 0.3) |
| $C_8$ | (0.55, 0.77, 1) | (0.33, 0.55, 0.77) | (0.33, 0.55, 0.77) | (0.11, 0.33, 0.55) |
| $C_9$ | (0.55, 0.77, 1) | (0.11, 0.33, 0.55) | (0.33, 0.55, 0.77) | (0, 0.11, 0.33) |
| $C_{10}$ | (0.7, 0.9, 1) | (0.5, 0.7, 0.9) | (0.5, 0.7, 0.9) | (0.1, 0.3, 0.5) |
| $C_{11}$ | (0.42, 0.71, 1) | (0.14, 0.42, 0.71) | (0.14, 0.42, 0.71) | (0.42, 0.71, 1) |
| $C_{12}$ | (0.55, 0.77, 1) | (0.11, 0.33, 0.55) | (0.33, 0.55, 0.77) | (0.33, 0.55, 0.77) |

*Step 7*: According to step 7, the weights ($W_i$) calculated via AHP and presented in Table 7, were used for the calculation of the weighted normalized fuzzy decision matrix (Table 11).

**Table 11.** Weighted normalized fuzzy decision matrix.

| Criteria | SO | WO | ST | WT |
|---|---|---|---|---|
| $C_1$ | (0.016, 0.018, 0.018) | (0.013, 0.016, 0.018) | (0.005, 0.009, 0.013) | (0.009, 0.013, 0.016) |
| $C_2$ | (0.028, 0.039, 0.050) | (0.017, 0.028, 0.039) | (0.006, 0.017, 0.028) | (0.017, 0.028, 0.039) |
| $C_3$ | (0.028, 0.039, 0.050) | (0.006, 0.017, 0.028) | (0.006, 0.017, 0.028) | (0.017, 0.028, 0.039) |
| $C_4$ | (0.041, 0.058, 0.075) | (0.025, 0.041, 0.058) | (0.000, 0.008, 0.025) | (0.000, 0.000, 0.008) |
| $C_5$ | (0.022, 0.024, 0.024) | (0.007, 0.012, 0.017) | (0.012, 0.017, 0.022) | (0.002, 0.007, 0.012) |
| $C_6$ | (0.112, 0.144, 0.160) | (0.016, 0.048, 0.080) | (0.080, 0.112, 0.144) | (0.016, 0.048, 0.080) |
| $C_7$ | (0.112, 0.144, 0.160) | (0.080, 0.112, 0.144) | (0.048, 0.080, 0.112) | (0.000, 0.016, 0.048) |
| $C_8$ | (0.061, 0.085, 0.110) | (0.036, 0.061, 0.085) | (0.036, 0.061, 0.085) | (0.012, 0.036, 0.061) |
| $C_9$ | (0.041, 0.058, 0.075) | (0.008, 0.025, 0.041) | (0.025, 0.041, 0.058) | (0.000, 0.008, 0.025) |
| $C_{10}$ | (0.161, 0.207, 0.230) | (0.115, 0.161, 0.207) | (0.115, 0.161, 0.207) | (0.023, 0.069, 0.115) |
| $C_{11}$ | (0.015, 0.025, 0.035) | (0.005, 0.015, 0.025) | (0.005, 0.015, 0.025) | (0.015, 0.025, 0.035) |
| $C_{12}$ | (0.007, 0.010, 0.013) | (0,001, 0,004, 0,007) | (0,004, 0,007, 0,010) | (0.004, 0.007, 0.010) |

*Step 8*: It is observed that the elements that construct the Table 11 are normalized positive triangular fuzzy numbers and their ranges belong to the closed interval [0, 1]. Thus, the fuzzy positive-ideal solution (FPIS, A*) and the fuzzy negative-ideal solution (FNIS, A-) are defined as A* = (1,1,1) and A- = (0,0,0) for benefit criterion, and A* = (0,0,0) and A- = (1,1,1) for cost criterion. In this study, all criteria are considered as benefit criteria.

*Step 9*: The distance of each alternative from the positive ideal solution FPIS, d(Aj, A*), and the negative ideal solution FNIS, d(Aj, A-), are presented below (Tables 12 and 13):

**Table 12.** Distances from the positive ideal solution (FPIS).

| Criteria | SO | WO | ST | WT |
|---|---|---|---|---|
| $C_1$ | 0.983 | 0.984 | 0.991 | 0.987 |
| $C_2$ | 0.961 | 0.973 | 0.984 | 0.973 |
| $C_3$ | 0.961 | 0.984 | 0.984 | 0.973 |
| $C_4$ | 0.942 | 0.959 | 0.989 | 0.997 |
| $C_5$ | 0.977 | 0.988 | 0.983 | 0.993 |
| $C_6$ | 0.862 | 0.952 | 0.888 | 0.952 |
| $C_7$ | 0.862 | 0,888 | 0.920 | 0.979 |
| $C_8$ | 0.915 | 0.940 | 0.940 | 0.964 |
| $C_9$ | 0.942 | 0.975 | 0.959 | 0.989 |
| $C_{10}$ | 0.801 | 0.840 | 0.840 | 0.932 |
| $C_{11}$ | 0.979 | 0.985 | 0.982 | 0.987 |
| $C_{12}$ | 0.990 | 0.996 | 0.993 | 0.993 |
| A* | 11.175 | 11.464 | 11.453 | 11.719 |

**Table 13.** Distances from the negative ideal solution (FNIS).

| Criteria | SO | WO | ST | WT |
|---|---|---|---|---|
| $C_1$ | 0.017 | 0.016 | 0.009 | 0.013 |
| $C_2$ | 0.040 | 0.029 | 0.019 | 0.029 |
| $C_3$ | 0.040 | 0.019 | 0.019 | 0.029 |
| $C_4$ | 0.060 | 0.043 | 0.015 | 0.005 |
| $C_5$ | 0.023 | 0.013 | 0.017 | 0.008 |
| $C_6$ | 0.140 | 0.055 | 0.115 | 0.055 |
| $C_7$ | 0.140 | 0.115 | 0.084 | 0.029 |
| $C_8$ | 0.087 | 0.064 | 0.064 | 0.041 |
| $C_9$ | 0.060 | 0.028 | 0.043 | 0.015 |
| $C_{10}$ | 0.201 | 0.165 | 0.165 | 0.079 |
| $C_{11}$ | 0.022 | 0.017 | 0.022 | 0.017 |
| $C_{12}$ | 0.010 | 0.005 | 0.008 | 0.008 |
| A- | 0.840 | 0.569 | 0.580 | 0.328 |

***Step 10***: Through the equation (FPIS, FNIS) and the application of the Euclidean distance of each alternative from $A^*$ and $A^-$, the closeness coefficient is calculated, enabling the alternatives ranking (Table 14). As a result, the closeness coefficient can indicate which alternative strategy is optimal for achieving an energy cooperation with Morocco and which are the crucial factors (Strengths, Weaknesses, Opportunities, Threats) that the EU should capitalize on.

**Table 14.** Alternative strategies ranking for the case of Morocco with use of closeness coefficient.

| Ranking | $A_i$ | Alternative Strategies | Closeness Coefficient |
|---|---|---|---|
| 1 | $A_1$ | SO | 0.07 |
| 3 | $A_2$ | WO | 0.047 |
| 2 | $A_3$ | ST | 0.048 |
| 4 | $A_4$ | WT | 0.027 |

The final ranking is described as $A_1 > A_3 > A_2 > A_4$ and highlights as the best, the alternative strategy $A_1$ "Strategy SO", since the closeness coefficient of the alternative $A_1$ is closer to one and thus the largest over others Pilot Application in Egypt: Results

In order to validate the results of the proposed methodology, it has been also applied to the case study of Egypt, so as to compare the strategies that are proposed in both countries, according to the decision maker judgments. Of course, the criteria weights as they are assessed in Step 4 of the proposed methodology through the AHP method, are the same for both case studies, since it captures the significance of the identified factors in terms of RES-E cooperation deployment in a host country.

Egypt represents the region's most populated energy market and is the largest oil and gas consumer in the continent. Of the total primary energy consumption in Egypt, 94% is from fossil fuels, while some energy comes from hydropower. [85]. This strong dependence on fossil fuels is mainly due to the strong subsidies on fossil energy utilities imposed so far, which caused a vulnerability of the country to socio-economic events, resulting in low levels of reliability and security of supply [86]. The natural decline of the deposits due to their continued use in the generation process, as well as the aging of the infrastructure and the inadequate generation and transmission capacity are some of the features of the country's energy sector [87]. In addition, domestic electricity demand is growing rapidly.

There has been a change in domestic energy policy in recent years as, in order to meet these challenges, the country recognizes the need to diversify the energy mix and improve energy security through renewables [88]. For this reason, the government has adopted the so-called "Egypt's Vision 2030", which includes various goals in the direction of sustainable development with a time horizon of 2030 [89]. In addition, Egypt, with its excellent prospects due to its geographical location, plans a brave increase in electricity from renewable energy beyond 2030. It is also worth mentioning that Egyptian Electricity Transmission Company (EETC) signed a framework agreement with Euro Africa Interconnector Company to connect the power grids of Egypt, Cyprus and Greece through Crete by a 2000-megawatts (MW) electricity interconnection, which, of course, is in favor of the need for cooperation mechanism establishment.

As far as the institutional framework is concerned, laws, regulations and implementation schemes already exist that support the wind and solar sectors. However, project developers are not encouraged to implement investments due to complex administrative procedures, including the unavailability of contractual documents for projects and multiple focal points for renewable energy deployment [90]. To overcome these risks, institutional roles should be further defined.

At the social level, Egypt can greatly benefit from the development of alternative sources of energy as the socio-economic benefits they bring are unquestionable. The country is familiar with RES technologies and especially with regards to wind energy, as actions have already been taken. However, the country's limited industrial development coupled with the imbalance between demand and supply of electricity, are factors that slow down potential cooperation with other countries as the growing domestic needs have to be firstly met [84].

Based on the Egypt profile, the decision maker assessed the alternative policy strategies at EU level to each one of the criteria (Table 15).

**Table 15.** Ratings of the alternative strategies by the DM for the case of Egypt.

| Criteria | SO | WO | ST | WT |
|---|---|---|---|---|
| $C_1$ | VG | G | MG | F |
| $C_2$ | MG | F | F | MG |
| $C_3$ | F | MG | P | F |
| $C_4$ | G | F | MG | MP |
| $C_5$ | VG | MG | F | F |
| $C_6$ | MG | MG | F | MP |
| $C_7$ | F | MG | MP | P |
| $C_8$ | MG | MP | F | MG |
| $C_9$ | MG | MP | MG | F |
| $C_{10}$ | G | MG | MG | VP |
| $C_{11}$ | MG | MG | F | P |
| $C_{12}$ | F | P | VP | MG |

After the implementation of the Fuzzy TOPSIS method to evaluate the alternative strategies in Egypt, the ranking is presented in Table 16.

**Table 16.** Alternative strategies ranking for the case of Egypt.

| Ranking | $A_i$ | Alternative Strategies | Closeness Coefficient |
|---|---|---|---|
| 1 | $A_1$ | SO | 0.0648 |
| 2 | $A_2$ | WO | 0.0547 |
| 3 | $A_3$ | ST | 0.0478 |
| 4 | $A_4$ | WT | 0.030 |

In the case of Egypt, the final ranking is described as $A_1 > A_2 > A_3 > A_4$ and proposed as the more preferable strategies, the SO and WO as resulted from the calculation of the closeness coefficient.

The following figure (Figure 7) illustrates the results of the implementation of the proposed methodology in the case study of Morocco and Egypt.

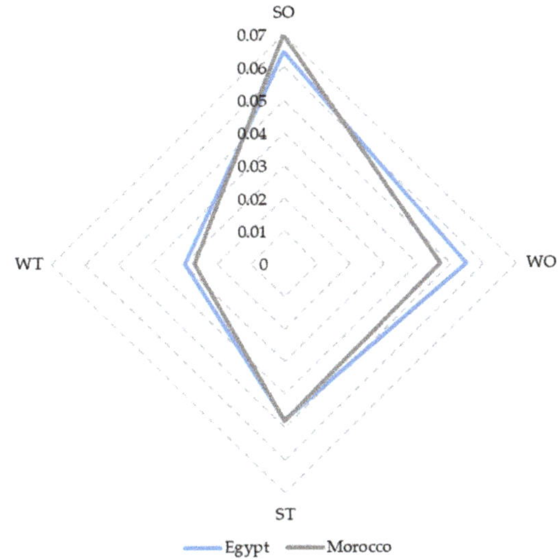


**Figure 7.** Results for Morocco and Egypt.

## 4. Discussion

To sum up, in this study the preferences of the decision-making team were determined through a SWOT analysis, while the AHP method was utilized to identify their priorities towards these preferences and set weights to the specific criteria. TOWS analysis and fuzzy TOPSIS method resulted in a final ranking of four alternative strategic policy directions. This combination collects and evaluates all the different aspects of the problem simultaneously, while it also provides a flexible environment for the decision maker, in order to dynamically formulate the judgments and to manage properly the diverse and conceptually different criteria. Given the SWOT analysis of a host country, the decision makers gain insights into the internal and external factors (current and future) that play a crucial role towards cooperation, while the use of linguistic variables significantly facilitate the decision making process, as it allows for flexible mapping of their preferences.

Based on the application of the AHP method, the criterion with the greatest significance to the specific problem is $C_{10}$ (RES-E capacity and potential). This criterion includes all the factors related to each country's potential and capability to develop RES and the progress achieved in the integration of technologies related to this field. Given that the EU's desire is to import renewable energy quantities from developing countries, while simultaneously to transfer expertise and make investments for the sustainable development of these countries, it is understood how important this criterion is for policy makers.

According to the final ranking of criteria, it is obvious that the criteria that cover areas, such as economy, entrepreneurship and investments are of particular value to the decision maker. If a host country does not perform well in these sectors, then a cross border energy cooperation with the EU is unlikely to happen. This is the reason why the criteria that represent these sectors, such as the RE regulatory and policy framework ($C_6$), the institutional framework for RES-E deployment ($C_7$), financial risks and uncertainty ($C_8$), the market structure of energy system ($C_4$) and investment facilitation ($C_9$), have a higher ranking compared to other criteria.

In general, according to the results from the pilot application in Morocco, it is possible to draw some very important conclusions in evaluating the possibility of implementing an EU–Morocco cross border cooperation within the field of renewable energy. The same stands also for the case of Egypt.

Using the proposed methodology, the overall rankings for all possible strategies were determined, and finally, SO policy strategy, which addresses strengths and opportunities, prevailed in the case of Morocco. This means that EU policy makers should focus on policy strategies that explore how the strengths could be used to benefit from external opportunities. An offensive strategy (ST) is placed second, which explores how policy makers could benefit from the strengths of Morocco to avoid or lessen (potential) external threats that may hinder the implementation of cooperation. This result reveals that according to a decision maker's opinion, the strengths of the country could easily come up against the potential threats. Morocco presents a very balanced picture on all three levels of analysis, as its energy sector is relatively well developed and targets are set for increasing the share of renewable energy in the future. It is the only African country that has electrical connection with Europe, while the domestic network is relatively integrated. At the same time, the appropriate energy laws on RES have been enacted, while the country's market is accessible to potential investors. The risk levels are relatively low, while good social conditions prevail and this fact facilitates the acceptance of new energy projects, as long as these materialize in the country. Finally, the more defensive policy strategies WT and WO are not so popular in the particular problem area. This means that it is not a priority for the decision maker to try to mitigate the internal weaknesses so as to exploit opportunities or to minimize any weaknesses to avoid potential threats.

On the other hand, the application of the proposed methodology in Egypt reveals that the most preferable strategies are the SO and WO, with the SO strategy that uses the internal strengths to take advantage of opportunities prevailing in both case study countries. The country already has the necessary legislation in place to support RES and the gradual penetration of RES into the energy reality, as the domestic market is largely liberalized, and the energy sector has seen slight growth in recent years. The country's economy is characterized by levels of volatility mainly due to monetary policy, while the investment sector has seen significant improvement through various reforms. However, the country has significant opportunities for RES development, as also described in its energy plan, which explains that the SO strategy dominates, followed by the WO which suggests using these opportunities to meet potential challenges. According to the decision makers evaluation, the country has a relatively manageable number of elements that may hinder potential energy co-operation. Thus, the existence of the ST and WT strategies in the last two ranking positions indicates that the country should rely mainly on its future potential for RES development and investment attractiveness. It seems that this country is eligible to develop cooperation with the EU, mainly because of future prospects and opportunities identified and not because of the advantages and strengths in the energy and investment sector during recent years.

## 5. Conclusions

This paper conducted a preference analysis based on the combination of SWOT analysis with two multi-criteria methods, AHP and fuzzy TOPSIS, in order to assess several factors that play a crucial role in the implementation of cross border cooperation in the field of RES and propose strategic directions towards the successful implementation. The methodology was applied for the case of Morocco and Egypt.

According to the results, Morocco performs well in most of the criteria and, thus the strategy that ranked first was the alternative SO with the focus to use strengths so as to exploit opportunities. On the other hand, the results from the application in Egypt reveal that the favorable strategic options that may boost the implementation of the EU–Egypt cooperation are the strategies that will try to exploit the opportunities of the country. This means that the current status of Egypt, compared to that of Morocco may not be so efficient to promote Egypt as a host country. However, future prospects seem very exploitable, and thus a defensive and an offensive strategy, based both on the exploitation of future opportunities, are ranked in the first two places.

The proposed decision-making approach is able to manage the uncertainty, inaccuracy and complexity of decisions, as they emerge from the various different and conflicting criteria. In order to ensure further improvement of the decision model, more criteria could be used to perform a more thorough analysis of the options available, as well as to include more information regarding the country concerned (e.g., Current Account Balance Index, the Energy Development Index, the Global Competitiveness Index etc.). The proposed methodological framework could also be applicable to more than one decision maker by adjusting and using the AHP and Fuzzy TOPSIS methods for group decision making. This will provide even greater accuracy and objectivity in the methodology, as it will contain different and often controversial judgments. Last but not least, this methodology could be applied to other potential host countries, such as the Western Balkans and Turkey and assist EU policy makers mapping the important factors, as well as hazards in the neighboring countries, that may emerge from a potential cooperation in the field of RES.

## 6. Patents

This section is not mandatory but may be added if there are patents resulting from the work reported in this manuscript.

**Author Contributions:** Conceptualization, A.P. and C.K.; Methodology, A.P., C.K. and G.A.; Writing—Original Draft Preparation, A.P. and G.A.; Writing—Review and Editing, C.K., H.D. and A.P.; Visualization, A.P. and G.A.; Supervision, H.D. All authors have read and agreed to the published version of the manuscript.

**Funding:** This research received no external funding.

**Conflicts of Interest:** The authors declare no conflict of interest.

## Abbreviations and Variable Definitions

The following abbreviations are used in this manuscript:

| | |
|---|---|
| AHP | Analytic Hierarchy Process |
| AMEE | Agence Marocaine pour l' Efficacité Énergétique |
| ANP | Analytic Network Process |
| ANRE | National Agency for Regulation of Electricity |
| BETTER | Bringing Europe and Third countries closer together through renewable Energies |
| BRI | Belt and Road Initiative |
| CNG | Compressed Natural Gas |
| CPF | Country Partnership Framework |
| CSP | Concentrated Solar Power |
| EC | European Commission |
| EETC | Egyptian Electricity Transmission Company |
| FDI | Foreign Direct Investment |
| Fuzzy TOPSIS | Fuzzy Technique for Order of Preference by Similarity to Ideal Solution |
| GHG | Greenhouse Gasses |
| GIS | Geographic Information System |
| IPCC | Intergovernmental Panel on Climate Change |
| IRESEN | Institute for Research in Solar Energy and Renewable Energies |
| MASEN | Moroccan Agency for Sustainable Energy |
| MCDM | Multi-Criteria Decision Making |
| NIS | Negative Ideal Solution |
| ONEE | Office National de l' Électricité et de l' Eau Potable |
| PIS | Positive Ideal Solution |
| RECAI | Renewable Energy Country Attractiveness Index |
| RED | Renewable Energy Directive |
| RES | Renewable Energy Sources |
| RES-E | RES electricity |
| SWOT | Strengths, Weaknesses, Opportunities, Threats |

The following variable definitions are used in this manuscript:

| | |
|---|---|
| $\widetilde{A}_n$ | The normalisation of the square matrix $\widetilde{A}$: |
| $\widetilde{D}$ | The fuzzy decision matrix |
| $d_i^-$ | The distance of alternatives from FNIS |
| $d_j^*$ | The distance of alternatives from FPIS |
| $\widetilde{R}$ | The normalized fuzzy decision matrix |
| $\widetilde{V}$ | The weighted normalized fuzzy decision matrix |
| $\widetilde{W}$ | The weights of the criteria |
| $\widetilde{A}$ | The square matrix that is developed from the pairwise comparison of the criterion i with the criterion j |
| C | The set of the evaluation criteria |
| CC | Closeness Coefficient |
| CI | Consistency Index |
| CR | Consistency Ratio |
| FNIS | The fuzzy negative ideal solution |
| FPIS | The fuzzy positive ideal solution |
| RI | Random Index |
| $\lambda_{max}$ | The principal eigen value of the matrix $\widetilde{A}$ |

## References


1. Doukas, H.; Nikas, A. Decision Support Models in Climate Policy. *Eur. J. Oper. Res.* **2020**, *280*, 1–24. [CrossRef]
2. European Commision. *Clean Energy for All Europeans*; COM(2016) 860 Final: Brussels, Belgium, 2016; p. 13.

3.  European Commision. *Directive (EU) 2018/2001 of the European Parliament and of the Council of 11 December 2018 on the Promotion of the Use of Energy from Renewable Sources*; Official Journal of the European Union: Luxembourg, 2018; p. 128.

4.  Klinge Jacobsen, H.; Pade, L.L.; Schröder, S.T.; Kitzing, L. Cooperation Mechanisms to Achieve EU Renewable Targets. *Renew. Energy* **2014**, *63*, 345–352. [CrossRef]

5.  Karakosta, C.; Marinakis, V.; Psarras, J. RES Cooperation Opportunities between EU and MENA Countries: The Case of Morocco. *Energy Strateg. Rev.* **2013**, *2*, 92–99. [CrossRef]

6.  Papapostolou, A.; Karakosta, C.; Nikas, A.; Psarras, J. Exploring Opportunities and Risks for RES-E Deployment under Cooperation Mechanisms between EU and Western Balkans: A Multi-Criteria Assessment. *Renew. Sustain. Energy Rev.* **2017**, *80*, 519–530. [CrossRef]

7.  Karakosta, C.; Marinakis, V.; Papapostolou, A.; Psarras, J. Benefits and costs sharing through RES electricity cooperation between Europe and third countries. In Proceeding of the 3rd International Conference on Energy Systems and Technologies, Cairo, Egypt, 16–19 February 2015.

8.  Papapostolou, A.; Karakosta, C.; Marinakis, V.; Flamos, A. Assessment of RES Cooperation Framework between the EU and North Africa. *Int. J. Energy Sect. Manag.* **2016**, *10*, 402–426. [CrossRef]

9.  Karakosta, C.; Papapostolou, A.; Dede, P.; Marinakis, V.; Psarras, J. Investigating EU-Turkey Renewable Cooperation Opportunities: A SWOT Analysis. *Int. J. Energy Sect. Manag.* **2016**, *10*, 337–362. [CrossRef]

10.  Papapostolou, A.; Karakosta, C.; Doukas, H. Analysis of Policy Scenarios for Achieving Renewable Energy Sources Targets: A Fuzzy TOPSIS Approach. *Energy Environ.* **2017**, *28*, 88–109. [CrossRef]

11.  Papapostolou, A.; Karakosta, C.; Kourti, K.-A.; Doukas, H.; Psarras, J. Supporting Europe's Energy Policy Towards a Decarbonised Energy System: A Comparative Assessment. *Sustainability* **2019**, *11*, 4010. [CrossRef]

12.  Cinelli, M.; Coles, S.R.; Kirwan, K. Analysis of the Potentials of Multi Criteria Decision Analysis Methods to Conduct Sustainability Assessment. *Ecol. Indic.* **2014**, *46*, 138–148. [CrossRef]

13.  Kumar, A.; Sah, B.; Singh, A.R.; Deng, Y.; He, X.; Kumar, P.; Bansal, R.C. A Review of Multi Criteria Decision Making (MCDM) towards Sustainable Renewable Energy Development. *Renew. Sustain. Energy Rev.* **2017**, *69*, 596–609. [CrossRef]

14.  Lee, H.-C.; Chang, C.-T. Comparative Analysis of MCDM Methods for Ranking Renewable Energy Sources in Taiwan. *Renew. Sustain. Energy Rev.* **2018**, *92*, 883–896. [CrossRef]

15.  Şengül, Ü.; Eren, M.; Eslamian Shiraz, S.; Gezder, V.; Şengül, A.B. Fuzzy TOPSIS Method for Ranking Renewable Energy Supply Systems in Turkey. *Renew. Energy* **2015**, *75*, 617–625. [CrossRef]

16.  Stein, E.W. A Comprehensive Multi-Criteria Model to Rank Electric Energy Production Technologies. *Renew. Sustain. Energy Rev.* **2013**, *22*, 640–654. [CrossRef]

17.  Wang, J.-J.; Jing, Y.-Y.; Zhang, C.-F.; Zhao, J.-H. Review on Multi-Criteria Decision Analysis Aid in Sustainable Energy Decision-Making. *Renew. Sustain. Energy Rev.* **2009**, *13*, 2263–2278. [CrossRef]

18.  Kousksou, T.; Allouhi, A.; Belattar, M.; Jamil, A.; El Rhafiki, T.; Zeraouli, Y. Morocco's Strategy for Energy Security and Low-Carbon Growth. *Energy* **2015**, *84*, 98–105. [CrossRef]

19.  Helms, M.M.; Nixon, J. Exploring SWOT Analysis—Where Are We Now? *J. Strateg. Manag.* **2010**, *3*, 215–251. [CrossRef]

20.  Chiu, A.S.; Yong, G. On the Industrial Ecology Potential in Asian Developing Countries. *J. Clean. Prod.* **2004**, *12*, 1037–1045. [CrossRef]

21.  Terrados, J.; Almonacid, G.; Hontoria, L. Regional Energy Planning through SWOT Analysis and Strategic Planning Tools. *Renew. Sustain. Energy Rev.* **2007**, *11*, 1275–1287. [CrossRef]

22.  Zharan, K.; Bongaerts, J.C. Decision-Making on the Integration of Renewable Energy in the Mining Industry: A Case Studies Analysis, a Cost Analysis and a SWOT Analysis. *J. Sustain. Min.* **2017**, *16*, 162–170. [CrossRef]

23.  Jaber, J.O.; Elkarmi, F.; Alasis, E.; Kostas, A. Employment of Renewable Energy in Jordan: Current Status, SWOT and Problem Analysis. *Renew. Sustain. Energy Rev.* **2015**, *49*, 490–499. [CrossRef]

24.  Chen, W.-M.; Kim, H.; Yamaguchi, H. Renewable Energy in Eastern Asia: Renewable Energy Policy Review and Comparative SWOT Analysis for Promoting Renewable Energy in Japan, South Korea, and Taiwan. *Energy Policy* **2014**, *74*, 319–329. [CrossRef]

25.  Dyson, R.G. Strategic Development and SWOT Analysis at the University of Warwick. *Eur. J. Oper. Res.* **2004**, *152*, 631–640. [CrossRef]

26. Lei, Y.; Lu, X.; Shi, M.; Wang, L.; Lv, H.; Chen, S.; Hu, C.; Yu, Q.; da Silveira, S.D.H. SWOT Analysis for the Development of Photovoltaic Solar Power in Africa in Comparison with China. *Environ. Impact Assess. Rev.* **2019**, *77*, 122–127. [CrossRef]

27. Agyekum, E.B.; Ansah, M.N.S.; Afornu, K.B. Nuclear Energy for Sustainable Development: SWOT Analysis on Ghana's Nuclear Agenda. *Energy Rep.* **2020**, *6*, 107–115. [CrossRef]

28. Kamran, M.; Fazal, M.R.; Mudassar, M. Towards Empowerment of the Renewable Energy Sector in Pakistan for Sustainable Energy Evolution: SWOT Analysis. *Renew. Energy* **2020**, *146*, 543–558. [CrossRef]

29. Igliński, B.; Iglińska, A.; Koziński, G.; Skrzatek, M.; Buczkowski, R. Wind Energy in Poland – History, Current State, Surveys, Renewable Energy Sources Act, SWOT Analysis. *Renew. Sustain. Energy Rev.* **2016**, *64*, 19–33. [CrossRef]

30. Ervural, C.B.; Zaim, S.; Demirel, O.F.; Aydin, Z.; Delen, D. An ANP and Fuzzy TOPSIS-Based SWOT Analysis for Turkey's Energy Planning. *Renew. Sustain. Energy Rev.* **2018**, *82*, 1538–1550. [CrossRef]

31. Wang, Y.; Xu, L.; Solangi, Y.A. Strategic Renewable Energy Resources Selection for Pakistan: Based on SWOT-Fuzzy AHP Approach. *Sustain. Cities Soc.* **2020**, *52*, 101861. [CrossRef]

32. Khan, M.I. Evaluating the Strategies of Compressed Natural Gas Industry Using an Integrated SWOT and MCDM Approach. *J. Clean. Prod.* **2018**, *172*, 1035–1052. [CrossRef]

33. Solangi, Y.A.; Tan, Q.; Mirjat, N.H.; Ali, S. Evaluating the Strategies for Sustainable Energy Planning in Pakistan: An Integrated SWOT-AHP and Fuzzy-TOPSIS Approach. *J. Clean. Prod.* **2019**, *236*, 117655. [CrossRef]

34. Velasquez, M.; Hester, P.T. An Analysis of Multi-Criteria Decision Making Methods. *Int. J. Oper. Res.* **2013**, *10*, 56–66.

35. Badri, M.A. A Combined AHP–GP Model for Quality Control Systems. *Int. J. Prod. Econ.* **2001**, *72*, 27–40. [CrossRef]

36. Wang, B.; Song, J.; Ren, J.; Li, K.; Duan, H.; Wang, X. Selecting Sustainable Energy Conversion Technologies for Agricultural Residues: A Fuzzy AHP-VIKOR Based Prioritization from Life Cycle Perspective. *Resour. Conserv. Recycl.* **2019**, *142*, 78–87. [CrossRef]

37. Ghimire, L.P.; Kim, Y. An Analysis on Barriers to Renewable Energy Development in the Context of Nepal Using AHP. *Renew. Energy* **2018**, *129*, 446–456. [CrossRef]

38. Ozdemir, S.; Sahin, G. Multi-Criteria Decision-Making in the Location Selection for a Solar PV Power Plant Using AHP. *Measurement* **2018**, *129*, 218–226. [CrossRef]

39. Colak, H.E.; Memisoglu, T.; Gercek, Y. Optimal Site Selection for Solar Photovoltaic (PV) Power Plants Using GIS and AHP: A Case Study of Malatya Province, Turkey. *Renew. Energy* **2020**, *149*, 565–576. [CrossRef]

40. Keeley, A.R.; Matsumoto, K. Relative Significance of Determinants of Foreign Direct Investment in Wind and Solar Energy in Developing Countries—AHP Analysis. *Energy Policy* **2018**, *123*, 337–348. [CrossRef]

41. Wu, Y.; Xu, C.; Ke, Y.; Tao, Y.; Li, X. Portfolio Optimization of Renewable Energy Projects under Type-2 Fuzzy Environment with Sustainability Perspective. *Comput. Ind. Eng.* **2019**, *133*, 69–82. [CrossRef]

42. Karakosta, C.; Psarras, J. Fuzzy TOPSIS approach for understanding a country's development priorities within the scope of climate technology transfer. In *Advances in Energy Research*; Acosta, M.J., Ed.; Nova Science Publishers: Hauppauge, NY, USA, 2012; Volume 9, pp. 123–149. ISBN 978-1-61470-485-0.

43. Hwang, C.L.; Yoon, K. *Multiple Attribute Decision Making: Methods and Applications a State-of-the-Art Survey*; Springer: Berlin/Heidelberg, Germany, 1981; ISBN 978-3-540-10558-9.

44. Zadeh, L.A. Fuzzy Sets. *Inf. Control* **1965**, *8*, 338–353. [CrossRef]

45. Kahraman, C.; Onar, S.C.; Oztaysi, B. Fuzzy Multicriteria Decision-Making: A Literature Review. *Int. J. Comput. Intell. Syst.* **2015**, *8*, 637–666. [CrossRef]

46. Bellman, R.E.; Zadeh, L.A. Decision-Making in a Fuzzy Environment. *Manag. Sci.* **1970**, *17*, B-141–B-164. [CrossRef]

47. Karakosta, C. *Multi-Criteria Decision Making Methods for Technology Transfer*; LAP LAMBERT Academic Publishing: Riga, Latvia, 2018; ISBN 978-613-9-86527-7.

48. Khalili-Damghani, K.; Sadi-Nezhad, S.; Tavana, M. Solving Multi-Period Project Selection Problems with Fuzzy Goal Programming Based on TOPSIS and a Fuzzy Preference Relation. *Inf. Sci.* **2013**, *252*, 42–61. [CrossRef]

49. Lima Junior, F.R.; Osiro, L.; Carpinetti, L.C.R. A Comparison between Fuzzy AHP and Fuzzy TOPSIS Methods to Supplier Selection. *Appl. Soft Comput.* **2014**, *21*, 194–209. [CrossRef]

50. Zhao, H.; Guo, S. Selecting Green Supplier of Thermal Power Equipment by Using a Hybrid MCDM Method for Sustainability. *Sustainability* **2014**, *6*, 217–235. [CrossRef]
51. Papadogeorgos, I.; Papapostolou, A.; Karakosta, C.; Doukas, H. Multicriteria assessment of alternative policy scenarios for achieving EU RES target by 2030. In *Strategic Innovative Marketing*; Kavoura, A., Sakas, D., Tomaras, P., Eds.; Springer Proceedings in Business and Economics; Springer: Cham, Switzerland, 2017.
52. Rani, P.; Mishra, A.R.; Mardani, A.; Cavallaro, F.; Alrasheedi, M.; Alrashidi, A. A Novel Approach to Extended Fuzzy TOPSIS Based on New Divergence Measures for Renewable Energy Sources Selection. *J. Clean. Prod.* **2020**, *257*, 120352. [CrossRef]
53. Çolak, M.; Kaya, İ. Prioritization of Renewable Energy Alternatives by Using an Integrated Fuzzy MCDM Model: A Real Case Application for Turkey. *Renew. Sustain. Energy Rev.* **2017**, *80*, 840–853. [CrossRef]
54. Karunathilake, H.; Hewage, K.; Mérida, W.; Sadiq, R. Renewable Energy Selection for Net-Zero Energy Communities: Life Cycle Based Decision Making under Uncertainty. *Renew. Energy* **2019**, *130*, 558–573. [CrossRef]
55. Ligus, M.; Peternek, P. Determination of Most Suitable Low-Emission Energy Technologies Development in Poland Using Integrated Fuzzy AHP-TOPSIS Method. *Energy Procedia* **2018**, *153*, 101–106. [CrossRef]
56. Proctor, T. *Strategic Marketing, an Introduction*, 1st ed.; Taylor & Francis Ltd.: London, UK, 2008.
57. Weihrich, H. The TOWS Matrix—A Tool for Situational Analysis. *Long Range Plan.* **1982**, *15*, 54–66. [CrossRef]
58. Chen, C.-T. Extensions of the TOPSIS for Group Decision-Making under Fuzzy Environment. *Fuzzy Sets Syst.* **2000**, *114*, 1–9. [CrossRef]
59. Chen, S.; Wei, Z. A Rational Consensus Model in Group Decision Making based on Linguistic Assessment Information. In Proceeding of the International Conference on Management and Service Science (MASS), Wuhan, China, 24–26 August 2010.
60. Alptekin, N. Integration of SWOT Analysis and TOPSIS Method in Strategic Decision Making Process. *Macrotheme Rev.* **2013**, *2*, 1–8.
61. Saaty, T.L. Decision Making with the Analytic Hierarchy Process. *Int. J. Serv. Sci.* **2008**, *1*, 83. [CrossRef]
62. World Future Council. *A Roadmap for 100% Renewable Energy in Morocco*; World Future Council: Hamburg, Germany, 2017.
63. Bouhal, T.; Agrouaz, Y.; Kousksou, T.; Allouhi, A.; El Rhafiki, T.; Jamil, A.; Bakkas, M. Technical Feasibility of a Sustainable Concentrated Solar Power in Morocco through an Energy Analysis. *Renew. Sustain. Energy Rev.* **2018**, *81*, 1087–1095. [CrossRef]
64. Lotfi, M.; Monteiro, C.; Shafie-khah, M.; Catalao, J.P.S. Evolution of Demand Response: A Historical Analysis of Legislation and Research Trends. In Proceeding of the 2018 Twentieth International Middle East Power Systems Conference (MEPCON), Cairo University, Cairo, Egypt, 18–20 December 2018.
65. Cantoni, R.; Rignall, K. Kingdom of the Sun: A Critical, Multiscalar Analysis of Morocco's Solar Energy Strategy. *Energy Res. Soc. Sci.* **2019**, *51*, 20–31. [CrossRef]
66. GIZ. No *Étude de l'effet sur l'Emploi d'un Régime D'autoproduction avec Facturation nette D'électricité (Loi 13-09 et loi 58-15) issue D'installations Photovoltaïques Connectées au Réseau de Basse Tension au Maroc*; Deutsche Gesellschaft für Internationale Zusammenarbeit (GIZ) GmbH: Eschborn, Germany, 2017.
67. OECD/IEA. Renewables 2018. Available online: www.iea.org/statistics/ (accessed on 12 March 2020).
68. Renewable Energy Solutions for the Mediterranean (RES4MED). Country Profile: Morocco 2018. Available online: https://www.res4med.org/wp-content/uploads/2018/06/Country-profile-Marocco-2.pdf (accessed on 12 March 2020).
69. Moore, S. Evaluating the Energy Security of Electricity Interdependence: Perspectives from Morocco. *Energy Res. Soc. Sci.* **2017**, *24*, 21–29. [CrossRef]
70. Nordea. Country Profile Morocco. Available online: https://www.nordeatrade.com/en/explore-new-market/morocco/investment (accessed on 3 March 2020).
71. IEA. *Energy Policies Beyond IEA Countries: Morocco 2019*; OECD: Paris, France, 2019; Available online: https://doi.org/10.1787/10f93028-en (accessed on 12 March 2020).
72. Schinke, B.; Klawitter, J. *Summary: Country Fact Sheet Morocco, Energy and Development at a Glance, Middle East North Africa Sustainable ELECtricity Trajectories (MENA-SELECT) Project*; Germanwatch: Bonnn, Germany, 2016.
73. OECD/IEA. World Energy Balances 2018. Available online: www.iea.org/statistics/ (accessed on 12 March 2020).

74. Schwerhoff, G.; Sy, M. Financing Renewable Energy in Africa—Key Challenge of the Sustainable Development Goals. *Renew. Sustain. Energy Rev.* **2017**, *75*, 393–401. [CrossRef]
75. Warren, B. Renewable Energy Country Attractiveness Index (RECAI). Available online: https://www.ey.com/en_uk/power-utilities/renewable-energy-country-attractiveness-index (accessed on 12 March 2020).
76. Schinko, T.; Komendantova, N. De-Risking Investment into Concentrated Solar Power in North Africa: Impacts on the Costs of Electricity Generation. *Renew. Energy* **2016**, *92*, 262–272. [CrossRef]
77. Royaume du Maroc Ministere de l'Industrie. Industrial Acceleration Plan 2014–2020. Available online: http://www.mcinet.gov.ma/en/content/industrial-acceleration-plan-2014-2020-0 (accessed on 12 March 2020).
78. Choukri, K.; Naddami, A.; Hayani, S. Renewable Energy in Emergent Countries: Lessons from Energy Transition in Morocco. *Energy. Sustain. Soc.* **2017**, *7*, 25. [CrossRef]
79. Schinko, T.; Bohm, S.; Komendantova, N.; Jamea, E.M.; Blohm, M. Morocco's Sustainable Energy Transition and the Role of Financing Costs: A Participatory Electricity System Modeling Approach. *Energy. Sustain. Soc.* **2019**, *9*, 1. [CrossRef]
80. Tramblay, Y.; Ruelland, D.; Bouaicha, R.; Servat, E. Projected climate change impacts on water resources in northern Morocco with an ensemble of regional climate models. Hydrology in a Changing World: Environmental and Human Dimensions. In Proceeding of the FRIEND-Water 2014, Montpellier, France, 7–10 October 2014. IAHS Publ. 363.
81. Simonneaux, V.; Cheggour, A.; Deschamps, C.; Mouillot, F.; Cerdan, O.; Le Bissonnais, Y. Land Use and Climate Change Effects on Soil Erosion in a Semi-Arid Mountainous Watershed (High Atlas, Morocco). *J. Arid Environ.* **2015**, *122*, 64–75. [CrossRef]
82. Kousksou, T.; Allouhi, A.; Belattar, M.; Jamil, A.; El Rhafiki, T.; Arid, A.; Zeraouli, Y. Renewable Energy Potential and National Policy Directions for Sustainable Development in Morocco. *Renew. Sustain. Energy Rev.* **2015**, *47*, 46–57. [CrossRef]
83. The World Bank. Morocco Overview. Available online: https://www.worldbank.org/en/country/morocco/overview#2 (accessed on 12 March 2020).
84. Trieb, F.; Hess, D.; Kern, J.; Fichter, T.; Moser, M.; Pfenning, U.; Caldez, N.; de la Rua, C.; Türk, A.; Frieden, D.; et al. *WP3:North Africa Case Study Final Report Bringing Europe and Third Countries Closer Together through Renewable Energies (BETTER) Project, European Commission Intelligent Energy for Europe*; Contract No: IEE/11/845/SI2.61637; European Commission: Brussels, Belgium, 2015.
85. Davies, M.; Elmatbouly, S.; El-Mazghouny, D.; Schellkens, G.; Ahmad, S. Developing Renewable Energy Projects: A Guide to Achieving Success in the Middle East. Available online: https://www.pwc.com/m1/en/publications/documents/developing-renewable-energy-projects-egypt.pdf (accessed on 3 April 2020).
86. Rady, Y.Y.; Rocco, M.V.; Serag-Eldin, M.A.; Colombo, E. Modelling for Power Generation Sector in Developing Countries: Case of Egypt. *Energy* **2018**, *165*, 198–209. [CrossRef]
87. Elkadeem, M.R.; Wang, S.; Azmy, A.M.; Atiya, E.G.; Ullah, Z.; Sharshir, S.W. A Systematic Decision-Making Approach for Planning and Assessment of Hybrid Renewable Energy-Based Microgrid with Techno-Economic Optimization: A Case Study on an Urban Community in Egypt. *Sustain. Cities Soc.* **2020**, *54*, 102013. [CrossRef]
88. Davies, M.; Hodge, B.; Ahmad, S.; Wang, Y. Developing Renewable Energy Projects: A Guide to Achieving Success in the Middle East. Available online: https://www.pwc.com/m1/en/publications/documents/eversheds-pwc-developing-renewable-energy-projects.pdf (accessed on 3 April 2020).
89. Bohl, D.K.; Hanna, T.; Scott, A.C.; Moyer, J.D.; Hedden, S.G. *Sustainable Development Goals Report: Egypt 2030*; Frederick S. Pardee Center for International Futures and United Nations Development Programme: Denver, CO, USA; New York, NY, USA, 2018.
90. IRENA. *Renewable Energy Outlook: Egypt, Executive Summary*; International Renewable Energy Agency: Abu Dhabi, United Arab Emirate, 2018.

*Article*

# Development and Validation of a Knowledge Management Questionnaire for Hospitals and Other Healthcare Organizations

**Ioanna Karamitri [1], Fotis Kitsios [2] and Michael A. Talias [1,*]**

[1] Healthcare Management Postgraduate Program, Open University Cyprus, P.O. Box 12794, Nicosia 2252, Cyprus; ioanna.karamitri@st.ouc.ac.cy

[2] Department of Applied Informatics, University of Macedonia, 156 Egnatia Street, 54636 Thessaloniki, Greece; kitsios@uom.gr

* Correspondence: michael.talias@ouc.ac.cy

Received: 25 February 2020; Accepted: 27 March 2020; Published: 31 March 2020

**Abstract:** Sustainable societies need to consider the connection between knowledge management (KM) and healthcare as a critical issue for social development. They need to investigate how to create knowledge and identify possible predictors of knowledge-sharing behavior that can support a hospital's sustainable knowledge-management strategy. KM strategies could help managers to increase the performance of hospitals and other healthcare organizations. The purpose of this paper is to present a valid and reliable questionnaire about KM in healthcare organizations. We develop a new knowledge-management questionnaire based on the use of an extensive literature review and health professionals' consensus. The Applied Knowledge Management Instrument (AKMI) questionnaire was pilot tested and retested on a small group of employees of healthcare organizations (n = 31). After the pilot process, a larger group of health professionals (n = 261) completed the questionnaire. Further investigation resulted in item reduction and verification of the dimensions of AKMI. Finally, we explore the psychometric properties of the developed tool. The developed questionnaire seems to be reliable, valid, and suitable to be used for studying the suggested nine dimensions of KM: perceptions of KM, intrinsic and extrinsic motivations, knowledge synthesis and sharing, cooperation, leadership, organizational culture, and barriers. The developed questionnaire can help policymakers and hospital administrators collect information about KM processes in healthcare organizations and this can result in higher performance of health organizations.

**Keywords:** knowledge management; knowledge creation; healthcare organizations; hospital; organizational learning; sustainable knowledge management; strategy; validation; questionnaire; quality of care

---

## 1. Introduction

Knowledge is a valuable resource for the growth of individuals and organizations. It represents a cognitive framework that makes possible the meaning and understanding of raw data and information [1] and sometimes leads to wisdom [2]. Scientists distinguish two types of knowledge, explicit and implicit [3–5]. Explicit knowledge can be expressed through words, numbers, or figures and represents the tip of an iceberg. Most of our knowledge is tacit, and it is hard to formulate and share. It is what Michael Polanyi [6] said: "We can know more than we can tell".

The cornerstone of knowledge creation and transfer theory was introduced by Nonaka and Takeuchi [5] with the SECI model. Since knowledge increases with interaction, it can be articulated and amplified in various entities where individuals cooperate, like businesses and other organizations, making them sustainable [7]. In the past, scientists insisted on the personal character of knowledge.

Next, many agreed that organizational culture exists, especially the heuristic knowledge that is developed by employees while working. Organizational knowledge is achievable when organizations sustain a spirit of cooperation, motivate their personnel, and encourage them to innovate, which means that they have competent management [8].

Management embraced knowledge and, around the 1990s, a contemporary business philosophy attracted the interest of executive officers, researchers, and scholars. In this way, the interest in knowledge management (KM) has grown and has been sustained. An increasing amount of digitized information is available because the decision-maker allows an organization to outperform its competitors. The complexity of modern business needs proper information to minimize errors and ensure future success [9], and the need for quality and best economic outcomes within the business strategy management framework [10].

The definition and conceptualization of knowledge management are not easily distinguished [11] mainly because of the two disjointed approaches that identify KM as technology-centered and people-centered. The first suggests that KM resembles information system management, which uses high technology to make information available and accessible at the right time for the users. The latter focuses on managing knowledge via human resource management practices [12]. The perspective of this article is human-orientated. Like other scientists, we believe that information and communication systems are tools for effective knowledge management and that attention should be focusing on the human, organizational and cultural aspects of knowledge management [13]. Healthcare organizations are examples of the balance between humans and engines. Even if the provision of health services relies on modern technologies, health professionals take the final decision for the diagnosis and treatment of the patient.

As Peter Drucker, the renowned professor, stated [14], "Hospitals are the most complex human organizations ever derived . . . and the fastest-growing in all developed countries". Even medium-sized hospitals occupy hundreds of employees from various scientific fields, educational backgrounds, socioeconomic status, and occasionally different cultures. Different groups of employees often have their regulations, perspectives, requirements, and accreditation. Still, they have to interact, cooperate, share information, transform it into knowledge, and perform efficiently to provide high-quality services to the patients and their caregivers. Consequently, it is difficult to share experiences and make comparisons between healthcare settings and other types of organization, and these should be studied independently via their social context and norms.

Healthcare agencies are late adopters of KM philosophy compared to the business sector [15]. Therefore, healthcare experts have just recently started to show interest in research for evaluating the existence and quality of a knowledge environment in hospitals. In 2015, we conducted a systematic review of knowledge management practices in healthcare settings. We accessed three databases (Medline, Cinahl, and Health Source: nursing/academic edition) for 10 years (1/1/2004-25/11/2014) and retrieved 604 articles, of which 20 articles were eligible for analysis. Most of the studies had a qualitative approach, and researchers collected data through interviews with a small number of individuals or focus groups. Details about preparation, analysis, and results of our systematic review are published [16]. We confirmed that quantitative research about KM in a healthcare settings is scarce, and there is a lack of an integrated self-administered questionnaire for health professionals who work in healthcare organizations.

Academics and practitioners from other scientific fields have developed appropriate questionnaires for KM [17]. Still, to our knowledge, no one has until now introduced a reliable quantitative tool that explores KM elements in healthcare organizations.

The purpose of this study is to develop and test a questionnaire to learn more about knowledge management in healthcare settings. We aim to create a tool that could explore attitudes, emotions, cognition, intention or behavior, and identify motivators of and barriers to employees about KM.

## 2. Methods

### *2.1. Questionnaire Design*

Rattray and Jones [18] claimed that researchers who design a questionnaire should use various resources, such as discussions with experts, proposals of participants, and an extensive literature review to increase the face and content validity. Before creating this questionnaire, we conducted a systematic review of the literature [16]. This review identified six critical elements of KM in healthcare settings: perceptions of KM, synthesis, dissemination, collaboration, means of KM, and leadership. Furthermore, it detected several barriers, which restrict the implementation of knowledge management practices. These findings stimulated us to attempt the development of a questionnaire, which we named the Applied Knowledge Management Instrument or AKMI. The word "AKMI" is similar to the English word acme, which has a Greek origin and means the highest point or peak. We chose this name to stress that effective knowledge management could lead hospitals and other healthcare units to top performance.

### *2.2. Ethical Issues*

The study protocol received approval by the scientific and the border committee of the General Hospital of Messinia, Greece. We composed a letter stating that the completion of the questionnaire is voluntary, and that we will protect the privacy of human subjects while collecting, analyzing, and reporting data by anonymity [19]. Furthermore, we clearly announced the purpose of the study, the significance of the contribution of each employee, and that the completion of the questionnaire will have a positive impact on the hospital and science in general [20]. A cover letter stated a brief definition of KM to clarify the term for those who were not familiar with it. In this way, we motivated potential participants to complete the questionnaire.

### *2.3. Research Tool*

### 2.3.1. Selection of the Factors

We selected the following factors of knowledge management for healthcare settings for analysis: perceptions about KM, intrinsic motives, extrinsic motives, knowledge synthesis, dissemination, cooperation, leadership, culture, and barriers. The items of the factors consisted of closed-ended statements, and participants completed AKMI by reporting their level of agreement on a five-point Likert scale. The first statement, for example, is, "Each hospital should implement KM politics." There is only one open-ended question, which asks, "In your opinions, which are the three most important barriers of knowledge management," to identify KM obstacles. Furthermore, the questionnaire had questions for job satisfaction, self-efficacy, and state anxiety, and six items for demographics. These questions were the last to diminish the possibility of drop-outs [18].

### 2.3.2. Perceptions about Knowledge Management (KM)

The theory of reasoned action assumes that there is a relationship between attitudes and volitional behavior [21]. As a result, a positive attitude or perception towards KM could lead to action like knowledge creation or sharing. Chang et al. [22] claimed that better comprehension of KM improves employees' performance at hospitals. Another study revealed that a positive attitude for KM could give a competitive advantage and can increase innovation [23]. Another supposition is the existence or absence of a correlation between positive perceptions about KM and self-efficacy, as it appears in a sample of librarians in Israel [17]. This factor consists of five items.

### 2.3.3. Motives

Motives activate individuals to fulfill their needs [24]. If the reward from a specific action is endogenous (e.g., feeling of satisfaction), the motivation is intrinsic, and if it is exogenous (e.g., financial compensation), the motive is extrinsic [25]. The exploration of the motivations that drive employees to knowledge creation and sharing is one of the main goals of our study. Two factors are needed to represent motives; one for the intrinsic features and one for the extrinsic characteristics. Each latent factor consists of four items.

At the intrinsic motives, the person draws satisfaction from other external rewards, like the challenge of completing a difficult task. It is hard but possible for managers to handle intrinsic motives [26]. Comprehension of human behavior could help managers. For instance, people seek purposes to fulfill their life, and a shared goal promotes collaborative relationships. On the contrary, the lack of use places a psychological burden on employees [27]. Intrinsic motives for participating in KM procedures could be the satisfaction of having and transmitting knowledge, and the joy of helping others [28].

The reinforcement theory suggests that individuals are motivated when their behavior is reinforced positively (with rewards) or negatively (with the reduction or removal of positive rewards) [29]. The theory has received some criticism, but tangible rewards are significant motivators for other methods as well [25]. Recorded extrinsic motives for KM are personal, professional, and financial rewards for participating in KM procedure, work safety and stability, and other ways of individual support. Effects of extrinsic and intrinsic motivation on employees' knowledge-sharing intentions were examined in Taiwan, using structural equation modeling. The sample was 172 employees of a big group of firms [30]. A comparison of these results with participants from a Greek public hospital could be rather interesting.

### 2.3.4. Knowledge Synthesis

Knowledge synthesis is a fundamental element of KM [31] and healthcare settings [14]. Results showed many ways of creating knowledge, such as interactions of colleagues [32], formal and informal meetings [33], and recorded evidence [34]. The synthesis will attempt to clarify which groups of employees are more involved in the process and whether it connects strongly with other factors like perceptions, culture [35], or leadership [36], in different environments besides hospitals.

### 2.3.5. Knowledge Sharing and Collaboration

Knowledge sharing through collaboration is fundamental for KM because it can be carried out by all employees, regardless of their ability to create knowledge. Next, we will examine whether this factor (using five items) relates to various motives. Furthermore, we will test if knowledge sharing correlates with personal and demographic characteristics [37]. The "openness" to cooperation will also be examined [38] with four items, as stated in the case that collaboration among different groups of health professionals increases the likelihood of innovation [39]. Pezeshki Rad et al. [40] designed a questionnaire for knowledge sharing at the Iranian Ministry of Agriculture that had some interesting questions, which we modified and adopted for this factor.

### 2.3.6. Leadership

Leaders have a significant impact on businesses and organizations. They are expected to ease access to information, encourage innovation, and empower employees to implement KM practices [41]. The way they act is fundamental to the success of knowledge sharing [42]. They should build a culture of knowledge [43], reinforce continuous learning, and create communication channels [22]. Leaders have the power to provide support and rewards [43]. Factor "leadership," which consists of three items, will be tested for correlations with culture, extrinsic motives, and self-efficacy, to clarify its impact on KM.

### 2.3.7. Culture

Knowledge culture represents the factor with the most items (nine items). Here, it is examined if a healthcare setting supports innovation, research, and cultivates a learning environment. Organizational culture is a broad term that refers to ideologies, practices, norms, and social behaviors. It gives integration and differentiation opportunities [44]. Sibbald et al. [45] found that leadership and organizational culture are instrumental in supporting knowledge management procedures in hospitals. We will also examine the relationship between corporate culture and perceptions of employees about KM.

### 2.3.8. Barriers

Even if administrators and employees might have the best intentions to create and share knowledge, there are often obstacles complicating their efforts. Most studies exploring the subject have a qualitative orientation [46–48]. Likewise, we chose an open-ended question of AKMI to reveal more barriers to implementing KM in hospitals. Still, we also entered three items for the quantitative part of the scale.

### 2.4. Pilot Study

We initially developed a pilot questionnaire of 38 questions. Then, we asked the opinion of three experts from the Hospital. The first expert was a medical doctor, who was the manager of a public Health Center, and responsible for the continuous education of physicians who undertake their internship. The second expert was the administrative manager of a General Hospital, an expert in Hospital Administration. The third expert was a registered nurse with a Master's degree in special education [49]. After the discussions, the number of questions increased to 64.

The extensive questionnaire of the 64 questions was pilot tested in a sample of 31 employees (physicians, nurses, midwives, health visitors, and administrative staff) who work at two public health centers and a public Physical Medicine and Rehabilitation Center. The participants (24 females and 7 males) had a mean of 19.82 years of working experience (SD 7.99) and a mean age of 45.42 (SD 6.72) years.

The questionnaire was completed for a second time 15 days later by the same group of people. Test-retest was measured with the intraclass correlation coefficient (ICC, two-way mixed model on absolute agreement) [50] for all the questions except the demographics. The results of ICC are interpreted according to the scores as follows: <0.40 poor, 0.40–0.49 adequate, 0.60–0.74 good, 0.75–1.00 excellent [51]. The ICC of our study was excellent (ICC average measures: mean 0.904, min 0.717, max 1.000). For the questions answered by the Likert scale, we measured Cronbach's Alpha coefficient with mean value equals to 0.905 [52]. Figure 1 shows the dimensions of AKMI.

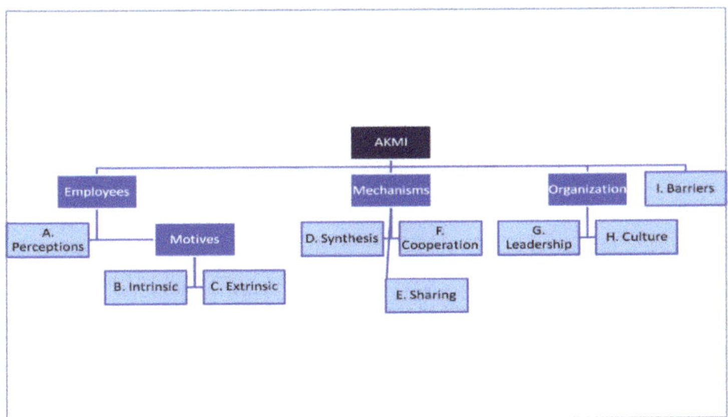


**Figure 1.** The architecture of Applied Knowledge Management Instrument (AKMI).

*2.5. Data Analysis*

Data were collected and entered on Microsoft Excel and analyzed with SPSS v22 (SPSS Inc, Chicago, IL, USA). Missing values were less than 3.2% for each item (overall missing values less than 4‰). We conducted Little's MCAR test, which indicated that values were missing completely at random. Therefore no entry was excluded from the analysis, and missing values were not replaced because sometimes imputation techniques for handling missing data result in biased estimates [53,54]. Reverse coding questions were recoded into different variables before further analysis.

To test whether data were appropriate for factor analysis, we measured the Kaiser–Meyer–Olkin (KMO) coefficient (KMO = 0.696). Furthermore, we carried out Barlett's test of sphericity, which showed a significant p-value <0.001. Both tests indicated that the dataset was suitable for factor analysis [55]. Furthermore, the sample size of the dataset was larger than 250, which is a prerequisite for obtaining reliable results [56].

## 3. Results

*3.1. Study Population*

This study took place from February to June of 2015 at General Hospital of Kalamata, which is a medium-sized public hospital with 300 beds, and the biggest (out of two) from a rural area of 200,000 inhabitants in Messinia, Greece. In 2015, the hospital had approximately 700 employees, of which 30% were males and 70% females. We asked 300 employees to participate in the study, and 261 employees agreed and completed the questionnaire (87% response rate). Even if there is no rule of thumb for the ideal sample size for testing a newly developed scale, a sample size of more than 200 people is acceptable [57].

Demographics of participants as regards gender, age, professional status, and working experience resembled the rest of the employees, who did not participate (Table 1). The completion time was 10–15 minutes.

**Table 1.** Demographic characteristics of the 261-employee sample.

| | | Frequency | Percent | Valid Percent | Cumulative Percent |
|---|---|---|---|---|---|
| Gender | | | | | |
| Valid | Male | 75 | 28.7 | 29.1 | 29.1 |
| | Female | 183 | 70.1 | 70.9 | 100 |
| | Total | 258 | 98.9 | 100 | |
| Missing | | | | | |
| Total | | | | | |
| Professional status | | | | | |
| Valid | Physician | 47 | 18 | 18 | 18 |
| | Nurse | 102 | 39.1 | 39.1 | 57.1 |
| | Administrative staff | 59 | 22.6 | 22.6 | 79.7 |
| | Paramedics | 21 | 8 | 8 | 87.7 |
| | Midwives - Health visitors | 7 | 2.7 | 2.7 | 90.4 |
| | Other | 25 | 9.6 | 9.6 | 100 |
| | Total | 261 | 100 | 100 | |
| Educational status (in years of study) | | | | | |
| | 9 years | 6 | 2.3 | 2.3 | 2.3 |
| | 12 years | 57 | 21.8 | 22.2 | 24.5 |
| | | 107 | 98.9 | 100 | |
| | 16 years (technological orientation) | 107 | 98.9 | | |
| Missing | | | | | |
| Total | | | | | |
| | **N** | **Min** | **Max** | **Mean** | **Std. Dev.** |
| Age | 254 | 21 | 62 | 44.61 | 8.18 |
| Overall Working experience | 257 | 0.4 | 36 | 18.94 | 8.76 |
| Working experience | 256 | 0.4 | 36 | 14.17 | 9.18 |

### 3.2. Validity

Face validity is a measure for the suitability of the project. It concerns the appropriateness, sensibility, or relevance of a test and its items and evaluates how it appears to the people who undertake it. Even if face validity seems an ambiguous term, it is essential for the success of a test or scale [58]. Many participants reported that our questionnaire was exciting and comprehensive. Additionally, they realized how effective knowledge management is and stated that they could participate more in knowledge sharing in the future. Content validity is a characteristic associated with the scale's adequacy for the measurement of the concept under consideration. It can only be checked subjectively through its approval by connoisseurs [59]. Our questionnaire was a subject of extended discussions at the pilot phase with three experienced health professionals with various educational and professional backgrounds, to ensure content validity. Additionally, we performed a factor analysis to establish construct validity [60].

### 3.3. Exploratory Factor Analysis

We conducted factor analysis, with extract method Alpha factoring, which resulted in 19 components with an eigenvalue greater than 1.0 that explained 52.8% of the variance. Due to low scoring, we removed a group of questions regarding "facilities" and nine more items. Most of the single items we excluded were reverse coded, and that means that they confused the participants. Following principal component factoring and varimax rotation, we repeated the analysis with a forced nine-factor solution. This time, the solution explained 56.87% of the variance. Table 2 illustrates results from factor analysis.

The estimate for the internal consistency of the entire questionnaire (Cronbach's alpha) was 0.802. Each dimension had the following Cronbach's $\alpha$: perceptions—0.724, intrinsic motivation—0.626, extrinsic motivation—0.739, knowledge synthesis—0.652, knowledge sharing—0.570, cooperation—0.567, leadership—0.717, culture—0.821, barriers—0.664. Median, interquartile range, and outliers of the results from the nine dimensions of AKMI are presented in Figure 2, and the final version of AKMI with preliminary results is shown in Table 3.

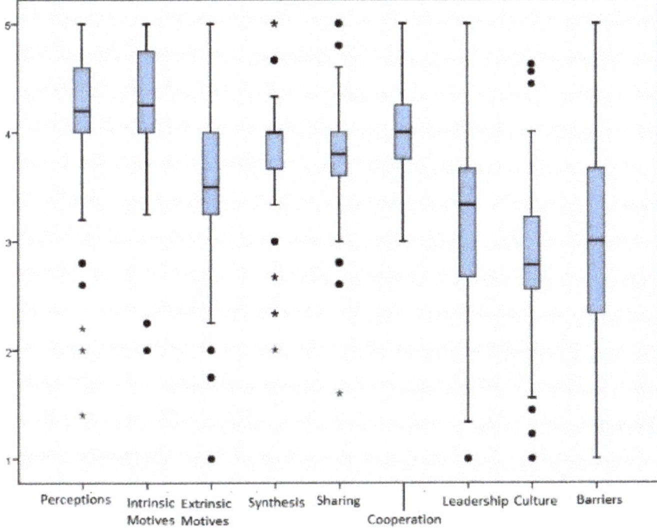


**Figure 2.** Boxplots of the nine dimensions.

**Table 2.** Exploratory factor analysis of the knowledge-management (KM) dimensions.

| Items | Communalities | Loadings | | | | | | | | |
| --- | --- | --- | --- | --- | --- | --- | --- | --- | --- | --- |
| | | 1 | 2 | 3 | 4 | 5 | 6 | 7 | 8 | 9 |
| Each hospital should implement KM politics | 0.772 | 0.854 | | | | | | | | |
| KM is essential for the performance of the hospital. | 0.830 | 0.900 | | | | | | | | |
| KM helps decrease errors. | 0.665 | 0.779 | | | | | | | | |
| Knowledge acquisition helps the individual's autonomy | 0.260 | 0.323 | | | | | | | | |
| Knowledge recording helps employees adapt when they are transferred to different departments of the hospital. | 0.453 | 0.272 | | | | | | | | |
| I feel content when I share my knowledge with others. | 0.579 | | 0.668 | | | | | | | |
| When I know something is useful for my colleagues, I inform them. It is a matter of principle. | 0.575 | | 0.675 | | | | | | | |
| I could participate in a seminar because I like knowledge even if I would not receive credit or a certificate | 0.433 | | 0.531 | | | | | | | |
| Knowledge acquisition gives me the power | 0.466 | | 0.381 | | | | | | | |
| When I share my knowledge, my colleagues respect me more. | 0.703 | | | 0.790 | | | | | | |
| When I share my knowledge, I bond with my colleagues. | 0.788 | | | 0.860 | | | | | | |
| When I help my colleagues, they help me, respectively. | 0.600 | | | 0.736 | | | | | | |
| I have higher chances of promotion where I possess knowledge. | 0.529 | | | 0.366 | | | | | | |
| I create knowledge through observation of the working environment. | 0.433 | | | | 0.438 | | | | | |
| Knowledge is created during group meetings. | 0.698 | | | | 0.512 | | | | | |
| Knowledge is created during group seminars. | 0.618 | | | | 0.371 | | | | | |
| Knowledge is shared during group meetings. | 0.454 | | | | | 0.488 | | | | |
| Knowledge is shared using electronic means (websites, wikis, forums). | 0.453 | | | | | 0.488 | | | | |
| I share knowledge with colleagues who are my friends. | 0.319 | | | | | 0.514 | | | | |
| I share knowledge with colleagues from my department. | 0.453 | | | | | 0.544 | | | | |
| I share knowledge with colleagues of other professional groups at the hospital. | 0.714 | | | | | 0.200 | | | | |
| I often cooperate with my colleagues to face a new situation | 0.359 | | | | | | 0.411 | | | |
| When I come across difficulties, I ask my colleagues. | 0.497 | | | | | | 0.487 | | | |
| When I know the work of others, it improves my performance | 0.319 | | | | | | 0.301 | | | |
| Cooperation when creating new knowledge reduces the anxiety of responsibility in the case of an error | 0.391 | | | | | | 0.241 | | | |
| My supervisor provides the required knowledge to solve problems. | 0.686 | | | | | | | 0.762 | | |
| My supervisor rewards people who share their knowledge. | 0.727 | | | | | | | 0.793 | | |
| Leadership creates channels of communication that help knowledge transfer. | 0.674 | | | | | | | 0.310 | | |
| At this hospital, there are KM strategies. | 0.487 | | | | | | | | 0.592 | |
| In this hospital, personnel are encouraged to innovate if they have a new idea. | 0.692 | | | | | | | | 0.778 | |
| This hospital supports research. | 0.668 | | | | | | | | 0.764 | |
| This hospital is a knowledge creation agency. | 0.589 | | | | | | | | 0.726 | |
| This cooperation strengthens the knowledge culture of my department. | 0.572 | | | | | | | | 0.275 | |
| In this hospital, there are commonly shared files to inform employees. | 0.484 | | | | | | | | 0.594 | |
| Leadership at this hospital has not understood the importance of KM (reverse coding). | 0.522 | | | | | | | | 0.680 | |
| There is no knowledge culture in this hospital (reverse coding) | 0.458 | | | | | | | | 0.634 | |
| Most colleagues share their knowledge freely. | | | | | | | | | 0.507 | |
| Hospitals' information system does not facilitate KM. | 0.514 | | | | | | | | | 0.409 |
| I have no access to useful information for my work. | 0.658 | | | | | | | | | 0.727 |
| I do not know very well where to find useful information for my work. | 0.725 | | | | | | | | | 0.838 |

**Table 3.** The final version of the Applied Knowledge Management Instrument.

| Item | Dimensions | Mean | Median | Factor's Mean | Cronbach Alpha |
|---|---|---|---|---|---|
| **A.** | **PERCEPTIONS** | | | | |
| 1 | Each hospital should implement KM politics. | 4.26 | 4.00 | | |
| 2 | KM is essential for the performance of the hospital. | 4.30 | 4.00 | 4.19 | 0.724 |
| 3 | KM helps decrease errors. | 4.27 | 4.00 | | |
| 5 | Knowledge acquisition helps the individual's autonomy. | 4.12 | 4.00 | | |
| 6 | Knowledge recording helps employees adapt when they are transferred to different departments of the hospital. | 3.99 | 4.00 | | |
| **B.** | **INTRINSIC MOTIVATION** | | | | |
| 7 | I feel content when I share my knowledge with others. | 4.27 | 4.00 | | |
| 8 | When I know something is useful for my colleagues, I inform them. It is a matter of principle. | 4.42 | 4.00 | 4.33 | 0.626 |
| 9 | I could participate in a seminar because I like knowledge even if I would not receive credit or a certificate of participation. | 4.29 | 4.00 | | |
| 12 | Knowledge acquisition gives me power. | 4.35 | 4.00 | | |
| **C.** | **EXTRINSIC MOTIVATION** | | | | |
| 13 | When I share my knowledge, my colleagues respect me more. | 3.56 | 4.00 | | |
| 14 | When I share my knowledge, I bond with my colleagues. | 3.64 | 4.00 | 3.55 | 0.739 |
| 15 | When I help my colleagues, they help me, respectively. | 3.59 | 4.00 | | |
| 17 | I have higher chances of promotion where I possess knowledge. | 3.39 | 4.00 | | |
| **D.** | **SYNTHESIS** | | | | |
| 18 | I create knowledge through observation of the working environment | 3.91 | 4.00 | | |
| 19 | Knowledge is created during group meetings. | 3.68 | 4.00 | 3.85 | 0.652 |
| 20 | Knowledge is created during group seminars. | 3.95 | 4.00 | | |
| **E.** | **SHARING** | | | | |
| 24 | Knowledge is shared during group meetings. | 3.65 | 4.00 | | |
| 25 | Knowledge is shared using electronic means (websites, wikis, forums). | 3.96 | 4.00 | | |
| 26 | I share knowledge with colleagues who are my friends | 3.44 | 4.00 | 3.74 | 0.570 |
| 27 | I share knowledge with colleagues from my department. | 3.91 | 4.00 | | |
| 28 | I share knowledge with colleagues of other professional groups at the hospital. | 3.76 | 4.00 | | |
| **F.** | **COOPERATION** | | | | |
| 30 | I often cooperate with my colleagues to face a new situation. | 3.95 | 4.00 | | |
| 31 | When I come across difficulties, I ask my colleagues. | 4.18 | 4.00 | | |
| 33 | When I know the work of the others, it improves my performance. | 3.97 | 4.00 | 4.04 | 0.567 |
| 34 | Cooperation when creating new knowledge reduces the anxiety of responsibility in case of an error. | 4.05 | 4.00 | | |
| **G.** | **LEADERSHIP** | | | | |
| 35 | My supervisor provides the required knowledge to solve problems. | 3.52 | 4.00 | | |
| 36 | My supervisor rewards people who share their knowledge. | 3.29 | 4.00 | 3.23 | 0.717 |
| 39 | Leadership creates channels of communication that help knowledge transfer. | 2.88 | 3.00 | | |
| **H.** | **CULTURE** | | | | |
| 37 | Leadership at this hospital has not understood the importance of KM (reverse coding). | 2.61 | 3.00 | | |
| 38 | At this hospital, there are KM strategies. | 4.42 | 3.00 | | |
| 40 | In this hospital, personnel is encouraged to innovate if they have a new idea. | 4.29 | 3.00 | | |
| 41 | This hospital supports research. | 4.35 | 3.00 | 2.83 | 0.821 |
| 42 | This hospital is a knowledge creation agency. | 2.72 | 3.00 | | |
| 43 | This cooperation strengthens the knowledge culture of my department. | 3.68 | 4.00 | | |
| 44 | There is no knowledge culture in this hospital (reverse coding). | 2.95 | 3.00 | | |
| 45 | In this hospital, there are commonly shared files to inform employees. | 2.71 | 3.00 | | |
| 23 | Most colleagues share their knowledge freely. | 2.76 | 3.00 | | |
| **I.** | **BARRIERS** | 2.61 | | | |
| 46 | The hospitals' information system does not facilitate KM. | 3.19 | 3.00 | | |
| 52 | I have no access to useful information for my work. | 2.81 | 3.00 | 2.95 | 0.644 |
| 53 | I do not know very well where to find useful information for my work. | 2.84 | 3.00 | | |

We further estimated the polychoric inter-correlations among the AKMI subscales and reported significant correlations between factors. Table 4 shows the estimated polychronic intercorrelations.

**Table 4.** Polychronic inter-correlations between factors.

| Factors | (1) | (2) | (3) | (4) | (5) | (6) | (7) | (8) | (9) |
|---|---|---|---|---|---|---|---|---|---|
| Perception (1) | 1.000 | | | | | | | | |
| Intrinsic motives (2) | 0.059 | 1.000 | | | | | | | |
| Extrinsic motives (3) | 0.110 | 1.054 *** | 1.000 | | | | | | |
| Synthesis (4) | 0.285 *** | 0.230 ** | 0.359 *** | 1.000 | | | | | |
| Sharing (5) | 0.201 * | 0.421 ** | 0.283 ** | 0.469 *** | 1.000 | | | | |
| Cooperation (6) | 0.280 ** | 0.752 ** | 0.660 *** | 0.526 *** | 0.617 *** | 1.000 | | | |
| Leadership (7) | −0.041 | 0.252 * | 0.098 | −0.059 | 0.101 | 0.096 | 1.000 | | |
| Culture (8) | −0.015 | 0.271 ** | 0.202 * | 0.018 | 0.151 | 0.056 | 0.949 *** | 1.000 | |
| Barrier (9) | 0.000 | −0.301 ** | −0.210 * | −0.098 | −0.223 | −0.173 ** | 0.501 *** | −0.486 *** | 1.000 |

Correlations are significant as follows: *** $p < 0.001$, ** $p < 0.05$, * $p < 0.01$.

## 4. Discussion

This research aimed to develop a questionnaire to understand the concepts of knowledge management and to investigate the organizational factors that affect all aspects of the knowledge creation process within hospitals.

Knowledge management is related to sustainability, organizational learning, knowledge transfer, quality of care and safety, type of motivations, and barriers, all of which will affect the level of service.

### 4.1. Knowledge Management and Sustainability

The application of knowledge management can lead to a sustainable healthcare system, and leaders can achieve the goals of their organizations [61]. It is important to note that the knowledge management process can be significantly related to improvements in the quality of healthcare as well as the organizational-level of social and economic outcomes, as stated by Popa [10,62]. Doctors may process the information related to the healthcare industry, and based on their experience and knowledge, can improve the quality of the system and the management of their patients. Moreover, patients can increase their knowledge from various sources like the internet, social media, and other medical staff. In this way, patients can determine or change their behavior and thoughts and demand the best possible service. The optimal management of the knowledge process affects the quality of a system.

Social stainability issues in healthcare facilities is another aspect which is explained by [63,64]. An organization with collaboration can apply knowledge management to share information to make healthcare organizations sustainable.

### 4.2. Knowledge Management and Human Resources

Knowledge is also regarded as organizational culture, skills, reputation, intuition, and codified theory that influences human behavior and thoughts [65,66]. There is also a concern about the current and future status of human resources management in healthcare organizations [67] and the impact of human resources information systems technology. Each organization will need to use HR practices that will balance evidence from data, its objectives, individual factors, and Human Resources Information systems. Organizations are becoming increasingly aware of the importance of employees in gaining and maintaining competitive advantage.

The competitiveness of a healthcare organization depends on the effectiveness of its knowledge management [62], and the knowledge-sharing process helps sustainable engagement in healthcare.

### 4.3. Knowledge Management and Organisational Learning

With knowledge management, healthcare leaders can understand how collective learning enhances the quality and safety improvement of hospitals. Organizations can support the process of internal learning if the goal is the improvement of their services. External knowledge acquisition often occurs through processes involving people. Knowledge management can help to reduce errors. For example, effective control is achieved using a clinical decision-support system. As a result, the potential reduction of medical errors can affect the improvement of healthcare delivery.

For example, research suggests that collective learning plays a role in improvement [68]. Specifically, cooperative learning is the process of gaining information which helps the capabilities in groups and organizations. Another process is collective learning, which has to do with the understanding and skills in groups and organizations [68,69]. Collective learning differs from individual learning because it requires individuals to analyze and interpret organizational experience [68].

The implementation of knowledge management can be thought of in two different ways [70,71]. The first is that there is a possibility that knowledge management to increase the autonomy of the medical staff by enhancing knowledge access. Knowledge sharing can lead to knowledge creation. On the other hand, controlling activities of the team can decrease collective intelligence. The excess of autonomy can encourage individuals to destabilize the organization, and there is a chance for them to act against the interests of the organization.

*4.4. Knowledge Management and the Developed Questionnaire*

Scientific interest in the various aspects of knowledge management can allow the connection of past results and the creation of knowledge. The findings and their implications should be addressed in the broadest context possible. Future research directions may also be highlighted. Perceptions of knowledge management were examined for another group of professionals, such as librarians in India [72] and other sectors, like construction and design companies in Spain [73]. Comparisons have been made between the perceptions of employees about knowledge management from small and large organizations in the United Kingdom [74]. Intrinsic and extrinsic motivations of KM were explored by researchers from various scientific fields [30,75]. There is still a debate in this field if external rewards can be considered as drivers for knowledge sharing, and our questionnaire aspires to clarify this issue. Knowledge creation, sharing, and cooperation are amongst the most researched topics in this area. However, in the healthcare sector, the focuses were mainly qualitatively analyzed [16], even if there are a small number of surveys, e.g., [37]. As regards leadership, studies have indicated individual styles of leadership to be significantly associated with the art of KM practices [76]. Zheng et al. [77] suggest that KM fully mediates the impact of organizational culture, and Leidner et al. [78] claim that organizational culture influences knowledge management initiatives. Based on these findings, we will subsequently create a model to determine the correlation structure of KM dimensions using a structural equation modeling procedure.

We think that self-efficacy plays an essential role in knowledge sharing. Until now, self-efficacy is mainly correlated with computer skills and knowledge-management systems [79] and less with occupational self-efficacy. With our dataset, we could check for significant connections between occupational self-efficacy and intentions to create or share knowledge.

The barriers of knowledge management procedure will be studied using the information we have collected with a closed and open-ended question. We asked health professionals to name the three most essential barriers according to their experience about the implementation of knowledge management in their organizations. The rationale of the task is to reveal existing barriers, especially in their working environment, and understand the correlations of barriers with the rest of the dimensions of the set-up questionnaire, e.g., leadership, and organizational culture.

The main advantage for the use of a specific knowledge-management instrument for healthcare units concerning a standard KM questionnaire is that the former takes into account the sui generis nature of the healthcare environment and the particular type of working relationships among health professionals. Additionally, the design of AKMI was done cautiously, with carefully examined methodological steps of an exhaustive literature review, pilot testing and retesting extended discussions with health professionals, and item reduction with factor analysis according to the main findings. The completion time was acceptable, and the dropouts were practically non-existent. Finally, participants spontaneously expressed their content after completed the questionnaire by stating that "this was their first step to actively participating in the knowledge-management process."

In terms of limitations, there are some caveats about specific dimensions of the questionnaire due to a just fair Cronbach's alpha score. Furthermore, our study does not permit premature generalization of the results obtained.

## 5. Conclusions

In summary, this paper describes the process of development and validation of a questionnaire with nine dimensions of knowledge management in healthcare organizations, perceptions of knowledge management, intrinsic and extrinsic motives, knowledge creation and sharing, cooperation, leadership, culture, and barriers. The introduction with AKMI makes a novel contribution to the study of KM in the area of healthcare organizations, adopting a social orientation at which employees and managers are the protagonists for successful KM in contrast to systems and sophisticated structures. Thus, AKMI has theoretical and practical implications. Universities may use the scale to explore knowledge management as a social process, at which people are the drivers of knowledge, smoothing the transitions from academia to practice. Similarly, it is useful to managers who need to know how they could motivate their personnel to engage in knowledge creation and sharing in an unimpeded way, in a working environment where innovation is supported.

Hospitals, as part of their operations, need to use knowledge-management systems to facilitate their operations' sustainably. Learning is an essential process, which is related to knowledge management [80]. The knowledge-creation process can lead to a sustainable competitive advantage process. However, few studies have empirically investigated how individual characteristics and organizational work practices influence knowledge sharing [81]. The knowledge creation process is vital for sustainability [82], and social media have an impact on this process [83]. Knowledge management enablers such as organizational structure, information technology (IT), strategy, and culture can be essential factors for the sustainability process of any healthcare organization. Different categories of healthcare employees have their role within sustainable operations, and human resources managers are encouraged to recruit people with the right qualifications to apply knowledge management.

**Author Contributions:** Conceptualization, I.K. and M.A.T.; methodology, F.K. and M.A.T.; software, I.K. and F.K.; data curation I.K. and F.K., I.K.; writing—original draft preparation, I.K.; writing—review and editing, M.A.T.; supervision, M.A.T.; All authors have read and agreed to the published version of the manuscript.

**Funding:** This research received no external funding.

**Acknowledgments:** The authors would like to express their gratitude to the participants of the survey.

**Conflicts of Interest:** The authors declare no conflict of interest.

## References

1. Zins, C. Conceptual approaches for defining data, information, and knowledge. *J. Am. Soc. Inf. Sci. Technol.* **2007**, *58*, 479–493. [CrossRef]
2. Ackoff, R.L. From Data to Wisdom. *J. Appl. Syst. Anal.* **1989**, *16*, 3–9.
3. Smith, E.A. The role of tacit and explicit knowledge in the workplace. *J. Knowl. Manag.* **2001**, *5*, 311–321. [CrossRef]
4. Collins, H. *Tacit and Explicit Knowledge*; University of Chicago Press: Chicago, IL, USA, 2010.
5. Nonaka, I.; Takeuchi, H. *The Knowledge-Creating Company: How Japanese Companies Create the Dynamics of Innovation*; Oxford University Press: New York, NY, USA, 1995.
6. Polanyi, M. *The Tacit Dimension*; Anchor Books: New York, NY, USA, 1967.
7. Sanguankaew, P.; Vathanophas Ractham, V. Bibliometric review of research on knowledge management and sustainability, 1994–2018. *Sustainability* **2019**, *11*, 4388. [CrossRef]
8. Tsoukas, H.; Vladimirou, E. What is organizational knowledge? *J. Manag. Stud.* **2001**, *38*, 973–993. [CrossRef]
9. Guptill, J. Knowledge management in health care. *J. Health Care Financ.* **2005**, *31*, 10–14.

10. Popa, I.; Ștefan, S.C. Modeling the Pathways of Knowledge Management Towards Social and Economic Outcomes of Health Organizations. *Int. J. Environ. Res. Public Health* **2019**, *16*, 1114. [CrossRef] [PubMed]

11. Begoña Lloria, M. A review of the main approaches to knowledge management. *Knowl. Manag. Res. Pract.* **2008**, *6*, 77–89. [CrossRef]

12. Hislop, D. *Knowledge Management in Organizations: A Critical Introduction*; Oxford University Press: Oxford, UK, 2013.

13. Šajeva, S. The analysis of key elements of socio-technical knowledge management system. *Econ. Manag.* **2010**, *15*, 765–774.

14. Drucker, P. *Managing in the Next Society*; Taylor & Francis Ltd.: Oxford, UK, 2007.

15. Kothari, A.; Hovanec, N.; Hastie, R.; Sibbald, S. Lessons from the business sector for successful knowledge management in health care: A systematic review. *BMC Health Serv. Res.* **2011**, *11*, 173. [CrossRef]

16. Karamitri, I.; Talias, M.A.; Bellali, T. Knowledge management practices in healthcare settings: A systematic review. *Int. J. Health Plann. Manag.* **2017**, *32*, 4–18. [CrossRef] [PubMed]

17. Aharony, N. Librarians' Attitudes toward Knowledge Management. *Coll. Res. Libr.* **2011**, *72*, 111–126. [CrossRef]

18. Rattray, J.; Jones, M.C. Essential elements of questionnaire design and development. *J. Clin. Nurs.* **2007**, *16*, 234–243. [CrossRef] [PubMed]

19. Bissett, A.F. Designing a questionnaire. Send a personal covering letter. *BMJ* **1994**, *308*, 202–203. [CrossRef]

20. Duncan, A.; Bonetti, D.; Clarkson, J.; Ramsay, C. Improving trial questionnaire response rates using behaviour change theory. *Trials* **2015**, *16*, P92. [CrossRef]

21. Fishbein, I.; Ajzen, I. *Belief, Attitude, Intention and Behaviour: An Introduction to Theory and Research*; Addison-Wesley: Boston, MA, USA, 1975.

22. Chang, Y.-Y.; Hsu, P.-F.; Li, M.-H.; Chang, C.-C. Performance evaluation of knowledge management among hospital employees. *Int. J. Health Care Qual. Assur.* **2011**, *24*, 348–365. [CrossRef]

23. Potgieter, A.; Dube, T.; Rensleigh, C. Knowledge management awareness in a research and development facility: Investigating employee perceptions. *S. Afr. J. Inf. Manag.* **2013**, *15*, 6. [CrossRef]

24. Maslow, A.H. A theory of human motivation. *Psychol. Rev.* **1943**, *50*, 370–396. [CrossRef]

25. Buchbinder, S.B.; Shanks, N.H. *Introduction to Health Care Management*, 2nd ed.; Jones & Bartlett Learning: Denver, CO, USA, 2011.

26. Abbas, J. Impact of total quality management on corporate sustainability through the mediating effect of knowledge management. *J. Clean. Prod.* **2020**, *244*, 118806. [CrossRef]

27. Thomas, K.W. *Intrinsic Motivation at Work: What Really Drives Employee Engagement*; Berrett-Koehler Publishers: San Francisco, CA, USA, 2009.

28. Bednarczyk, T.K. Human resources and motivation in knowledge management. *Manag. Prod. Eng. Rev.* **2010**, *1*, 6.

29. Skinner, B.F. *Science and Human Behavior*; New Impression edition; Free Press: New York, NY, USA, 1965.

30. Lin, H.-F. Effects of extrinsic and intrinsic motivation on employee knowledge sharing intentions. *J. Inf. Sci.* **2007**, *33*, 135–149. [CrossRef]

31. Colnar, S.; Dimovski, V.; Bogataj, D. Knowledge Management and the Sustainable Development of Social Work. *Sustainability* **2019**, *11*, 6374. [CrossRef]

32. Ghosh, B.; Scott, J.E. Comparing knowledge management in health-care and technical support organizations. *IEEE Trans. Inf. Technol. Biomed.* **2005**, *9*, 162–168. [CrossRef] [PubMed]

33. Orzano, A.J.; Ohman-Strickland, P.A.; Patel, M. What can family medicine practices do to facilitate knowledge management? *Health Care Manag. Rev.* **2008**, *33*, 216–224. [CrossRef] [PubMed]

34. Gerrish, K.; McDonnell, A.; Nolan, M.; Guillaume, L.; Kirshbaum, M.; Tod, A. The role of advanced practice nurses in knowledge brokering as a means of promoting evidence-based practice among clinical nurses. *J. Adv. Nurs.* **2011**, *67*, 2004–2014. [CrossRef] [PubMed]

35. Auernhammer, J.; Hall, H. Organizational culture in knowledge creation, creativity and innovation: Towards the Freiraum model. *J. Inf. Sci.* **2013**, *40*, 154–166. [CrossRef]

36. Kumar, M. Effect of Leadership Behaviors on Knowledge Creation in Indian Organizations. *Delhi Bus. Rev.* **2013**, *14*, 25.

37. Lee, H.S.; Hong, S.A. Factors affecting hospital employees' knowledge sharing intention and behavior, and innovation behavior. *Osong Public Health Res. Perspect.* **2014**, *5*, 148–155. [CrossRef]

38. Nembhard, I.M. All teach, all learn, all improve?: The role of interorganizational learning in quality improvement collaboratives. *Health Care Manag. Rev.* **2012**, *37*, 154–164. [CrossRef]

39. Williams, P.M. Integration of health and social care: A case of learning and knowledge management. *Health Soc. Care Community* **2012**, *20*, 550–560. [CrossRef]

40. Pezeshki rad, G.; Alizadeh, N.; Zamani Miandashti, N.; Shabanali Fami, H. factors influencing knowledge sharing among personnel of agricultural extension and education organization in iranian ministry of jihad-e agriculture. *J. Agric. Sci. Technol.* **2011**, *13*, 491–501.

41. Nguyen, H.; Mohamed, S. Leadership behaviors, organizational culture and knowledge management practices: An empirical investigation. *J. Manag. Dev.* **2011**, *30*, 206–221. [CrossRef]

42. Liu, F.-C.; Cheng, K.-L.; Chao, M.; Tseng, H.-M. Team innovation climate and knowledge sharing among healthcare managers: Mediating effects of altruistic intentions. *Chang. Gung Med. J.* **2012**, *35*, 408–419. [CrossRef] [PubMed]

43. Sánchez-Polo, M.T.; Cegarra-Navarro, J.G. Implementing Knowledge Management Practices in Hospital-in-the-Home Units. *J. Nurs. Care Qual.* **2008**, *23*, 18–22. [CrossRef]

44. Alavi, M.; Kayworth, T.R.; Leidner, D.E. An Empirical Examination of the Influence of Organizational Culture on Knowledge Management Practices. *J. Manag. Inf. Syst.* **2005**, *22*, 191–224. [CrossRef]

45. Sibbald, S.L.; Wathen, C.N.; Kothari, A. An empirically based model for knowledge management in health care organizations. *Health Care Manag. Rev.* **2016**, *41*, 64–74. [CrossRef]

46. Karamat, J.; Shurong, T.; Ahmad, N.; Waheed, A.; Khan, S. Barriers to knowledge management in the health sector of Pakistan. *Sustainability* **2018**, *10*, 4155. [CrossRef]

47. Dobbins, M.; DeCorby, K.; Twiddy, T. A knowledge transfer strategy for public health decision makers. *Worldviews Evid. Based Nurs.* **2004**, *1*, 120–128. [CrossRef]

48. Lin, C.; Tan, B.; Chang, S. An exploratory model of knowledge flow barriers within healthcare organizations. *Inf. Manag.* **2008**, *45*, 331–339. [CrossRef]

49. Woodward, C.A. Questionnaire construction and question writing for research in medical education. *Med. Educ.* **1988**, *22*, 345–363. [CrossRef]

50. Shrout, P.E.; Fleiss, J.L. Intraclass correlations: Uses in assessing rater reliability. *Psychol. Bull.* **1979**, *86*, 420–428. [CrossRef] [PubMed]

51. Cicchetti, D.V. Guidelines, criteria, and rules of thumb for evaluating normed and standardized assessment instruments in psychology. *Psychol. Assess.* **1994**, *6*, 284–290. [CrossRef]

52. Cronbach, L.J. Coefficient alpha and the internal structure of tests. *Psychometrika* **1951**, *16*, 297–334. [CrossRef]

53. Donders, A.R.T.; van der Heijden, G.J.M.G.; Stijnen, T.; Moons, K.G.M. Review: A gentle introduction to imputation of missing values. *J. Clin. Epidemiol.* **2006**, *59*, 1087–1091. [CrossRef] [PubMed]

54. Tabachnick, B.G.; Fidell, L.S. *Using Multivariate Statistics*, 5th ed.; Pearson: London, UK, 2006.

55. Hammer, A.; Ernstmann, N.; Ommen, O.; Wirtz, M.; Manser, T.; Pfeiffer, Y.; Pfaff, H. Psychometric properties of the Hospital Survey on Patient Safety Culture for hospital management (HSOPS_M). *Bmc Health Serv. Res.* **2011**, *11*, 165. [CrossRef] [PubMed]

56. MacCallum, R.C.; Widaman, K.F.; Zhang, S.; Hong, S. Sample size in factor analysis. *Psychol. Methods* **1999**, *4*, 84–99. [CrossRef]

57. Streiner, D.L.; Kottner, J. Recommendations for reporting the results of studies of instrument and scale development and testing. *J. Adv. Nurs.* **2014**, *70*, 1970–1979. [CrossRef]

58. Weiner, I.B.; Craighead, W.E. *The Corsini Encyclopedia of Psychology*, 4th ed.; John Wiley & Sons: Hoboken, NJ, USA, 2010; Volume 2.

59. Gotzamani Katerina, D. An empirical study of the ISO 9000 standards' contribution towards total quality management. *Int. J. Oper. Prod. Manag.* **2001**, *21*, 1326–1342. [CrossRef]

60. Windisch, W.; Freidel, K.; Schucher, B.; Baumann, H.; Wiebel, M.; Matthys, H.; Petermann, F. The Severe Respiratory Insufficiency (SRI) Questionnaire: A specific measure of health-related quality of life in patients receiving home mechanical ventilation. *J. Clin. Epidemiol.* **2003**, *56*, 752–759. [CrossRef]

61. Karamat, J.; Shurong, T.; Ahmad, N.; Afridi, S.; Khan, S.; Mahmood, K. Promoting healthcare sustainability in developing countries: Analysis of knowledge management drivers in public and private hospitals of Pakistan. *Int. J. Environ. Res. Public Health* **2019**, *16*, 508. [CrossRef]

62. Stefan, S.C.; Popa, I.; Dobrin, C.O. Towards a model of sustainable competitiveness of health organizations. *Sustainability* **2016**, *8*, 464. [CrossRef]

63. Djukic, A.; Marić, J. Towards Socially sustainable Healthcare Facilities – the Role of Evidence-based Design in Regeneration of Existing Hospitals in Serbia. *Procedia Environ. Sci.* **2017**, *38*, 256–263. [CrossRef]
64. Capolongo, S.; Gola, M.; Di Noia, M.; Nickolova, M.; Nachiero, D.; Rebecchi, A.; Settimo, G.; Vittori, G.; Buffoli, M. Social sustainability in healthcare facilities: A rating tool for analysing and improving social aspects in environments of care. *Annali Dell'Istituto Superiore Sanita* **2016**, *52*, 15–23.
65. Hall, R.; Andriani, P. Managing knowledge associated with innovation. *J. Bus. Res.* **2003**, *56*, 145–152. [CrossRef]
66. Abubakar, A.M.; Elrehail, H.; Alatailat, M.A.; Elçi, A. Knowledge management, decision-making style and organizational performance. *J. Innov. Knowl.* **2019**, *4*, 104–114. [CrossRef]
67. Tursunbayeva, A. Human resource technology disruptions and their implications for human resources management in healthcare organizations. *Bmc Health Serv. Res.* **2019**, *19*, 268. [CrossRef] [PubMed]
68. Singer, S.; Benzer, J.; Hamdan, S. Improving health care quality and safety: The role of collective learning. *J. Healthc. Leadersh.* **2015**, *7*, 91–107. [CrossRef] [PubMed]
69. Levitt, B.; March, J.G. Organizational Learning. *Annu. Rev. Sociol.* **1988**, *14*, 319–338. [CrossRef]
70. Khalil, C.; Dudezert, A. Knowledge management systems, Autonomy and control: How to regulate? A case-study in an industrial company. In *Materiality, Rules and Regulation*; Palgrave Macmillan: London, UK, 2015; pp. 143–153.
71. Gagné, M.; Tian, A.W.; Soo, C.; Zhang, B.; Ho, K.S.B.; Hosszu, K. Different motivations for knowledge sharing and hiding: The role of motivating work design. *J. Organ. Behav.* **2019**, *40*, 783–799. [CrossRef]
72. Steyn, P.D.; Du Toit, A.S.A. Perceptions on the use of a corporate business incubator to enhance knowledge management at Eskom. *S. Afr. J. Econ. Manag. Sci.* **2013**, *10*, 33–50. [CrossRef]
73. Forcada, N.; Fuertes, A.; Gangolells, M.; Casals, M.; Macarulla, M. Knowledge management perceptions in construction and design companies. *Autom. Constr.* **2013**, *29*, 83–91. [CrossRef]
74. McAdam, R. SME and large organisation perceptions of knowledge management: Comparisons and contrasts. *J. Knowl. Manag.* **2001**, *5*, 231–241. [CrossRef]
75. Harder, M. How Do Rewards and Management Styles Influence the Motivation to Share Knowledge? 2008. Available online: https://ssrn.com/abstract=1098881 (accessed on 31 March 2020).
76. Singh Sanjay, K. Role of leadership in knowledge management: A study. *J. Knowl. Manag.* **2008**, *12*, 3–15. [CrossRef]
77. Zheng, W.; Yang, B.; McLean, G.N. Linking organizational culture, structure, strategy, and organizational effectiveness: Mediating role of knowledge management. *J. Bus. Res.* **2010**, *63*, 763–771. [CrossRef]
78. Dorothy, L.; Maryam, A.; Timothy, K. The Role of Culture in Knowledge Management: A Case Study of Two Global Firms. *Int. J. E-Collab.* **2006**, *2*, 17–40. [CrossRef]
79. Lai, J.-Y. How reward, computer self-efficacy, and perceived power security affect knowledge management systems success: An empirical investigation in high-tech companies. *J. Am. Soc. Inf. Sci. Technol.* **2009**, *60*, 332–347. [CrossRef]
80. Choo, A.S.; Linderman, K.W.; Schroeder, R.G. Method and psychological effects on learning behaviors and knowledge creation in quality improvement projects. *Manag. Sci.* **2007**, *53*, 437–450. [CrossRef]
81. Kitsios, F.; Kamariotou, M.; Talias, M.A. Corporate Sustainability Strategies and Decision Support Methods: A Bibliometric Analysis. *Sustainability* **2020**, *12*, 521. [CrossRef]
82. Yu, C.; Zhang, Z.; Lin, C.; Wu, Y.J. Knowledge creation process and sustainable competitive advantage: The role of technological innovation capabilities. *Sustainability* **2017**, *9*, 2280. [CrossRef]
83. Parveen, F.; Jaafar Noor, I.; Ainin, S. Social media's impact on organizational performance and entrepreneurial orientation in organizations. *Manag. Decis.* **2016**, *54*, 2208–2234. [CrossRef]

*Article*

# Environmental Sustainability Assessment of Spatial Entities with Anthropogenic Activities-Evaluation of Existing Methods

**Despoina Aktsoglou * and Georgios Gaidajis**

Laboratory of Environmental Management and Industrial Ecology, Department of Production Engineering and Management, Democritus University of Thrace, 67100 Xanthi, Greece; geogai@pme.duth.gr
* Correspondence: daktsog@pme.duth.gr; Tel.: +30-25410-79356

Received: 19 February 2020; Accepted: 27 March 2020; Published: 29 March 2020

**Abstract:** The aim of this study is to evaluate the existing available methods that can potentially be applied to anthropogenic spatial entities to access their environmental sustainability. The paper offers an overview of existing sustainability assessment methods, discusses their adequacy, and evaluates their efficiency to assess the environmental sustainability of anthropogenic spatial entities. From a list of sixty-one (61) sustainability assessment methods for spatial entities with anthropogenic activities that had been identified and examined, thirteen (13) methods were selected to be assessed based on specific exclusion criteria set. The thirteen methods were further classified into four categories, namely, (1) Indicators/Indices, (2) Resource Availability Assessment, (3) Material and Energy Flow Analysis, and (4) Life-Cycle Assessment, and then these methods were evaluated using specific evaluation criteria. The "Resource Availability Assessment" category, and particularly the "Ecological Footprint" method, was indicated as the most appropriate method to assess the environmental sustainability of anthropogenic spatial entities.

**Keywords:** environmental sustainability; sustainable development; environmental assessment; anthropogenic spatial entities

## 1. Introduction

The growth of the human population enhances the extensive use and consumption of materials, while the existing lifestyle and established consumer patterns lead the planet and its population to an unstable situation without possible reversal [1]. In other words, the rate of use and consumption of resources and material, especially in developed countries, cannot be maintained without causing significant damage to the environment, society, and economy. Therefore, keeping anthropogenic systems sustainable is of crucial importance; but how do we define, maintain, and finally measure sustainability?

According to Graymore et al. [2], "sustainability is an essential goal for planning and natural resource management at all spatial scales", as it requires the population to live within the limits of the supporting systems, ensuring equitable sharing of resources and opportunities for this and future generations [3].

Moreover, the measurement and the assessment of sustainability is a particularly complex process due to the wide range of issues and the complexity of the systems involved. As mentioned by Gasparatos et al. [4], "the sustainability assessment does the difficult job to discover, study and suggest solutions for a large and heterogeneous set of issues that concern the stakeholders and extend to different spatial and temporal scales".

According to Ness et al. [5], "The purpose of sustainability assessment is to provide decision-makers with an evaluation of global to local integrated nature-society systems in short and long term perspectives in order to assist them to determine which actions should or should not be taken in an attempt to make society sustainable". It is clear from the above definition that sustainability assessment mainly evolves as a decision-making tool [6].

During the last decades, strategic planning and natural resource management are increasingly attempted at anthropogenic spatial entity scale, which refers to the spatial scale below a nation and usually includes a province or a municipality or parts of them that are under a specific management scheme or other specific authority (ex. National park). This spatial scale is the most appropriate for this purpose as it is "at this scale where ecological functioning and human activities most intensely interact" [7] and "the governance for planning, coordinating and assessing actions towards sustainable development" is effective at the specific scale [8].

There are hundreds of sustainability assessment methods [9]. Several studies [4,5,9–14] have attempted to summarize relevant methods. However, "we are still far from agreeing on how to define, plan and measure the progress towards sustainability" [10].

Most of the terms of sustainability are associated principally with the environment [15], since environmental protection is essential for sustainable development [16]. Therefore, the environmental aspects of sustainability are at the forefront [14], and the concept of environmental sustainability assessment integrates the environmental component in the decision-making process, providing deeper and more formalized knowledge [13]. Angelakoglou and Gaidajis [14] define an environmental sustainability assessment method as "a method which can provide quantitative information that can potentially help to assess the environmental sustainability".

Considering the above, a literature review of available sustainability assessment methods at anthropogenic spatial entity scale was implemented by the authors of this article, in order to assess the methods' adequacy. The literature reviews include (1) the methodology of the review and a brief description of the selected methods, (2) the criteria of the assessment and the reasoning behind their selection, and finally (3) the results of the methods' assessment.

## 2. Methodology of Literature Search

The literature search was implemented utilizing appropriate keywords relevant to the notions of sustainability, sustainability assessment, regional/local environmental sustainability assessment, etc., to specific search engines. As a result, articles with similar themes that had been published in international scientific journals were selected.

The methods emerged from the initial "filtering process" were thoroughly analyzed. Due to the wide range of the particular research area, a set of exclusion criteria was set and several methods were eliminated based on these criteria. In Table 1, the exclusion criteria and the excluded methods are summarized.

**Table 1.** Exclusion criteria of initially selected methods.

|  | Exclusion Criterion | Methods Excluded and a Brief Description of their Exclusion | References |
|---|---|---|---|
| 1 | Proven quality of references | (1) The method "Two Synthetic Environmental Indices" was excluded due to insufficient data for its analysis. | [11] |
| 2 | Ability to simultaneously evaluate various anthropogenic activities | (2) The method "Sustainability/Environmental Rating Systems" was excluded because it is focused on the evaluation of Construction Industry. | [9] |
|  |  | (3) The method "Environmental Performance Index" was excluded because it is focused on the evaluation of human health. | [11] |

**Table 1.** *Cont.*

| | Exclusion Criterion | Methods Excluded and a Brief Description of their Exclusion | References |
|---|---|---|---|
| 3 | Holistic evaluation at local level (region, municipality) | Methods 4 to 19 were excluded because they are developed to assess specific projects: | |
| | | (4) Environmental Quality Index; | [11] |
| | | (5) Cost–Benefit Analysis; | [5,9,10,17] |
| | | (6) Multicriteria Analysis; | [5,9,10,17] |
| | | (7) Full Cost Accounting; | [10] |
| | | (8) Sustainability Assessment Modeling; | [10] |
| | | (9) Environmental, Social, and Economic Impact Analysis; | [9] |
| | | (10) Analysis Network Process; | [9] |
| | | (11) Environmental Impact Assessment; | [5,17] |
| | | (12)EU Sustainability Impact Assessment; | [17] |
| | | (13)Strategic Environmental Assessment; | [9,17] |
| | | (14) Material Intensity per Service Unit; | [5] |
| | | (15) Risk Analysis; | [5] |
| | | (16) Conceptual Modeling; | [5] |
| | | (17) System Dynamics; | [5] |
| | | (18) Uncertainty Analysis; | [5] |
| | | (19) Vulnerability analysis. | [5] |
| | | Methods 20 to 29 were excluded because they are only implemented on a national scale: | |
| | | (20) UNCSD 58; | [5] |
| | | (21) Sustainable National Income; | [5] |
| | | (22) Adjusted Net Saving (Genuine Saving); | [5] |
| | | (23) Wellbeing Index; | [5] |
| | | (24) Genuine Progress Indicator and Index of Sustainable Economic Welfare; | [5,17] |
| | | (25) Human Development Index; | [5,17] |
| | | (26) Environmental Sustainability Index; | [11] |
| | | (27) Environment Sustainability Index; | [11] |
| | | (28) Environmental Policy Performance Indicator; | [11] |
| | | (29) Ecosystem Health Assessment. | [2] |
| 4 | Focus on the sustainable aspect of environmental performance | Methods 30 to 39 were excluded because they mainly assess public concern about the environmental impact of projects or activities: | |
| | | (30) Concern about Environmental Problems; | [11] |
| | | (31) Index of Environmental Friendliness; | [11] |
| | | (32) Environmental Vulnerability Index; | [11] |
| | | (33) Market prices; | [10] |
| | | (34) Benefit transfer; | [10] |
| | | (35) Choice modeling; | [10] |
| | | (36) Hedonic pricing; | [9,10] |
| | | (37) Travel Cost Method; | [9,10] |
| | | (38) Contingent Valuation Method; | [9,10] |
| | | (39) Community Impact Evaluation. | [9] |
| 5 | Basic idea analysis (when it comes to a family of methods resulting from the extension, improvement, or segmentation of an original method) | Methods 40 to 47 were excluded because they belong to the same family of methods with the method "Ecological Footprint": | |
| | | (40) Carrying Capacity; | [18] |
| | | (41) Natural Resource Availability; | [19] |
| | | (42) Carbon Footprint; | [12] |
| | | (43) Fossil Fuel Sustainability Index; | [11] |
| | | (44) Green Gas Inventory; | [20] |
| | | (45) Eco-Index Methodology; | [21] |
| | | (46) Sustainable Process Index; | [22] |
| | | (47) Energy Footprint. | [23] |
| | | (48) The method "Ecological Network Analysis" was excluded because it is developed in the same basic idea with the method "Physical Input–Output Tables". | [13] |

To sum up the above, from a list of sixty-one (61) examined methods that assess environmental sustainability of a spatial entity with anthropogenic activities, forty-eight (48) were excluded and thirteen (13) methods were finally selected (Table 3). Although not exhaustive, this list is, in the authors' opinion, extremely satisfactory.

In order to facilitate their analysis, the thirteen selected methods were classified based on their particular characteristics and their basic idea in four (4) categories, namely, "Indicators/Indices", "Resource Availability Assessment", "Material and Energy Flow Analysis", and "Life-Cycle Assessment" (see Table 3). The specific categorization is based on the categorization proposed by Angelakoglou and Gaidajis [14] and is widely acceptable in the economic sectors [24].

**Table 2.** The categorization of the selected methods and their brief description.

| | Method | Description of Method |
|---|---|---|
| | *Indicators/Indices* | |
| 1.1 | Sustainable Development Indicators (SDIs) | The SDIs consist a specific range of indicators for sustainable development, which have been developed according to the Driving Force-pressure-state-impact-response (DPSIR) framework, in order to support the stakeholders to evaluate the effectiveness of the policy on the way to sustainable development [8,25]. |
| 1.2 | Environmental Pressure Indicators (EPIs) | The EPIs have been developed by Eurostat and consist of sixty (60) indicators, six (6) for each of the ten (10) policy areas according to the 5th Environmental Action Program [26]. |
| 1.3 | The Dashboard of Sustainability (DoS) | The DoS method is a mathematical and graphical tool, designed to integrate the complex implications of sustainability and to support the decision-making process at nation level with the generation of brief evaluations. The tool evaluates indicators relative to environmental protection, economic development and social improvement [27]. |
| 1.4 | Quality of Life (QoL) | The method is based on trends and conditions related to indicators such as crime, participation in cultural activities, health, education, income, unemployment, water quality, air pollution and the proportion of unstructured areas, assessing the relevant areas of "Quality of Life" [28]. |
| | *Resource Availability Assessment* | |
| 2.1 | Ecological Footprint (EF) | The Ecological Footprint [29] expresses «the theoretical area (in global hectares) which is used by humans to produce the resources they consume, and to absorb the waste generated (including CO2 emissions from energy consumption)". |
| 2.2 | Water Footprint (WF) | The WF method [30] is based on the calculation of the "total volume of fresh water required to meet the direct and/or indirect needs of the entity under consideration». |
| 2.3 | Wellbeing Assessment (WA) | The method has been developed by the World Conservation Union for its use at various levels of spatial entities. This is a holistic approach to evaluating sustainability using plenty of indicators, covering all parts of the entity [31]. |
| | *Material and Energy Flow Analysis* | |
| 3.1 | Material Flow Analysis (MFA) | The MFA method is used to determine the material and energy balance of an entity. This method is mostly implemented at national level due to the easy access to the required data and the existence of a methodological framework developed by Eurostat [32]. |
| 3.2 | Substance Flow Analysis (SFA) | The SFA method [33,34] aims at the control of the flows of substances (chemicals and/or compounds) that contain significant levels of concern about their impact on ecological and human health in their production and use. |
| 3.3 | Physical Input–Output Tables (PIOT) | The methods study the direct and indirect flows of an entity, applying the principle of mass conservation. Especially, the PIOT considers the environment as a source of raw materials and a "sink" of the residuals of the production processes of an economy [35]. |

**Table 3.** The categorization of the selected methods and their brief description.

| | Method | Description of Method |
|---|---|---|
| | | *Material and Energy Flow Analysis* |
| 3.4 | Emergy Analysis (EMA) | The EMA [36] method is used to measure "the work previously done by nature and/or man that contributed to the realization of a product or service". The energy required is expressed as the sum of the individual types of energy, expressed as a final form of energy, usually solar energy (expressed in emjoules). |
| 3.5 | Exergy Analysis (EXA) | The EXA method [37] is used to measure "the maximum equivalent mechanical work that can be derived from a system when it tends to a thermodynamic equilibrium state compared with a reference system. The application of the method allows the definition and evaluation of the flows that contain exergy (so it is possible to be further exploited) or where it is completely lost (so it has to be further analyzed) [38]. |
| | | *Life-Cycle Assessment* |
| 4.1 | Life-Cycle Sustainability Analysis (LCSA) | The Life-Cycle Assessment (LCA) is mostly applied to assess and evaluate product sustainability. However, Guinée et al. [39] proposed a new framework, namely Life-Cycle Sustainability Analysis, which extends the scope of analysis from product-related to economic issues, including an intermediate level, like anthropogenic spatial entities. |

## 3. Assessment of Selected Methods

The scope of the evaluation of the selected 13 methods is the selection of the most appropriate method for assessing the environmental sustainability of anthropogenic spatial entities. Furthermore, the evaluation aims to document whether there is a necessity to develop new methods (or/and improving the available ones). Within that framework, the thirteen (13) methods, mentioned above, were assessed using predefined evaluation criteria.

Each method has specific advantages and disadvantages associated with its particular characteristics. However, the general features that an efficient method of evaluating sustainability must, as described (a) in the literature [4,24,40], (b) in the definition of sustainability, and (c) as defined by the requirements of the end-user, i.e., addressing issues related to its applicability and usability [2], do the following [14]:

- Take into account the particular spatial characteristics of the anthropogenic spatial entity under consideration and be able to evaluate its progress over time.
- Enhance decision-making on promoting desired actions that improve sustainability and the possibility of adding new activities within the administrative boundaries of the spatial entity.
- Establish sustainability reference benchmarks.
- Ensure an adequate balance between the level of complexity and the coverage of key sustainability issues.
- Emphasize clearly and minimize assumptions and weaknesses that arise during its development.
- Be appropriate for comparisons.
- Ensure that it can be modified to incorporate other aspects of sustainability or that it can be combined with other methods to implement a more comprehensive assessment.

Therefore, based on the above characteristics that a sustainability assessment method should have, five criteria listed in Table 4 were selected to evaluate the 13 methods selected. More specifically, the methods were evaluated with the application of a zero to three scale, where zero indicates the minimum and three the maximum satisfaction of the criterion. For each criterion, three questions were developed based on the characteristics previously mentioned. If the examined method satisfies the issue raised by the question (answer to the question "Yes"), then it receives one point. Alternatively, no point is attributed to the method (answer to the question 'No'). Therefore, the maximum score for each method can be 15 points (five criteria and three questions for each criterion). Higher score implies higher satisfaction of all evaluation criteria, and consequently higher suitability for assessing environmental sustainability.

At this point, it should be mentioned that the authors recognize a certain degree of subjectivity in the evaluation process, as the satisfaction of the criteria for each of the 13 methods was judged not by a focus group but according to the analysis of the methods made by the authors and the corresponding literature [2], which referred to similar criteria for evaluating a subset of the specific methods. Among the future objectives of the authors is the more objective evaluation of the aforementioned methods with the assistance of an experts group.

## 4. Results and Discussion

The evaluation of the methods is presented in detail in Table 4.

The methods belonging to the Second Category, i.e., "Resource Availability Assessment Methods", exhibit the highest average score in terms of satisfaction of all the evaluation criteria (11 points out of the maximum 15 points). More specifically, the Method 2.1 "Ecological Footprint (EF)" of the Second Category of Methods exhibits the highest degree of satisfaction for all criteria (13 points out of the maximum 15 points), followed by the Method 2.3 "Wellbeing Assessment (WA)" of the same Category (11/15), the Method 1.1 "Sustainable Development Indicator Method", and the Method 1.4 "Quality of Life" of the First Category of Methods. The EF method was the highest rated method with a total score of 13/15. Particularly, the Ecological Footprint has the ability to assess environmental sustainability holistically (Criterion 1 score: 3/3), is easy to use (Criterion 4 score: 3/3), and integrates spatial and temporal characteristics of the anthropogenic spatial entity under study (Criterion 5 score: 3/3). All Methods of Category 2 satisfy to a maximum degree Criterion 5 ("Integration of spatial/temporal characteristics") and extremely well Criterion 1 ("Ability to access environmental sustainability holistically"), satisfying therefore the spatial and temporal expansion of assessment and the holistic approach of the assessment.

The assessment of all categories and methods identified some issues that need improvement, such as the ability of methods to incorporate new activities within the environmental boundaries of the anthropogenic spatial entity under study (Q.2.2: Score 0). Moreover, few methods include specific thresholds in order to evaluate quantitatively whether the performance is sustainable (Q.3.2.: Score 4).

An ideal method of evaluating sustainability should take into account all the parameters simultaneously [5]. However, the development of such a method, due to the nature of the concept of sustainability, involves a high degree of complexity, which in the case of anthropogenic spatial entities translates into high costs and time.

A new methodological framework in assessing environmental sustainability for spatial entities with anthropogenic activities that will be able to take advantages of the existing methods and deal with their challenges and their drawbacks should be developed. The scope of the framework will be to constitute a tool of particular importance for the decision-making process at the local level [14,17], giving answers to the issues highlighted above.

**Table 4.** Evaluation of the environmental sustainability assessment methods under examination.

| Evaluation Criteria | Categorization of Methods | | | | | | | | | | | | | Criterion Average | Question Score |
| | 1 Indicators/Indices | | | | 2 Resource Availability Assessment | | | 3 Material and Energy Flow Analysis | | | | | 4 Life-Cycle Assessment | | |
| | 1.1 SDI | 1.2 EPI | 1.3 DoS | 1.4 QoL | 2.1 EF | 2.2 WF | 2.3 WA | 3.1 MFA | 3.2 SFA | 3.3 PIOT | 3.4 EmA | 3.5 ExA | 4.1 LCSA | | |
|---|---|---|---|---|---|---|---|---|---|---|---|---|---|---|---|
| *Criterion 1: Ability to assess environmental sustainability holistically* | | | | | | | | | | | | | | 2.3 | |
| Q.1.1: Do methods assess more than one sector? | Y | Y | Y | Y | Y | Y | Y | Y | Y | Y | Y | Y | Y | | 13 |
| Q.1.2: Do methods assess an adequate number of environmental issues? | Y | Y | N | Y | Y | N | Y | Y | N | N | N | N | Y | | 7 |
| Q.1.3: Do methods promote energy and resource efficiency? | Y | N | N | N | Y | Y | Y | Y | Y | Y | Y | Y | Y | | 10 |
| **Score per criterion** | 3 | 2 | 1 | 2 | 3 | 2 | 3 | 3 | 2 | 2 | 2 | 2 | 3 | | |
| **Average Score of each category** | | | 2 | | | 2.7 | | | | 2.2 | | | 3 | | |
| *Criterion 2: Ability to help decision making* | | | | | | | | | | | | | | 1.3 | |
| Q.2.1: Can methods communicate their results to public? | Y | Y | Y | Y | Y | Y | Y | N | N | N | N | N | N | | 7 |
| Q.2.2: Can methods answer to the potential addition of a new activity; | N | N | N | N | N | N | N | N | N | N | N | N | N | | 0 |
| Q.2.3: Can methods identify specific environmental "hot spots" of the spatial entity? | Y | Y | Y | Y | Y | Y | Y | Y | Y | N | N | Y | N | | 10 |
| **Score per criterion** | 2 | 2 | 2 | 2 | 2 | 2 | 2 | 1 | 1 | 0 | 0 | 1 | 0 | | |
| **Average Score of each category** | | | 2 | | | 2 | | | | 0.6 | | | 0 | | |
| *Criterion 3: Potential for benchmarking* | | | | | | | | | | | | | | 1.3 | |
| Q.3.1: Can methods aggregate the results into single scores? | N | Y | N | N | Y | Y | Y | N | N | Y | Y | Y | Y | | 8 |
| Q.3.2: Do methods include specific thresholds/targets of sustainable performance? | Y | N | N | Y | Y | N | N | N | N | N | N | Y | N | | 4 |
| Q.3.3: Can methods be applied/updated to compare overall sustainability? | Y | N | Y | Y | N | N | Y | N | N | Y | N | N | N | | 5 |
| **Score per criterion** | 2 | 1 | 1 | 2 | 2 | 1 | 2 | 0 | 0 | 2 | 1 | 2 | 1 | | |
| **Average Score of each category** | | | 1.5 | | | 1.7 | | | | 1 | | | 1 | | |

**Table 4.** *Cont.*

| Evaluation Criteria | Categorization of Methods | | | | | | | | | | | | | Criterion Average | Question Score |
|---|---|---|---|---|---|---|---|---|---|---|---|---|---|---|---|
| | 1 Indicators/Indices | | | | 2 Resource Availability Assessment | | | 3 Material and Energy Flow Analysis | | | | | 4 Life-Cycle Assessment | | |
| | 1.1 SDI | 1.2 EPI | 1.3 DoS | 1.4 QoL | 2.1 EF | 2.2 WF | 2.3 WA | 3.1 MFA | 3.2 SFA | 3.3 PIOT | 3.4 EmA | 3.5 ExA | 4.1 LCSA | | |
| *Criterion 4:Applicability and ease of use* | | | | | | | | | | | | | | 1.7 | |
| Q.4.1: Can methods be easily applied by nonexperts? | Y | Y | Y | Y | Y | N | Y | N | N | N | N | N | N | | 6 |
| Q.4.2: Can methods be easily applied by local government (data/cost involved)? | Y | Y | Y | N | Y | N | N | Y | N | N | N | N | N | | 5 |
| Q.4.3: Do methods include clear guidelines of implementation (freely available)? | N | Y | Y | Y | Y | Y | Y | Y | Y | N | Y | Y | Y | | 11 |
| Score per criterion | 2 | 3 | 3 | 2 | 3 | 1 | 2 | 2 | 1 | 0 | 1 | 1 | 1 | | |
| Average Score of each category | | | 2.5 | | | 2 | | | | 1 | | | 1 | | |
| *Criterion 5:Integration of spatial and temporal characteristics* | | | | | | | | | | | | | | 2.1 | |
| Q.5.1: Do methods integrate physical and anthropogenic characteristics? | N | N | N | Y | Y | Y | N | Y | Y | Y | Y | Y | Y | | 9 |
| Q.5.2: Do methods assess environmental sustainability at local level? | Y | N | N | Y | Y | Y | Y | Y | Y | Y | Y | Y | N | | 10 |
| Q.5.3: Are methods able to evaluate progress over time? | Y | N | Y | Y | Y | Y | Y | N | N | N | N | Y | N | | 7 |
| Score per criterion | 2 | 0 | 1 | 3 | 3 | 3 | 3 | 2 | 2 | 2 | 2 | 3 | 1 | | |
| Average Score of each category | | | 1.5 | | | 3 | | | | 2.2 | | | 1 | | |
| *Total Score of each method* | 11 | 8 | 8 | 11 | 13 | 9 | 11 | 8 | 6 | 6 | 6 | 9 | 6 | | |
| *Average Score of each category* | | | 9.5 | | | 11 | | | | 7 | | | 6 | 8.6 | |
| Y: YES and N: NO | | | | | | | | | | | | | | | |

## 5. Conclusions

The scope of this article is to summarize and evaluate the available methods for assessing the environmental sustainability of spatial entities with anthropogenic activities (for example, municipalities, regions, protected areas, and national parks). To achieve this, a literature search of existing methods for assessing the environmental sustainability of anthropogenic spatial entities was carried out. From the total of 61 methods that were found and reviewed, 48 were excluded based on specific exclusion criteria related to the proven quality of the references, the ability to simultaneously evaluate different issues of anthropogenic activity, the holistic evaluation at local level, and the focus on the sustainable aspect of environmental performance.

The 13 finally selected methods were classified according to their particular characteristics into categories, and the methods were evaluated with specific criteria such as the ability to holistically assess environmental sustainability, the ability to support the decision-making process, the ability to perform benchmarking, the applicability and ease of use, and the integration of spatial and temporal characteristics of the entity under study.

The evaluation of available methods, based on the above criteria, indicated that the methods of the "Resource Availability Assessment" category satisfy to a significant extent all the criteria set, whereas a method of the above category, and more specifically the "Ecological Footprint Method" satisfied to a maximum degree the criteria set and was considered therefore as the most appropriate method for assessing the environmental sustainability of a spatial entity with anthropogenic activities.

Further research is necessary in order to indentify methods that are able to assess economic and social sustainability, hence a more holistic inventory of sustainability assessment methods for spatial entities with anthropogenic activities will be developed. The evaluation process followed provides a basis to address the effectiveness of the methods, to develop evaluation criteria, and to select the most suitable method in any case. Future work by the authors will be undertaken to improve the above-mentioned issues.

Results presented in this study can be further utilized to improve current methods and/or develop new ones. The authors are planning to take advantage of these findings and develop an environmental sustainability assessment framework for spatial entities that will be based on the most appropriate method as mentioned, thus the "Ecological Footprint", which will be able to deal with the most shortcomings of the existing methods.

**Author Contributions:** All authors contributed equally to the text. All authors have read and agreed to the published version of the manuscript.

**Funding:** This research received no external funding.

**Conflicts of Interest:** The authors declare no conflict of interest.

## References

1. Allenby, B.R.; Graedel, T.E. *Industrial Ecology*, 2nd ed.; Pearson Prentice Hall: Upper Saddle River, NJ, USA, 2002; pp. 25–27.
2. Graymore, M.L.M.; Sipe, N.G.; Rickson, R.E. Regional sustainability: How useful are current tools of sustainability assessment at the regional scale? *Ecol. Econ.* **2008**, *67*, 362–372. [CrossRef]
3. Graymore, M. The Journey to Sustainability: Small Regions, Sustainable Carrying Capacity and Sustainability Assessment Methods. Ph.D. Thesis, Australian School of Environmental Studies, Griffith University, Brisbane, Australia, 2005.
4. Gasparatos, A.; El-Haram, M.; Horner, M. A critical review of reductionist approaches for assessing the progress towards sustainability. *Environ. Impact Assess. Rev.* **2008**, *289*, 286–311. [CrossRef]
5. Ness, B.; Urbel_Piirsalu, E.; Andeberg, S.; Olsson, L. Categorising tools FOS sustainability assessment. *Ecol. Econ.* **2007**, *60*, 498–508. [CrossRef]
6. Pope, J.; Annandale, D.; Morrison Sanders, A. Conceptualizing sustainability assessment. *Environ. Impact Assess. Rev.* **2004**, *24*, 595–616. [CrossRef]

7.  Coehlo, P.; Mascarenhas, A.; Vaz, P.; Beja, I.; Dores, A.; Ramos, T.B. A Methodological Framework for Indicators of Sustainable Development at the Regional Scale: The Case of the Portuguese Region (Algarve). In Proceedings of the 12th Annual International Sustainable Development Research Conference 2006, Hong Kong, China, 6–8 April 2006; University of Hong Kong, in Association with ERP Environment: Hong Kong, China, 2006.

8.  Mascarenhas, A.; Coelho, P.; Subtil, E.; Ramos, T.B. The role of common local indicators in regional sustainability assessment. *Ecol. Indic.* **2010**, *10*, 646–656. [CrossRef]

9.  Poveda, C.A.; Lipsett, M.G. A review of sustainability assessment and sustainability/environmental rating systems and credit weighting tools. *J. Sustain. Dev.* **2011**, *4*, 36–55. [CrossRef]

10.  Gasparatos, A.; Scolobig, A. Choosing the most appropriate sustainability assessment tool. *Ecol. Econ.* **2012**, *80*, 1–7. [CrossRef]

11.  Singh, R.K.; Murty, H.R.; Gupta, S.K.; Dikshit, A.K. An overview of sustainability assessment methodologies. *Ecol. Indic.* **2012**, *15*, 281–299. [CrossRef]

12.  Cucek, L.; Klemes, J.J.; Kravanja, Z. A review of Footprint analysis tools for monitoring impacts on sustainability. *J. Clean. Prod.* **2012**, *34*, 9–20. [CrossRef]

13.  Loiseau, E.; Junqua, G.; Roux, P.H.; Bellon-Maurel, V. Environmental assessment of a territory: An overview of existing tools and methods. *J. Environ. Manag.* **2012**, *112*, 213–225. [CrossRef]

14.  Angelakoglou, K.; Gaidajis, G. A review of methods contributing to the assessment of the environmental sustainability of industrial systems. *J. Clean. Prod.* **2015**, *108*, 725–747. [CrossRef]

15.  Glavic, P.; Lukman, R. Review of sustainability terms and their definitions. *J. Clean. Prod.* **2007**, *15*, 1875–1885. [CrossRef]

16.  Collin, R.M.; Collin, R.W. *Encyclopedia of Sustainability*; ABC-CLIO/Greenwood Press: Santa Barbara, CA, USA, 2010; Volume 2, pp. 101–113.

17.  Sala, S.; Ciuffo, B.; Nijkamp, P. A systemic framework for sustainability assessment. *Ecol. Econ.* **2015**, *119*, 314–325. [CrossRef]

18.  Lane, M.C. The Carrying Capacity Imperative: Assessing Regional Carrying Capacity Methodologies for Sustainable Land-use Planning. In Proceedings of the 53rd Annual Meeting of the International Society for the Systems Sciences, Brisbane, Queensland, Australia, 12–17 July 2009; University of Queensland: Queensland, Australia, 2009.

19.  Graymore, M.L.; Sipe, N.G.; Rickson, R.E. Sustaining Human Carrying Capacity: A tool for regional sustainability assessment. *Ecol. Econ.* **2010**, *69*, 459–468. [CrossRef]

20.  Pulselli, F.; Ciampalini, F.; Leipert, C.; Tiezzi, E. Integrating methods for environmental sustainability: The SPIn-Eco Project in the Province of Siena (Italy). *J. Environ. Manag.* **2008**, *86*, 332–341. [CrossRef] [PubMed]

21.  Chambers, N.; Simmons, C.; Wackernagel, M. Chapter 9: Assessing Impact of organizations and Services. In *Sharing Natures's Interest: Ecological Footprints as an Indicator of Sustainability*; Earthscan: London, UK, 2000; p. 145.

22.  Kettl, K.H.; Niemetz, N.; Sandor, N.; Eder, M.; Heckl, I.; Narodoslawsky, M. Regional Optimizer (RegiOpt)-Sustainable energy technology network solution for regions. *Comput. Aided Chem. Eng.* **2011**, *29*, 1959–1963.

23.  Global Footprint Network (Glossary 2009). Available online: www.footprintnetwork.org (accessed on 3 February 2017).

24.  OECD 2009. Chapter 3: Tracking Performance: Indicators of Sustainable Manufacturing. In *Eco-innovation in Industry- Enabling Green Growth*; Organization for Economic Co-Operation and Development (OECD): Paris, France, 2009; pp. 95–144.

25.  Scipioni, A.; Mazzi, A.; Zuliani, F.; Mason, M. The ISO 14031 standard to guide the urban sustainability measurement process: An Italian experience. *J. Clean. Prod.* **2008**, *16*, 1247–1257. [CrossRef]

26.  Lammers, P.E.M.; Gilbert, A.J. *Towards Environmental Pressure Indicators for the EU: Indicator Definition*; EUROSTAT: Brussels, Belgium, 1999.

27.  Scipioni, A.; Mazzi, A.; Mason, M.; Manzardo, A. The Dashboard of Sustainability to measure the local urban sustainable development: The case study of Padua Municipality. *Ecol. Indic.* **2009**, *99*, 364–380. [CrossRef]

28.  Gatt, L. *Quality of Life in New Zealand's Six Largest Cities*; Auckland City Council: Auckland, New Zealand, 2001; pp. 1–122.

29. Wackernagel, M.; Rees, W.E. *Our Ecological Footprint: Reducing Human Impact on the Earth. New Catalyst Bioregional Series*; New Society Publishers: Gabriola Island, BC, Canada; Philadelphia, PA, USA, 1996; Volume 9, pp. 1–160.

30. Hoekstra, A.Y.; Chapagain, A.K.; Aldaya, M.M.; Mekonnen, M.M. *Water Footprint Manual—State of the Art 2009*; Water Footprint Network: Enschede, The Netherlands, 2009; pp. 1–127.

31. Guijt, I.; Moiseev, A.; Prescott-Allen, R. *Resource Kit for Sustainability Assessment: Part A*; IUCN-The World Conversation Union: Gland, Switzerland; Cambridge, UK, 2001.

32. Eurostat 2001. *Economy-wide Material Flow Accounts and Derived Indicators, A Methodological Guide*; Statistical Office of the European Union: Luxembourg, 2001.

33. Bruner, P.H. Substance flow analysis—A key tool for effective resource management. *J. Ind. Ecol.* **2012**, *16*, 293–295. [CrossRef]

34. Huang, C.L.; Vause, J.; Ma, H.W.; Yu, C.P. Using material/substance flow analysis to support sustainable development assessment: A literature review and outlook. *Resour. Conserv. Recycl.* **2012**, *68*, 104–116. [CrossRef]

35. Giljum, S.; Hubacek, K. Conceptual foundations and applications of physical input-output tables. In *Handbook of Input-Output Economics in Industrial Ecology, Eco-Efficiency in Industry and Science*; Suh, S., Ed.; University of Minnesota: Minneapolis, MN, USA, 2009; pp. 61–75.

36. Brown, M.; Ulgiati, S. Energy quality, emergy, and transformity: H.T. Odum's contributions to quantifying and understanding systems. *Ecol. Model.* **2004**, *178*, 201–213. [CrossRef]

37. Rosen, M.; Dincer, I. Exergy as the confluence of energy, environment and sustainable development. *Int. J. Exergy* **2001**, *1*, 3–13. [CrossRef]

38. Apaiah, R.K.; Linnemann, A.R.; van der Kooi, H.J. Exergy analysis: A tool to study the sustainability of food supply chains. *Food Res. Int.* **2006**, *39*, 1–11. [CrossRef]

39. Guinèe, J.B.; Heijungs, R.; Huppes, G.; Zamangni, A.; Masoni, P.; Buonamici, R.; Ekvall, T.; Rydberg, T. Life cycle assessment: Past, present, and future. *Environ. Sci. Technol.* **2011**, *45*, 90–96. [CrossRef]

40. Bond, A.J.; Morrison-Saunders, A. Re-evaluating sustainability assessment: Aligning the vision and the practice. *Environ. Impact Assess. Rev.* **2011**, *11*, 1–7. [CrossRef]

*Article*

# A Conceptual Framework to Evaluate the Environmental Sustainability Performance of Mining Industrial Facilities

**Komninos Angelakoglou * and Georgios Gaidajis**

Laboratory of Environmental Management and Industrial Ecology, Department of Production Engineering and Management, School of Engineering, Democritus University of Thrace, 67100 Xanthi, Greece; geogai@pme.duth.gr
* Correspondence: kangelak@pme.duth.gr; Tel.: +30-25410-79356

Received: 22 February 2020; Accepted: 6 March 2020; Published: 10 March 2020

**Abstract:** The aim of this study is to strengthen the capacity of mining industries to assess and improve their environmental sustainability performance through the introduction of a relevant framework. Specific assessment categories and respective indicators were selected according to predefined steps. Sustainability threshold values were identified for each indicator to enable the comparison of the facility's performance with a sustainability reference value. The application of the framework results in the extraction of an Environmental Sustainability Assessment of Mining Industries Index ($I_{ESAMI}$). The framework was applied to evaluate a mining facility in Greece, with a view to improve its applicability in parallel. The final score of environmental sustainability for the examined facility was 3.0 points ($I_{ESAMI}$ = 3.0 points), indicating significant room for improvement where the company should aim to further enhance its sustainability performance.

**Keywords:** sustainability assessment; index; sustainable mining; threshold values; sustainability indicators

---

## 1. Introduction

Business and industrial activity are being reformed to cope with the needs and challenges of sustainable development. Environmental responsibility is moving beyond being just a legal obligation—it also stands out as a good business practice through the expansion of markets and the improvement of sales [1]. Managers face the challenge to deliver better corporate sustainability strategies [2], whereas external agents are increasingly paying attention on the concept of sustainable development [3]. While the reasoning behind the need of companies to contribute to sustainable development has been extensively analyzed in literature, relatively less progress has been made in developing integrated approaches for sustainability evaluations [4]. As a result, during the last few years, a significant number of studies attempted to strengthen sustainability evaluations at the corporate level by providing both generic and specific recommendations and guidelines [5–8].

This field of research is especially critical for mining industries which are inherently disruptive to the environment. The interdependence of mining activity and sustainable development is reflected on relevant initiatives (such as the Global Mining Initiative, Towards Sustainable Mining commitment of the Mining Association of Canada, the Sustainable Mining Initiative by Federation of Indian Mineral Industries, etc.) that attempt to set a common framework to promote responsible mining. In parallel, there is a growing interest among the academic community regarding issues associated with mining and sustainable development [9]. Mining industries usually exhibit commitment to the environment through the adoption of environmentally responsible practices [10] and sustainability reporting, with a view to balance negative impacts and reduce opposition by local communities [11].

Corporate sustainability reporting, especially through the adoption of the Global Reporting Initiative (GRI) guidelines, is now considered a common practice for measuring, reporting, and comparing sustainability performance [12].

Despite this background, there are many criticisms regarding the relationship between mining activity and environmental sustainability, and the literature argues that there is still significant room for improvement. Although social and environmental reporting is becoming increasingly sophisticated in the mining industry, there is a lack of uniformity that hinders the progress toward measuring corporate social responsibility and sustainable goals [13]. Belkhir et al. [14] assessed whether the GRI impacts environmental sustainability in terms of $CO_2$ emissions and found no correlation between GRI-reporting and sustainability improvement, a result that calls for the re-examination of the effectiveness of corporate social responsibility strategies. Another issue that has been raised in the literature is the ability to compare the sustainability performance of firms, even from the same sector, which remains problematic [12]. The mining corporations' framework for measuring and reporting sustainability progress needs to be changed in order to reflect more accurate and meaningful information [15]. Tost et al. [16] argue that the mining industry is at risk of failing societal expectations regarding climate change and falling behind from other industries on natural capital considerations.

According to Lopez et al. [6], research on corporate sustainability performance need to focus on the standardization of measurements, whereas stakeholders should apply indicators measured at wider scales. It is necessary to develop and implement effective tools and methodologies to support decision making, taking into account the complexity of sustainability problems [17]. One of the biggest challenges at the moment is closing the gap between theory and practice. Despite the fact that many researchers have been working on developing sustainability assessment methods and tools, relatively few of these are applied by manufacturing companies [18].

Serving this challenge, the aim of the specific study is to provide a practical framework that is able to strengthen the evaluation and monitoring of the environmental sustainability of mining industrial facilities. The proposed framework capitalizes the results and proposals of an extensive literature review we have conducted in a previous work [19]. In this work, 48 methods were identified and clustered into six categories (individual/set of indicators, composite indices, socially responsible investment indices, material and energy flow analysis, life cycle analysis, and environmental accounting), extracting in parallel their key attributes. These categories were further evaluated based on five criteria—(a) ability to promote actions of improvement, (b) ability to help decision making, (c) potential for benchmarking, (d) applicability and ease of use, and (e) integration of wider spatial and temporal characteristics. This analysis highlighted key recommendations that can help improve the efficiency and applicability of environmental sustainability evaluations of industrial systems. More specifically, we found out that an industrial facility should be assessed both in terms of performance and concern and provide environmental sustainability threshold values for every indicator applied. An effective environmental sustainability assessment method should take into account the spatial characteristics of the examined industrial systems and assess the progress towards sustainability over time.

This paper consolidates key findings from our previous work [20–26] (i.e., proposed environmental sustainability assessment categories—building upon the principles of industrial ecology, criteria for selecting indicators to assess industrial facilities, a proposed normalization method combining categorical scale and distance to a reference), but takes one big step further by integrating all information into an applicable framework that focuses on mining industry and that was tested in a mining facility in Greece to examine its utility (closing the gap between theory and practice).

## 2. Method

The proposed framework was developed building upon a standard methodology for constructing composite indicators. A theory-driven (top-down) over data-driven (bottom-up) approach was adopted to ensure that environmental sustainability will be efficiently assessed through the selection of proper indicators. The methodology applied consists of 10 steps (Figure 1) that were defined by taking into account available guidelines for the construction and use of composite indicators [27] and the recommendations from the analysis of 48 sustainability assessment methods as described above [19].

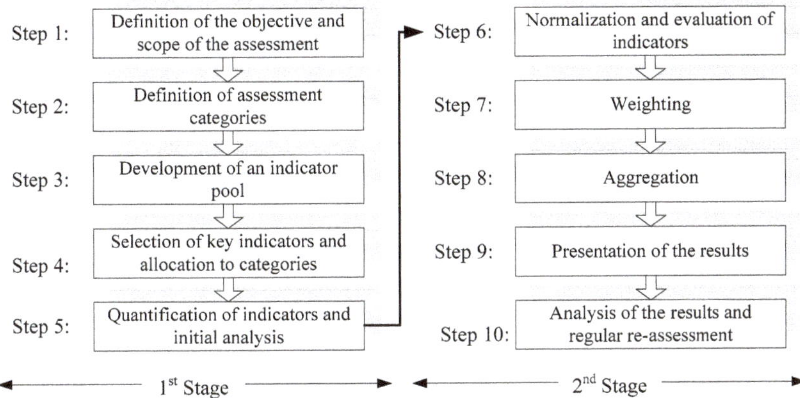


**Figure 1.** Methodology followed to construct the proposed assessment framework.

The 10 steps are divided into two stages of implementation in order to accommodate the gradual application of the proposed framework. The first stage comprises steps 1–5 and includes the minimum actions to be conducted by a mining industry on a facility level, who wish to acquire an initial overview of its performance related to environmental sustainability aspects. This stage is addressed to industries that have little time and resources at their disposal. The implementation of the first stage provides the data required for the assessment of the environmental sustainability. The second stage comprises the steps 6–10 and includes all actions required for the effective assessment of the environmental sustainability of the facility. The implementation of the second stage results in the development of a final comprehensive environmental score and the identification of environmental "hot spots" that call for improvement.

In the following sections, the 10 steps comprising the conceptual framework are presented. Each step contains both generic implementation guidelines and more specific instructions. Generic implementation guidelines can be used to improve existing assessment methods applied by industries, whereas specific instructions provide all the information necessary for the direct assessment of environmental sustainability. As a result, the utility of the proposed framework is twofold. It can act both as a path to develop new assessment methods (or improve existing ones) and as a ready to be used tool.

### 2.1. Definition of the Objective and Scope of the Assessment

The first step taken in order to develop the framework was to define the objective of the assessment. Indicative objectives include the identification of weaknesses and the development of improvement measures, comparison with other industrial facilities, monitoring the performance of industrial facility over time, and enhancing sustainability reports. Additionally it is necessary to define the scope of the assessment. The reference year and the boundaries of the industrial facility under examination must be selected in order to serve the objectives of the assessment. Significant factors to consider among others are the availability of data, the budget and time available to perform the assessment. The identification of the target groups to whom the results of the assessment will be addressed can further increase the effectiveness of the framework. Ecological organizations, local residents, and control bodies favor methods that follow more eco-centric approaches.

The objective of the proposed framework is to enable the evaluation and monitoring of the environmental sustainability performance of the examined facility, thus strengthening corporate decision making on a higher level. This choice is based on the fact that key strategic decisions on sustainability-related issues are mostly taken on an upper-management level. The full implementation of the proposed framework is addressed primarily to mining industrial facilities that are already in operation. The scope of the assessment includes the whole supply chain from raw material extraction to final processing (cradle to gate approach). This approach is essential to capture environmental impacts deriving from different life cycle stages of the operation of the examined facility. The time reference for the assessment is one year (annual implementation/results) to serve compatibility with annual reports and facilitate data mining.

The reason for conducting the assessment on the industrial facility level (rather than corporate level) is that it enables the identification of unsustainable industrial processes/practices at the source while taking into account specific spatial characteristics of the facility under examination. Consequently, the results of the assessment are expected to better facilitate decision making and the identification of amelioration actions. The proposed methodology is focusing at the moment only on the environmental aspect of sustainability but can be adapted to include economical and social aspects.

*2.2. Definition of Assessment Categories*

The next step was to identify various sub-categories that compose the examined phenomenon, which in our case is environmental sustainability [27]. In other words, the parameters need to be considered for efficiently assessing the environmental sustainability of the system under examination had to be identified. Sustainability assessments should be based on certain principles to enhance their effectiveness [28]. Additionally, every sub-category must have a clear purpose that serves the goal of the evaluation. To ensure that the most important issues in terms of environmental sustainability will be assessed, the principles of industrial ecology (IE) were utilized to develop the specific framework. The authors have summarized and examined the principles of IE in a previous work [21]. Additionally, the environmental assessment categories proposed by the Global Reporting Initiative [15] and further recommendations by mining experts were taken into account. As a result of this process, a list of eight key categories to assess the environmental sustainability was developed (Table 1).

Table 1. Proposed environmental sustainability assessment categories.

| S/N | Name of the Assessment Category |
|:---:|:---:|
| 1 | Mineral resources and materials |
| 2 | Emissions and waste |
| 3 | Energy and water |
| 4 | Suppliers and environmental performance |
| 5 | Biodiversity |
| 6 | Land use and rehabilitation |
| 7 | Impact on the environmental and climate change |
| 8 | Impact on human health |

*2.3. Development of an Indicator Pool*

Following the definition of assessment categories, an indicator pool was developed including indicators that can potentially be utilized to assess the eight environmental sustainability categories. A filtering procedure according to predefined criteria was undertaken to narrow down the vast number of potential indicators that can be included in the pool [29]. Two screening criteria were selected by the authors in order to identify potential indicators that can be utilized in the proposed framework—(a) the indicator must be included in at least one of the available environmental sustainability methods reviewed [19] and (b) the indicator must satisfy the principles of IE to ensure that sustainable actions highlighted by IE will be promoted [22]. According to that criteria set, we developed a list of 108 indicators. Indicators with similar name or common indicators expressed in different units appear once in the list so the actual number of indicators analyzed is much higher.

*2.4. Selection of Key Indicators and Allocation to Categories*

The final key indicators to be integrated into the framework were selected from the pool with the application of five criteria [26]—(a) be easily understandable to various stakeholders, (b) be easily measured and are cost-effective in terms of data collection, (c) can cover long-term issues and be applied in multiple scales, (d) support decision making and promote desired behavior, and (e) be potentially comparable. All indicators included in the indicator pool (Step 3) were ranked according to these criteria with the utilization of a 0–4 scoring scale per criterion, resulting in a final score from 0–20 points per indicator (where 20 indicates maximum performance). The indicators were scored by a panel of five experts from both academia and industry. The analytical process followed to select the final core indicators is the following:

1. Distribution of the indicators included in the indicator pool to one of the eight proposed assessment categories.
2. Selection of the indicators with the highest score. In case two indicators served the same purpose, the one with the highest score was selected. Extra care was given to not include indicators that assess the same environmental parameter (double counting).
3. Configure, if necessary, the scope and the unit of measurement of indicators in order to improve the effectiveness of evaluation and comply with the principles of sustainable development. In many cases, absolute units (e.g., kWh consumed) had to be altered into relative ones (e.g., kWh consumed per tons of product) to strengthen decision making.
4. Consultation with the environmental department of the mining facility selected as a case study (see Section 3) to implement the proposed framework. In this way, indicators that present a particular interest for the mining facility have been integrated or excluded by the framework.

Resulting from this process, 19 final core indicators of environmental sustainability were selected that constitute the basis of the proposed framework (Table 2). The fact that the indicators were selected according to a structured and clear process increases the efficiency of the evaluation and ensures that most important parameters in terms of environmental sustainability are measured without increasing too much the time and resources required for the implementation. However, additional case specific indicators may be integrated by the user, if deemed necessary.

*2.5. Quantification of Indicators and Initial Analysis*

During this step, the evaluator must collect all data needed for the efficient quantification of the indicators and perform an initial analysis of the results. The initial analysis should include a validity check in order to examine whether the indicator values are reasonable and identify any errors in data collection [24]. If data are available from previous years, comparing the results and trying to justify differences is encouraged. In Section 3, the framework was implemented in a mining facility in order to examine its applicability and usefulness.

*2.6. Normalization and Evaluation of Indicators*

Normalization of indicators is necessary if we want to proceed to data aggregation [30]. Indicators above or below the mean, min-max, z-score, distance to a reference, and categorical scales are some examples of the normalization methods available [27]. A sensitivity analysis of normalization methods indicated that the distance to a reference method is the most suitable choice for sustainability performance evaluations in industry [30]. Based on this finding, we propose a hybrid normalization procedure, combining the categorical scale and the distance to a reference approaches. Distance to a reference is applied to compare the value of an indicator to a reference value whereas categorical scale assigns a score to every indicator using a numerical or qualitative scale. The same approach has been successfully applied to normalize and assess key indicators utilized for the evaluation of sustainable water consumption and management of industrial facilities [26].

**Table 2.** Final core environmental sustainability indicators selected to be included in the framework (ENi: EN stands for environmental, whereas i is the serial numbering).

| S/N | Name of Indicator |
|---|---|
| | *Category 1: Mineral resources and materials* |
| EN1 | Total consumption of resources and materials and quantity of products produced (in absolute and relative units) |
| EN2 | Utilization of mining waste |
| | *Category 2: Emissions and waste* |
| EN3 | Air emissions by type and total (in absolute and relative units) |
| EN4 | Liquid waste by type and total (in absolute and relative units) |
| EN5 | Mining waste to be deposited by type and total (in absolute and relative units) |
| | *Category 3: Energy and water* |
| EN6 | Energy consumption by type and total (in absolute and relative units) |
| EN7 | Total industrial water consumption (in absolute and relative units) |
| EN8 | Percentage of water that is recycled or reused as compared to total consumption |
| EN9 | Availability of water resources at local level |
| | *Category 4: Suppliers and environmental performance* |
| EN10 | Environmental assessment of suppliers and contractors |
| EN11 | Initiatives to enhance environmental performance, accountability and equity |
| | *Category 5: Biodiversity* |
| EN12 | Numbers of species included in the IUCN red list or national lists of protected species and are affected by the facility's activities, classified by the level of extinction risk |
| EN13 | Identification and mitigation actions on biodiversity |
| | *Category 6: Land use and rehabilitation* |
| EN14 | Total area restored to total disruption |
| | *Category 7: Impact on the environment and climate change* |
| EN15 | Global warming potential |
| EN16 | Number and description of environmental accidents |
| EN17 | Hazard risk of utilized materials, emissions and waste |
| | *Category 8: Impact on human health* |
| EN18 | Ambient air quality of the industrial site |
| EN19 | Risk of failure of infrastructure |

The reference value serves as the starting criterion to assign sustainability scores with the application of a 5-point semi-qualitative scale (Very High (5), High (4), Medium (3), Low {2), and Very Low (1)). Specific sustainability reference values were identified for each indicator to enable the comparison of the examined industry's performance with a sustainability reference point (see Appendix A). In our case, sustainability reference values reflect (a) either target values set by EU or international organizations, (b) either values derived from the analysis of the performance of international mining projects, (c) either expert's estimations and proposals from international scientific literature, or (d) either corporate objectives.

The pros and cons of the proposed normalization procedure were summarized and are presented in Table 3. The identification of commonly accepted reference values for every indicator was found to be a time and effort intensive process especially for uncommon indicators. To cope with this challenge, internal targets and expert judgments can be applied until a more concrete reference value is available. These targets must be evaluated and modified regularly to reduce subjectivity. The basic goal of the specific procedure is to aggregate indicators into one single sub-index. Loss of information during normalization can be balanced if the initial analysis (Step 5) has been carefully conducted [24].

**Table 3.** Pros and cons of the proposed normalization method [23].

| Pros | Cons |
| --- | --- |
| + Enables the aggregation of indicators into a composite environmental sustainability index while in parallel reflecting a distance from an environmentally sustainable performance per indicator.<br>+ It can assess both quantitative and qualitative indicators.<br>+ The process of finding a reference point generates value in terms of environmental sustainability knowledge.<br>+ No extensive data for a great number of industries and/or data over many consecutive years are required in contrast with common normalization methods.<br>+ Results are easily apprehendable, even by non-experts.<br>+ "Best in class" industry still receives a better score than other industries but not necessarily the maximum score if the sustainability criteria are not met.<br>+ If common indicators and respective reference points are applied, it can be used to benchmark the environmental sustainability performance of different facilities. | − The process of finding well accepted reference points for every indicator is time and effort intensive.<br>− A lot of information is lost due to the fact that a discrete five-point scale is applied, which may lead to accumulation of scores into the same cluster (many industries with the same score).<br>− The selection of a reference point and a scoring scale entails significant level of subjectivity (thus uncertainty of the results).<br>− It may be too hard for industries to gain a high score during the first years of implementation. |

## 2.7. Weighting

Weighting (expression of how important is a parameter compared to another) is a particularly significant step during the development of a composite indicator [31]. As in the case of a normalization process, there is not a commonly accepted way of data weighting [32]. Weights selected by the analyst is not necessarily a bad practice; however, it is very likely to have negative consequences regarding the acceptance of the results [33]. On the other hand, weights result from statistical methods, may be even less acceptable from the perspective of decision-making and policy development toward sustainability, as insignificant political parameters can receive high scores, while innovative approaches on sustainable development may not even taken into consideration [32].

In the proposed framework, we suggest equal weights to be attributed to all indicators and assessment categories. The specific decision is based on two basic reasons. The indicators of the framework were selected using a concrete procedure to ensure that key environmental issues are examined with maximum efficiency. Every indicator serves a different aspect of environmental sustainability, and all issues must be taken into account if we want to move towards sustainable development. Equal weights discourage industries from merely focusing on the improvement of the indicators with the higher weights. This approach serves better the sustainability notion according to which the performance of a system should be assessed taking into account various parameters (holistic approach). Second, it should be taken into account that the proposed framework attempts to assess industrial facilities regardless their special characteristics (e.g., size). If weights were to be adopted, these would have to be adapted to specific types, sizes, and spatial characteristics of the facility, which would significantly increase the complexity and uncertainty of the results, especially if used for benchmarking purposes.

## 2.8. Aggregation

The utilization of a high number of indicators might be problematic for the efficient communication of the environmental sustainability to the senior management of the industry and the general public. Since all indicators are expressed in a common quantitative scale (1–5 points), the extraction of individual sub-indices, and a final single index of environmental sustainability is possible. More specifically, by applying the proposed framework, the following environmental sustainability scores can be extracted:

- Per assessment category—Eight sub-indices:

- ○ I1: Mineral resources and materials index
- ○ I2: Emissions and waste index
- ○ I3: Energy and water index
- ○ I4: Suppliers and environmental performance index
- ○ I5: Biodiversity index
- ○ I6: Land use and rehabilitation index
- ○ I7: Impact on the environmental and climate change index
- ○ I8: Impact on human health index

- Total— One final index of environmental sustainability:

  - ○ $I_{ESAMI}$: Environmental Sustainability Assessment of Mining Industries index

The sub-indices per assessment category result from the average score of the indicators that compose the category:

$$I_X = \frac{\sum_{i=1}^{n} w_i}{n},$$
(1)

where x is the examined category ($x \in \mathbb{Z}\{1,\ldots,8\}$), $w_i$ is the score of the indicator i, and n is the number of indicators included in the assessment category x. Sub-indices Ix can help industries identify and analyze those categories where there is a great potential for improving their environmental sustainability. The final index of environmental sustainability is estimated from the average score of the eight assessment categories:

$$I_{ESAMI} = \frac{\sum I_X}{8} \text{ where } x \in \mathbb{Z}\{1,\ldots,8\} .$$
({2)

The extraction of a final overall score enables the efficient communication of the results and comparison with other facilities provides an overview of the environmental sustainability of the examined facility and allows the regular re-evaluation of the progress achieved through the years.

*2.9. Presentation of the Results*

The presentation of the results is an issue that should not be neglected and depends on many factors such as the target audience [34]. The proposed framework provides significant feedback (indicators utilization, extraction of sub-indices, finding reference values, quantitative scores, etc.) for the efficient presentation of the results. A number of key techniques for enhancing the presentation of the results include: the development of summarized tables of results, the development of trend charts per indicator for consecutive years, the development of graphs depicting the scores per sub-indices, the presentation of key results to websites, leaflets, and others. The results from the implementation of the framework could serve as an important means of strengthening sustainability reports, corporate social responsibility reports, environmental impact studies, and relevant presentations at meetings and conferences.

*2.10. Analysis of theRresults and Regular Re-Assessment*

The final step refers to the analysis and interpretation of the results. The analysis of the results is the step with the highest impact on the facility under examination since their interpretation will lead to successful decision making and the formulation of strategies for improving environmental sustainability. Consequently it should be performed with great attention. A successful analysis of results should be able to answer the following key questions:

Analysis per assessment category: Which categories present the lowest score and what factors (indicators) caused this? In this case it is particular useful to return to the results of step 5 in order to explore in greater depths the indicators that received the lowest scores.

Total analysis: What is the final score of environmental sustainability and how it can be improved? The conduction of a sensitivity analysis to quantify the alternative actions of improvement is highly encouraged at this stage. For the easier analysis of the results, the following general rule of interpretation is proposed depending on the rating of the index:

- Score: 1–1.5 points: very low sustainability performance
- Score: 1.5–2.5 points: low sustainability performance
- Score: 2.5–3.5 points: moderate sustainability performance
- Score: 3.5–4.5 points: high sustainability performance
- Score: 4.5–5 points: very high sustainability performance

As already mentioned, the utility of the proposed framework is twofold. It can act both as a path to develop new assessment methods (or improve existing ones) and as a ready to be used tool by industries who wish to assess their environmental sustainability. In the second case, the user can utilize the analytical results of the implementation of each step described before (proposed assessment categories, indicator pool, implementation guide, core indicators etc.). In this way, industries only need to apply steps 1, 5, 6, 8, 9, and 10, which reduces the time and cost of implementation.

## 3. Case Study—Evaluation of a Mining Facility in Greece

The proposed framework was implemented in a mining industrial facility in Greece in order to examine its applicability and usefulness. The system under examination was selected with a view to represent facilities of relatively high complexity of material and energy balances and significant environmental concerns. The steps described in the previous section were applied by taking into account the following assumptions.

The objective defined was to assess the overall environmental sustainability performance (current state of operation) and identify key weak spots that call for improvement. The scope of the assessment included the entire mining and metallurgical installations of the mines in the area that were examined as an entity to reflect the overall sustainability performance of the mining activity of the company.

The environmental sustainability performance was assessed using the eight proposed thematic categories and the 19 core environmental sustainability indicators of the framework. Indicators EN3, EN4, and EN19 were not evaluated due to the lack of relevant data because the facility was in the process of developing the necessary procedures for their quantification during the conduction of this study.

The data required for the assessment of each indicator were acquired from the following sources: a) published sustainability reports of the company, b) published data from the Association of Mining Enterprises in Greece, c) raw data from the environmental monitoring program of the company, and d) published data from technical reports and Environmental Impact Assessment studies. The normalization and evaluation of the indicators was carried out using the proposed method and respective sustainability reference values presented in Appendix A. It was further decided that all indicators and categories are of the same significance (no weights were attributed). Specific data and calculations will not be presented in detail due to confidentiality reasons and space restrictions. For a better understanding of the results though, a number of particular features of the case study are clarified below that have significantly influenced the evaluation.

A number of sites within the facility were still under development and/or at the stage of preparing the start of production, resulting in the consumption of high quantities of raw materials, energy, and water without the corresponding production of marketable products. That led to an abnormal correlation between the quantity of run of mine (ROM) ore and the marketable products with the inputs (raw materials, energy, water) and outputs (gaseous emissions, liquid waste, solid waste). As a result, indicators assessed in relative units (per ton of ROM ore) exhibited low or very low performance even when there has been an annual reduction in absolute terms. Typical examples are the indicators EN5, EN6, and EN15 related to the deposition of mining waste, energy consumption and global warming potential respectively. The production process is expected to be normalized during the full development of the examined facility, and thus, the assessment will better capture the efforts of the company to improve its sustainability performance.

*3.1. Results PerIindicator and Assessment Category*

The results per indicator and per assessment category are summarized in Table 4 and Figure 2 respectively. The score of the assessment categories ranged from 1.0 ($I_2$—missions and waste) to 5.0 points ($I_8$—human health impacts) exhibiting a noticeable fluctuation per indicator and category. The case study received the highest score in 6 out of 16 indicators evaluated (indicators 2, 8, 11, 13, 16, and 18) and the minimum score to 6 (indicators 1, 5, 6, 7, 12, and 15). The peculiarities of the year of examination, discussed in the previous section, do not allow the extraction of safe conclusions for the assessment categories $I_1$, $I_2$, and $I_3$. It is characteristic that for the specific year, the mass balance of inputs and outputs was negative while the output intensity (tons of outputs produced per ton of ROM ore mined) was higher than that input intensity (tons of inputs consumed per ton of ROM ore mined).

**Table 4.** Environmental sustainability performance per indicator.

| S/N | Name of indicator | Performance | Score |
|---|---|---|---|
| | *Category 1: mineral resources and materials* | | |
| EN1 | Total consumption of resources and materials and quantity of products produced | Very Low | 1.0 |
| EN2 | Utilization of mining waste | Very High | 5.0 |
| | *Score of Category 1:* | | *3.0/5.0* |
| | *Category 2: emissions and waste* | | |
| EN3 | Air emissions by type and total | No data available | |
| EN4 | Liquid waste by type and total | No data available | |
| EN5 | Mining waste to be deposited by type and total (in absolute and relative units) | Very Low | 1.0 |
| | *Score of Category 2:* | | *1.0/5.0* |
| | *Category 3: energy and water* | | |
| EN6 | Energy consumption by type and total | Very Low | 1.0 |
| EN7 | Total industrial water consumption | Very Low | 1.0 |
| EN8 | Percentage of water that is recycled or reused as compared to total consumption | Very High | 5.0 |
| EN9 | Availability of water resources at local level | Moderate | 3.0 |
| | *Score of Category 3:* | | *2.5/5.0* |
| | *Category 4: suppliers and environmental performance* | | |
| EN10 | Environmental assessment of suppliers and contractors | Moderate | 3.0 |
| EN11 | Initiatives to enhance environmental performance, accountability and equity | Very High | 5.0 |
| | *Score of Category 4:* | | *4.0/5.0* |
| | *Category 5: biodiversity* | | |
| EN12 | Numbers of species included in the IUCN red list or national lists of protected species and are affected by the facility's activities, classified by the level of extinction risk | Very Low | 1.0 |
| EN13 | Identification and mitigation actions on biodiversity | Very High | 5.0 |
| | *Score of Category 5:* | | *3.0/5.0* |
| | *Category 6: land use and rehabilitation* | | |
| EN14 | Total area restored to total disruption | Low | 2.0 |
| | *Score of Category 6:* | | *2.0/5.0* |
| | *Category 7: impact on the environment and climate change* | | |
| EN15 | Global warming potential | Very Low | 1.0 |
| EN16 | Number and description of environmental accidents | Very High | 5.0 |
| EN17 | Hazard risk of utilized materials, emissions and waste | High | 4.0/5.0 |
| | *Score of Category 7:* | | *3.3/5.0* |
| | *Category 8: impact on human health* | | |
| EN18 | Ambient air quality of the industrial site | Very High | 5.0 |
| EN19 | Risk of failure of infrastructure | No data available | |
| | *Score of Category 8:* | | *5.0/5.0* |
| | **Total Environmental Sustainability Score I(ESAMI):** | | **3.0/5.0** |

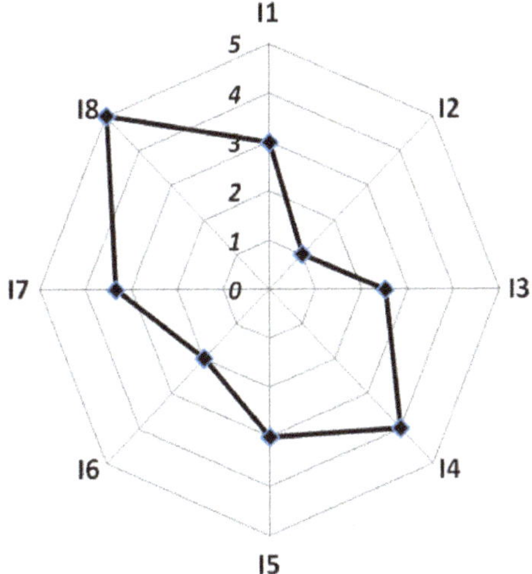

| | |
|---|---|
| I₁ | Mineral resources and materials |
| I₂ | Emissions and waste |
| I₃ | Energy and water |
| I₄ | Suppliers and environmental performance |
| I₅ | Biodiversity |
| I₆ | Land use and rehabilitation |
| I₇ | Impact on the environmental and climate change |
| I₈ | Impact on human health |

**Figure 2.** Environmental sustainability performance per category.

The examined facility is characterized by very high rates of recovery/reuse of mining waste (EN2 = 5.0 points) and water (EN8 = 5.0 points), thus balancing the negative impact of development and restoration works during this stage. The facility management has undertaken a significant number of initiatives to enhance its environmental performance, accountability and equity, whereas takes into account the environmental performance of its contractors ($I_4$ = 4.0 points), supporting the view that industries with particular environmental concerns and pressures, tend to account more on environmental management related issues.

The facility is characterized by moderate performance regarding biodiversity protection ($I_5$ = 3.0 points). The high concern due to the significant biodiversity of the area (EN12 = 1.0 points) is offset by mitigation actions on biodiversity (EN13 = 5.0 points) as the company has taken a significant number of relevant initiatives in line with best available practices for biodiversity protection. At this stage (development of sub-projects), the case study is characterized by low performance in land use and rehabilitation ($I_6$ = 2.0 points) as the majority of the disturbed area has not been restored. The score in the specific category is expected to improve over time as rehabilitation works are in progress.

The assessment categories related to environmental and climate change impacts ($I_7$ = 3.3 points) and human health impacts ($I_8$ = 5.0 points) exhibited moderate and very high performance respectively. This result indicates that the operation of the facility affects the environment and human health in an acceptable way, in case an environmental accident does not occur. An exception is indicator EN15 regarding global warming potential, which received a minimum score as the equivalent carbon dioxide emissions are calculated on the basis of the energy consumption per ton of ROM ore. Carbon dioxide emissions in absolute units ($\approx$57,000 tons) are at normal levels for heavy industries and much lower that energy and chemical industries. Reagents utilized in the production process, although mostly hazardous, represent a very small proportion of the overall input–output balance and with proper management the risk of large-scale accidents is minimized. On the contrary, the produced products and materials left over after the separation process present a significant risk due to the combination of their quantities and risk.

*3.2. Total Results – The $I_{ESAMI}$ Index*

According to the results from the implementation of the proposed framework, the final score of environmental sustainability for the examined facility was 3.0 points ($I_{ESAMI}$ = 3.0 points). The score is considered satisfactory (moderate environmental sustainability) if we take into account the fact that the system under examination is characterized by very high environmental and social concerns and the particularities of the operation phase during the assessment year.

The minimization of environmental impacts is one of the biggest challenges of the mining industry. The proposed framework follows a more eco-centric approach according to which the facility examined should continually improve the efficiency of its processes, set long-term goals (e.g., complete land rehabilitation), and take into account concerns in which one cannot directly intervene (e.g., biodiversity of the area, availability of water in the area, etc.).

The assessment and analysis of the core indicators indicated significant room for improvement where the company should aim to further enhance its sustainability performance. A number of suggestions have been developed taking into account the possible ways to increase the score of the indicators with the lowest performance.

- Reduce solid waste deposited annually by finding ways to exploit/reuse them and/or innovative techniques for their management.
- Strengthen energy saving efforts and reduce the dependency on utilization of oil. In this context, a) examine the viability of meeting specific energy needs (e.g. offices) using renewable energy sources, b) place simple control and energy saving mechanisms (e.g. sensors, photocells, led lighting), and c) continuously monitor energy management and consumption indicators.
- Continuously monitor indicators on water management/consumption and the elaboration of an analytical water balance for each sub-project within the facility and as a whole.
- Develop concrete procedures for the environmental assessment of the suppliers.
- Develop concrete procedures for estimating the risk of failure of infrastructures.

The structure of the framework and the assessment process presented a high level of flexibility that allowed its implementation with minimum resources and time available. Thus, one of the key objectives of the framework which is providing scalable levels of difficulty was achieved, thus allowing facilities of varying capacity to be able to implement it.

The procedure of quantifying and scoring some of the indicators presented great added value for the facility since, in several cases, it raised issues (and respective concerns) that were not taken into account until then. The need to develop analytical balances of materials/energy/water/waste and the extraction of composite indicators provided a general overview of the environmental performance of the facility, indicating areas of high concern. The complete implementation of the framework required the cooperation of different sections of the industry something that helps to detect broader issues and data gaps. Results obtained during the implementation of the steps could be used in parallel with the development/update of sustainability and corporate responsibility reports.

## 4. Concluding Remarks and Future Research

This study introduced a conceptual framework that is able to strengthen the evaluation and monitoring of the environmental sustainability of mining industrial facilities. The framework can be either utilized as a guide by industries who wish to enhance the way they assess environmental sustainability issues and/or as a stand-alone tool of assessment. It consists of 10 well-defined steps that include specific guidelines and tips that allow for its gradual implementation.

The framework attempts to adopt a more proactive approach, aiming to assess the actions that lead to reduced/increased environmental impact rather than the impacts themselves. The specific approach strengthens the probability that the results will be accepted from external relevant agents and auditors. The framework covers environmental issues in a wider spatial and temporal frame in accordance with the targets of sustainable development.

The reason for conducting the assessment on the industrial facility level (rather than corporate level) is that it enables the identification of unsustainable industrial processes at the source while taking into account specific spatial characteristics of the facility under examination. Consequently, the results of the assessment are expected to better facilitate the identification of amelioration actions. The proposed methodology is focused at the moment only on the environmental aspect of sustainability but can be adapted to include the economic and social aspects.

Linking environmental sustainability indicators to a sustainability reference point can provide a meaningful sustainability performance based on a distance-to-target approach. To cope with this challenge, a hybrid normalization procedure combining the categorical scale and the distance to a reference approach was introduced in the framework. The combination of these two methods enables the aggregation of indicators into a composite environmental sustainability index, whereas in parallel, it reflects a distance from an environmentally sustainable performance-target per indicator. These targets (and thus the assessment of the industry) must be re-evaluated and modified regularly due to the dynamic nature of environment.

The implementation of the proposed methodological framework indicated a number of particular points of interest that can further enhance its applicability. The categories that were harder to be assessed (in terms of data and time needed) were Category 1: mineral resources and materials and Category 2: emissions and waste. The development of an analytical input-output inventory and its reduction in the same units entails the conduction of transformations that can complicate the calculations and affect the reliability of the results. Therefore, it is particular useful to develop a conversion factor inventory to ensure reliable estimations. Despite the effort to minimize the data required for the assessment, the framework still requires information related to a number of different activities within the facility. The development of a standardized data acquisition form will reduce the time of implementation and will facilitate the exchange of information among the various departments of the facility.

The authors are planning to continuously improve the effectiveness of the proposed framework by updating sustainability reference points and developing supporting software that will enable its fast and reliable implementation. Additionally the framework should be implemented into a significant number of industrial facilities. The specific feedback will provide the opportunity to examine issues such as the framework's ability to predict future performance and concerns and the examination of interrelations among the proposed indicators. Other key issues that need to be examined in the future include the potential usability of the framework to assess economic and social aspects of sustainability, its combination with other methods to assess different types of systems, and its applicability to develop relevant sustainability eco-labeling schemes.

Industries and organizations in general will be expected to assess and report their environmental sustainability in the near future. The ideas and steps described in this study can help in developing a common methodological framework that can be applied by mining industries and promote environmental sustainability.

**Author Contributions:** All authors contributed equally to the text. All authors have read and agreed to the published version of the manuscript

**Funding:** This research received no external funding.

**Conflicts of Interest:** The authors declare no conflict of interest.

# Appendix A



**Table A1.** Normalization and assessment procedure per indicator included in the proposed framework.

| Indicator | Calculation | Identification of a Sustainability Reference Value | Normalization and Assessment |
|---|---|---|---|
| **EN1** *Total consumption of resources and materials and quantity of products produced* | (1) Calculation and recording of all inputs necessary for the production process (water and fuels not included); (2) Convert all inputs to a common unit of measurement and sum up results; (3) Define a functional unit (tons of run of mine (ROM) ore is proposed); (4) Divide total input consumption with functional unit and find resource/material intensity; (5) Repeat steps 1–4 for outputs (including products) to find output intensity and examine mass balance (inputs-outputs). | An annual reduction of resource/material intensity over 2% can be set as a reference value for VH sustainability (and/or a total reduction of over 20% in comparison with a past reference year). These values were selected taking into account EU's targets for strengthening resource efficiency in industries by 2030 [35]. | VL: Increase $Xi_t$ >2% and/or $Xy_t$ >20% <br> L: Increase $Xi_t$ from 0-2% and/or $Xy_t$ from 0-20% <br> M: Decrease $Xi_t$ from 0-1% and/or $Xy_t$ from 0-10% <br> H: Decrease $Xi_t$ from 1-2% and/or $Xy_t$ from 10-20% <br> VH: Decrease $Xi_t$ >2% and/or $Xy_t$ >20% <br> Where $Xi_t$ is the percentage of increase/decrease of resource/material intensity for year t compared to year t-1, and $Xy_t$ is the respective total increase/decrease compared to a reference year. |
| **EN2** *Utilization of mining waste* | Reuse percentage is calculated according to the following formula $$\frac{\text{Tailings} + \text{WR (reused)}}{\text{ROM+WR−concentrates produced}}\ \%$$ WR: Waste Rock | Mining waste reuse over 40% can be set as a reference value for VH sustainability. This percentage assures the significant reduction of tailings and rock waste disposal; whereas the target value can be achieved since the company can use tailings/rock wastes of previous years. | VL: $Xi_t \leq 10\%$ <br> L: $10\% < Xi_t \leq 20\%$ <br> M: $20\% < Xi_t \leq 30\%$ <br> H: $30\% < Xi_t \leq 40\%$ <br> VH: $Xi_t > 40\%$ <br> Where $Xi_t$ is the percentage of reuse of mining waste for year t. |
| **EN3** *Air emissions by type and total* | Dust emissions and intensity of dust emissions in the atmospheric environment are calculated according to the formulas <br> - Dust emissions in the air (ton) = $Ci \times Ri \times Oi$ <br> - Intensity (g/ton) = Dust emission in the air/ROM ore <br> *where* <br> C: average concentration (mg/m$^3$) <br> R: average emission rate from the source (m$^3$/hr) <br> O: unit operation time (hrs) <br> i: year under examination <br> Dust definition: Particulate matter of all aerodynamic diameters emitted in the atmosphere after the application of anti-pollution technology (i.e., cyclones, bag filters, etc.). | An annual reduction of dust emissions intensity over 2% can be set as a reference value for VH sustainability (and/or a total reduction of over 20% in comparison with a past reference year). These values were selected in accordance with EN1 indicator target values for the reduction of inputs/outputs and are taking into account the strategic 40% target value for the reduction of greenhouse gases of EU until 2030. | VL: Increase $Xi_t$ >2% and/or $Xy_t$ >20% <br> L: Increase $Xi_t$ from 0-2% and/or $Xy_t$ from 0-20% <br> M: Decrease $Xi_t$ from 0-1% and/or $Xy_t$ from 0-10% <br> H: Decrease $Xi_t$ from 1–2% and/or $Xy_t$ from 10–20% <br> VH: Decrease $Xi_t$ >2% and/or $Xy_t$ >20% <br> Where $Xi_t$ is the percentage of increase/decrease of emissions intensity for year t compared to year t-1, and $Xy_t$ is the respective total increase/decrease compared to a reference year. |
| **EN4** *Liquid waste by type and total* | The total amount of water pollutants is calculated according the following formula: $$\text{Total pollution load} = \sum_{i=1}^{i=n} Ci \times Q$$ Where: <br> C: average concentration of pollutant in the treated effluent discharged to a natural receiver (sea, river, lake, etc. in µg/m$^3$) <br> Q: volume of treated effluent discharged to natural receivers(m$^3$) <br> i: pollutant <br> n: number of pollutants under examination | An annual reduction of pollutant load discharged to natural receivers (i.e., sea, river, lake, etc.) over 1.5% can be set as a reference value for VH sustainability (and/or a total reduction of over15% in comparison with a reference year). These values were selected taking into account the maximum potential average water saving using the existing technology [36] and are in accordance with EN1 indicator target values for the reduction of inputs/outputs. | VL: Increase $Xi_t$ > 1.5% and/or $Xy_t$ > 15% <br> L: Increase $Xi_t$ from 0–1.5% and/or $Xy_t$ from 0–15% <br> M: Decrease $Xi_t$ from 0–0.5% and/or $Xy_t$ from 0-5% <br> H: Decrease $Xi_t$ from 0.5–1.5% and/or $Xy_t$ from 5–15% <br> VH: Decrease $Xi_t$ > 1.5% and/or $Xy_t$ > 15% <br> Where $Xi_t$ is the percentage of increase/decrease of water pollutants for year t compared to year t-1, and $Xy_t$ is the respective total increase/decrease compared to a reference year. |

**Table A1.** *Cont.*

| Indicator | Calculation | Identification of a Sustainability Reference Value | Normalization and Assessment |
|---|---|---|---|
| EN5 Mining waste to be deposited by type and total | Intensity of mining waste (tones/ ROM ore tones) is calculated according to the following formula: $$\frac{(\text{Tailings} + \text{Waste Rock produced}) - (\text{Tailings} + \text{Waste Rock waste reused})}{\text{Quantity of ROM ore}}$$ | An annual reduction of mining waste intensity over 1% can be set as a reference value for VH sustainability (and/or a total reduction of over 10% in comparison with a reference year). These values were selected taking into account the EU targets for the efficient resource use and are in accordance with EN1 indicator target values for the reduction of inputs/outputs. | VL: Increase $Xi_t$ > 1% and/or $Xy_t$ > 10% <br> L: Increase $Xi_t$ from 0–1% and/or $Xy_t$ from 0–10% <br> M: Decrease $Xi_t$ from 0–0.5% and/or $Xy_t$ from 0–5% <br> H: Decrease $Xi_t$ from 0.5–1% and/or $Xy_t$ from 5–10% <br> VH: Decrease $Xi_t$ > 1% and/or $Xy_t$ > 10% <br> Where $Xi_t$ is the percentage of increase/decrease of mining waste intensity for year t compared to year t-1, and $Xy_t$ is the respective total increase/decrease compared to a reference year. |
| EN6 Energy consumption by type and total | (1) Calculation of the energy consumption of the industrial facility per type (electricity, diesel oil, etc.) and in total (suggested conversion factors 1 kWh electricity = 3.6 MJ, 1 Lt. diesel = 38.7 MJ); <br> (2) Define a functional unit (tons of ROM ore is proposed),; <br> (3) Calculation of energy intensity by dividing the total consumption with the selected functional unit. | An annual reduction of energy intensity over 2% can be set as a reference value for VH sustainability (and/or a total reduction of over 20% in comparison with a past reference year). These values were selected taking into account the suggested target of the United Nations Industrial Development Organization [37] and are in accordance with EN1 indicator target values for the reduction of inputs/outputs. | VL: Increase $Xi_t$ > 2% and/or $Xy_t$ > 20% <br> L: Increase $Xi_t$ from 0–2% and/or $Xy_t$ from 0-20% <br> M: Decrease $Xi_t$ from 0–1% and/or $Xy_t$ from 0-10% <br> H: Decrease $Xi_t$ from 1–2% and/or $Xy_t$ from 10–20% <br> VH: Decrease $Xi_t$ > 2% and/or $Xy_t$ > 20% <br> Where $Xi_t$ is the percentage of increase/decrease of energy intensity for year t compared to year t-1, and $Xy_t$ is the respective total increase/decrease compared to a reference year. |
| EN7 Total industrial water consumption | (1) Calculation of the total quantity of mine waters entering the Water Treatment Plant (WTP in); <br> (2) Deduction of the treated mine waters from the WTP discharged to natural receivers (WTP out); <br> (3) Calculation of the total water consumption in the industrial facilities, (Make up water) = (WTP in)– (WTP out); <br> (4) Define a functional unit (tons of ROM ore is proposed); <br> (5) Calculation of the water use intensity by dividing the total consumption with the selected functional unit. | An annual reduction of water use intensity over 2% can be set as a reference value for VH sustainability (and/or a total reduction of over 20% in comparison with a past reference year). These values were selected taking into account the recorded water use intensity reduction for several industries [38] and are in accordance with EN1 indicator target values for the reduction of inputs/output. | VL: Increase $Xi_t$ >2% and/or $Xy_t$ >20% <br> L: Increase $Xi_t$ from 0–2% and/or $Xy_t$ from 0-20% <br> M: Decrease $Xi_t$ from 0–1% and/or $Xy_t$ from 0-10% <br> H: Decrease $Xi_t$ from 1–2% and/or $Xy_t$ from 10–20% <br> VH: Decrease $Xi_t$ >2% and/or $Xy_t$ >20% <br> Where $Xi_t$ is the percentage of increase/decrease of water use intensity for year t compared to year t-1, and $Xy_t$ is the respective total increase/decrease compared to a reference year. |
| EN8 Percentage of water that is recycled or reused as compared to total consumption | Final calculation method is differentiated on a case-by-case basis and depends on the special characteristics of the examined facility. For this case study the annual rainfall was taken into account whereas the water inputs and outputs to the WTP during the days when treatment capability was exceeded and therefore untreated mine water was discharged to the sea were not included in the estimations. | A zero percentage of water reuse/recycling can be set as a reference value for VL sustainability whereas industries that reuse/recycle over 30% of their water needs are characterized by VH sustainability. | VL: $X_t = 0$ <br> L: 0% < $X_t$ ≤ 10% <br> M: 10% < $X_t$ ≤ 20% <br> H: 20% < $X_t$ ≤ 30% <br> VH: $X_t$ > 30% <br> Where $X_t$ is the percentage of water that is reused/recycled for year t. |

**Table A1.** *Cont.*

| Indicator | Calculation | Identification of a Sustainability Reference Value | Normalization and Assessment |
|---|---|---|---|
| EN9 Availability of water resources at local level | This indicator is calculated according to the Overall Water Risk for the region that can be extracted from the database AQUEDUCT of the World Resources Institute (WRI) and the respective interactive map (http://www.wri.org/our-work/project/aqueduct/aqueduct-atlas/). Alternatively, water resources availability at a local level can be calculated taking into account the annual rainfall (mm) for the examined year. | A score higher than four (4) points that expresses very high risk-concern regarding water resources protection in the area can be set as a reference value for VL sustainability (following the scale of WRI scoreboard). A total annual rainfall higher than 1,200 mm can be set as a reference value for VH sustainability. Normalization of rainfall was conducted according to the range of average rainfall in Europe (1940-1995) so as to exclude areas that are characterized by extreme values (e.g. tropical forests). | VL: 4–5 points and/or $X_t \leq 200mm$ L: 3–4 points and/or $200mm < X_t \leq 400mm$ M: 2–3 points and/or $400mm < X_t \leq 800mm$ H: 1–2 points and/or $800mm < X_t \leq 1,200mm$ VH: 0–1 points and/or $i_t > 1,200mm$ According to the results-score of the AQUEDUCT database of WRI and/or in combination with the regional rainfall, where $X_t$ is the total annual rainfall for year t. |
| EN10 Environmental assessment of suppliers and contractors | The environmental performance of suppliers and contractors of the company for the year under examination is calculated with the utilization of a relevant questionnaire developed by the company or an external agent. The questionnaire should cover broader issues of environmental performance and should include five-scale closed type questions to facilitate implementation. | An average score higher than 4.5 points (maximum performance in all categories) can be set as a reference value for VH sustainability. | VL: $X_t \leq 1.5$ L: $1.5 < X_t \leq 2.5$ M: $2.5 < X_t \leq 3.5$ H: $3.5 < X_t \leq 4.5$ VH: $X_t > 4.5$ Where $X_t$ is the average score of the answers of the questionnaire for year t. |
| EN11 Initiatives to enhance environmental performance, accountability and equity | Suggested initiatives: (1) Adoption of an Environmental Management System (EMS); (2) Environmental certification by independent entities; (3) Development and publication of a Corporate Social Responsibility and Sustainability Report; (4) Publication of environmental data; (5) Synergies to enhance environmental performance, exchange wastes, common use of equipment and infrastructure; (6) Other initiatives. | The adoption of at least five initiatives to enhance environmental performance, accountability and equity can be set as a reference value for VH sustainability. Suggested initiatives were selected to reflect different degree of implementation difficulty and to be complementary. | VL: At least one/zero initiative L: At least two initiatives M: At least three initiatives H: At least four initiatives VH: At least five initiatives Assessment according to the number of initiatives adopted during year t. |
| EN12 Numbers of species included in the IUCN red list or national lists of protected species and are affected by the facility's activities | Identification and recording of the number of species included in the IUCN red list or national lists of protected species and are affected by the facility's activities, classified by the level of extinction risk according to relevant directives and standards. | The operation of the facility at an environment where there are only species of limited concern and as a result risks from non-reversible consequences are minimized can be set as a reference value for VH sustainability. | VL: Presence of critically endangered species (CR) or species of priority of Annex II of the Directive 92/43/EC or species of the category SPEC 1* L: Presence of endangered species (EN) or species of Annex II and IV of the Directive 92/43/EC or species of the category SPEC 2* M: Presence of vulnerable species (VU) or species of the category SPEC 3* H: Presence of near threatened species (NT) VH: Presence of species of limited concern (LC) Assessment according to the classification of species per extinction risk level that are affected by the facility's operation. |

**Table A1.** *Cont.*

| Indicator | Calculation | Identification of a Sustainability Reference Value | Normalization and Assessment |
|---|---|---|---|
| EN13 Identification and mitigation actions on biodiversity | Suggested actions according to literature and best available techniques: (1) Development of relevant observatory; (2) Vulnerability study of affected species; (3) Action plan for climate change and impacts on biodiversity; (4) Technical works for the protection of species (e.g., fences, opening alternative roads); (5) Other actions. | A number of at least four actions to mitigate the impacts of the facility's activity on biodiversity can be set as a reference value for VH sustainability. Suggested actions were selected to reflect different degree of implementation difficulty and to be complementary. | VL: No actions L: At least one action M: At least two actions H: At least three actions VH: At least four actions Assessment according to the number of actions/initiatives taken by the examined facility during year t. |
| EN14 Total area restored to total disruption | Calculation of the total area that has been restored or is under restoration expressed as a percentage of the total disrupted area for the year under examination. | The restoration of almost all the area (95%) that has been disrupted by the company can be set as a reference value for VH sustainability. | VL: $X_t < 5\%$ L: $5\% \leq X_t < 35\%$ M: $35\% \leq X_t \leq 65\%$ H: $65\% < X_t \leq 95\%$ VH: $X_t > 95\%$ Where $X_t$ is the percentage of the total are restored to the total area disrupted for year t. |
| EN15 Global warming potential | (1) Classification and quantification of the facility's energy consumption per energy/fuel type; (2) Calculation of equivalent $CO_2$ emissions using national or local transformation factors per type and use. The utilization of factors that take into account the fuel lifecycle is suggested; (3) Sum of the results and calculation of the total $CO_2$ emissions due to energy consumption; (4) Define a functional unit (tons of ROM ore is proposed); (5) Calculation of the carbon footprint intensity by dividing the total emissions with the selected functional unit. | An annual reduction of carbon footprint intensity over 2.5% can be set as a reference value for VH sustainability (and/or a total reduction of over 20% in comparison with a past reference year). These values were selected taking into account the new strategic target of EU [39] for the reduction of greenhouse gases emissions over 40% until 2030 (in comparison with 1990 emissions). | VL: Increase $Xi_t >2.5\%$ and/or $Xy_t >20\%$ L: Increase $Xi_t$ from 0–2.5% and/or $Xy_t$ from 0–20% M: Decrease $Xi_t$ from 0–1.25% and/or $Xy_t$ from 0–10% H: Decrease $Xi_t$ from 1.25–2.5% and/or $Xy_t$ from 10–20% VH: Decrease $Xi_t >2.5\%$ and/or $Xy_t >20\%$ Where $Xi_t$ is the percentage of increase/decrease of global warming potential (carbon footprint intensity) for year t compared to year t-1, and $Xy_t$ is the respective total increase/decrease compared to a reference year. |
| EN16 Number and description of environmental accidents | Recording of the number and detailed description of the environmental accidents occurred during the year of examination. The severity and impact of the accidents must be taken into account in the evaluation. | The complete absence of environmental accidents during the year of examination can be set as a reference value for VH sustainability. | VL: Extended environmental accident (severe impacts) L: Increase in number and/or severity of environmental accidents M: Decrease in number and/or severity of environmental accidents H: Small scale environmental accidents (minimal impacts) VH: Zero environmental accidents (no impacts) Assessment according to the number, extend and severity of the environmental accidents occurred compared to the previous year. |

**Table A1.** *Cont.*

| Indicator | Calculation | Identification of a Sustainability Reference Value | Normalization and Assessment |
|---|---|---|---|
| **EN17** Hazard risk of utilized materials, emissions and waste | (1) Utilization of the inputs/outputs list of EN1, expansion of the list, if needed, with waste and emissions from EN3–EN5; (2) Calculation of the contribution of every entry in the total mass balance (separate inputs/outputs); (3) Evaluate every entry according to its hazard risk with a grading scale of 1–3 points, where 1 refers to inert material and 3 to hazardous materials. Suggested grading methods: a) Material Safety Data Sheets, b) according to Scorecard guide for chemicals, c) according to other regulations (i.e., SEVESO directive); (4) Multiplication of the contribution of every entry with its hazard risk score and calculation of the average score of all data. | An average hazard risk score lower than 1.5 can be set as a reference value for VH sustainability (the majority of the inputs/outputs mass does not present a critical hazard risk). | VL: $\Sigma(Xi_t \times Ei_t)/100 \geq 2.5$ L: $2.5 > \Sigma(Xi_t \times Ei_t)/100 \geq 2.0$ M: $2.0 > \Sigma(Xi_t \times Ei_t)/100 \geq 1.75$ H: $1.75 > \Sigma(Xi_t \times Ei_t)/100 \geq 1.5$ VH: $\Sigma(Xi_t \times Ei_t)/100 < 1.5$ Where $\Sigma (Xi_t \times Ei_t)$ is the average hazard risk score of inputs/ outputs of the facility for the year t, Xi is the percentage of contribution of the input/output i to the mass balance and Ei is the respective hazard risk score of the input/output i. |
| **EN18** Ambient air quality of the industrial site | (1) Evaluation of the ambient air quality for the year under examination according to the suggested gradation of the European Environment Agency and the database Airbase for the main air pollutants ($PM_{10}$, $PM_{2.5}$, $NO_2$, $SO_2$, CO); (2) Scoring according to the suggested limit values, (3) Calculation of the average score for the five pollutants. | An average score higher than 4.5 can be set as a reference value for VH sustainability (low concentrations for all pollutants). Normalization of the reference values follows the suggested gradation of the European Environmental Agency, in accordance with the limit values for human and ecosystems health. | VL: $X_t \leq 1.5$ L: $1.5 < X_t \leq 2.5$ M: $2.5 < X_t \leq 3.5$ H: $3.5 < X_t \leq 4.5$ VH: $X_t > 4.5$ Where Xt is the average score for ambient air quality for the year t. |
| **EN19** Risk of failure of infrastructure (i.e. tailings dam stability, underground mines stability, etc.) | (1) Assessment of the potential accidents (due to technical failure and/or human error) including accidents due to natural disasters or terrorism. Large Scale Industrial Accidents are the accidents that lead to large fires, explosions, leakage of toxic substances or their combination, and their impacts go beyond the limits of the industrial facility; (2) Estimation of the impact radius in km of the potential accidents taking into account the worst-case scenario; (3) Recording of inhabited areas within the impact radius and of other facilities that can worsen the extent and/or the severity of the impacts (domino effect); (4) Analysis of the results and short description of the preventive measures. | The limitation of impacts within the industrial facility (zero impact radius) can be set as a reference value for VH sustainability. This value is attributed to industrial facilities that use non-hazardous substances and/or their industrial activity is of very low hazard risk. On the contrary, industries whose impacts have a broader scale radius are characterized by low sustainability for human health. | VL: Large scale impact radius (>20km) L: Broad scale impact radius (1–20 km) – inhabited areas M: Broad scale impact radius (1–20 km) – non inhabited areas H: Accidents may happen only during the transfer of hazardous substances VH: Accidents may happen only within the industrial facility According to the results of risk analysis. If infrastructure is designed to withstand extreme phenomena with a return period > 5 times their expected life cycle then the performance is VH. |

## References


1. Curkovic, S.; Sroufe, R. A literature review and taxonomy of environmentally responsible manufacturing. *Am. J. Ind. Bus. Manag.* **2016**, *6*, 323–346. [CrossRef]
2. Goyal, P.; Rahman, Z.; Kazmi, A.A. Corporate sustainability performance and firm performance research. *Manag. Decis.* **2013**, *51*, 361–379. [CrossRef]
3. Barkemeyer, R.; Holt, D.; Preuss, L.; Tsang, S. What happened to the 'development' in sustainable development? Business guidelines two decades after Brundtland. *Sustain. Dev.* **2014**, *22*, 15–32. [CrossRef]
4. Maas, K.; Schaltegger, S.; Crutzen, N. Integrating corporate sustainability assessment, management accounting, control and reporting. *J. Clean. Prod.* **2016**, *136*, 237–248. [CrossRef]
5. Nicolaescu, E.; Alpopi, C.; Zaharia, C. Measuring Corporate Sustainability Performance. *Sustainability* **2015**, *7*, 851–865. [CrossRef]
6. Lopez, R.A.; Ceballos, J.D.; Montiel, I. Deconstructing corporate sustainability: a comparison of different stakeholder metrics. *J. Clean. Prod.* **2016**, *136*, 5–17. [CrossRef]
7. Bergman, M.M.; Bergman, Z.; Berger, L. An Empirical Exploration, Typology, and Definition of Corporate Sustainability. *Sustainability* **2017**, *9*, 753. [CrossRef]
8. Villeneuve, C.; Tremblay, D.; Riffon, O.; Lanmafankpotin, G.Y.; Bouchard, S. A Systemic Tool and Process for Sustainaibility Assessment. *Sustainability* **2017**, *9*, 1909. [CrossRef]
9. Moran, C.J.; Lodhia, S.; Kunz, N.C.; Huisingh, D. Sustainability in mining, minerals and energy: new processes, pathways and human interaction for cautiously optimistic future. *J. Clean. Prod.* **2014**, *84*, 1–15. [CrossRef]
10. Vintro, C.; Sanmiquel, L.; Freijo, M. Environmental sustainability in the mining sector: evidence from Catalan companies. *J. Clean. Prod.* **2014**, *84*, 155–163. [CrossRef]
11. Kazakidis, V.; Gaidajis, G.; Angelakoglou, K. Evaluation of environmental and social parameters of a gold-mining project at the prefeasibility stage: A case study. *Glob. Perspect. Eng. Manag.* **2013**, *2*, 93–104.
12. Boiral, O.; Henri, J.F. Is sustainability performance comparable? A study of GRI report of mining organizations. *Bus. Soc.* **2017**, *56*, 283–317. [CrossRef]
13. Jenkins, H.; Yakovleva, N. Corporate social responsibility in the mining industry: Exploring trends in social and environmental disclosure. *J. Clean. Prod.* **2006**, *14*, 271–284. [CrossRef]
14. Belkhir, L.; Bernard, S.; Abdelgadir, S. Does GRI reporting impact environmental sustainability? A cross-industry analysis of $CO_2$ emissions and performance between GRI-reporting and non-reporting companies. *Manag. Environ. Qual.* **2017**, *28*, 138–155. [CrossRef]
15. Fonseca, A.; McAllister, M.L.; Fitzpatrick, P. Sustainability reporting among mining corporation: a constructive critique of the GRI approach. *J. Clean. Prod.* **2014**, *84*, 70–83. [CrossRef]
16. Tost, M.; Hitch, M.; Chandurkar, V.; Moser, P.; Feiel, S. The state of environmental sustainability considerations in mining. *J. Clean. Prod.* **2018**, *182*, 969–977. [CrossRef]
17. Gonzalez, E.D.R.S.; Sarkis, J.; Huisingh, D.; Huatuco, L.H.; Maculan, N.; Torres, J.R.M.; Almeida, C.M.V.B. Making real progress toward more sustainable societies using decision support models and tools: introduction to the special volume. *J. Clean. Prod.* **2015**, *105*, 1–13. [CrossRef]
18. Moldavska, A.; Welo, T. On the applicability of sustainability assessment tools in manufacturing. *Procepedia Cirp* **2015**, *29*, 621–626. [CrossRef]
19. Angelakoglou, K.; Gaidajis, G. A review of methods contributing to the assessment of the environmental sustainability of industrial systems. *J. Clean. Prod.* **2015**, *108*, 725–747. [CrossRef]
20. Angelakoglou, K.; Gaidajis, G. Assessing the progress of mining industry towards sustainability—In need of new methodological frameworks. In Proceedings of the 6th International Conference on Sustainable Development in the Minerals Industry-SDIMI, Milos, Greece, 30 June–3 July 2013.
21. Angelakoglou, K.; Gaidajis, G. Utilization of Industrial Ecology Principles for assessing the environmental sustainability of industries. In Proceedings of the 1st National Conference: Economics of Natural Resources and Environment—Climate Change-ENVECON, Volos, Greece, 27 March 2014.
22. Angelakoglou, K.; Gaidajis, G. Selection of indicators for assessing the environmental sustainability of industrial facilities and processes. In Proceedings of the 4th International Symposium on Green Chemistry, Health and Development, Kos Island, Greece, 24–26 September 2014.

23. Angelakoglou, K.; Gaidajis, G. Moving from mere quantification to meaningful evaluation of environmental sustainability indicators in industry. In Proceedings of the 4th International Symposium on Green Chemistry, Health and Development, Kos Island, Greece, 24–26 September 2014.

24. Angelakoglou, K. Development of a Methodological Framework for Assessing the Environmental Sustainability of Industrial facilities and Processes. Ph.D. Thesis, Democritus University of Thrace, Xanthi, Greece, 2015. (in Greek).

25. Angelakoglou, K.; Gaidajis, G. ENSAI-index: A new methodological framework for assessing the environmental sustainability of industrial facilities. In Proceedings of the ISIE joint 12th Socio-Economic Metabolism section conference and 5th Asia-Pacific conference, Nagoya, Japan, 28–30 September 2016.

26. Gaidajis, G.; Angelakoglou, K. Sustainability of industrial facilities through water indicators. *Environ. Process.* **2016**, *3*, 91–103. [CrossRef]

27. OECD. *Handbook on Constructing Composite Indicators. Methodology and User Guide. Part I. Constructing a Composite Indicator*; OECD: Paris, France, 2008.

28. Pope, J.; Annandale, D.; Morrison Sanders, A. Conceptualizing sustainability assessment. *Environ. Impact Assess. Revis.* **2004**, *24*, 595–616. [CrossRef]

29. Keeble, J.J.; Topiol, S.; Berkeley, S. Using indicators to measure sustainability performance at corporate and project level. *J. Bus. Ethics* **2003**, *44*, 149–158. [CrossRef]

30. Zhou, L.; Tokos, H.; Krajnc, D.; Yang, Y. Sustainability performance evaluation in industry by composite sustainability index. *Clean Technol. Environ. Policy* **2012**, *14*, 789–803. [CrossRef]

31. Singh, R.K.; Murty, H.R.; Gupta, S.K.; Dikshit, A.K. An overview of sustainability assessment methodologies. *Ecol. Indic.* **2012**, *15*, 281–299. [CrossRef]

32. Bohringer, C.; Jochem, P.E.P. Measuring the immeasurable—A survey of sustainability indices. *Ecol. Econ.* **2007**, *63*, 1–8. [CrossRef]

33. Gasparatos, A. Embedded value systems in sustainability assessment tools and their implications. *J. Environ. Manag.* **2010**, *91*, 1613–1622. [CrossRef] [PubMed]

34. Becker, J. Making sustainable development evaluations work. *Sustain. Dev.* **2014**, *12*, 200–211. [CrossRef]

35. SPIRE. Sustainable Process Industry through Resource and Energy Efficiency. Available online: http://www.spire2030.eu/uploads/Modules/Publications/spire-roadmap_december_2013_pbp.pdf (accessed on 25 October 2019).

36. Benito, P.; Mugdal, S.; Dias, D.; Jean-Baptiste, V.; Kong, M.A.; Inman, D.; Muro, M. Water Efficiency Standards. In *Bio Intelligence Service and Cranfield University, Report for European Commission (DG Environment)*; Bio Intelligence Service: Paris, France, 2009.

37. UNIDO. Industrial Development Report 2011. In *Industrial Energy Efficiency for Sustainable Wealth Creation*; United Nations Industrial Development Organizations: Vienna, Austria, 2011; UNIDO: ID No.: 442 (2011); ISBN 978-92-1-106448-3.

38. Florke, M.; Alcamo, J. *European Outlook on Water Use*; Final Report (2004), EEA/RNC/03/007; Center for Environmental Systems Research, University of Kassel: Kassel, Germany, 2004.

39. European Commission, 2030 Climate and Energy Framework. Available online: https://ec.europa.eu/clima/policies/strategies/2030_en (accessed on 25 October 2019).

 *sustainability*

Article

# A HAZOP with MCDM Based Risk-Assessment Approach: Focusing on the Deviations with Economic/Health/Environmental Impacts in a Process Industry

Panagiotis K. Marhavilas *, Michail Filippidis, Georgios K. Koulinas and Dimitrios E. Koulouriotis

Department of Production and Management Engineering, Democritus-University of Thrace, 67132 Xanthi, Greece; michalisfilippidis@gmail.com (M.F.); gkoulina@pme.duth.gr (G.K.K.); jimk@pme.duth.gr (D.E.K.)
* Correspondence: marhavil@pme.duth.gr

Received: 12 December 2019; Accepted: 28 January 2020; Published: 30 January 2020

**Abstract:** A joint-analysis by the use of (i) the multi-criteria decision-making (MCDM) technique of Typical-Analytical-Hierarchy-Process (T_AHP) and Fuzzy-Analytical-Hierarchy-Process (F_AHP) with (ii) the Hazard and Operability (HAZOP) method respectively, was conducted in a sour-crude-oil process industry (SCOPI), focusing on the deviations with economic/health/environmental impact. Consequently, the intention of this paper is dual; that means the first one is the presentation and application of the traditional HAZOP method on a SCOPI, and the second one the illustration and usage of the combined HAZOP_TAHP/FAHP framework in the previous referred industry, via the resulted hazards with economical, health and also environmental impacts, unveiling the sustainable dimension of this approach. The choice of the particular SCOPI for executing the HAZOP process is greatly powerful for the implementation of this approach, due to the plentiful hazards that the process includes, and due to its sophisticated design, as well. Originally, the conventional HAZOP study was carried out for finding out the potential destructive causes of abnormal circumstances in the SCOPI's establishment. Subsequently, the defined (by HAZOP) hazards with particular economic, health and environmental impacts were elaborated by the T_AHP and F_AHP techniques in order to broaden the HAZOP results and prioritize the risks in the worksite of the SCOPI. It is worthwhile mentioning that this joint-analysis can afford a reliable base to enhance procedure safety and upgrade the occupational health safety's level (OHS). Likewise, it might be a constructive means for the decision-makers to: (i) evaluate the urgent situation of the restricted resources' investment, in measures of preventing particular deviations (like the ones with economical, health and environmental consequences), and (ii) to endorse the sustainable growth of this industry, taking into account that a significant part of literature utilized the issue of sustainability as a frame for the development of OHS RAA techniques.

**Keywords:** crude-oil; sour gas; hydrogen sulfide; crude stabilization; hazard identification; HAZOP study; risk analysis; operability study; process safety; risk-matrix; DMRA; multi-criteria decision-making; Analytical-Hierarchy-Process; Typical-AHP; Fuzzy-AHP

---

## 1. Introduction

Although the chemical process industries (CPI) are among the safest workplaces in the world [1–3], mainly due to the severe procedures applied towards hazards, the public sense about them is that they are extremely hazardous with people fearing for their lives and/or health, and on the other hand, for their belongings and for the environment as well. Even so, throughout the last twenty years, the risk management and process safety actions were unfolded to a remarkable level, so that disasters like Flixborough (England, 1974), Seveso (Italy, 1976), Bhopal (India, 1984), Phillips (Pasadena/Texas, USA, 1989), AZF explosion (Toulouse/France, 2001) and Jilin (China, 2005) start to become lighter from memory. Above and beyond, a SCOPI can be greatly valuable for the local economies due to: (i) the work opportunities and (ii) the broad financial benefits it provides. Hence, even though there is always an existing likelihood for a catastrophe commencing, on the other side, disasters could be eliminated by proper risk management.

The Hazard and Operability analysis (abbreviated as HAZOP) emerged with the purpose of recognizing potential hazards in establishments that operate using extremely harmful materials. The major care was to abolish every source that can probably lead to a serious accident, such as explosions, fires and toxic-release [4]. Nevertheless, with the passage of time, HAZOP's handling was expanded to various other kinds of services because of its capability, not only to recognize hazards, but also to identify functional deviations from the preferred state. Thus, HAZOP technique has been chosen, among other actions, for road safety actions [5], hazard identification in photovoltaic installations [6], and diagnostic systems in medicine [7]. The miscellany of fields, that used HAZOP, shows that this methodology has been constituted as a dominant technique to meliorate numerous types of systems.

On the other hand, the Analytical-Hierarchy-Process (AHP), introduced by [8] was recognized as one of the more frequently used and proficient multi-criteria decision-making methods. Many applications (which employed AHP) were used for assessing health and safety risk factors such as the pioneer study of Henderson and Dutta (1992) [9], in which ergonomic factors were assessed. Other interesting applications were the multi-criteria-based occupational health safety (OHS) risk-evaluation method proposed by Badri et al. (2012) [10] and the study of Aminbakhsh et al. (2013) [11] which employed Typical-AHP (T_AHP) for assigning priorities to safety risks in construction sites.

Over and above, one of the highly significant features of risk management is the hazard identification, while a significant means is the HAZOP method. According to Mayes and Kilsby (1989) [12], HAZOP is a greatly versatile technique for hazards' identification in the chemical engineering industry, and has been confirmed, over the years, to be one of the most essential tools in order to ensure: (i) the safety of the employees in a plant, (ii) the environment, and naturally (iii) the probity of the facility itself [2]. Besides, taking into account that HAZOP's major intention is hazards' identification, it is assumed as the most excellent PHA technique of producing data for consequent quantitative-analysis techniques like QRA or FTA [2,13]. What is more, as the motivation for all industries (and the same for a SCOPI) is the turnover's raise, HAZOP analysis has additionally the considerable potentiality of identifying (i) not only the hazards for health and safety, but in addition (ii) the functional establishment's troubles, and (iii) the deviations with economic and environmental impacts, as well.

Nonetheless, although being efficient and well-organized for hazards' identification and risk-ranking in hazardous systems, HAZOP has its weaknesses; it is susceptible to ambiguous decisions, and, additionally, it has a variety of restrictions, like the subsequent: (a) it considers equivalent weights for the risk-factors (i.e., "low-probability/high-consequence" and "high-probability/low-consequence" hazards are approximately equivalent as far as the risk-ranking is concerned), and (ii) it utilizes accurate data, which are seldom available, particularly in the SCOPI's establishments [14,15].

On the other hand, risk prioritization (by T_AHP and F_AHP) can certainly help a manager: (i) to define the most urgent risk factors and implement the most significant safety measures, in order to attain the best allotment of a limited budget in risk-diminishing projects, and (ii) to degrade and/or eradicate the most critical risk-deviations (determined by HAZOP) with economic/health/environmental impacts by fulfilling the most imperative safety measures. In other words, improved distribution of the constricted funds: (a) can decrease support and financial costs, and (b) can normally allow managers to have available resources for further risk-diminishing in the enterprise [16,17].

This work concentrates (i) on an extensive HAZOP methodology with the DMRA tool (i.e., decision-matrix risk-assessment), combined with the Typical-Analytical-Hierarchy-Process

(T_AHP) and also with the Fuzzy-Analytical-Hierarchy-Process (F_AHP), and (ii) on the jointed HAZOP_TAHP/FAHP approach, executed as a case study in a SCOPI's establishment. Accordingly, the paper's intentions are the following: (i) the presentation and application of the HAZOP process in a specific SCOPI, (ii) the execution of the combined HAZOP_TAHP/FAHP framework on this SCOPI (focusing on the defined hazards that have economic, health and environmental impact), and additionally (iii) to disclose that the proposed HAZOP_TAHP/FAHP scheme can be a valuable technique for the decision-makers to meliorate OHS and process safety, and also to support the sustainable-operation of this SCOPI, given that a considerable part of literature utilized the matter of sustainability as a frame for the progress of OHS-RAA techniques [18].

In particular, the pathway toward accomplishing the sustainable development targets in a corporation, is achieved by intensifying amalgamation among different fields, and is associated with various key topics, like: (i) sustainable environment (that concerns the dynamics between human activities and ecosystem, including energy, chemistry, climate change, etc.), (ii) sustainable living (that means the achievement of health, carbon neutral and zero waste communities, including among other aspects, industrial production and consumption, buildings and infrastructure, supply chain and transportation, etc.), and (iii) sustainable technology (i.e., technologies that improve sustainability in all aspects) [19] (ICST&D, 2019).

Principally, a separate technique could not accomplish the best RAA performance in the industries' workplaces, and consequently, the development of an integrated RAA technique, which will combine a well-considered selection of comprehensive techniques (like HAZOP, DMRA, T_AHP, F_AHP, etc.) would allow industries (by the joint-analysis) to attain significant results in risk assessment [20–22].

The paper is structured by the subsequent parts: (1) introduction, (2) theoretical background, which concentrates on the conventional HAZOP method, on the DMRA risk-ranking technique and on the T_AHP/F_AHP processes, (3) the technical illustration of a particular sour-crude-oil process industry (SCOPI), (4) the development of the proposed HAZOP_TAHP/FAHP approach and its application, as a case study, on the precise SCOPI, and (5) the discussion of the foremost points and conclusions.

## 2. Theoretical Background

### 2.1. The Hazard and Operability (HAZOP) Procedure

HAZOP study was introduced, in the decade of 1960, through the "crucial examination" procedure of ICI (the Imperial Chemical Industries); while a decade afterward it was officially published as a restricted procedure for defining deviations from the design target. More explicitly, the HOC/ICI (Heavy Organic Chemicals) tried to design a construction for producing acetone and phenol from cumene, and so that, the responsible management team identified hazards or deviations from the normal operation. Thus, the HAZOP process (or another one similar to the current standardized method) appeared.

Furthermore, Lawley, in his interesting article of 1974, described the required principles to accomplish HAZOP analysis/studies due to the rising difficulty of the novel technique that might not be comprehensively explained by the normal standards [23,24], while the first primary directions concerning HAZOP were published in 1977 [25], and thus, it was recommended as a method to be applied in the process industries in order to recognize hazards and prepare safety actions.

The most important issue of the HAZOP technique is the detection of such design features that are susceptible to deviations which may result in a state of affairs within the study's intentions or aims. Normally, the design goal for a process is composite with some features that are subtle and its description is demanding, and on the other hand, the HAZOP study's team frequently identifies the design's goal plainly by choosing process factors from a checklist with no complete consideration of all key features of the design's goal [26]. To continue, Markowski and Siuta (2018) [27] developed a fuzzy-relied HAZOP for PHA of LNG storage services, where various indices are set up for managing the study's effectiveness and its quality as well. A widespread literature review for HAZOP technique has been recently incorporated in the work of Marhavilas et al. (2019) [2], while in the review of RAA techniques by Dallat et al. (2019) [28], some HAZOP derivatives are additionally referred.

Despite its usefulness, the HAZOP study has its own weakness and it shares limitations with other process hazard-analysis techniques [29]. More explicitly, Baybutt [29] presents a detailed critique wherein restrictions of HAZOP coming from the people executing the HAZOP study, in association with the deviations, the initiating events, the design intent, the technical coverage, the guide words, and the documentation, are discussed exhaustively. Moreover, Sultana et al. (2019) [3] notice that complex systems could not be suitably analyzed by conventional hazard-analysis tools (like the HAZOP one) and/or plain reliability-analysis techniques (like the FMEA one), so other techniques, such as the systematic hazard-analysis technique of STPA (System Theoretical Process Analysis) can be complementary to conventional HAZOP.

Recently, Animah and Shafiee (2020) [30] in their analysis exemplify several of the HAZOP's advantages (like its helpfulness in cases of risks which are complicated to be quantified in the LNG sector), and on the other hand, the disadvantages (like its incapability to appraise the risks appeared in a multi-component system, wherein diverse parts interrelate with each other).

The HAZOP technique needs the joint efforts of a team that is responsible for the system's management and is separated into the next stages: preparation, organizing, and execution/documentation [31,32].

Several factors are taken into consideration for the proper preparation of the HAZOP process, and the specification of its steps, like the consequent of the interrogative words: "how", "why", "where", "when", and "who" [32–35].

HAZOP technique cannot be performed at the conceptual stage of a project, due to the lack of any comprehensive P&ID, i.e., any "Piping and Instrumentation Diagram" at that moment. A hazard appraisal has to be achieved for identifying possible hazards existing in the process. Through the design phase, wherein its basic concept is entirely defined, the P&IDs are becoming available in order to provide adequate information for the deviations' investigation step of the study. In addition, HAZOP process is an extra handy process for accomplishing a comprehensive appraisal, earlier than installations' commencement [25].

Initially, HAZOP was considered appropriate only for new installations, but shortly after, was recognized as a valuable tool for existing establishments as well, because of the introduction of new regulations (such as SEVESO-directive), which require the performing of a periodic hazard-analysis in a facility, and due to the increasing safety-consciousness in industry. The possible hazards in an existing establishment cannot be calculated or recognized at the commencement of its functioning because the elaborated processes could change noticeably over time. In the case that a hazard might be present in the process, the existing HAZOP process should be revised, according to Baybutt (2013) [36]. Taking into account the article of Frank and Whittle (2001) [37], three types of HAZOP revalidation exist, such as, support, revision and replacement.

The study's definition has an evident consequence on the analysis's content, and also settles the criteria that the team's leader will use in order to gather the team-experts for conducting the HAZOP P-study. The project-leader in collaboration with the HAZOP's team-leader will determine the two key-topics of the process; initially, the details of its intention (i.e., why the study is performed), scope (i.e., what is included in the study) and objectives (i.e., what is to be considered), and secondarily, the efficient HAZOP team to realize the specific objectives (more information in the guidelines for defining HAZOP studies, by CCPS (1995)) [38].

Furthermore, the study's scope will comprise the whole equipment/tools which are used in the installation. Eventually, the intentions must always contain the identification of hazards in the establishment and the possible operating troubles. The hazards, according to Qureshi, can be safety/health and/or environmental ones, which might come from, or create an impact, both on- and off-site [39].

There are various types of hazards in a variety of installations that the HAZOP process might take place (for instance, fire, explosion, physical explosions, chemical explosions, toxic release, reactivity, etc.). Several of the potential hazards which must be recognized, particularly in process-installations that manage exceedingly hazardous materials, have been published by CCPS [40] in a catalog. This one constitutes a guidelines-list of the objectives as they are the hazards that need to be determined in the HAZOP analysis [41].

A HAZOP team involving: (a) more than seven (7) members, is not creative enough (mainly, due to communication problems), and (b) fewer than five (5) members, typically does not cover the essential knowledge-field for ensuring the execution of a thorough study. Consequently, the best size for the HAZOP's team must be equal to five, six or seven members [42].

The HAZOP analysis is a structured technique of defining forthcoming hazards. Hence, one critical point for the success of the effort is the study's planning into manageable parts, suitable for analyzing. The appropriate partition into nodes, with the right sizes, is a complicated job, so the practice can be considered as a subject of an art, due to the fact there is no aided node-management, but on the other hand, it relies completely on the judgment of the HAZOP team leader [43].

Chemical processes incorporate a broad range of interrelated components and auxiliary devices to implement the establishment's design intention (for instance, the crude-oil stabilization), so the equipment splitting into various nodes could be thorny due to their interconnectivity in the specific process industry. Thus, there are two views, which the HAZOP team's leader has to consider, when he decides the selection of the nodes, i.e., the node-parts ought to be equivalent as far as the size is concerned, and the involved equipment of a node ought to share the similar design and aim [2].

The members of the HAZOP team can recognize hazards that originate from various situations, such as: (a) deviations, that influence more than one-node (for instance a "flooding", or an "electric-power failure"), (b) situations, which also influence more than one-node (like "human factors", "piping and vessel physical configuration" and others), and (c) multiple-failure scenarios [42].

The HAZOP execution sub-stage of the study, associated with a division of the system into nodes, with particular interconnectivity and operation, follows the arrangement of nodes. The method is relied on the handling of specific guidewords, like the subsequent: (i) "No", (ii) "More" and (iii) "Less", in amalgamation with a variety of parameters, like the following: (i) "Flow" (ii) "Pressure" and (iii) "Temperature", in order to generate deviations from the regular functioning. The HAZOP team investigates the potential hazards of a node and proceeds to the identification of specific causes or/and consequences, constituting a scenario with safeguards which can detect/prevent and control/alleviate the hazardous circumstances. Ultimately, when the safeguards are inadequate to appropriately manipulate the specific problem, subsequently, the HAZOP team's members will be obliged to give its advices [41].

Even though, HAZOP process "knows" how to adapt the diverse requirements, there are some critical and most frequently used guidewords [44], as follows:

- "No" (alternatively: none/not), that means, nothing of the aim is realized and nothing else happens, as well.
- "More" (alternatively: more of/higher), that means, the design's target is accomplished in such a way which is quantitatively greater.
- "Less" (alternatively: less of/lower), that means, the design's target is implemented by a quantitative reduce of the parameter it is associated to.
- "As well as", that means, the design's aim is succeeded in parallel with something else (e.g., an extra action).
- "Part of", that means, the design's aims are achieved, only partially.
- "Reverse", that means, the logical opposite of the design's goals really happens.
- "Other than", that means, nothing of the intent is realized, and on the other hand, something entirely unusual happens.

The combination of these guidewords with process factors (or parameters) defines particular deviations from the design's goals.

### 2.2. Risk Ranking

It is a valuable procedure for prioritizing the recommendations resulted by HAZOP, and it is attained by the evaluation of the severity (S) and probability (P) of hazardous event sequences, in order to produce risk-outcomes which are illustrated in a table (e.g., matrix), implementing in this way the DMRA (i.e., decision-matrix risk-assessment, or risk-matrix) technique as an efficient tool for risk-estimations [45].

As soon as the hazards have been defined, the issue of designating the ratings of severity (S) and probability (P) must be arranged. It is essential to note that frequency estimations and consequence estimations are successfully determined by experienced risk–managers. Thus, the risk (R) is produced by establishing the risk–matrix according to relation of R-S x P [46].

In the recent work of Marhavilas et al. (2019) [2], the reader could find: (i) the necessary risk-ranking produced by the combination of the severity/reasons-frequency grades, according to the guidelines of IEC (2003) [47], and (ii) the resulted risk-matrix, according to Alaei (2014) [13].

### 2.3. The Typical- and Fuzzy-Analytical-Hierarchy-Process

During the past, there are many studies that use multi-criteria decision-making methods in cooperation with risk management methods, widely used in process industries. However, these studies used only Typical-AHP or Fuzzy-AHP, separately, and no comparative analysis was performed about the specific circumstances under which using fuzzy extensions of AHP could benefit the whole analysis.

More specifically, Abdelgawad and Fayek (2010) [48] used a Fuzzy-AHP approach to address the limitations of the traditional "failure mode and effect analysis" (FMEA) for managing risks in construction, and Zheng et al. (2012) [49] employed the Fuzzy-AHP method with trapezoidal fuzzy numbers for safety assessments in hot and humid environments.

Additionally, Basahel and Taylan (2016) [50] proposed a method for finding, assessing and evaluating safety risk factors at the construction sites (in Saudi Arabia), while Pour and Gheorghe (2017) [51] applied Fuzzy Risk Assessment (FRA) and AHP for determining and prioritizing the aggregate risk of oil and natural gas drilling process.

What is more, Gul (2018) [52] presented a significant review of OHS-risk assessment studies, via multi-criteria decision-making-based (MCDM) techniques. He incorporates in his work the fuzzy-side of MCDM techniques concerning OHS risk-assessment, and also refers to a structured methodology that uses prioritization (through T_AHP) in HAZOP analysis (presented by Othman et al. (2016) [14]).

In the article of Zhen et al. (2018) [53], a modified Fuzzy-AHP approach, used for importance degree of assessment of human, operational and organizational risk influencing factors at offshore maintenance work, and Fattahi and Khalilzadeh (2018) [54] developed an approach for evaluating risks in steel industry using FMEA, Fuzzy-AHP and also Fuzzy Multi-Objective-Optimization based on the Ratio-Analysis (MULTIMOORA) methods.

**The Typical-AHP (T_AHP):** The method defines the problem as a hierarchy and calculates the priorities (local and global) of the problem's factors (criteria and sub-criteria). The process' inputs are the decision-maker's judgments, namely the relative significance of each factor over another while performing pair-wise comparisons. The output result is a ranking of factors in descending order of their importance. The decision-maker must express the relative importance of the factors using a standard scale (Table 1) in order to transform the qualitative estimations of importance to numerical values.

One of the most important advantages of the T_AHP method is the integrated calculations for the decision-maker's consistency of judgments. Thus, the index CR (Consistency Ratio) is evaluated in order to check that the expert has a stable axiom system and is not assigning priorities "randomly". A pair-wise comparison matrix is characterized as consistent if the corresponding CR index is less than 10%. This characteristic improves the reliability of the results.

**Table 1.** The scale used to transform qualitative estimations to numerical values according to Saaty (1990) [8].

| Linguistic Variable<br>(Expressing the Significance of Factor I over Factor J to the Objective) | Value |
|---|---|
| "Equal" | 1 |
| "Equal to Moderate" | 2 |
| "Moderate" | 3 |
| "Moderate to Strong" | 4 |
| "Strong" | 5 |
| "Strong to Very Strong" | 6 |
| "Very Strong" | 7 |
| "Very Strong to Extremely" | 8 |
| "Extremely" | 9 |

**The Fuzzy AHP (F_AHP):** Besides the proven efficiency of the Typical-AHP (T_AHP), it was criticized for not taking into consideration that there is uncertainty to the decision-maker's judgments. For this reason, the Fuzzy Extended AHP proposed by Chang (1996) [55] using Extent Analysis Method was combined with Triangular Fuzzy Numbers (TFNs) to better correspond the expert's choices. The fuzzy extended AHP is a very popular AHP alternative because of its simplicity and its proven efficiency. In our approach, the decision-maker's judgments are inserted and assigned to numerical values using a scale, proposed by Lamata (2004) [56], and are illustrated in Table 2. Note that each linguistic variable corresponds to a value of the standard Saaty's scale and then to a Triangular Fuzzy Number (TFN).

**Table 2.** Linguistic variables to Saaty's scale and Triangular Fuzzy Numbers (TFNs) according to Lamata (2004) [56].

| Linguistic Variable<br>(Expressing the Significance of Factor I over Factor J to the Objective) | Saaty's Scale | Triangular Number |
|---|---|---|
| "Equal" | 1 | [1,1,1] |
| "Equal to Moderate" | 2 | [1,2,3] |
| "Moderate" | 3 | [2,3,4] |
| "Moderate to Strong" | 4 | [3,4,5] |
| "Strong" | 5 | [4,5,6] |
| "Strong to Very Strong" | 6 | [5,6,7] |
| "Very Strong" | 7 | [6,7,8] |
| "Very Strong to Extremely" | 8 | [7,8,9] |
| "Extremely" | 9 | [8,9,9] |

A Triangular Fuzzy Number $T = (t1, t2, t3)$ has as membership-function the one of $\mu_M(x)$ defined by Zimmermann (2001) [57].

$$\mu_m(x) = \begin{cases} 0, & x \leq t1 \\ \frac{x-t1}{t2-t1}, & t1 \leq x \leq t2 \\ \frac{t3-x}{t3-t2}, & t2 \leq x \leq t3 \\ 0, & x \geq t3 \end{cases} \tag{1}$$

Two TFNs $T_1 = (t1_1, t2_1, t3_1)$ and $T_2 = (t1_2, t2_2, t3_2)$ can be summed, subtracted, and multiplied with the operational laws described by Chang (1996) [55]. According to the process of the FEAHP, for a set of objects $O = \left\{o_1, o_2, \ldots, o_n\right\}$, and a goal set $GS = \{g_1, g_2, \ldots, g_n\}$, the extent analysis is applied resulting to $m$ values for each object:

$$T_{g_i}^1, T_{g_i}^2, \ldots, T_{g_i}^m, i = 1, 2, \ldots, n \tag{2}$$

where all $T_{g_i}^j$ $(j = 1, 2, \ldots, m)$ are TFNs. The value of the fuzzy synthetic-extent, for each $i_{th}$ object, is described by:

$$FSE_i = \sum_{j=1}^{m} T_{g_i}^j \otimes \left[\sum_{i=1}^{n} \sum_{j=1}^{m} T_{g_i}^j\right]^{-1} \tag{3}$$

These fuzzy numbers' output is transformed to crisp numbers using a defuzzification process. In the present study, we perform an average value approach, according to the work of Zimmermann (2001) [57], where a fuzzy number TFN $T = (t1, t2, t3)$ is altered into a crisp number of CRN as:

$$CRN(M) = (l + m + u)/3 \tag{4}$$

### 3. Presentation of the SCOPI

In this section, we give a technical illustration of a particular sour-crude-oil process industry (SCOPI), which has been chosen as a case study in this work, and additionally (a) constitutes the binder between the manuscript's sections concerning HAZOP, T_AHP and F_AHP techniques, and (b) could represent something in general, as far as the benefits of the application of the cooperative HAZOP_T AHP/F AHP framework on SCOPIs and other chemical industries, is concerned.

**Depiction of the plant:** A SCOPI performs a series of chemical and mechanical processes, which are applied on the sour-crude-oil (SCO), i.e., the one including hydrogen-sulfides and coming from the oil-production platform, in order to produce marketable crude-oil, incorporating the next units:

(i) The unit for the desalinization and stabilization of crude-oil: It desalinates and stabilizes approximately of 100 m$^3$/hr crude-oil, coming from the platform, while the produced sour-gas is of ~70,000 m$^3$/day.

(ii) The unit of the gas-compression, essential for compressing the sour-gas from stabilizer, and merging this with the stream of sour gas coming from the platform after it passes through a vessel, called three-phase separator that separates the well-fluids into: (a) gas and (b) two liquid kinds (oil and water). The cooling process of the two streams is realized by heat-exchanger units used to cool and/or condensate process streams with ambient air as the cooling medium rather than water, while later through a three-phase separator, there is a separation of the water and the "NGL".

(iii) The unit for "NGL"-stripping, wherein the produced "NGL" is stripped of several volatile hydrocarbons and hydrogen-sulfide (H$_2$S), before its mixture with the stabilized crude-oil.

Moreover, there is a variety of processes which happen, like the subsequent:

(i) The steam of 20.7 bar is directed to heat-exchangers in order to provide the essential energy for the "NGL"-stripping and also the crude stabilization

(ii) The water is mixed with the SCO to improve the desalinization process

(iii) The compressed dehydrated air to supply control air-actuated valves' functioning

(iv) The inaction of vessels or pipelines is achieved by occasional use of Nitrogen

The input-stream of the SCO arriving at the plant (through an 8″ pipe), has a pressure of ~20 bars, a temperature equal to the sea's one (i.e., presents a seasonal fluctuation from 14 °C (winter) to 25 °C (summer)), and constitutes a mixture that contains: (a) saturated volatile hydrocarbons, (b) a variety of volatile compounds (e.g., Hydrogen-Sulfide/ Nitrogen/Carbon-dioxide), (c) heavier compounds and (d) medium volatility hydrocarbons. The unprocessed SCO that inserts in the unit also includes dissolved sodium, salt-water, magnesium-chlorides, and calcium in tiny amounts of water. The salt-water is scattered in the crude-oil in the form of little balanced droplets, and due to their diameter, the droplets could be divided by an electrical field or by gravity in calm conditions. A characteristic laboratory analysis of the inward SCO is illustrated in Table 3.

**Table 3.** The input-composition of Sour-Crude-oil (SCO).

| Volatile—Low Molecular Weight | | | | | | | | | | | |
|---|---|---|---|---|---|---|---|---|---|---|---|
| **Component** | $N_2$ | $H_2S$ | $CO_2$ | COS | $CH_3SH$ | $C_2H_5SH$ | $CS_2$ | $C_1$ | $C_2$ | $C_3$ | $iC_4$ | $nC_4$ |
| **MOLE %** | 0.01 | 21.89 | 0.68 | 0.01 | 0.03 | 0.01 | 0.04 | 0.65 | 1.06 | 2.99 | 1.24 | 3.32 |

| Medium Volatile—Medium Molecular Weight | | | | | |
|---|---|---|---|---|---|
| **Component** | $iC_5$ | $nC_5$ | $C_6$ | $C_7$ | $C_8$ | $C_9$ |
| **MOLE %** | 2.88 | 2.64 | 4.75 | 3.73 | 3.52 | 3.41 |

| Heavy Components | | |
|---|---|---|
| **Component** | 150–200 °C | 200–350 °C | Boiling Point > 350 °C |
| **MOLE %** | 5.17 | 7.35 | 34.61 |

**The intention of the establishment:** The unit has been designed to: (i) take away (from the SCO), water, hydrogen-sulfide, suspended particles, salts and a great amount of the volatile hydrocarbons, and (ii) transform SCO into stabilized crude-oil, which is proper and safer for the storage/transportation and additional treating in a refinery plant, and on the other side, has specific features as follows: (a) Vapor pressure-reid less than 12 PSI, (b) $H_2S$ content less than 15 ppm, (c) water content around zero, as it is possible, and (d) salt content less than 28 mg/lt. The subtraction of the dissolved hydrogen-sulfide gas is compulsory, due to its volatile, toxic and flammable features that could make transportation and storage considerably more unsafe. The existence of salts in the final product is dangerous because it can enhance the corrosion in the involved transportation pipelines and in the storage equipment. What is more, there are chloride salts, for instance magnesium-chlorides, which form the enormously corrosive hydrochloric-acid. Furthermore, the avoidance of the problematical function in the sour-gas handling unit is achieved by the separation of the produced sour-gas from its liquefiable contents, while the recovered "NGL" ought to be splitted from the containing $H_2S$, due to the fact of a hazardous-spiking occurring in the crude-oil.

Likewise, stabilizers decrease the stored crude-oil volatility aiming that the final product has specific features as follows: (a) Vapor pressure-reid less than 12 PSI, (b) $H_2S$ content less than 15 ppm, (c) water content around zero, as it is possible, and (d) salt content less than 28 mg per L. The subtraction of the dissolved $H_2S$ gas is compulsory, due to its volatile, toxic and flammable features that could make transportation and storage considerably more unsafe. The presence of salts in the final product is dangerous because it can enhance corrosion in the involved steel pipelines for transfer or transportation and in the storage tanks and vessels. What is more, there are chloride salts, for instance magnesium-chlorides, which form an acidic environmental damage-mechanism, the Hydrochloric (HCL) Acid Corrosion.

Furthermore, the avoidance of the problematical function in the sour-gas handling-unit is achieved by the separation of the produced sour-gas from its liquefiable contents, while the recovered "NGL" ought to be splitted from the containing $H_2S$, due to the fact of a hazardous-spiking occurring in the crude-oil.

**The equipment of the establishment:** In Table 4 we present synoptically the installation's equipment that is structured with heat-exchangers, compressors, vessels, pumps and control/relief valves, while an analytical description of their functioning, with extensive details and further information, is presented in the thesis of Filippidis (2017) [58] and the article of Marhavilas et al. (2019) [2]. Moreover, in Figure 1 we illustrate the process flow diagram (PFD) concerning the SCOPI's functioning, which is located in the north part of Greece. Besides, the PFD drawing depicts by dashed-line frames, the necessary HAZOP nodes (i.e., simpler parts) in order to perform, in the consequent sections, the HAZOP analysis.

**Table 4.** A synoptic presentation of the equipment of SCOPI.

| Nr | Parts of the Equipment | Basic Technical Clarifications |
|---|---|---|
| 1 | **Vessels** | **V-101:** A three-phase separator vessel that releases sour-gas is from the sour-crude-oil |
| | | **ME-101:** A desalinization vessel that separates the salts from the sour-crude-oil |
| | | **V-102:** A crude stabilizer column |
| | | **V-103:** A scrubber vessel that retains water which is contained in the gas output from the top of V-102 |
| | | **V-106:** A vessel that acts as a buffer between the potable water intake and the crude-oil pipeline injection point. |
| | | **V-107:** A "N G L" (Natural Gas Liquids) separator which collects the "N G L" from the cooled gas stream from E-104 |
| | | **V-108:** A condensate stripper column |
| | | **V-110:** A three-phase separator |
| 2 | **Heat–Exchangers** | *There are two cooling heat-exchangers using sea-water and one using fans. There are also two heating heat-exchangers that use 20.7 bar steam as a source of energy. There is also a heat recovery heat-exchanger.* |
| | | **E-101:** A heat-exchanger which uses steam to heat the sour-crude-oil in V-102 |
| | | **E-102:** A heat-recovery heat-exchanger that utilizes the energy from the stabilized crude-oil leaving V-102 to heat the incoming sour-crude-oil entering the unit |
| | | **E-103:** A cooling heat-exchanger that uses sea-water to achieve its purpose |
| | | **E-104:** An air-cooled heat-exchanger that is used to cool the stream of sour-gas originating from C-101 and V-101 |
| | | **E-105:** A heat-exchanger that uses steam to heat the sour-crude-oil in V-108 |
| | | **E-106:** A water cooled heat-exchanger using sea-water as a cooling medium |
| 3 | **Pumps** | **P-101:** A centrifugal pump to move the stabilized crude-oil from the bottom of V-102 through E-102 and E-103 to storage |
| | | **P-102:** A reciprocating pump to inject potable water in the sour-crude-oil stream |
| | | **P-103:** A centrifugal pump to move the sour "N G L" from the bottom of V-107 to V-108 for stripping |
| 4 | **Compressors** | **C-101:** A two-stage double action reciprocating gas compressor |
| 5 | **Control–Valves** | *There are three types of control valves, i.e., pressure, flow and level valves that control the respective element. They are pneumatic valves that draw their power from compressed air supplied from a utilities unit. If the air supply is lost, for whatever reason, the control valves will assume a predetermined position, either closed or open, to ensure the safety of the installation.* |
| | | **Pressure Control Valves:** PV-101, PV-102A, PV-102B, PV-103A, PV-103B, PV-113, PV-117, PV-118 |
| | | **Flow Control Valves:** FV-107, FV-110, FV-124, FV-127 |
| | | **Level Control Valves:** LV-101, LV-102, LV-104, LV-106, LV-109, LV-114, LV-120, LV-121 |
| 6 | **Relief–Valves** | *There are several pressure relief-valves and alarm-switches to ensure the safe operation of the facility.* |
| | | **PSV**103-104, **PSV**105–106, **PSV**107, **PSV**108-109, **PSV**110-111, **PSV**115-118, **PSV**116-117, **PSV**121-123, **PSV**126-133, **PSV**127-128, **PSV**129-132, **PSV**134-135, **PSV**138, PSV139, **PSV**140, **PSV**141 |

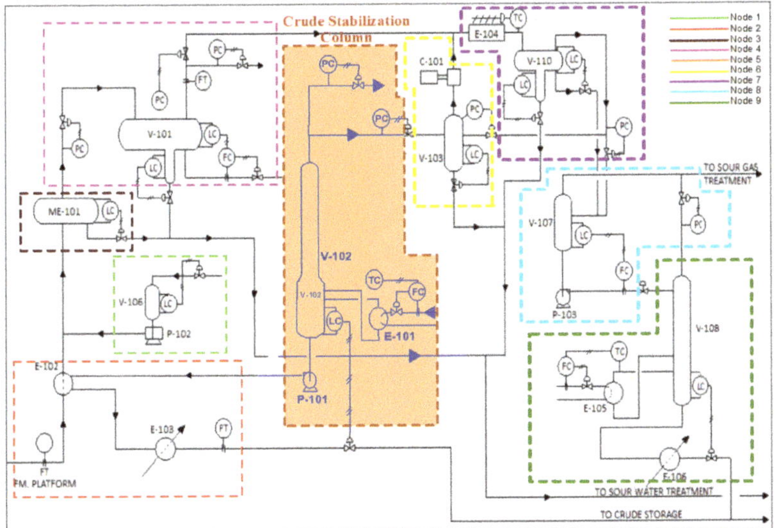



**Figure 1.** The Unit Process Flow Diagram (PFD) concerning the SCOPI's functioning.

## 4. Case Study: Application on a SCOPI Plant

### 4.1. The Procedural Framework of HAZOP_TAHP/FAHP

In Figure 2, we illustrate the flowchart of a RAA process (as a part of the risk management) after the incorporation of a HAZOP_TAHP/FAHP framework and the DMRA risk-ranking tool.

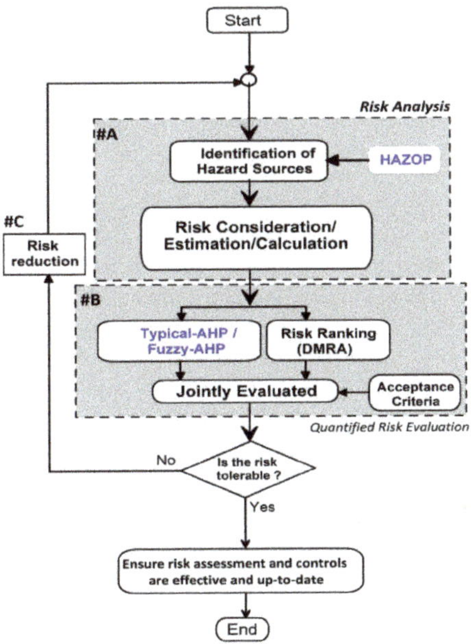



**Figure 2.** The flow diagram of a RAA process (as a part of the risk management) after the incorporation of a HAZOP_TAHP/FAHP framework.

This methodological framework is composed of three (3) separate sub-phases: the risk-analysis (#A) and risk-evaluation (#B), constituting the risk-assessment stage, and also the risk reduction/control (#C). In particular, the first sub-process, which includes the identification of hazard sources and the risk-consideration, incorporates HAZOP analysis, while the second sub-process considers the combination and joint-evaluation of Typical-AHP/Fuzzy-AHP and also DMRA, as a portion of QRE (i.e., quantitative risk-evaluation). The T_AHP/F_AHP process is applied for prioritizing the risk-factors of the project. Eventually, the third module #C incorporates safety-related decision-making. Besides, it is worthwhile mentioning, that the creation of this drawing was based on the safety-topics/guidelines of the works by Høj and Kröger (2002), Marhavilas (2015) and ISO/IEC (1999, 2009) [45,59–61].

*4.2. Application of the HAZOP Process*

The systematic HAZOP study was executed in a SCOPI, which is placed in NE-Greece, by the members of the HAZOP team which consisted of two unit-operators, e.g., the plant-operation administrator and the maintenance-department manager. The installation-operation superintendent is a capable chemical engineer with substantial capability in CPIs, whereas the maintenance-department superintendent is a mechanical engineer. The plant-operators have a significant ability for operating the SCOPI's units, and important experience on preceding incidents. The HAZOP sessions were accomplished through a six-month period, including entirely twelve sessions. The first two sessions included the explanation of the SCOPI process to the team, and also the selection of the HAZOP-nodes. Below, it is collocated the essential steps for implementing the HAZOP process, as follows:

**(1) Selection of the nodes and intentions:** Taking into consideration the equipment's operation and their interconnectivity, wholly nine nodes were chosen, such as:

(i)    **"Node–1"**, which is the "Water-injection system"
(ii)    **"Node–2"**, which is the "Reheating/cooling exchangers"
(iii)    **"Node–3"**, that means, the "Desalting process unit"
(iv)    **"Node–4"**, i.e., the "Low-Pressure L.P. separator"
(v)    **"Node–5"**, which is the "Crude stabilizer column"
(vi)    **"Node–6"**, that means the "Gas compressor"
(vii)    **"Node–7"**, e.g., the "Compressor gas cooling and liquids recovery"
(viii) **"Node–8"**, i.e., the "Condensate stripper column" that feeds vessel and "Pump P103"
(ix)    **"Node–9"**, the "Condensate s t r i p p e r and cooling unit"

**(2) Guideword and safe limits of the process**: The central idea of HAZOP method is the identification: (a) of deviations (i.e., hazards) from the normal and prospective (by the design) safe-state, and also (ii) the crucial process parameters, like: "Composition", "Flow", "Level", "Phase", "Pressure" and "Temperature". Moreover, the proper guidewords concerning the deviation from the safe operation are: "Less", "More", "No", "Other than" and "Reverse", which are combined with the previous referred process factors (parameters) in order to define the deviation from the conditions of a safe situation.

**(3) The identification of the causes and the hazards**: This step is realized after the previously referred two steps, whence the HAZOP team identifies the root causes for any hazard that is associated with a particular node.

**(4) The determination of consequences**: At this step, the members of the HAZOP team define the consequences of any hazard in association with safety, environmental, and economical aspects.

**(5) The identification of safeguards**: At this step, the HAZOP team documents, in a HAZOP-worksheet, the related safeguards in order to recognize the hazards and prevent the consequences.

**(6) Recommendations**: At this step, the necessary recommendations about decreasing the risk-level are discussed and analyzed by HAZOP's team members and are recorded in the worksheet of HAZOP.

It is worthwhile to mention, that the risk-matrix used in RAA, will have the structure presented by Table 4, involved in the paper of Marhavilas et al. (2019) [2]. Thus, (i) the probability of the recognized hazards and (ii) the failure-rate of ordinary control-equipment (like level/pressure-transmitters), were specified by the usage of an information-base of preceding incidents in SCOPI-units with equivalent equipment. On the other side, the evaluation of the hazards severity was performed with the accordance of the team's members and their individual decision. Besides, Table 5 illustrates the "Guidewords" and "Elements", which were used to generate the deviations produced by the HAZOP procedure.

**Table 5.** Depiction of "Guidewords" and "Elements" of the HAZOP study.

| Elements | | | | | | |
|---|---|---|---|---|---|---|
| Level | Flow | Pressure | Temperature | Composition | Phase | Containment |
| **Guidewords** | | | | | | |
| More | | Less | | Reverse | No | Other than |

At the beginning of the study, we depicted by drawing colored dashed-line frames, the existence of the previously referred nine nodes, as structural parts of the unit's 'process-flow-diagram' (PFD) presented in Figure 1.

The first node (Node-1) which is the water-injection system, was analyzed (by HAZOP) in association with the general process parameters (or elements) of "Level", "Pressure", "Flow" and "Containment".

Node-2, i.e., the preheating and cooling exchangers, was analyzed taking into account the elements "Flow", "Pressure", "Temperature", "Composition" and "Containment".

The HAZOP analysis of the installation's desalting vessel (i.e., Node-3) was achieved by using the elements "Level", "Flow", "Pressure", "Temperature", "Composition" and "Containment".

Besides, the HAZOP analysis of the low-pressure separator vessel (of Node-4) was conducted in association with the process parameters: "Level", "Flow", "Pressure", "Phase" and "Containment".

Node-5, which is the crude stabilization column-group of equipment, was HAZOP-analyzed taking into consideration the elements of "Composition", "Containment", "Flow", "Level", "Pressure" and "Temperature".

What is more, Node-6 (i.e., the 'gas compressor station' and its supplementary scrubber vessel) was analyzed taking into account the elements of "Containment", "Flow", "Level", "Phase", "Pressure", and "Temperature".

In addition, the HAZOP analysis of the compressor gas cooling and liquid-recovery system (in Node-7), was performed by the elements "Containment", "Flow", "Level", "Pressure" and "Temperature".

Moreover, Node-8, i.e., the condensate stripper column that feeds vessel V-107 and pump P103, was HAZOP-analyzed in association with the elements of "Containment", "Flow", "Level", and "Pressure".

Finally, the HAZOP analysis in Node-9, which is the condensate stripper and cooling system of the facility, was applied by the elements "Containment", "Flow", "Level", "Pressure" and "Temperature".

Taking into consideration the sizable SCOPI's plant, the large number of the HAZOP-nodes, and their operability as well, we concentrated, via this article, on a considerable node of the SCOPI's plant, as far as the risk is concerned. More explicitly, we selected Node-5, for applying (as an instance) the conventional HAZOP-DMRA technique, in order to identify the likely fault causes due to irregular deviations. Hence, for this node, which is shown in Figure 3, we operate and implement HAZOP analysis using the elements "Level", "Flow", "Pressure", "Temperature", "Composition" and "Containment" having as final outcomes the ones shown in Tables A1–A6, correspondingly. We elucidate, that the conventional HAZOP/DMRA technique was also performed for the rest nodes in order to achieve the full application of the HAZOP-DMRA scheme in the specific SCOPI.

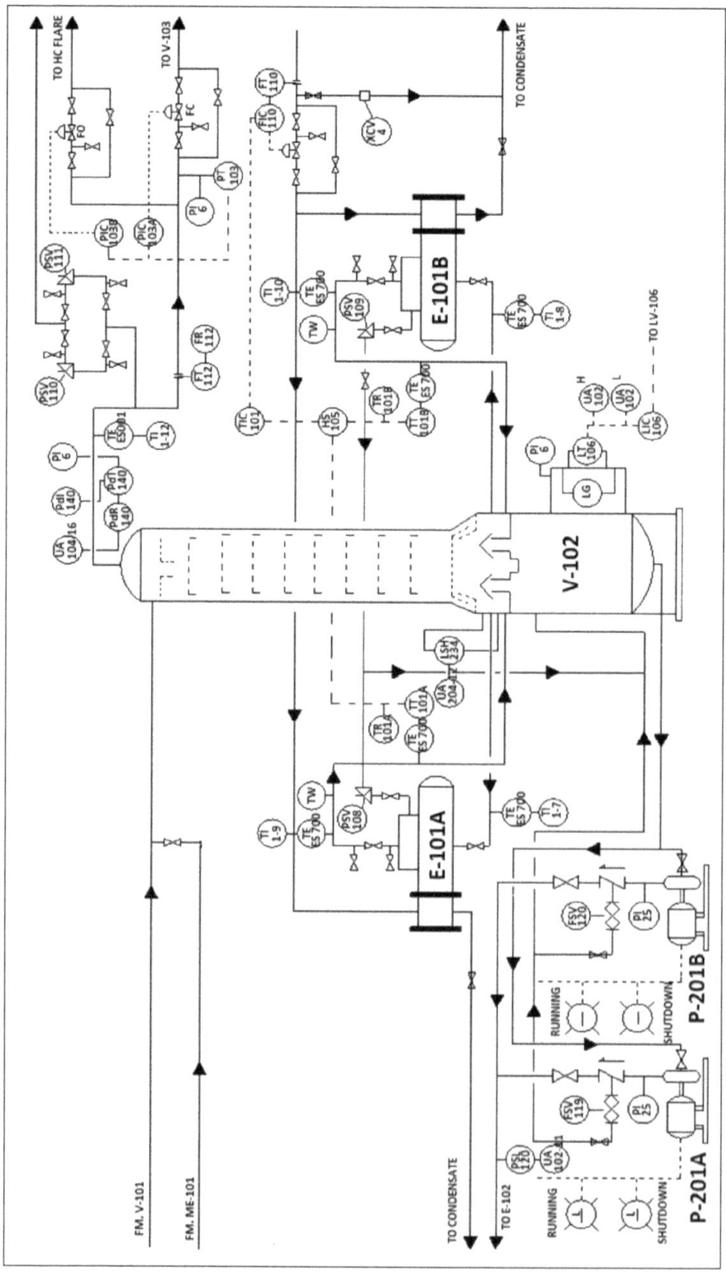

**Figure 3.** The crude stabilization column of "Node–5".

In the tables originated from the HAZOP study, all the plausible sources and hazards due to a particular deviation have been defined, and with the intention that HAZOP study forms a useful guide for providing a list of priorities to the maintenance/inspection staff. Many of them are ordinary for multiple deviations and some additional analysis was essential to determine distinctive hazards, which have been categorized into the next three classes: The first one corresponds to '**hazards to human life and the environment**' and has been analyzed thoroughly in the work of Marhavilas et al. (2019) [2]. The second one, presents the '**operability hazards**' i.e., the ones associated with the product quality and economic operation of the establishment and will be analyzed comprehensively in a forthcoming work. Finally, the third category includes the '**hazards with both economic and health (or environmental) impact**', which means the hazards that influence the economical performance of the establishment and concurrently cause a risk to human life and/or to the environment, and they are depicted in Table 6.

**Table 6.** Deviations (or hazards) that include "economic and health and/or environmental" impacts.

| Nr | Deviations/Hazards |
|:---:|:---:|
| "1" | Higher temperature crude to tanks due to more oil flow |
| "2" | Stabilized crude-oil containing $H_2S$ due to leak in E–102 |
| "3" | No crude stabilization due to column flooding |
| "4" | Volatile hydrocarbons to crude storage tanks from V–102 |
| "5" | Stabilized crude containing $H_2S$ due gas from V–102 |
| "6" | C–101 damage due to overheating |
| "7" | Hydrate formation in V–110 |
| "8" | "NGL" release in the sea |

Several serious deviations (given by Table 6) are the '**H2S releasing**' and also the '**hydrocarbons releasing**', taking into consideration the impacts of bad-breathing that might have happened to people, producing substantial problems concerning health. What is more, when **hydrocarbons** are taken into consideration, these would be liquid and/or gas. The discrimination can be hard due to the volatility of these components. In addition, "**NGL**" can be in liquid form, inside a vessel under pressure, and be evaporated nearly immediately, when it is exposed to atmospheric conditions. An additional important aspect, which has financial effects on the plant's operations, is the **crude-oil storage,** when this specific kind of crude-oil does have proper quality specifications (for instance, due to enhanced **H2S** content, and/or because of the existence of water/salt content).

*4.3. Application of the Typical-AHP and Fuzzy-AHP Processes*

In light of the outcomes of the prior HAZOP/DMRA scheme (shown by Table 6), we continue to the subsequent stage, e.g., the execution of the Typical-AHP and Fuzzy-AHP processes, on the SCOPI's plant (as a case study), by the usage of the hazards depicted in Table 6, so as to entirely implement the HAZOP_TAHP/FAHP approach.

The problem of assessing eight hazard factors with health and environmental impacts in a SCOPI is modeled by a hierarchy in which the goal is safety's prevention while the first level consists of eight hazard factors (Figure 4).

It is worthwhile to mention that, the decision-maker fills the pair-wise comparison matrix presented in Table 7 with numbers that correspond to his preferences about the risks' relative importance, as described by linguistic variables (Tables 1 and 2). Firstly, the Typical-AHP is applied to calculate each hazard's weight and construct the ranking. An important element of the Typical-AHP calculation is the consistency-ratio (CR) estimation, which, in a case that is below 10% means that the judgments of the expert are consistent and the ranking is reliable. Note that, in the present case the consistency-ratio is found to be CR ≤ 10%.

**Figure 4.** The hierarchy.

**Table 7.** The expert's judgments serving as input on both Typical-AHP and Fuzzy-AHP.

| Safety | [01] | [02] | [03] | [04] | [05] | [06] | [07] | [08] |
|--------|------|------|------|------|------|------|------|------|
| [01] | 1.00 | 3.00 | 2.00 | 3.00 | 6.00 | 0.50 | 3.00 | 0.50 |
| [02] |  | 1.00 | 0.33 | 0.20 | 2.00 | 0.17 | 0.50 | 0.25 |
| [03] |  |  | 1.00 | 0.33 | 3.00 | 0.20 | 0.50 | 0.25 |
| [04] |  |  |  | 1.00 | 5.00 | 0.50 | 2.00 | 0.33 |
| [05] |  |  |  |  | 1.00 | 0.14 | 0.25 | 0.13 |
| [06] |  |  |  |  |  | 1.00 | 4.00 | 0.33 |
| [07] |  |  |  |  |  |  | 1.00 | 0.50 |
| [08] |  |  |  |  |  |  |  | 1.00 |

The weights for each hazard calculated by Typical-AHP and F_AHP, are illustrated in Table 8 and also the rankings of hazards extracted by each method, in descending order of importance. It is worth mentioning that both methods use as input the same judgments pair-wise comparison matrix.

**Table 8.** Weights and rankings derived with both Typical- and Fuzzy-AHP.

| Weights for the Hazards (Typical-AHP) | | Final Ranking (Typical-AHP) | | Weights for the Hazards (F_AHP) | | Final Ranking (F_AHP) | |
|---|---|---|---|---|---|---|---|
| [01] | 16.50% | [08] | 28.17% | [01] | 18.00% | [08] | 25.29% |
| [02] | 3.92% | [06] | 22.53% | [02] | 4.70% | [06] | 25.11% |
| [03] | 6.19% | [01] | 16.50% | [03] | 8.51% | [01] | 18.00% |
| [04] | 12.22% | [04] | 12.22% | [04] | 16.14% | [04] | 16.14% |
| [05] | 2.38% | [07] | 8.08% | [05] | 2.66% | [07] | 10.26% |
| [06] | 22.53% | [03] | 6.19% | [06] | 25.11% | [03] | 8.51% |
| [07] | 8.08% | [02] | 3.92% | [07] | 10.26% | [02] | 4.70% |
| [08] | 28.17% | [05] | 2.38% | [08] | 25.29% | [05] | 2.66% |

According to the Typical-AHP calculations, the factors "NGL release in the sea" (08) and "C-101 damage due to overheating" (06) are more important since their weights are 28.17% and 22.53%, respectively. The second more important group consists of the factor "Higher temperature crude to tanks due to more oil flow" (01) with weight of 16.5%, and the (04) "Volatile hydrocarbons to crude storage tanks from V-102" with 12.22% score. The third more important group includes hazards with small differences on their weights but they are responsible for the 20.57% (cumulatively) of the total risk. The risk factors belonging to this group are "Hydrate formation in V-110" (07), "No crude stabilization due to column flooding" (03), "Stabilized crude containing $H_2S$ due to leak in E-102" (02), and "Stabilized crude containing $H_2S$ due gas from V-102" (05).

The calculation with Fuzzy-AHP, results to a same ranking of the hazards, as shown in the last column of Table 8. However, the distance of the weights between the first factor [08] and the second factor [06] is dramatically reduced, meaning that the F_AHP can better describe the uncertainty of the expert's choices. Additionally, it is observed that except the ranked—as more important—hazards, the scores of all the other hazards are increased, and especially the financial impact hazards ([06], [01], [03]) are responsible (cumulatively) for the 51.62% of the total risk instead of the 45.22% found by T_AHP. In addition, the scores of the hazards with impact on the workers' health are higher than these computed by Typical-AHP, namely the factor [04] is weighted to 16.14% and [07] to 10.26%. This finding is of great interest for business managers and safety executives who, usually, are concerned more about workers' safety and health issues, as well as, about business profits, rather than impacts on the environment. Additionally, the above findings could be used for prioritizing investments in health and safety prevention measurements.

These results verify the findings of the recent study of Chan et al. (2019) [62] wherein the circumstances under which the F_AHP is preferred over Typical-AHP are stated. More specifically, in our case, quantitative difference was observed while the pair-wise comparison matrix is highly consistent, and the importance of hazards is relatively close to one another. Due to the presence of the above conditions, the F_AHP is used, having as a result the efficiency's rising of the proposed approach, since the proven ability of Fuzzy-AHP to handle uncertainty is necessary for applications such as risk assessment in industry where ambiguous conclusions might affect health and safety of employees.

In the beginning of this article, the execution of the conventional HAZOP study is attained, to identify the potential consequences and causes which correspond to irregular conditions. Thus, several serious hazards/deviations (depicted in Table 6) are the following: "hydrocarbons releasing", "$H_2S$ releasing", and "NGL releasing", with regards to affecting breathing that may happen to people, causing substantial problems to their health. In association with the results of the HAZOP technique, on the one hand, and also the utilization of the DMRA for risk-ranking, on the other side, the T_AHP and the F_AHP procedure is then applied consecutively, to expand HAZOP study by prioritizing the risks in the SCOPI's establishment. The choice of the explicit SCOPI for executing the HAZOP study is greatly valuable for the implementation of this approach, due to a variety of hazards included in the process, and because of its innovative design.

The hazards' recognition (by HAZOP study) denotes that we could systematically assess and judge the SCO process. This technique can be, consequently, considered as an efficient method, for hazards' identification and prediction, and it may enhance the safety levels, obstruct accidents and increase the reliability of systems, through the decrease of operational and environmental problems [13]. The specific SCOPI, under consideration in this article, was proved to be trustworthy and safe, at such a level that right establishment inspections and maintenance are imposed. The arranged approach and systematic analysis that HAZOP technique provides, is a brilliant tool for judging the SCOPI's parts' operation. In general, the equipment of chemical industries (and of SCOPIs as well) is extremely expensive, because of the unique technical and quality specifications, and due to the fact, that such equipment is regularly custom-engineered to particular applications. Therefore, any act of changing (entirely or partially) the equipment, creates significant economic consequences to the corporation.

To be brief, the examined SCOPI's plant, despite its lengthy lifetime, presents tolerable risk in its functioning, because of the appropriate inspection and maintenance [2].

The main results derived by applying the T_AHP and F_AHP are the following: (i) The outcomes of the T_AHP and F_AHP process, agree (between them) on the final results as far as the hazards' ranking is concerned; (ii) The most important source of risk is the deviation Nr [08], which has an environmental impact. This finding is of great value given the fact, that both at the state/national level (with the development of a legislative framework for environmental impacts) and also at the international level (with global transnational agreements on the environmental impact of industrial activity), the international community is becoming more and more aware of environmental issues (e.g., $CO_2$ footprint, climate change, industrial pollutants, etc.); (iii) F_AHP enhances (in comparison with T_AHP) the hazards' importance factors which have impacts on the financial consequences on the company (e.g., deviations [06], [01], [03], [02]), as well as the impact on employee health (e.g., deviations [04], [07]). This may be of greater interest to business manager and safety executives, who are firstly concerned with the safety/health of workers, and the economical factor of their business, and less with the impact on the environment.

In the present approach, we used F_AHP and T_AHP, so as to rank hazards, while exploiting the expertise of the decision-maker. The hazards' ranking using F_AHP remains the same with the ranking using T_AHP, but the scores for each hazard are quite different, and the relative distances between them are smaller/larger, respectively. These different scores are very important, in case of spending budget, to safety measures, as they can lead to over/under-investing in measures for a hazard, resulting to lower effectiveness in prevention of accidents. Additionally, as stated in Chan et al. (2019) [62], given that there are present *"judgments with dominant preference and equal important elements"* in our case, the total effectiveness of the approach could be benefited by using the Fuzzy-AHP.

## 5. Discussion and Conclusions

This work concentrates on the spreading out of the HAZOP process (i.e., one of the most frequently applied in industry) and more particularly, on the widespread presentation: (i) of the conventional HAZOP study, and (ii) of an extended HAZOP methodology with DMRA, combined with the Typical-Analytical-Hierarchy-Process (T_AHP) and also with the Fuzzy-Analytical-Hierarchy-Process (F_AHP). Additionally, the cooperative HAZOP_TAHP/FAHP scheme is performed as a unified approach to identify and prioritize crucial points and impending hazards in a SCOPI (i.e., a case study), which is remarkably important, since it combines many hazards, like gases, toxic gases and flammable liquids as well. Hence, this is the reason for choosing the SCO chemical process as a case study.

This article provides a framework for systematically evaluating economical, health and also environmental impacts, and prioritizing the risks by hazards or deviations from the normal operation of a specific chemical process (i.e., a SCOPI). In particular, an individual technique cannot attain optimal RAA results in the workplaces, and future points of view should focus on the concurrent application of diverse RAA methods. Consequently, the development of an integrated RAA technique, which will combine a well-considered selection of wide-ranging techniques (like HAZOP, DMRA, T_AHP, F_AHP, etc.) would make able industries achieve substantial results in risk assessment [20–22].

Besides, this study is an effort to downgrade the negative aspects of conventional HAZOP through the jointed HAZOP_TAHP/FAHP approach. In the literature, the improvement of HAZOP methodology is underway, so that various techniques have been incorporated in traditional HAZOP studies, like the subsequent combinations: (a) HAZO–DMRA that presented by Alaei et al. (2014) [13], (b) Fuzzy-DMRA by Gul and Guneri (2016) [63], (c) HAZOP-AHP by Othman et al. (2016) [14] and Marhavilas et al. (2019) [2], (d) HAZOP-dynamic FTA by Guo and Kang (2015) [64], (e) Fuzzy-based HAZOP by Ahn and Chang (2016) [65], (f) Fuzzy multi-attribute HAZOP by Cheraghi et al. (2019) [15], (g) HAZOP-Fuzzy TOPSIS by Cheraghi et al. (2019) [15], and (h) HAZOP–Fuzzy AHP by Marhavilas et al. (2020) [66].

It is worth mentioning that for the first time, through this work, we designate the weight of risk-factors and prioritize the hazards by using together T_AHP and F_AHP in the environment of

HAZOP. Likewise, the usage of the HAZOP_TAHP/FAHP framework on a SCOPI shows that this combined technique displays a more distinct environment, as far as the hazards' ranking is concerned, compared to conventional HAZOP.

What is more, the Fuzzy-extended AHP was applied in the present study because of its usability, since only the expert has to fill in a pair-wise comparison matrix with relative importance of each hazard of the analysis over another. Next, the assignment of linguistic values (to Triangular Fuzzy Numbers) rises the efficiency of the analysis, as the uncertainty of the expert's judgments is taken into account, which leads to a more reliable ranking in applications such as health and safety hazards analysis, due to the impacts on the environment and workers.

It is noteworthy that by focusing on the deviations (illustrated in Table 6 and defined by HAZOP study) with economic/health/environmental impacts: (i) we broaden, through this article, the outcomes of our previous work, which have already integrated HAZOP with AHP [2], and (ii) we unveil the sustainable dimension of the HAZOP_TAHP/FAHP approach, having as the main gain, the improvement of the sustainable living in the specific SCOPI, taking into account that a significant part of literature [18] utilized the issue of sustainability as a frame for the development of OHS RAA techniques.

In fact, the idea of sustainability was used as a frame to develop OHS RAA methods, and the greater part of OHS subjects are associated with enterprises' compliance with legislation requests, while a considerable part of literature integrated OHS issues into the concept of social responsibility of organizations, which should go beyond the law by adopting voluntary OHS standards [18]. Furthermore, the implementation of the methodology that developed within this article could be useful to reorganize the production processes of some industries in order to reduce operational risk and to lead sustainable manufacturing. To this logic, the concept of sustainability is utilized as a frame for the development of OHS RAA techniques.

Therefore, this is the new contribution of this piece of writing (in comparison with previous work), and also one of its main aims. Likewise, an extra intention is to disclose that the proposed HAZOP_TAHP/FAHP framework could be a valuable means for decision-makers to meliorate OHS and process safety, as well, and to support the sustainable operation of this SCOPI.

We have the opinion that the usage of the new alternative framework, which is presented in this paper, would help industries in the achievement of occupational risk prevention. In future work, we are planning (a) the development of additional RAA frameworks (including more stochastic/deterministic techniques), and (b) the application on OHSS of SCOPIs and other industries. This means that according to Zheng and Liu (2009) [67], there is a potentiality for the combination of more different techniques (like neural networks, Markov chains, grey model, Bayesian networks, etc.) with traditional HAZOP.

In conclusion, the joint-analysis (by HAZOP_TAHP/FAHP) in this article can afford a reliable base to enhance process safety and meliorate OHS. Similarly, it might be a constructive means for the decision-makers to: (a) evaluate the urgent situation of investment's restricted resources in measures preventing particular deviations (like the ones with economic, health and environmental consequences), and (b) to endorse the sustainable growth of this industry. Besides, the novelty and key points of this study or, in other words, its contributions to "new knowledge" as far as health and safety science are concerned, could be focused on: (i) the development of an alternative beneficial RAA process (i.e., the HAZOP_TAHP/FAHP one), (ii) the improvement of RAA processes due to combination of separate RAA techniques with real data of deviations (hazards), (iii) the comparison of their outcome risk-estimation results (T_AHP vs. F_AHP), (iv) the application of this new technique (as a case study) on a SCOPI, which presents significant scientific interest, and (v) the enrichment of the scientific literature with new results for the OHSS and principally in the field of sour-crude-oil processes.

**Author Contributions:** Conceptualization, P.K.M., D.E.K.; data curation, P.K.M., M.F., G.K.K.; formal analysis, P.K.M., D.E.K.; investigation, P.K.M., G.K.K., D.E.K.; methodology, P.K.M., D.E.K.; writing—original draft, P.K.M., M.F., G.K.K.; writing—review and editing, P.K.M., M.F., G.K.K., D.E.K. All authors have read and agreed to the published version of the manuscript.

**Funding:** This research received no external funding.

**Conflicts of Interest:** We declare no conflict of interest.

## Acronyms/Abbreviations

| | |
|---|---|
| AHP | "Analytical Hierarchy-Process" |
| BDV | "Blowdown Valve" |
| CCPS | "Center for Chemical Process Safety" |
| CIA | "Chemical Industries Association" |
| CPI | "Chemical Process Industry" |
| DMRA | "Decision-Matrix Risk-Assessment" |
| EPSC | "European Process Safety Centre" |
| ESD | "Emergency Shutdown" |
| ESDV | "Emergency Shutdown Valve" |
| F_AHP | "Fuzzy-Analytical-Hierarchy-Process" |
| FEAHP | "Fuzzy Extended Analytical Hierarchy-Process" |
| FEED | "Front-End Engineering Design" |
| FMEA | "Failure Mode and Effect Analysis" |
| FTA | "Fault Tree Analysis" |
| HAZOP | "Hazard and Operability" |
| ICI | "Imperial Chemical Industries" |
| ICST&D | "International Conference on Sustainable Technology and Development" |
| IEC | "International Electrotechnical Commission" |
| LAHH | "Level Alarm High High" |
| LALL | "Level Alarm Low Low" |
| LNG | "Liquefied Natural Gas" |
| MCDM | "Multi-Criteria Decision-Making" |
| NGL | "Natural Gas Liquids" |
| OHSS | "Occupational Health and Safety System" |
| P&IDs | "Piping and Instrumentation Diagrams" |
| PHA | "Preliminary-Hazard-Analysis" |
| PFD | "Process Flow Diagrams" |
| PIC | "Pressure-Indicator-Controller" |
| PPE | "Personal-Protective-Equipment" |
| PSV | "Pressure-Safety-Valve" |
| PT | "Pressure-Transmitter" |
| PV | "Pressure-Valve" |
| QRE | "Quantified-Risk-Evaluation" |
| QRA | "Quantitative Risk Assessment" |
| RA | "Risk Assessment" |
| RAA | "Risk Analysis and Assessment" |
| SCO | "Sour-crude-oil" |
| SCOPI | "Sour-crude-oil Process Industry" |
| SIL | "Safety Integrity Level" |
| SMS | "Safety Management System" |
| STPA | "System Theoretical Process Analysis" |
| T_AHP | "Typical-Analytical-Hierarchy-Process" |
| TFN | "Triangular Fuzzy Number" |

## Appendix A

The appendix contains the tables which illustrate the outcomes from the implementation of HAZOP analysis using the elements "Level", "Flow", "Pressure", "Temperature", "Composition" and "Containment", concerning "Node-5".

**Table A1.** The "Level"-Element of "Node-5".

| Node | 5 | | | | | | | |
|---|---|---|---|---|---|---|---|---|
| **Element** | **Level** | | | | | | | |
| **Guideword** | **Deviation** | **Possible Cause** | **Consequences** | **Safeguards** | **Risk Matrix** | | | **Recommendations** |
| | | | | | **S** | **L** | **R** | |
| More | Higher level in V-102 bottom | 1. P-101 shutdown | | - 1.1 High Level Alarm LT-106<br>- 1.2 Spare P-101<br>- 1.3 LSH-134 alarm | 4 | 2 | AC | None |
| | | 2. LT-106 failure | - No production to storage | - 2.1 LSH-134 alarm | 4 | 3 | AC | None |
| | | 3. Oil line to storage blocked | - Column flooding, poor $H_2S$ stripping | - 3.1 LSH-134 alarm<br>- 3.2 High Level Alarm LT-106 | 4 | 3 | AC | None |
| Less | Lower level in V-102 bottom | 4. LT-106 failure | - Gas to storage<br>- $H_2S$ release in the tanks | - 4.1 PdT high alarm | 2 | 3 | 1 | Install an LSL alarm at V-102 |
| | Lower level in V-102 chimney tray | 5. Low oil input to V-102 | - No feed to E-101<br>- Poor crude stabilization | - 5.1 Crude from LV-106 sampling every 4 hours | 4 | 2 | AC | None |

**Table A2.** The "Flow"-Element of "Node-5".

| Node | 5 | | | | | | | |
|---|---|---|---|---|---|---|---|---|
| Element | Flow | | | | | | | |
| Guideword | Deviation | Possible Cause | Consequences | Safeguards | Risk Matrix | | | Recommendations |
| | | | | | S | L | R | |
| More | More crude-oil input flow | 6. FV-107 bypass open | - Poor crude stabilization<br>- Crude to V-103. Possible damage to C-101 | None | 4 | 4 | NSR | None |
| | More crude vapors output from E-101 | 7. Increased temperature in E-101 | - Pressure increase<br>- Pipeline fracture<br>- $H_2S$ release<br>- Oil spill<br>- Fire Hazard<br>- Explosion Hazard | - 7.1 PSV-108/109 will open<br>- 7.1 $H_2S$ alarm<br>- 7.2 Explosive alarm | 2 | 3 | 1 | None |
| | | 8. High volatility oil from platform | | | 2 | 4 | AC | None |
| | More steam flow to E-101 | 9. TT-101A/B Failure | - Damage to E-101<br>- Loss of non-volatile hydrocarbons through V-103.<br>- Increased steam consumption | None | 4 | 3 | AC | Install TSH alarm in E-101 vapor exit |
| | More gas flow to V-103 | 10. High $H_2S$ content of oil | - Pressure increase<br>- Pipeline fracture<br>- $H_2S$ release<br>- Oil spill<br>- Fire Hazard<br>- Explosion Hazard | - 9.1 PSV-210/211 will open<br>- 9.2 PV-103B will open to flare<br>- 9.3 $H_2S$ alarm<br>- 9.4 Explosive alarm | 2 | 4 | AC | None |
| | | 11. Excessive vapor production in E-101 | - Pressure increase<br>- Loss of non-volatile hydrocarbons through V-103<br>- Pipeline fracture<br>- $H_2S$ release<br>- Oil spill<br>- Fire Hazard<br>- Explosion Hazard | - 10.1 PSV-210/211 will open<br>- 10.2 PV-103B will open to flare<br>- 10.3 $H_2S$ alarm<br>- 10.4 Explosive alarm<br>- 10.5 PdT alarm | 2 | 3 | 1 | None |

**Table A2.** *Cont.*

| Node | 5 | | | | | | | |
|---|---|---|---|---|---|---|---|---|
| Element | Flow | | | | | | | |
| Guideword | Deviation | Possible Cause | Consequences | Safeguards | Risk Matrix | | | Recommendations |
| | | | | | S | L | R | |
| Less | Less crude-oil input flow | 12. Low oil flow from platform | - Column disks not working properly. Poor stabilization | - 12.1 Crude from LV-106 sampling every 4 hours | 4 | 2 | AC | None |
| | | 13. FT-107 failure | | | 4 | 3 | AC | None |
| | Less crude-oil vapors output flow from E-101 | 14. TT-101 failure | - Poor stabilization | | 3 | 3 | AC | Install TSL alarm at E-101 vapor output |
| | | 15. Non-volatile oil from platform | | | | | | None |
| | Less steam flow to E-101 | 16. TT-101 failure | | | | | | None |
| | Less stabilized oil flow to P-101 | 17. Low oil flow from platform | - Pump cavitation | None | 4 | 3 | AC | Install a FSL alarm at P-101 suction |
| Reverse | Reverse flow of gas to FV-107 | 18. Pipeline fracture | - Pipeline fracture<br>- H$_2$S release<br>- Oil spill<br>- Fire Hazard<br>- Explosion Hazard | - 18.1 H$_2$S alarm<br>- 18.2 Explosive alarm | 2 | 4 | AC | None |
| No | No crude-oil input flow | 19. No oil flow from platform | - No production | - 19.1 PSL-120 alarm | 4 | 3 | AC | None |
| | No steam flow to E-101 | 20. FT-110 Failure | - No crude stabilization | - 20.1 TR-201 will record the temperature drop | 3 | 3 | AC | Install a TSL alarm at E-101 vapor exit |
| | | 21. Steam production facility failure | | None | 3 | 4 | AC | None |
| | | 22. Blocked E-101 tubes | | - 22.1 Spare E-101 | 3 | 3 | AC | None |
| | No stabilized crude flow | 23. P-101 shutdown | No Production | - 23.1 PSL-120 P-101 Discharge<br>- 23.2 Spare P-101 | 4 | 3 | AC | None |
| | | 24. Pipeline to storage blocked | | None | 4 | 3 | AC | None |

**Table A3.** The "Pressure"-Element of "Node-5".

| Node | 5 | | | | | | | |
|---|---|---|---|---|---|---|---|---|
| Element | Pressure | | | | | | | |
| Guideword | Deviation | Possible Cause | Consequences | Safeguards | Risk Matrix | | | Recommendations |
| | | | | | S | L | R | |
| | More pressure in crude input line | 25. Column tray disks clogged | - No crude stabilization | None | 3 | 3 | AC | None |
| More | More pressure in gas line | 26. PT-103 failure | - Pipeline fracture<br>- $H_2S$ release<br>- Oil spill<br>- Fire Hazard<br>- Explosion Hazard | - 26.1 PSV 110/111 will open to flare<br>- 26.2 $H_2S$ alarm<br>- 26.3 Explosive alarm | 2 | 3 | 1 | None |
| | | 27. Line to V-103 blocked | - Pipeline fracture<br>- $H_2S$ release<br>- Oil spill<br>- Fire Hazard<br>- Explosion Hazard | - 27.1 PSV 110/111 will open to flare<br>- 27.2 $H_2S$ alarm<br>- 27.3 Explosive alarm<br>- 27.4 PV-103B will open to flare | 3 | 3 | AC | None |
| | More pressure in pipeline to P-101 | 28. PT-103 Failure | - Pump seal failure<br>- Pipeline fracture<br>- $H_2S$ release<br>- Oil spill<br>- Fire Hazard<br>- Explosion Hazard | None | 3 | 3 | AC | None |
| | More pressure in E-101 | 29. Line from E-101 to V-102 blocked | - No crude stabilization<br>- E-101 fracture<br>- $H_2S$ release<br>- Oil spill-Fire Hazard<br>- Explosion Hazard | - 29.1 PSV-109 will open to flare | 3 | 3 | AC | None |
| | | 30. Steam to Crude leak | - Increased V-102 pressure<br>- Crude to storage increased water content | - 30.1 PSV-109 will open to flare | 3 | 3 | AC | During shutdown check E-101 integrity |

**Table A3.** *Cont.*

| Node | 5 | | | | | | | |
|---|---|---|---|---|---|---|---|---|
| **Element** | **Pressure** | | | | | | | |
| **Guideword** | **Deviation** | **Possible Cause** | **Consequences** | **Safeguards** | **Risk Matrix** | | | **Recommendations** |
| | | | | | **S** | **L** | **R** | |
| Less | Less pressure in crude input line | 31. PT-102 failure | - No production | None | 4 | 3 | AC | None |
| | | 32. Pipeline fracture | - Pipeline fracture<br>- $H_2S$ release<br>- Oil spill<br>- Fire Hazard<br>- Explosion Hazard | - 32.2 $H_2S$ alarm<br>- 32.3 Explosive alarm | 2 | 4 | AC | None |
| | Less pressure in pipeline to P-101 | 34. PT-102 failure | - Pump Cavitation | None | 3 | 3 | AC | None |
| | Less pressure in gas pipeline | 35. Pipeline fracture | - Pipeline fracture<br>- $H_2S$ release<br>- Oil spill<br>- Fire Hazard<br>- Explosion Hazard | - 32.2 $H_2S$ alarm<br>- 32.3 Explosive alarm | 2 | 4 | AC | None |

Table A4. The "Temperature"-Element of "Node-5".

| Node | 5 | | | | | | | |
|------|---|---|---|---|---|---|---|---|
| **Element** | **Temperature** | | | | | | | |
| **Guideword** | **Deviation** | **Possible Cause** | **Consequences** | **Safeguards** | **Risk Matrix** | | | **Recommendations** |
| | | | | | **S** | **L** | **R** | |
| More | Higher temperature in E-101 exit | 36. FT-110 failure | - Steam waste<br>- More hydrocarbon evaporation and escape to Sour Water | - 36.1 TR will record temperature increase | 3 | 3 | AC | Install a TSH alarm at E-101 output |
| | | 37. TT-101 failure | | None | 3 | 3 | AC | |
| | | 38. Steam leakage to crude in E-101 | - Increased pressure in V-102<br>- Steam Waste | - 38.1 PSV-109 will open | 3 | 4 | AC | |
| Less | Low temperature in E-101 exit | 39. FT-110 failure | - Poor Crude stabilization | - 36.1 TR will record temperature increase | 3 | 3 | AC | None |
| | | 40. TT-101 failure | | None | 3 | 3 | AC | None |

Table A5. The "Composition"-Element of "Node-5".

| Node | 5 | | | | | | | |
|------|---|---|---|---|---|---|---|---|
| **Element** | **Composition** | | | | | | | |
| **Guideword** | **Deviation** | **Possible Cause** | **Consequences** | **Safeguards** | **Risk Matrix** | | | **Recommendations** |
| | | | | | **S** | **L** | **R** | |
| Other than | Composition in Crude to storage other than required | 41. Low steam flow to E-101 | - Increased $H_2S$ in crude to storage<br>- $H_2S$ release in the tanks | - 41.1 Crude to storage sampling every 4 hours | 3 | 3 | AC | Install a TSL alarm in E-101 output |
| | | 42. Low crude input | | | 3 | 2 | 1 | None |
| | | 43. High crude input | | | 3 | 4 | AC | None |
| | | 44.Asphaltenes on Column disk | | - 44.1 High Pressure differential PdT-140 | 3 | 3 | AC | None |

**Table A6.** The "Containment"-Element of "Node-5".

| Node | 5 | | | | | | | |
|---|---|---|---|---|---|---|---|---|
| **Element** | **Containment** | | | | | | | |
| **Guideword** | **Deviation** | **Possible Cause** | **Consequences** | **Safeguards** | **Risk Matrix** | | | **Recommendations** |
| | | | | | **S** | **L** | **R** | |
| No | No Containment | 45. Fractured Vessel | - Oil spill<br>- Sour water spill | - 45.1 H$_2$S Alarm<br>- 45.2 Explosives Alarm | 2 | 3 | 1 | Install a PSL alarm at V-101 |
| | | 46. Valve opened to atmosphere | - H$_2$S release<br>- Fire hazard<br>- Explosion hazard | | | | | Allow only qualified personnel to operate valves |

## References


1.  Lees, F.P. *Loss Prevention in the Process Industries: Hazard Identification, Assessment and Control*; Reed: Oxford, UK, 1980.
2.  Marhavilas, P.K.; Filippidis, M.; Koulinas, G.K.; Koulouriotis, D.E. The integration of HAZOP study with risk-matrix and the analytical-hierarchy process for identifying critical control-points and prioritizing risks in industry—A case study. *J. Loss Prev. Process. Ind.* **2019**. [CrossRef]
3.  Sultana, S.; Okoh, P.; Haugen, S.; Vinnem, J.E. Hazard analysis: Application of STPA to ship-to-ship transfer of LNG. *J. Loss Prev. Process. Ind.* **2019**, *60*, 241–252. [CrossRef]
4.  Swann, C.; Preston, M. Twenty-five years of HAZOPs. *J. Loss Prev. Process. Ind.* **1995**, *8*, 349–353. [CrossRef]
5.  Jagtman, H.; Hale, A.; Heijer, T. A support tool for identifying evaluation issues of road safety measures. *Reliab. Eng. Syst. Saf.* **2005**, *90*, 206–216. [CrossRef]
6.  Ftenakis, V.M.; Trammell, S.R. Reference Guide for Hazard Analysis in PV Facilities. Available online: https://www.researchgate.net/publication/228611604_Reference_Guide_for_Hazard_Analysis_in_PV_Facilities (accessed on 29 January 2020).
7.  Chudleigh, M.F. Hazard analysis of a computer based medical diagnostic system. *Comput. Methods Programs Biomed.* **1994**, *44*, 45–54. [CrossRef]
8.  Saaty, T.L. How to make a decision: The analytic hierarchy process. *Eur. J. Oper. Res.* **1994**, *44*, 45–54. [CrossRef]
9.  Henderson, R.D.; Dutta, S.P. Use of the analytic hierarchy process in ergonomic analysis. *Int. J. Ind. Ergon.* **1990**. [CrossRef]
10. Badri, A.; Nadeau, S.; Gbodossou, A. Proposal of a risk-factor-based analytical approach for integrating occupational health and safety into project risk evaluation. *Accid. Anal. Prev.* **2012**, *48*, 223–234. [CrossRef]
11. Aminbakhsh, S.; Gündüz, M.; Sonmez, R. Safety risk assessment using analytic hierarchy process (AHP) during planning and budgeting of construction projects. *J. Saf. Res.* **2013**, *46*, 99–105. [CrossRef]
12. Mayes, T.; Kilsby, D.C. The use of HAZOP hazard analysis to identify critical control points for the microbiological safety of food. *Food Qual. Prefer.* **1989**, *1*, 53–57. [CrossRef]
13. Alaei, R.; Mansoori, S.A.A.; Moghaddam, A.H.; Mansoori, S.M.; Mansoori, N. Safety assessment approach of hazard and operability (HAZOP) for sulfur recovery unit Claus reaction furnace package; blower; heat exchanger equipment in South Pars gas processing plant. *J. Nat. Gas Sci. Eng.* **2014**, *20*, 271–284. [CrossRef]
14. Othman, M.R.; Idris, R.; Hassim, M.H.; Ibrahim, W.H.W. Prioritizing HAZOP analysis using analytic hierarchy process (AHP). *Clean Technol. Environ. Policy* **2016**, *18*, 1345–1360. [CrossRef]
15. Cheraghi, M.; Baladeh, A.E.; Khakzad, N. A fuzzy multi-attribute HAZOP technique (FMA-HAZOP): Application to gas wellhead facilities. *Saf. Sci.* **2019**, *114*, 12–22. [CrossRef]
16. Koulinas, G.K.; Marhavilas, P.K.; Demesouka, O.E.; Vavatsikos, A.P.; Koulouriotis, D.E. Risk Analysis and Assessment in the worksites using the Fuzzy-Analytical Hierarchy Process and a Quantitative Technique—A case study for the Greek Construction sector. *Saf. Sci.* **2019**. [CrossRef]
17. Koulinas, G.K.; Demesouka, O.E.; Marhavilas, P.K.; Vavatsikos, A.P.; Koulouriotis, D.E. Risk Assessment Using Fuzzy TOPSIS and PRAT for Sustainable Engineering Projects. *Sustainability* **2019**, *11*, 615. [CrossRef]
18. Marhavilas, P.K.; Koulouriotis, D.E.; Nikolaou, I.; Tsotoulidou, S. International Occupational Health and Safety Management-Systems Standards as a Frame for the Sustainability: Mapping the Territory. *Sustainability* **2018**, *10*, 3663. [CrossRef]
19. ICST&D. In Proceedings of the International Conference on Sustainable Technology and Development, Shenzhen, China, 22–25 August 2020; Available online: https://www.elsevier.com/events/conferences/international-conference-on-sustainable-technology-and-development (accessed on 9 December 2019).
20. Marhavilas, P.; Koulouriotis, D. A combined usage of stochastic and quantitative risk assessment methods in the worksites: Application on an electric power provider. *Reliab. Eng. Syst. Saf.* **2012**, *97*, 36–46. [CrossRef]
21. Marhavilas, P.; Koulouriotis, D. Developing a new alternative risk assessment framework in the work sites by including a stochastic and a deterministic process: A case study for the Greek Public Electric Power Provider. *Saf. Sci.* **2012**, *50*, 448–462. [CrossRef]
22. Marhavilas, P.; Koulouriotis, D.; Spartalis, S. Harmonic analysis of occupational-accident time-series as a part of the quantified risk evaluation in worksites: Application on electric power industry and construction sector. *Reliab. Eng. Syst. Saf.* **2013**, *112*, 8–25. [CrossRef]

23. Lawley, H.G. Operability Studies and Hazard Analysis. *Chem. Eng. Prog.* **1974**, *4*, 105–116.
24. Lawley, H.G. Size up plant hazards this way, hydrocarb. *Process* **1976**, *55*, 247–261.
25. CIA. *A Guide to Hazard and Operability Studies*; Imperial Chemical Industries and Chemical Industries Associations: London, UK, 1977.
26. Baybutt, P. On the completeness of scenario identification in process hazard analysis (PHA). *J. Loss Prev. Process. Ind.* **2018**, *55*, 492–499. [CrossRef]
27. Markowski, A.S.; Siuta, D. Fuzzy logic approach for identifying representative accident scenarios. *J. Loss Prev. Process. Ind.* **2018**, *56*, 414–423. [CrossRef]
28. Dallat, C.; Salmon, P.M.; Goode, N. Risky systems versus risky people: To what extent do risk assessment methods consider the systems approach to accident causation? A review of the literature. *Saf. Sci.* **2019**, *119*, 266–279. [CrossRef]
29. Baybutt, P. A critique of the Hazard and Operability (HAZOP) study. *J. Loss Prev. Process. Ind.* **2015**, *33*, 52–58. [CrossRef]
30. Animah, I.; Shafiee, M. Application of risk analysis in the liquefied natural gas (LNG) sector: An overview. *J. Loss Prev. Process. Ind.* **2020**, *63*, 103980. [CrossRef]
31. Baladeh, A.E.; Cheraghi, M.; Khakzad, N. A multi-objective model to optimal selection of safety measures in oil and gas facilities. *Process Saf. Environ.* **2019**. [CrossRef]
32. Kletz, T. *HAZOP and HAZAN: Identifying and Assesing Process Industy Hazards*; Institution of Chemical Engineers: Rugby, UK; Sydney, Australia, 1999.
33. Andow, P.K. Improvement of Operator Reliability Using Expert Systems. *Reliab. Eng.* **1986**, *14*, 309–319. [CrossRef]
34. Mentzer, R.A.; Zhang, J.; Xu, W.; Mannan, M.S. Mannan, What Does "Safe" Look and Feel Like? *J. Loss Prev. Process. Ind.* **2014**, *32*, 265–275. [CrossRef]
35. Herbert, I.L. Learning the lessons—Retrospective HAZOPs. *Soc. Pet. Eng. J.* **2011**. [CrossRef]
36. Baybutt, P. The role of people and human factors in performing process hazard analysis and layers of protection analysis. *J. Loss Prev. Process. Ind.* **2013**, *26*, 1352–1365. [CrossRef]
37. Frank, W.L.; Whittle, D.K. *Revalidating Process Hazards Analyses*; American Institute of Chemical Engineers: New York, NY, USA, 2001.
38. Center for Chemical Process Safety (CCPS). *Guidelines for Process Safety Documentation*; American Institute of Chemical Engineers (AICE): New York, NY, USA, 1995.
39. Qureshi, A. The role of hazard and operability study in risk analysis of major hazard plant. *J. Loss Prev. Process. Ind.* **1988**, *1*, 104–109. [CrossRef]
40. Center for Chemical Process Safety (CCPS). *Layer of Protection Analysis-Simplified Process Risk Assessment*; American Institute of Chemical Engineers: New York, NY, USA, 2001; ISBN 978-0-8169-0811-0.
41. Center for Chemical Process Safety (CCPS). *Guidelines for Hazard Evaluation Procedures*, 3rd ed.; American Institute of Chemical Engineers: New York, NY, USA, 2008.
42. European Process Safety Centre (EPSC). *HAZOP: Guide to Best Practice*; Institute of Chemical Engineers: Rugby, UK, 2000.
43. Dunjó, J.; Fthenakis, V.; Vílchez, J.A.; Arnaldos, J. Hazard and operability (HAZOP) analysis. A literature review. *J. Hazard. Mater.* **2009**, *173*, 19–32.
44. Macdonald, D.; Mackay, S. *Practical HAZOPs, Trips and Alarms*; Newnes: Oxford, UK, 2004.
45. Marhavilas, P.K. Risk Assessment Techniques in the Worksites of Occupational Health-Safety Systems with Emphasis on Industries and Constructions. Ph.D. Thesis, Department of Production and Management Engineering, Democritus University of Thrace, Xanthi, Greece, 2015. Available online: http://hdl.handle.net/10442/hedi/35612 (accessed on 28 November 2019).
46. Reniers, G.; Dullaert, W.; Ale, B.; Soudan, K. The use of current risk analysis tools evaluated towards preventing external domino accidents. *J. Loss Prev. Process. Ind.* **2005**, *18*, 119–126. [CrossRef]
47. IEC. *Functional Safety: Safety Instrumented Systems for the Process Sector*; IEC61511; International Electrotechnical Commission: Geneva, Switzerland, 2003.
48. AbdelGawad, M.; Fayek, A.R. Risk Management in the Construction Industry Using Combined Fuzzy FMEA and Fuzzy AHP. *J. Constr. Eng. Manag.* **2010**, *136*, 1028–1036. [CrossRef]
49. Zheng, G.; Zhu, N.; Tian, Z.; Chen, Y.; Sun, B. Application of a trapezoidal fuzzy AHP method for work safety evaluation and early warning rating of hot and humid environments. *Saf. Sci.* **2012**, *50*, 228–239. [CrossRef]

50. Basahel, A.; Taylan, O. Using fuzzy ahp and fuzzy topsis approaches for assessing safety conditions at worksites in construction industry. *Int. J. Saf. Secur. Eng.* **2016**, *6*, 728–745. [CrossRef]
51. Pour, F.S.A.; Gheorghe, A. Risk assessment of oil and natural gas drilling process by employing fuzzy sets and analytical hierarchy process (AHP). In Proceedings of the 2017 International Annual Conference of the American Society for Engineering Management (ASEM), Huntsville, AL, USA, 18–21 October 2017; Available online: https://www.scopus.com/inward/record.uri?eid=2-s2.0-85040090723&partnerID=40&md5=78216ee82ff871e798e203527a169c2f (accessed on 29 January 2020).
52. Gul, M. A review of occupational health and safety risk assessment approaches based on multi-criteria decision-making methods and their fuzzy versions. *Hum. Ecol. Risk Assessment: Int. J.* **2018**, *24*, 1723–1760. [CrossRef]
53. Zhen, X.; Vinnem, J.E.; Peng, C.; Huang, Y. Quantitative risk modelling of maintenance work on major offshore process equipment. *J. Loss Prev. Process. Ind.* **2018**, *56*, 430–443. [CrossRef]
54. Fattahi, R.; Khalilzadeh, M. Risk evaluation using a novel hybrid method based on FMEA, extended MULTIMOORA, and AHP methods under fuzzy environment. *Saf. Sci.* **2018**, *102*, 290–300. [CrossRef]
55. Chang, D.-Y. Applications of the extent analysis method on fuzzy AHP. *Eur. J. Oper. Res.* **1996**, *95*, 649–655. [CrossRef]
56. Lamata, M.T. Ranking of alternatives with ordered weighted averaging operators. *Int. J. Intell. Syst.* **2004**, *19*, 473–482. [CrossRef]
57. Zimmermann, H.J. *Fuzzy Set Theory—And Its Applications*; Springer: Berlin, Germany, 2001.
58. Filippidis, M. *Hazard and Operability Study of a Sour Crude-Oil Processing Plant*; Department of Production and Management Engineering, Democritus University of Thrace (DUTH): Xanthi, Greece, 2017.
59. Høj, N.P.; Kroger, W. Risk analyses of transportation on road and railway from a European Perspective. *Saf. Sci.* **2002**, *40*, 337–357. [CrossRef]
60. ISO. ISO/IEC Guide 51, Safety Aspects—Guidelines for Their Inclusion in Standards. Available online: https://www.iso.org/standard/32893.html (accessed on 29 January 2019).
61. ISO. ISO/IEC Guide 73, Risk Management-Vocabulary. Available online: https://www.iso.org/standard/44651.html (accessed on 29 January 2019).
62. Chan, H.K.; Sun, X.; Chung, S.-H. When should fuzzy analytic hierarchy process be used instead of analytic hierarchy process? *Decis. Support Syst.* **2019**, *125*. [CrossRef]
63. Gul, M.; Guneri, A.F. A fuzzy multi criteria risk assessment based on decision matrix technique: A case study for aluminum industry. *J. Loss Prev. Process. Ind.* **2016**, *40*, 89–100. [CrossRef]
64. Guo, L.; Kang, J. An extended HAZOP analysis approach with dynamic fault tree. *J. Loss Prev. Process. Ind.* **2015**, *38*, 224–232. [CrossRef]
65. Ahn, J.; Chang, D. Fuzzy-based HAZOP study for process industry. *J. Hazard. Mater.* **2016**, *317*, 303–311. [CrossRef]
66. Marhavilas, P.K.; Filippidis, M.; Koulinas, G.K.; Koulouriotis, D.E. An expanded HAZOP-study with fuzzy-AHP (XPA-HAZOP technique): Application in a sour crude-oil processing plant. *Saf. Sci.* **2020**, *124*, 104590. [CrossRef]
67. Zheng, X.; Liu, M. An overview of accident forecasting methodologies. *J. Loss Prev. Process. Ind.* **2009**, *22*, 484–491. [CrossRef]

*Article*

# Sustainable Decision-Making in Road Development: Analysis of Road Preservation Policies

**Angie Ruiz * and Jose Guevara**

Department of Civil and Environmental Engineering, Universidad de Los Andes, Bogotá 5101, Colombia; ja.guevara915@uniandes.edu.co
* Correspondence: al.ruiz10@uniandes.edu.co

Received: 29 November 2019; Accepted: 7 January 2020; Published: 24 January 2020

**Abstract:** Road infrastructure in appropriate conditions is a requirement for the development of any country. The formulation of policies oriented to preserve road networks responds to political, socio-economic, and environmental interests. Through a hybrid methodology that integrates system dynamics (SD) and analytic hierarchical process (AHP) approaches, this paper compares some strategies employed in the development of sustainable road maintenance policies in Colombia. Using a hypothetical case study of a national road network, a set of maintenance policy alternatives are evaluated through the SD model in order to analyze the evolution of road conditions, and quantify costs and emissions. Then, a multi-criteria evaluation is performed applying the AHP methodology. Results show that in the Colombian context, decision-making processes regarding maintenance policies are highly influenced by economic factors, which lead to short-term strategies such as performing corrective maintenance over predictive maintenance. However, further analysis demonstrates that predictive maintenance allows the road network to remain in good conditions. Simultaneously, roadways in adequate conditions contribute to mitigate the environmental impact, because $CO_2$ emissions are directly related to the interventions performed to preserve these roads. The proposed methodology can be used as a support tool to formulate maintenance policies that consider the long-term effects at the technical, environmental, and economic levels.

**Keywords:** road network; sustainability; $CO_2$ emissions; system dynamics; analytic hierarchical process

---

## 1. Introduction

Road infrastructure is one of the basic facilities that serve social and economic purposes in a country. It is associated with the development capacity and competitiveness of any region, as it facilitates the transport of goods and passengers, the transfer of information, and ensures access to basic services, which are necessary conditions in the modern economy [1]. The deterioration of this type of infrastructure significantly affects economic activities, the environment, and quality of life of all the inhabitants of a country [2]. For these reasons, an efficient infrastructure network must be a priority for governments, especially those of developing countries.

In recent decades, governments and highway agencies have recognized the need to improve planning and management processes in the infrastructure sector; the adoption of small measures can generate significant changes and benefits in the population [3]. The infrastructure network is a large and complex system that requires the development of effective maintenance programs in order to adequately allocate resources between preservation, maintenance and rehabilitation (M&R) activities [4]. The development of infrastructure responds to the interests and needs of society and its environment, and is important for the economic well-being of countries; for this reason, it is necessary to make optimal decisions regarding which strategies to use, and how and when to implement them in order to adjust to the availability of economic and natural resources. This translates into a decision-making problem, since all the policies or strategies adopted when distributing resources must be supported by socioeconomic and environmental factors.

A variety of methods have been employed to address this problem. Most studies have used optimization techniques to determine how to distribute resources. For example, González, Dueñas, Sánchez, and Medaglia defined a minimum-cost reconstruction strategy for a partially destroyed infrastructure network through an Interdependent Network Design Problem (INDP) based approach and heuristic methods [5]. Similarly, Fwa and Farhan proposed an optimization analysis using the Pareto efficiency concept to achieve an equitable allocation of budget by maintaining a minimum performance in all criteria considered [6]. Some researchers have employed life-cycle analysis (LCA) to support optimization models in order to develop a management system that allows decision makers to preserve a healthy pavement network and minimize costs while meeting all types of agency constraints, such as budget limitations [7,8].

Other authors have proposed the use of genetic algorithms in road maintenance planning; Herabat and Tangphaisankun developed a multi-objective optimization model using constraint-based genetic algorithms to support the decision-making process of highway maintenance management in Thailand [9]. Chan, Fwa, and Tan used genetic algorithms to develop a model (PAVENET) that serves as an optimization technique for the road-maintenance planning problem at a network level [10]. Other methods such as linear-programming and nonlinear-programming have been used in optimization analysis [11]. Also, complex systems theory or dynamic system analysis have been incorporated in the pavement management process. Some authors have developed system dynamics simulation models to evaluate the performance of different road maintenance policies over time, concluding that authorities should give priority to preventive maintenance over corrective maintenance [12–14]. Other researchers have resorted to hybrid approaches; for example, Moazami, Bhabhan, and Unhandy proposed a model combining the analytic hierarchy process and fuzzy logic to establish priorities between projects in pavement rehabilitation and maintenance processes, considering both the structural performance and the total cost over the life cycle [15].

It is expected that a resources manager will be able to analyze the benefits for the community and compare them with benefits in other aspects [16]. In addition to economic restrictions, the environmental impact should be considered when making decisions on investment in road infrastructure. Regardless of the type of pavement, road infrastructure interventions involve construction, maintenance, and rehabilitation processes [17], which consume significant amounts of non-renewable resources [18] and generate large amounts of energy, greenhouse gas emissions, and waste [19]. In recent years, public concern about emissions and environmental pollution has increased. For this reason, there have been advances in sustainable technologies that reduce the environmental impacts of road development [20,21]. Similarly, the environmental component has been considered as a fundamental aspect in the pavement management system. Egilmez and Tatari proposed a dynamic modeling approach to analyze the US highway system and evaluate policies designed to reduce carbon emissions by 2050 [22]. Jullien, Dauvergne, and Cerezo also used the LCA methodology in the environmental assessment of road construction and maintenance policies for both asphalt concrete and cement concrete roads [23].

The construction and maintenance processes involved in pavement interventions depend largely on multiple factors, due to their great complexity. This makes it difficult to implement policies related to the road network, which are aimed at promoting sustainability in the network. Consequently, examining the relationship between road infrastructure, the economic sector, and the environment requires approaches capable of incorporating multiple variables from different perspectives. For this reason, this study aims to analyze maintenance policies in the road infrastructure sector in Colombia, in order to make them sustainable. Explicitly, it seeks to evaluate strategies in the allocation of resources corresponding to the construction, maintenance, and rehabilitation of roads considering technical, economic, and environmental factors through decision-making analysis.

Specifically, this work seeks to design a hybrid methodology, integrating system dynamics (SD) and the analytic hierarchy process (AHP), which allows decision makers to evaluate different maintenance policies in a road network. Through a hypothetical case study and application of the methodology, it is intended to answer the following questions: Which factors affect decision-making in budget distribution for road infrastructure? How do these decisions impact the service life of the infrastructure? Using the simulation results, a comparison and evaluation of different maintenance policies are made. Then, a set of recommendations and observations regarding the strategies are presented with the ultimate goal of improving the conditions of the roads during their life cycle.

## 2. Methodology

In order to accomplish the objectives of this work, the methodology shown in Figure 1 is proposed. The first step is to describe the situation and structure it as a decision problem, identifying the main aspects, actors, and variables involved in the problem. Then, a hypothetical case study is described in order to present the data employed in this research. The third step is to formulate a system dynamics model so as to analyze the identified problem through the lens of the hypothetical case study data. The SD approach was chosen due to the complex relationships between the variables involved and the influence of time on the behavior of the system analyzed. Based on the SD analysis, maintenance policy alternatives are discussed through the AHP approach. This methodology is employed to define priorities between multiple objectives and evaluate the performance of policy alternatives [24]. Finally, a global performance score is applied in order to rank the policy alternatives and in this way choose the one that allows the road network to remain in adequate conditions considering the economic and environmental aspects.

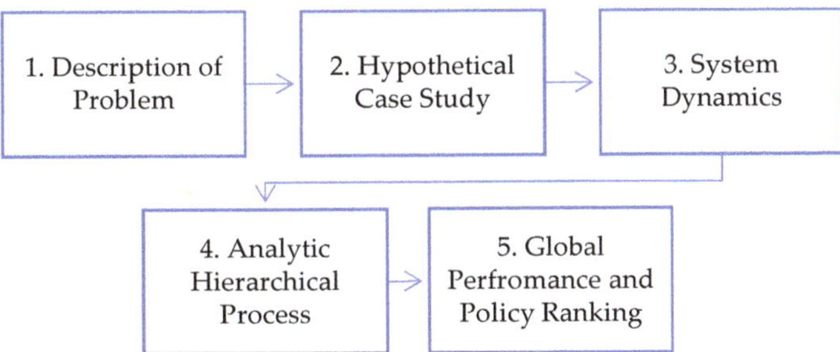


**Figure 1.** Methodology.

The methodology proposed can be described as a hybrid approach, where SD and AHP are integrated. Hybrid models combine continuous and discrete behaviors in a system [25], which adjust to the current study, since there are continuous variables, such as the pavement deterioration process and discrete events like the decisions made by the actors involved. On the one hand, SD simulation models allow the comprehension of the dynamic behavior of systems that exhibit non-linear behaviors, feedback mechanisms or anti-intuitive patterns [26]. On the other hand, the AHP is a methodology used to perform multi-attribute analysis, in which priorities between multiple objectives are defined and the performance of different alternatives is evaluated [24]. In addition, it allows decision makers to incorporate their experience in the analysis of the model results, complementing them, and thus overcoming some limitations of the simulation.

## 2.1. Description of the Problem

A road infrastructure system in excellent condition is one of the main requirements to reach a high level of socioeconomic development for any country. According to the World Economic Forum Report [27], there is a direct relationship between the gross domestic product (GDP) of countries and the kilometers of roads per inhabitant, and the quality of the roads. This study focuses on the Colombian road infrastructure sector, since it is currently one of the sectors with the greatest challenges for the country. In Colombia, the road infrastructure gap is particularly serious, which is reflected in the high logistics costs compared to countries with similar economies [28]. In addition, it highlights the lack of long-term strategic planning in the sector as one of the main challenges for the country.

For this reason, it is necessary to take measures to improve the way that the government and agencies manage the road network. This requires making decisions regarding what type of strategies should be implemented in the formulation of maintenance policies in order to optimally allocate the resources available to expand, maintain, and rehabilitate the road network. Specifically, the objective is to select the best maintenance policy for the preservation of a national road network, 17,201 km-roadway in length, which is described through the formulation of a hypothetical case study.

These decisions must be made considering several factors involved in the problem, such as technical, environmental, and socio-economic aspects. It is also important to consider the different actors involved such as road agencies, regulators, policy makers, and users. The amount of interactions implies a significant number of variables, such as road deterioration processes, types and frequency of interventions (construction, maintenance, and rehabilitation), road demand, as well as emission factors and costs associated with the interventions.

## 2.2. Hypothetical Case Study

Due to a lack of reliable historical data in Colombia, a hypothetical case study is used to evaluate different strategies in road maintenance policies within the country's road network. Accordingly, this study employs historical data obtained from the Chilean road network as reported by Ruiz and Guevara [29]. Although the Colombian and Chilean highway systems are not exactly the same, this jurisdiction was selected because of its similarities with the Colombian context in terms of number of roadways, pavement types, traffic volume, road growth, and overall system conditions [30,31]. Consequently, for this study, we proposed to evaluate a national road network with an initial length of 17,201 km-roadway. Initially, 58% of the roadways were in good condition, 22% in fair condition, and 20% in poor condition. The objective is to see how the network responds to the application of different maintenance policies over a period of ten years.

Table 1 shows the main characteristics of the hypothetical road network employed in this study. This information comprises four types of parameters and is particularly useful to develop the SD model described in the following section. System condition variables describe the highway system in terms of the main road categories. Time-based variables refer to pavement growth and deterioration processes. Emission-based parameters indicate the amount of $CO_2$ generated as a result of construction, maintenance, and rehabilitation activities. Cost-related information indicates the expenditures associated with the pavement interventions under analysis.

**Table 1.** Case study data.

|  | Main Variables and Parameters | Value | Unit | Comment |
|---|---|---|---|---|
| System Condition variables | Total roadways | 17,201 | km-roadway | Values adopted from historical data reports [32] |
|  | Good condition roads | 58 | % |  |
|  | Fair condition roads | 22 | % |  |
|  | Poor condition roads | 20 | % |  |
|  | Annual road growth | 4.2 | % |  |
| Time-Based Variables | Average time to fix roads | 1.547 | Years | Calibrated to fit historical values [32] |
|  | Time reconstruction | 7.396 | Years |  |
|  | Time structural failures | 8 | Years |  |
|  | Time superficial failures | 1.524 | Years |  |
|  | Time to rehabilitate roads | 1.945 | Years |  |
| Emission-sbased variables | Emission factor gray construction | 172,400 | $kgCO_2$/km-roadway | Values reported for typical roads [8,33] |
|  | Emission factor green flexible | 134,800 | $kgCO_2$/km-roadway |  |
|  | Emission factor green reconstruction | 153,600 | $kgCO_2$/km-roadway |  |
|  | Emission factor reconstruction | 191,500 | $kgCO_2$/km-roadway |  |
|  | Emission factor green maintenance | 12,700 | $kgCO_2$/km-roadway |  |
|  | Emission factor green rehabilitation | 62,950 | $kgCO_2$/km-roadway |  |
|  | Emission factor maintenance | 14,110 | $kgCO_2$/km-roadway |  |
|  | Emission factor rehabilitation | 86,230 | $kgCO_2$/km-roadway |  |
| Cost-Based Variables | Unit cost construction | 504,463 | USD[1]/km-roadway | Values based on maintenance contracts [34–36] |
|  | Unit cost maintenance | 198,043 | USD/km-roadway |  |
|  | Unit cost rehabilitation | 350,428 | USD/km-roadway |  |
|  | Unit cost reconstruction | 534,155 | USD/km-roadway |  |
|  | Unit cost green construction | 366,493 | USD/km-roadway |  |
|  | Unit cost green maintenance | 194,082 | USD/km-roadway |  |
|  | Unit cost green rehabilitation | 221,938 | USD/km-roadway |  |
|  | Unit cost green reconstruction | 398,747 | USD/km-roadway |  |

[1] USD: United States Dollar.

## 2.3. System Dynamics

SD is a methodology to study and understand complex and changing systems over time [37]. This methodology is based on analyzing the behavior of systems over time based on the relationships and dependencies between their components [38]. SD allows researchers to analyze problems from a global perspective, since it allows the identification of short and long-term effects on all components of the system, thus reducing the fear of uncertainty of taking long-term measures and reducing unanticipated side effects [38]. For this reason, this methodology is extensively used in solving engineering problems.

Specifically, researchers have employed SD to examine the effects of maintenance practices on highway system conditions. They analyzed how deferred maintenance affects road network performance in the long-term [13,14]. Additionally, SD has also been used for the analysis of environmental issues. For example, Mallick et al. [39] proposed a model to evaluate the impact of pavement network intervention on climate change. Similarly, other authors have studied the generation of $CO_2$ emissions [40–42] and analyzed strategies to reduce energy consumption rates [22,39].

The present study uses a previously developed SD model [29]. This model considers the deterioration phenomena of road networks, the different types of activities related to road preservation (construction, maintenance, and rehabilitation), as well as costs and $CO_2$ emissions associated with them. In addition, the model incorporates sustainable measures in all three types of activities, that is, the intervention of the road network can be carried out by conventional (gray) or sustainable (green) procedures. A sustainable measure is understood as any material or technique that contributes to the mitigation of $CO_2$ emissions.

In this study, the model was modified. The available budget constraint was added to the formulation of the pavement deterioration model, which means that the variables associated with construction and maintenance rates change from exogenous parameters to variables directly depending on the budget. This new restriction gives rise to new feedback loops in the system, significantly affecting the results obtained for the main outputs, which are the overall conditions of the road network, expenditures, and emissions associated with all types of preservation interventions. A complete description of its variables and formulations can be found in the Supplementary Materials.

### 2.3.1. Causal Loop Diagram (CLD)

To formulate the equations between budget variables and maintenance activities, it is necessary to identify what type of relationships and feedbacks exist between them. Feedbacks can be reinforcing or balancing loops, labeled with the letter R and B, respectively. Reinforcing loops represent interactions that promote their own growth, and balancing loops refer to mechanisms that help the system reach an equilibrium condition [38].

Figure 2 exposes the developed CLD, which shows that the system's behavior is highly influenced by the gap between the desired condition of the road network and the actual condition, which is a result of the deterioration processes of the pavement. With the purpose of reducing this gap, conventional construction, maintenance, and rehabilitation activities are implemented. As a first reaction to the deterioration, rehabilitation measures are carried out, thus improving the condition of the road network (loop B1). The execution of these types of activities consumes part of the resources allocated to the preservation of the road network, reducing the resources available for maintenance and construction activities. That is, the more rehabilitation activities are carried out, the smaller the gap between the desired and actual condition of the system, and also, the smaller the budget available for implementing maintenance activities (loop R1). Similarly, the execution of both types of activities (M&R) decreases the resources to build new roads (loop B2).

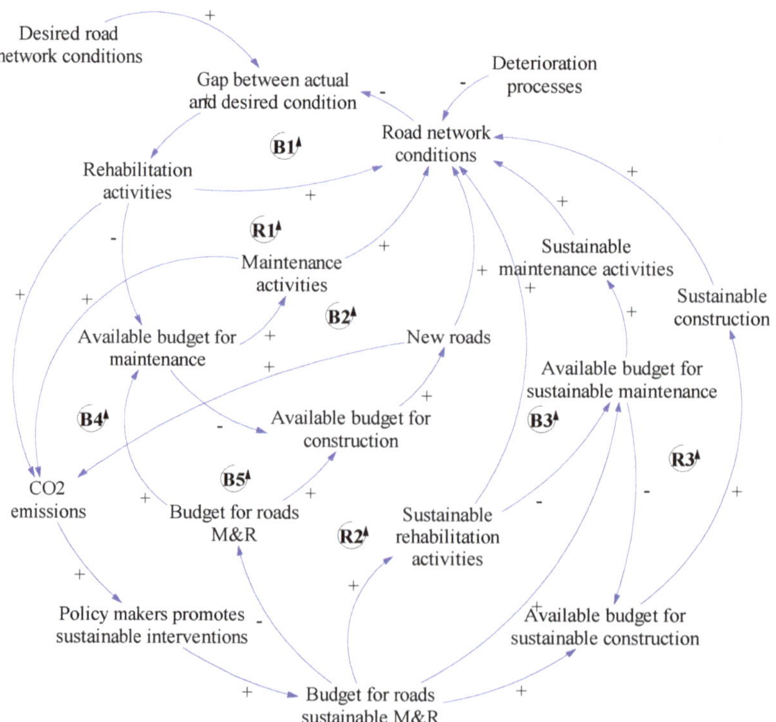


**Figure 2.** Causal loop diagram (CLD). M&R: maintenance and rehabilitation.

In addition to budget consumption, implementing this type of activity generates $CO_2$ emissions, which increase the concern of policy makers regarding environmental impact. As a measure to mitigate the environmental impact, agencies and governments allocate part of the road preservation budget to promote the implementation of sustainable alternatives in the construction, maintenance, and rehabilitation processes. This decreases the budget available for conventional construction, maintenance, and rehabilitation activities (loops B4 and B5). On the other hand, the implementation of sustainable alternatives also contributes to the improvement of the conditions of the road network, reducing the gap between the actual and desired conditions. This can be seen in loop R2, where the execution of sustainable rehabilitation activities is reinforced. In sustainable activities, the same dynamic as in conventional loops occurs, that is, the execution of sustainable rehabilitation activities reduces the resources available for sustainable construction and maintenance. These interactions are represented by loops R3 and B3, respectively.

### 2.3.2. Model Description

The model has three main components: technical, economic, and environmental components. The first one is composed of two aging chains (gray aging chain and green aging chain), which recreate the road network condition over time. The economic and environmental aspects are represented by the variables, budget for road preservation and $CO_2$ emissions, as shown in Figure 3. Following the relationships identified in the CLD, there is a feedback loop between the technical and economic components. On the one hand, the budget for road preservation determines the ability of governments to intervene in the road network; at the same time, the execution of these interventions decreases the available budget, resulting in a balance loop between the two components. On the other hand, the environmental component is directly related to the technical component, since $CO_2$ emissions are associated with the interventions performed.

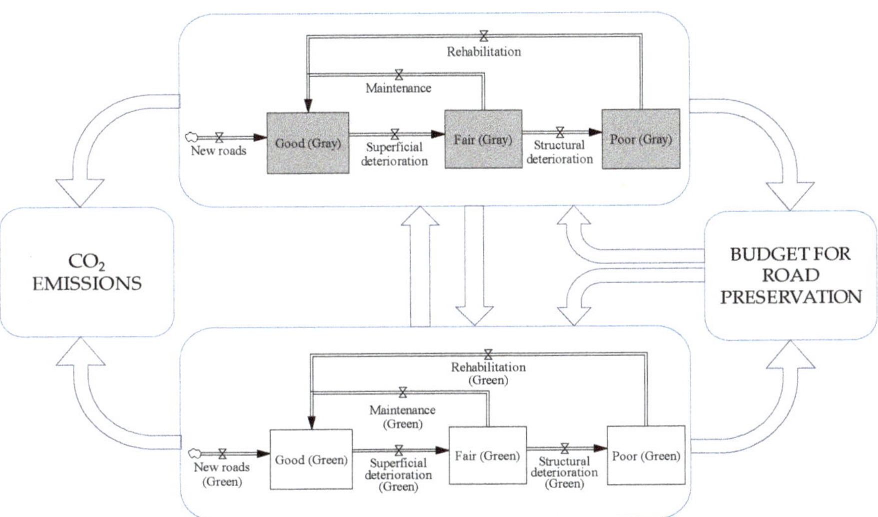


**Figure 3.** Model components.

The model simulates the processes of deterioration, maintenance, and rehabilitation of the national road network, which are represented through the gray aging chain and green aging chain. Three stocks compose each chain—good, fair, and poor—and are measured through km-roadways. The gray aging chain represents these processes for all the roadways that were intervened using conventional materials and techniques, and the green aging chain represents the roadways built or preserved through sustainable alternatives. The two chains have the same dynamics for the processes of deterioration and maintenance of roadways. These dynamics are presented in Figure 4, which shows the variables involved in the aging chain and their relationships.

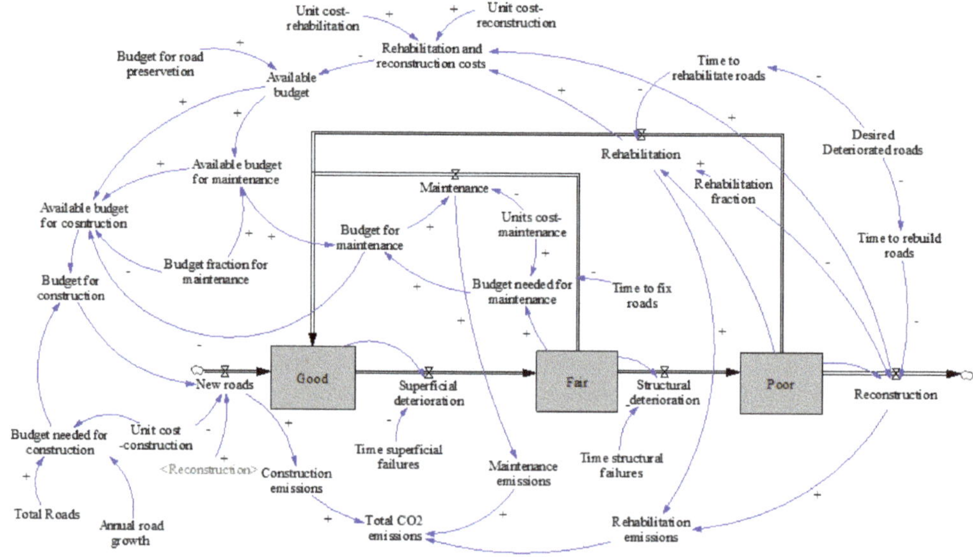

**Figure 4.** Stock and flow diagram: aging chains.

As shown in Figure 4, the stocks are connected through rates, which are represented by double-line arrows with a valve attached to them. The new roads rate denotes the roadways that are built each year and are expected to be in good condition. On the other hand, superficial failure and structural failure rates, correspond to the processes that lead roadways to a fair and poor condition, respectively. Both rates are directly related to the kilometers of roadways in each stock and the average age of roadways. The average times for superficial and structural failure occurrence were calculated using partial calibration methods [43], as shown in Table 1.

However, the roadway condition does not depend exclusively on the deterioration process, but also on the interventions performed. In addition to the construction of new roads, two types of interventions are considered: the first is called maintenance activities, which take roadways from fair condition to good condition. Likewise, the roadways in poor condition are intervened to take them to good condition through rehabilitation and reconstruction activities. There are two highly influential factors in the number of interventions carried out: the budget available for the preservation of roads and the government's goals regarding the desired condition of the network. The model assumes that the government gives priority to rehabilitation over maintenance, that is, according to the difference between the desired and actual condition of the network, the government executes the required rehabilitation and reconstruction activities.

After these interventions are performed, the costs associated with these types of activities are quantified and compared to the available budget for road preservation. Intervention costs are calculated employing the unit costs reported in Table 1. If there are available resources, a percentage of this is allocated to execute maintenance activities. The maintenance rate is directly related to the available budget and the budget needed to maintain the stock of roadways in fair conditions. If the budget needed for maintenance is less than the available one, the remaining resources become part of the available budget to build more roadways. Finally, the construction of new roads depends on the resources available after executing both types of interventions, maintenance and rehabilitation activities.

Additionally, $CO_2$ emissions associated with road network preservation are calculated. This is done using emission factors (measured in kg of $CO_2$/km-roadway) associated with each type of intervention that were obtained from the literature [8,33] (Table 1). $CO_2$ emissions are the variable that represents the environmental aspect in the model. In order to mitigate environmental impacts, sustainable alternatives are incorporated in the model. Therefore, the variables of percentage of sustainable construction, sustainable maintenance, and sustainable rehabilitation are added, which represent the proportion of interventions that will be carried out using sustainable or green activities. In other words, other flows of interventions arise (new green roads, green maintenance, and green rehabilitation), and behave exactly as the ones explained above, but employ green alternatives, and these new flows create the green aging chain. All roadways (i.e., those that are part of both the gray and green chain), are intervened using conventional or sustainable techniques. This means that a road that was built gray can remain gray or turn green and vice versa, and this creates flows between the two chains as shown in Figure 3.

### 2.3.3. SD Model Validation

System dynamics modelers have developed several tests for validation purposes which seek to find flaws in the model formula and ensure its efficiency to fulfill the objective for which it was built [38]. Once the objective and scope of the model are established, the next step is to formulate it. Dimensional consistency and structural testing tests were carried out during and at the end of this formulation process. To ensure dimensional consistency, all units of measure for each variable were specified as the model was built. The objective of evaluating the structure of the model is to verify that it is consistent with the real system and functional for the purpose of this project. The development of the model structure was done through stock and flow formulations that were used in other studies that analyzed road networks [12–14,44], where first-order aging chains were employed in order to endogenously capture the pavement deterioration process. In the structure evaluation, it is also important to verify that physical laws are not violated, therefore in this case, the formula used guarantees that stocks and flows remain at positive values under any circumstance.

To carry out the integration error test, a time step and integration method were chosen in such a way that if the method was changed or the time step was reduced by half, the simulation results would not change significantly. Additionally, behavior anomaly tests were performed; for this, the relationships between construction and maintenance rates and the budget were modified or deleted. Anomalous behaviors arise when the feedback loop governed by the budget is deleted; this indicates the importance of including these relationships. Behavior reproduction tests were also performed; the purpose of this test is to evaluate the model's capacity to represent historical data through indicators such as the coefficient of determination ($R^2$), the root mean square error (RMSE), and Theil's inequalities [38]. The results of these indicators suggest that the model successfully recreates historical data trends.

Additionally, parameter assessment, sensitivity, and extreme condition tests [38] were carried out in Vensim DSS (i.e., software for SD modeling). The parameter values were estimated using partial model calibration from numerical data, as presented in Table 1 [43]. This helped to confirm that model results were consistent with previous studies on pavement deterioration and system dynamics [12–14,44]. The sensitivity analysis was performed through Monte Carlo simulation procedures; for each exogenous parameter, a set of possible values was established. All sets follow a triangular distribution, where the most likely value is the value previously estimated through partial model calibration. The results showed that the model exhibited a logical behavior for any value of a set of values, and that the mean of the simulations exhibits trends similar to historical data. Four scenarios of extreme conditions were evaluated: accelerated growth of the road network, null growth of the road network, accelerated deterioration processes with slow M&R interventions, and slow deterioration processes with efficient M&R interventions. In all four scenarios, the system responded satisfactorily and consistently with the changes made.

### 2.4. AHP

The performance of systems such as road networks depends largely on the ability of countries and organizations to choose correct alternatives consistent with their contexts so that they align with their objectives. Making the right decisions is a critical factor since these processes are involved in complex contexts, affected by both quantitative and qualitative variables. In the allocation of resources in any sector, optimization methods are often used to choose between different alternatives as explained in the introduction. However, these types of methodologies are usually based on quantitative indicators, leaving aside influential factors in decision-making, such as the perceptions and preferences of decision makers. For this reason, the AHP methodology is widely used in solving engineering problems [44]. For example, Thanki, Govindan, and Thakkar [45] employed the methodology to evaluate the influence of lean and green paradigms on the overall performance of small and medium-sized enterprises. Similarly, Rosa and Haddad [46] evaluated the sustainability of existing buildings using AHP. Other authors have used the methodology as a support tool to prioritize and categorize challenges and to develop appropriate frameworks in project development [47,48]. The road infrastructure sector is no stranger to this methodology, having used it for decades in the prioritization of maintenance policies [49].

The AHP is a multi-criteria model for decision-making developed by Thomas L. Saaty [50]. The main purpose of this methodology is to select the best alternative from a set of possible alternatives based on several criteria and variables, setting priorities among them [51]. For this case, the set of alternatives corresponds to eight maintenance policies that will be described in the next section. This methodology allows decision makers to incorporate the opinions and experiences of experts in the decision-making processes, and for this reason it has been extensively used in scenarios and problems where human perceptions and junctions have significant influence [52]. The application of this methodology follows five major stages: description of policy alternatives, description of the decision problem, development of pairwise comparison matrices, estimation of weights, and finally results analysis [51]. More details regarding each stage are given in the next sections.

### 2.4.1. Description of Policy Alternatives

A maintenance policy is defined as a coordinated set of activities that seek to provide safe pavement and operations, taking full advantage of available public funds [53]. The set of activities includes all interventions related to planning, design, construction, maintenance, and evaluation of roads. The formulation of maintenance policies implies high-level decisions related to the planning, policy, and budget of the entire network [54]. Typically, to prepare a maintenance plan it is necessary to select the set of activities to be carried out, as well as the frequency of implementation. Since resources are limited, in most cases it is not possible to execute all the activities required to preserve roads in the desired conditions. This means that agencies and governments are forced to give priority to some activities. For example, most governments usually give priority to rehabilitation (corrective maintenance) over maintenance (predictive maintenance), although it has been shown that prevention activities bring greater benefits [55].

As mentioned above, sustainable alternatives have emerged in the road infrastructure sector. This means that in addition to choosing between intervention categories (construction, maintenance, rehabilitation, and reconstruction), policy makers also have to make decisions regarding the techniques for interventions. They have to choose if the roads are intervened using conventional techniques, sustainable techniques or both. In addition, once the set of interventions and priorities between them are established, agencies must estimate the time of interventions and the extent of repairs [56]. For the purposes of this study, several maintenance plan alternatives are compared. These alternatives are the result of the combination of different decision variables: types of activities, priorities between the types of activities, and the frequency of execution of the activities.

Regarding the frequency of interventions, two types of maintenance activities are considered: predictive maintenance and corrective maintenance [56]. Predictive maintenance is performed according to the condition of the system; usually it considers a minimum or maximum value of an indicator. In this case, the indicator corresponds to the percentage of roads in adequate conditions (good and fair). Additionally, this type of maintenance can be carried out giving priority to maintenance (intervention of roadways in fair condition) over rehabilitation (intervention of roadways in poor condition) or vice versa. On the other hand, corrective maintenance is performed every time a system failure occurs, which means only rehabilitation activities are carried out.

The inclusion of sustainable or green activities can be done by promoting sustainability in construction or sustainability in M&R. The first one involves using sustainable activities in the construction of a certain percentage of new roadways, and the second one promotes the use of sustainable activities in the maintenance and rehabilitation of a certain percentage of existing roadways. Based on that, Table 2 shows eight maintenance policy alternatives that were designed through the combination of the following decision variables: type of intervention (predictive or corrective), priority between maintenance and rehabilitation, and the inclusion of sustainable practices in construction or M&R operations. For the last decision variable, different percentage values were evaluated, however, not all of them are shown in this study because differences between simulation results were not significant.

**Table 2.** Maintenance policies alternatives.

| Policy Alternative | Type of Intervention | Priority Rule: Maintenance vs. Rehabilitation | Sustainable Practices | |
|---|---|---|---|---|
| | | | Green Intervention | Percentage of Implementation |
| Policy 1 (P1) | Predictive | Rehabilitation over Maintenance | N.A.[1] | 0% |
| Policy 2 (P2) | Predictive | Rehabilitation over Maintenance | Construction | 60% |
| Policy 3 (P3) | Predictive | Rehabilitation over Maintenance | M&R | 60% |
| Policy 4 (P4) | Predictive | Maintenance over Rehabilitation | Construction | 60% |
| Policy 5 (P5) | Predictive | Maintenance over Rehabilitation | M&R | 60% |
| Policy 6 (P6) | Predictive | Maintenance over Rehabilitation | M&R | 100% |
| Policy 7 (P7) | Corrective | N.A. | Construction | 60% |
| Policy 8 (P8) | Corrective | N.A. | M&R | 60% |

[1] N.A.: not applicable.

### 2.4.2. Description of the Decision Problem

One of the main principles of the AHP methodology is to define the decision problem as a hierarchical structure, where the upper level corresponds to the global goal. For the construction of hierarchical structures, it is necessary to identify the global goal, the criteria and sub criteria, as well as the set of alternatives [57]. The global goal in this case corresponds to selecting the best policy for the maintenance of the road network. Additionally, based on the SD model outputs, four criteria were defined for the evaluation of the maintenance policy alternatives: growth of the road network, technical performance, costs, and environmental impact. The first criteria represents the km-roadways built during the analysis period and the second one represents the percentage of roadways in each condition (good, fair, and poor condition) over time. Environmental impact and costs refer to tons of $CO_2$ emissions and the net present value (NPV) of the costs associated exclusively with all types of interventions carried out every year. It is important to mention that user costs like the ones derived from travel time and vehicle operating costs, are not considered in this study. To develop the hierarchical structures, the evaluation criteria were classified according to their impact on the performance of the alternatives, resulting in the development of two hierarchical structures: benefits hierarchy and risks hierarchy, as shown in Figure 5.

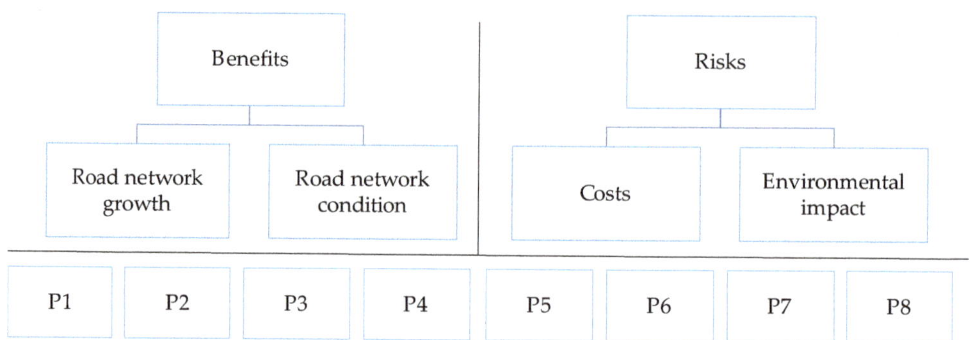


**Figure 5.** Benefits and risks hierarchies.

### 2.4.3. Pairwise Comparison Matrices and Establishment of Weights

AHP uses pairwise comparison matrices at each level to establish preferences between criteria and alternatives. To elaborate the matrices it is necessary to compare each element with respect to the other elements of the same level; this comparison can be done in multiple ways [44]. However, for the purposes of this study, the comparisons were carried out using the scale of relative importance suggested by Saaty. This scale consists of assigning a value from one to nine, according to the importance of an alternative with respect to another as shown in Table 3 [58]. This qualification system has proven to be very useful to incorporate judgments and personal values in a logical way.

**Table 3.** Saaty's relative importance scale.

| Scale | Numerical Rating | Reciprocal |
| --- | --- | --- |
| Extremely important | 9 | 1/9 |
| Very strongly to extremely important | 8 | 1/8 |
| Very strongly important | 7 | 1/7 |
| Strongly to very strongly important | 6 | 1/6 |
| Strongly important | 5 | 1/5 |
| Moderately to strongly important | 4 | 1/4 |
| Moderately important | 3 | 1/3 |
| Equally to moderately important | 2 | 1/2 |
| Equally important | 1 | 1 |

These comparisons are converted into numerical values through mathematical models, with the objective of estimating relative weights between the criteria and assigning probabilities to the alternatives [57]. This probability refers to the odds that each alternative has to fulfill the global goal [49]. The mathematical calculation is based on calculating the eigenvector of each matrix, ensuring certain consistency in the evaluation of judgments [59].

Judgements of the group of experts are represented by matrix $A_{nxn}$, where $a_{ij}$ is the relative importance of element $i$ compared to element $j$, and $n$ is the number of elements that are compared [57].

$$A = \begin{bmatrix} 1 & \cdots & a_{1n} \\ \vdots & \ddots & \vdots \\ \frac{1}{a_{1n}} & \cdots & 1 \end{bmatrix} \tag{1}$$

Once matrix A is defined, it is normalized, which is done by dividing the elements of each column of the matrix by the sum of all the elements of that column [49]. Then, the relative weight of each element is calculated through the eigenvector; the values in the eigenvector represent the contribution of each criterion to the global goal [49]. The next step is to ensure there is no inconsistency in the data collected. For this, Saaty [58] suggested employing the consistency rate (CR), which depends on the consistency index (CI) and the random consistency index (RI), and stated that a matrix is consistent if:

$$CR = \frac{CI}{RI} \leq 10\% \qquad (2)$$

CI depends on the maximum eigenvalue ($\lambda_{max}$) and the number of elements considered for the matrix development ($n$), and is described by the following expression:

$$CI = \frac{\lambda_{max} - n}{n - 1} \qquad (3)$$

On the one hand, the maximum eigenvalue is the sum of the products of each element of the eigenvector and the total value of the corresponding column in matrix A. On the other hand, the RI is determined by the number of elements ($n$) [58], as shown in Table 4.

**Table 4.** Random consistency index (RI) value [58].

| n | 1 | 2 | 3 | 4 | 5 | 6 | 7 | 8 | 9 | 10 |
|---|---|---|---|---|---|---|---|---|---|---|
| RI | 0 | 0 | 0.58 | 0.9 | 1.12 | 1.24 | 1.32 | 1.41 | 1.45 | 1.51 |

*2.5. Global Performance*

To complete the evaluation of the policy alternatives, a global performance (GP) index is built, which aims to integrate the results obtained for both hierarchies. Relative weights are established between the hierarchies ($W_B$ and $W_R$) to grant a rating to each alternative according to the following expression:

$$GP = W_B (Benefits) - W_R (Risks) \qquad (4)$$

## 3. Limitations

*3.1. SD Model Limitations*

The SD model described in this study has multiple limitations. First, the aging chain that describes the pavement deterioration process is based exclusively on information of the road network condition (Pavement Condition Index (PCI) or International Roughness Index (IRI) data), assuming the average time of occurrence of failures and the average time to carry out interventions as constant parameters. The behavior of these variables in reality is not constant; on the contrary, it responds to external factors such as climatic factors, volume and weight of vehicles, geotechnical conditions, properties of materials, etc. [60]. Considering these factors in the simulations would allow the model to recreate deterioration processes in a more realistic way.

Second, because of the lack of official information, both the analysis of costs and environmental impact is limited to two variables: capital costs and $CO_2$ emissions. The unit costs used to calculate capital costs are assumed constant during the analysis period and do not consider effects derived from market dynamics. Similarly, the emission factors used to quantify emissions were calculated from values reported in the literature [8,33] and not from official information. A complete analysis of the economic and environmental aspects implies the study of a significant number of variables that are not considered in this study.

Additionally, the implementation of sustainable measures in the preservation of the road network depends exclusively on how the available budget is distributed. Although economic resources are usually the main constraint in network management, it is important to consider other aspects, such as government incentives, availability and ease of adoption of measures, knowledge about measures' performance, etc. [61].

### 3.2. AHP Limitations

The AHP approach employed in this paper is limited in two important ways. First, expert interviews rely on maintenance policies derived from a hypothetical SD model that does not exactly reflect road conditions in Colombia. This is due to the lack of reliable information about the Colombian road network. Most of the pavement-related data in Colombia remains fragmented across different regions and have not been reported on a continuous basis over the last decade. In order to overcome this limitation, standard SD formulations have been employed and respondents have verified model results before conducting the AHP interviews. However, more good-quality data on Colombian pavements (or in any other jurisdiction) are required so that future studies can further explore this issue.

Second, AHP assumes that evaluation criteria are independent. However, this limitation is intended to be overcome by using SD as a complementing methodology [62]. Additionally, AHP has been strongly criticized for the poor accuracy of the methodology in interpreting the opinions of experts [63]. This is because AHP requires accurate judgments from experts, which leaves aside the uncertainty associated with decision-making processes. On the one hand, although the rating scales are unified, similar concepts can be interpreted differently by the group of respondents. To overcome this limitation, further studies can employ improved methods such as fuzzy AHP, which uses diffuse numbers to characterize the variables involved and thus to represent the imprecise nature of human cognition [63].

## 4. Results

### 4.1. SD Model Results

The main outputs of the SD model for each of the policies are shown below. First, three graphs that exhibit the evolution of the road network condition during the analysis period are presented. Figure 6 shows the percentages of km-roadway in Good, Fair and Poor condition over time. Next, Table 5 is presented, which summarizes the results for the other criteria (km-roadway built, costs and $CO_2$ emissions). Unlike the condition of the road network, results for the other criteria are not exhibited over time. Average values were estimated with the purpose of facilitating comparisons and the analysis; an average annual value was calculated for the growth of the road network and the emissions generated, and the Equivalent Annual Cost (EAC) was estimated for showing cost results. EAC values are negative in order to differentiate them from savings (positive) because they only represent expenditures associated with capital investments.

**Table 5.** System dynamics (SD) model results for road network growth, costs, and emissions.

| Policies | Roads Built (km-Roadway/Year) | $CO_2$ Emissions (Ton/Year) | EAC [1] (Millions of USD) |
|---|---|---|---|
| P1 | 37,392 | 37,334 | −$ 3.343 |
| P2 | 41,503 | 37,659 | −$ 3.343 |
| P3 | 42,615 | 34,841 | −$ 3.343 |
| P4 | 50,446 | 42,473 | −$ 3.219 |
| P5 | 24,106 | 33,063 | −$ 3.343 |
| P6 | 61,416 | 41,472 | −$ 2.727 |
| P7 | 112,960 | 54,136 | −$ 1.988 |
| P8 | 85,344 | 45,895 | −$ 1.705 |

[1] EAC: equivalent annual cost.

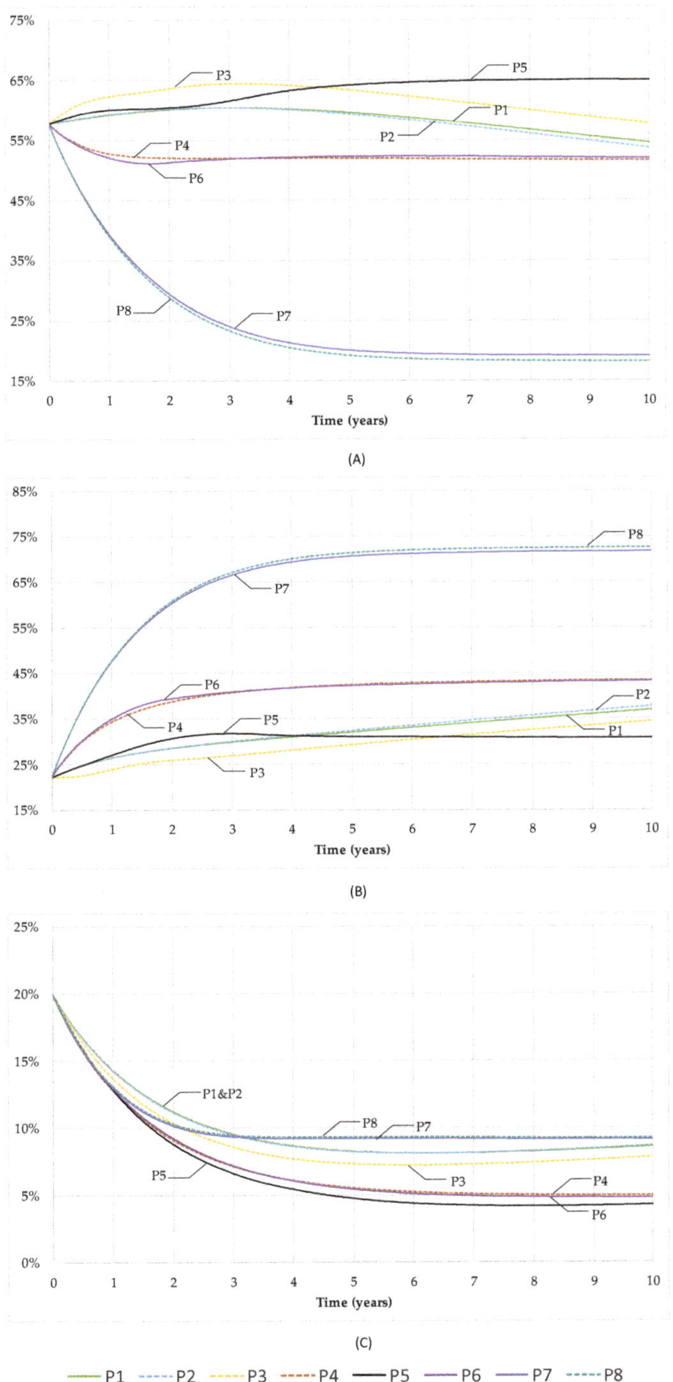



**Figure 6.** SD results for road network conditions: (**A**) good condition; (**B**) fair condition; (**C**) poor condition.

### 4.2. AHP Results

Twelve Colombian experts were interviewed for the application of the AHP methodology, as shown in Table 6. The sample size is consistent with the number of experts employed in AHP studies [64]. This is because the sample size does not matter, as long as the experts are a representative sample and can provide consistent information [24]. As previously stated, one of the objectives of using the AHP methodology was to complement the SD approach. Before carrying out the multi-attribute evaluation, the SD model and the hypothetical case study were examined through the lens of the group of experts. In each interview, respondents recognized the lack of reliable information in the Colombian context regarding the condition of the country's road network. As a result, they accepted the use of a hypothetical case study for the evaluation of maintenance policies. Additionally, after observing the results of the SD model, they confirmed that the hypothetical case study was appropriate for the Colombian highway system.

**Table 6.** Brief profile of experts interviewed.

| Expert | Industry Type |
| --- | --- |
| Expert 1 | Infrastructure Consultant/Designer |
| Expert 2 | Infrastructure Contractor |
| Expert 3 | Infrastructure Contractor |
| Expert 4 | Infrastructure Contractor |
| Expert 5 | Professor and Pavement Expert |
| Expert 6 | Infrastructure Contractor |
| Expert 7 | Infrastructure Consultant/Designer |
| Expert 8 | Infrastructure Consultant/Designer |
| Expert 9 | Professor and Pavement Expert |
| Expert 10 | Infrastructure Contractor |
| Expert 11 | Professor and Pavement Expert |
| Expert 12 | Infrastructure Consultant/Designer |

Once the SD model and the case study data studied were approved by the experts, the multi-attribute analysis was performed. Each participant was asked to establish relative importance between the criteria of each hierarchy and between the performances of each alternative, applying the Saaty scale. This resulted in the creation of six pairwise comparison matrices, three for each hierarchy. For the benefits hierarchy, three matrices were filled: one to establish priorities between the road network growth and the road network condition criteria, and two to establish scores between the performance of the alternatives for each criterion. For the risks hierarchy, the same matrices were developed for the costs and environmental impact criteria.

To process and weigh the opinions collected, AHP online software was used [65]. Regarding the benefits hierarchy, it was observed that experts gave priority to the condition of the network over its growth, granting local relative weights of 66.7% and 33.3% between these two criteria. To ensure the consistency of the results, the CR of each criterion was calculated: road network growth CR = 5.5% and road network condition CR = 1.6%. Additionally, the consensus of the group was estimated and qualified as "very high", with values higher than 90%. Employing the relative weights, the following policy ranking was obtained: P5, P7, P3, P6, P1, P2, P8, and P4. Policy 5 (P5) had the best performance in this hierarchy; it was concluded that this was because it allowed the road network to have better conditions over time, as based on values presented in Table 7. In second place was policy 7 (P7), which despite having a poor performance in the road network condition criterion, had an excellent performance regarding the road network growth; it presented the highest value of km-roadway built annually. Additionally, the sensitivity analysis of this hierarchy showed that a change in the weight of the road network condition criterion from 66.7% to 50.5% would change the ranking between P5 and P7, making P7 the top policy in the ranking.

**Table 7.** Benefits hierarchy results.

| Level 0 | Level 1 | Priority | P1 | P2 | P3 | P4 | P5 | P6 | P7 | P8 |
|---|---|---|---|---|---|---|---|---|---|---|
| Benefits Hierarchy | Road Network Growth | 33.30% | 0.038 | 0.040 | 0.045 | 0.072 | 0.021 | 0.128 | 0.382 | 0.274 |
| | Road Network Condition | 66.70% | 0.147 | 0.147 | 0.179 | 0.110 | 0.262 | 0.108 | 0.024 | 0.024 |
| | | **100%** | **0.111** | **0.111** | **0.134** | **0.097** | **0.182** | **0.114** | **0.143** | **0.107** |

On the other hand, in the risks hierarchy, experts considered costs as significantly more important than $CO_2$ emissions, with relative weights of 85.7% and 14.3%, respectively (Table 8). As was done for the benefits hierarchy, the CR was determined for each criterion, with a value of 1.1% for both criteria. In this case, the ranking obtained was: P1, P2, P3, P4, P5, P7, P6, and P8, where the first place is the alternative that had higher costs and generated more emissions. In terms of planning and policies, the most convenient alternative was the one with the worst performance in this hierarchy, namely policy 8 (P8) was the one that contributed the most to the global goal. This ranking was dominated by costs, due to the great importance given by experts, resulting in the most convenient alternative being the least expensive, independently of its environmental impact. The sensitivity analysis of this hierarchy indicates that by reducing the weight of costs to 70%, the positions in the ranking of policies P2 and P4 would change; however, significant variations between the priorities were required so that there were changes in the top and bottom of the ranking.

**Table 8.** Risks hierarchy results.

| Level 0 | Level 1 | Priority | P1 | P2 | P3 | P4 | P5 | P6 | P7 | P8 |
|---|---|---|---|---|---|---|---|---|---|---|
| Risks Hierarchy | Costs | 85.70% | 0.178 | 0.178 | 0.178 | 0.167 | 0.178 | 0.064 | 0.032 | 0.024 |
| | Environmental Impact | 14.300% | 0.063 | 0.063 | 0.051 | 0.118 | 0.042 | 0.115 | 0.345 | 0.203 |
| | | **100%** | **0.162** | **0.162** | **0.160** | **0.160** | **0.159** | **0.071** | **0.077** | **0.049** |

*4.3. Global Performance and Ranking*

To combine the results of both hierarchies, the global performance score for each alternative was calculated. A 60% weight was assigned to the benefits and 40% to the risks, giving rise to the following expression for the global performance score:

$$GP = 0.2(RNGrowth) + 0.4(RNCondition) - 0.34(Costs) - 0.06(Env.Impact) \qquad (5)$$

Table 9 presents the GP results for each policy alternative. In order to make comparisons easier, the last column shows a normalized GP value that was calculated in order to turn all GP results into positive values.

**Table 9.** Global performance (GP) results.

| Alternative | Benefits Hierarchy (0.6) | | Risks Hierarchy (0.4) | | GP | Normalized GP |
|---|---|---|---|---|---|---|
| | Growth (0.333) | Condition (0.667) | Costs (0.857) | Emission (0.143) | | |
| P1 | 0.038 | 0.147 | 0.178 | 0.063 | 0.002 | 0.73 |
| P2 | 0.040 | 0.147 | 0.178 | 0.063 | 0.002 | 0.78 |
| P3 | 0.045 | 0.179 | 0.178 | 0.051 | 0.017 | 2.22 |
| P4 | 0.072 | 0.110 | 0.167 | 0.118 | −0.006 | 0.00 |
| P5 | 0.021 | 0.262 | 0.178 | 0.042 | 0.046 | 5.13 |
| P6 | 0.128 | 0.108 | 0.064 | 0.115 | 0.040 | 4.57 |
| P7 | 0.382 | 0.024 | 0.032 | 0.345 | 0.055 | 6.08 |
| P8 | 0.274 | 0.024 | 0.024 | 0.203 | 0.045 | 5.020 |

The ranking of policies according to global performance is: P7, P5, P8, P6, P3, P2, P1, and P4. That is, for the Colombian experts, the best policy alternative was the one that proposed a corrective maintenance, mainly concerned with the rehabilitation of roads in poor condition. This result is consistent with the reality of Colombia and other countries. This was demonstrated in other studies that exhibited how governments and agencies usually perform short-term measures, promoting corrective maintenance over preventive maintenance [55,66,67]. Additionally, this policy promotes the use of sustainable measures in the rehabilitation of existing roads and not in the construction of new ones. However, adopting sustainable alternatives in the rehabilitation interventions does not contribute to a reduction of emissions. This is because the generation of emissions not only depends exclusively on the type of activities employed, but also is associated with the amount of interventions executed.

### 4.4. Sensitivity Analysis

AHP results were highly influenced by the experts' opinions. That is, the ranking of policies can easily vary if the perceptions of experts change. For this reason, a sensitivity analysis of the global performance was performed, varying the relative importance between benefits and risks. Figure 7 presents the variation of the policy rankings according to the relative weight of the risks. It shows that a decrease in the weight of risks from 40% to 32% would modify the ranking, leaving policy 5 in first place, characterizing it as a policy that is more expensive than policy 7 and one that allows road networks to keep better conditions since it gives priority to maintenance over rehabilitation. On the other hand, an increase in the weight of risks from 40% to 57% would position policy 8 at the top of the rankings. Independently of the weight of risks, P7 and P8 remain in the first place of the ranking. This demonstrates that the evaluation and selection of maintenance policies are governed by economic factors in the Colombian context.

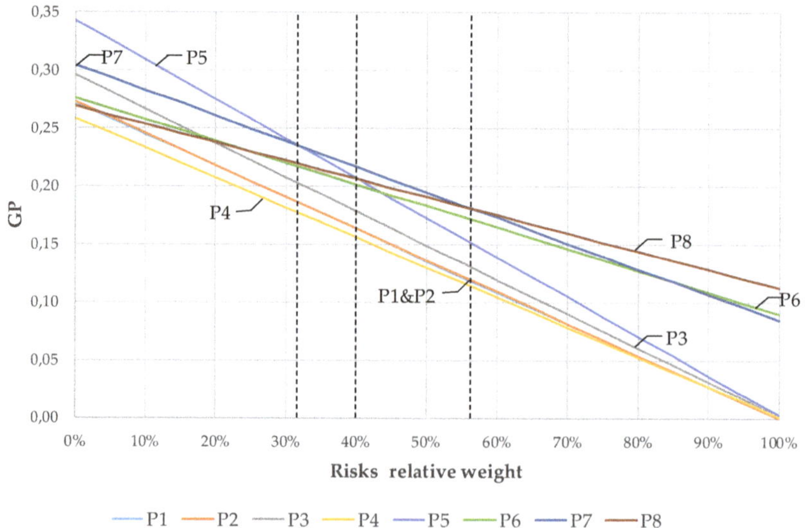


**Figure 7.** Global performance sensitivity analysis.

### 4.5. Scenario Analysis

One of the objectives of the study was to integrate technical, environmental, and economic aspects in the evaluation of maintenance strategies. For this reason, the analysis of an alternative scenario was proposed, in which the relative weights of the risk hierarchy were modified, with the purpose of giving less priority to the costs and thus increasing the importance given to emissions. For this, an alternative scenario was proposed, in which decision makers have a slightly more environmentally friendly perspective and their decisions are not based only on costs. The benefits hierarchy's weights are kept the same, since the relative importance between the criteria did not show a difference that could significantly affect the ranking. Results when the relative weight of costs is reduced from 86% to 60% are shown below in Table 10.

**Table 10.** Global performance results for the alternative scenario.

| Alternative | Benefits Hierarchy (0.6) | | Risks Hierarchy (0.4) | | GP | Normalized GP |
|---|---|---|---|---|---|---|
| | Growth (0.333) | Condition (0.667) | Costs (0.6) | Emission (0.4) | | |
| P1 | 0.038 | 0.147 | 0.178 | 0.063 | 0.014 | 1.409 |
| P2 | 0.040 | 0.147 | 0.178 | 0.063 | 0.014 | 1.447 |
| P3 | 0.045 | 0.179 | 0.178 | 0.051 | 0.030 | 3.015 |
| P4 | 0.072 | 0.110 | 0.167 | 0.118 | -0.001 | 0.000 |
| P5 | 0.021 | 0.262 | 0.178 | 0.042 | 0.060 | 6.017 |
| P6 | 0.128 | 0.108 | 0.064 | 0.115 | 0.035 | 3.537 |
| P7 | 0.382 | 0.024 | 0.032 | 0.345 | 0.023 | 2.357 |
| P8 | 0.274 | 0.024 | 0.024 | 0.203 | 0.026 | 2.668 |

The ranking for this scenario is as follows: P5, P6, P3, P8, P7, P2, P1, and P4. A sensitivity analysis was performed for the GP in this scenario, which showed that if the relative weight of risks is lower than 63%, the best alternative is policy 5, otherwise, the best alternative is policy 6. Both policies give priority to maintenance over rehabilitation and promote the use of sustainable techniques in the M&R of existing roads.

## 5. Discussion

Based on the results presented before, the combination of SD and AHP approaches allowed the authors to examine the main factors associated with the sustainable development of a road network. It also offered the opportunity to examine multiple pavement-related decisions from a strategic perspective by considering several decision-making criteria. Consequently, this paper provides a hybrid approach capable of examining the underlying social, environmental, and economic factors of road development.

Certainly, as discussed in previous sections, the proposed methodology has multiple limitations. However, the authors have implemented specific analytical approaches so as to increase the validity and reliability of this study. For instance, although it is clear that the SD model is based on a hypothetical case study, the stock and flow formulations employed in such a model have been used by several researchers to analyze multiple road networks worldwide. This means that model equations are not exclusively dependent on the historical data of the hypothetical case study; but rather respond to managerial and physical mechanisms that are common to multiple road networks in several jurisdictions at an international level. In line with that, although the SD model outcomes do not exactly reproduce the behavior of the Colombian road infrastructure sector, all the interviewed experts approved such results and agreed with them as a good estimation for the Colombian context. In this sense, the SD model helped to create a hypothetical reality upon which experts' judgements could be examined so as to better understand how to improve the Colombian road network in a sustainable way.

Results suggest that in the Colombian context, decision-making regarding the formulation of road maintenance policies is highly influenced by costs. AHP results showed that those alternatives that focus on corrective maintenance (rehabilitation and reconstruction), such as policies 7 and 8, have a better GP score. This is because these alternatives have the lowest implementation costs and the highest average km-roadway built. Although rehabilitation activities are the most expensive, the costs do not increase significantly since maintenance activities are not carried out. This generates two important consequences: First, there are more resources available to build roads and second, the existing roads deteriorate faster, increasing the number of roads in fair condition (Figure 6B). Results reflect the feedback between the budget and the different types of interventions; specifically, not spending on maintenance activities increases the construction of new roads. The construction of roads contributes to the quality of the road network, as this increases the kilometers of roadways. However, these new roads deteriorate rapidly because predictive maintenance of those in fair conditions is not executed. That is, Colombian experts seek that roads do not reach the worst conditions (i.e., poor condition), but they do not show an interest in keeping the road network in excellent condition; as a consequence, most kilometers of the road network remain in fair condition. This type of policy increases road connectivity in a country in terms of the number of highways but neglects their quality.

Overall, this work has several pavement management and sustainable policy implications. Results suggest that emissions not only depend on the type of road-related activities but also on the frequency of the pavement-related interventions. For example, policy 1 promotes the use of conventional activities for every type of intervention and has a smaller environmental impact than policies 4, 6, 7, and 8, which promote the use of sustainable activities in construction or M&R stages. Therefore, no goals are achieved at the environmental level if an appropriate maintenance policy, in terms of frequency, is not developed.

The proposed hybrid methodology also indicates that if a country is committed to reducing $CO_2$ emissions, it needs to develop maintenance policies that are environmentally friendly and at the same time allow the road network to remain in adequate conditions. The better the condition of the road network, the better the service for users, providing security and comfort to them. In this regard, results suggest that policies that promote the use of green activities in the M&R of existing roads (policies 3, 5, 6 and 8) have higher possibilities of fulfilling environmental goals compared to those that promote sustainable activities in the construction stage (policies 2 and 4).

However, despite the clear relationship between a good maintenance policy and a sustainable road network, decision-making processes within the Colombian road infrastructure sector seem to prioritize new road construction over maintenance and rehabilitation interventions. Although this increases the number of paved roads in the country, it highlights the fact that even when it is known that predictive maintenance is the best alternative, it is not always used to formulate policies.

## 6. Conclusions

This study evaluates strategies used in road maintenance policies, simultaneously considering technical, economic, and environmental factors. This work demonstrates that the environmental impact, specifically the $CO_2$ emissions generated, and the expenditures associated with the growth and preservation of the road network are directly interrelated with the condition of the network. Through a hybrid approach, relationships between deterioration phenomena, construction, and preservation procedures were identified, as well as their connections with budgetary expenditures and $CO_2$ emissions. The application of the methodology reveals that, in Colombia, decision-making processes seem to prioritize rehabilitation activities (i.e., corrective over predictive maintenance). This is because cost remains the factor that governs decisions related to the development of maintenance policies. This shows that $CO_2$ emissions have not received enough attention and that the economic component plays a very crucial role in respect to maintenance policy development within the road infrastructure sector.

Furthermore, through the scenario analysis, it was shown that the road network is maintained in better conditions if predictive maintenance is adopted, where priority is given to maintenance over rehabilitation. Additionally, it was demonstrated that only promoting the implementation of sustainable activities does not contribute to mitigating the environmental impact, since emissions not only depend on the type of activities but also on the number of interventions carried out. This means that promoting corrective maintenance does not contribute to the reduction of $CO_2$ emissions because of the amount and complexity of interventions needed to rehabilitate roads. The analysis carried out in this work also exhibits that, within the framework of an appropriate predictive maintenance policy, greater reductions in $CO_2$ can be achieved if policies focus on promoting sustainable M&R interventions instead of sustainable construction.

Overall, through this study, the proposed methodology has proven to be beneficial in terms of examining the different factors associated with developing a sustainable road network. The SD model allowed researchers to analyze the relationships between highway system conditions and M&R operations. It also provided a platform to quantify the influence of budgetary constraints over different pavement preservation policies. Additionally, it facilitated the estimation of the amount of $CO_2$ emissions generated in multiple policy-based scenarios and the application of the AHP methodology allowed the authors to gain insight into the multiple strategies related to preserving a highway system in good conditions with limited environmental impacts. Since, in many cases, such strategies depend on individual decisions, the analysis of experts' interviews helped to better understand several policy-based alternatives focused on social, environmental, and economic factors.

This study contributes to the literature in several different ways. First, it presents an SD model that allows researchers to analyze the development of road networks in a global and integrated way. Whereas the literature offers multiple examples of SD models within the road infrastructure sector, this study provides a simulation tool capable of recreating the feedback mechanisms associated with the evolution of the road network condition, considering the deterioration processes and preservation interventions, and describing the role played by the limited resources in these interactions. Since the model employs traditional system dynamics formulations, it can be adapted to reflect the specific conditions of any road network worldwide with the help of further studies. Second, the methodology proposed in this study can be used as a support tool in decision-making processes in road development. Although multiple authors have implemented AHP in multiple contexts, there is little evidence of combining AHP with SD for the road infrastructure sector. The integration of these two approaches, therefore, allows decision makers to analyze how the implementation of specific strategies affects road network performance, the environment, and available budget in the long term. Third, because results suggest that decision makers may tend to prioritize corrective over predictive maintenance, the proposed hybrid approach helps to generate awareness about the effects of formulating road maintenance policies without considering the long-term effects at the technical, environmental, and economic levels. Future studies can strengthen this contribution by exploiting the graphical and interactive procedures associated with the implementation of SD and AHP.

Although the study has made multiple contributions to the fields of sustainability and decision-making, it has some limitations imposed by its scope and assumptions, as discussed in previous sections. For instance, the SD model is based on mechanisms related to construction and M&R activities. It does not incorporate variables related to user costs, climatic factors, vehicle data, and geotechnical conditions. Also, due to the lack of reliable data in Colombia, the AHP methodology was applied by considering a hypothetical case study as the basis for decision-making processes. Additionally, AHP results only apply to a specific situation or context. In case of implementing the proposed methodology in other settings, policymakers must establish their own criteria to obtain priorities that fit their context, which means that AHP evaluation must be performed again for the analysis of new policy alternatives.

Further research is required in multiple fields to overcome the limitations in this study. In order to improve the decision-making processes associated with road network preservation, it is necessary to integrate experts from various areas, such as paving engineering, marketing, infrastructure systems management, material science, environmental impact, etc. In this way, a model that analyzes all the involved sectors at the same level of detail, understands all the factors that affect the deterioration phenomena, and includes user costs in the analysis could be developed. Also, understanding the dynamics in the economic sector could achieve a better quantification and evaluation of costs. Finally, studies related to sustainable activities and environmental impact are required to enrich the model, as well as the analysis of measures that allow the adoption of such innovations. The methodology requires several improvements; however, results of this study show that it is valuable to evaluate maintenance policies and identify key strategies, taking into account the relationships between technical, economic, and environmental factors.

**Supplementary Materials:** Models' variables and formulations can be found in the Supplemental Data. The following are available online at http://www.mdpi.com/2071-1050/12/3/872/s1, System Dynamics Equations: Equations (S1)–(S110).

**Author Contributions:** The authors confirm contribution to the paper as follows: conceptualization, methodology, investigation, data collection, validation, analysis and interpretation of results, draft manuscript preparation: A.R. and J.G. All authors have read and agreed to the published version of the manuscript.

**Funding:** This research received no external funding.

**Acknowledgments:** We would like to thank the Department of Civil and Environmental Engineering at Universidad de los Andes (Bogota, Colombia) for their valuable support and help in offering the resources needed to conduct this study.

**Conflicts of Interest:** The authors declare no conflicts of interest.

## References

1. World Economic Forum the Global Competitiveness Report 2018—Reports—World Economic Forum. Available online: http://reports.weforum.org/global-competitiveness-report-2018/chapter-3-benchmarking-competitiveness-in-the-fourth-industrial-revolution-introducing-the-global-competitiveness-index-4-0/ (accessed on 27 November 2019).
2. Queiroz, C.; Haas, R.; Cai, Y. National Economic Development and Prosperity Related to Paved Road Infrastructure. In Proceedings of the Tranportation Research Record, Washington DC, USA, 9–13 January 1994.
3. Augeri, M.G.; Colombrita, R.; Greco, S.; Lo Certo, A.; Matarazzo, B.; Slowinski, R. Dominance-Based Rough Set Approach to Budget Allocation in Highway Maintenance Activities. *J. Infrastruct. Syst.* **2011**, *17*, 77–85. [CrossRef]
4. Bjornsson, H.C.; De La Garza, J.M.; Nasir, M.J. A Decision Support System for Road Maintenance Budget Allocation. In Proceedings of the Eighth International Conference on Computing in Civil and Building Engineering, Stanford, CA, USA, 14–16 August 2000.
5. González, A.D.; Dueñas-Osorio, L.; Sánchez-Silva, M.; Medaglia, A.L. The Interdependent Network Design Problem for Optimal Infrastructure System Restoration. *Comput. Civ. Infrastruct. Eng.* **2016**, *31*, 334–350. [CrossRef]
6. Fwa, T.F.; Farhan, J. Optimal Multiasset Maintenance Budget Allocation in Highway Asset Management. *J. Transp. Eng.* **2012**, *138*, 1179–1187. [CrossRef]
7. Zhang, H.; Keoleian, G.A.; Lepech, M.D. Network-Level Pavement Asset Management System Integrated with Life-Cycle Analysis and Life-Cycle Optimization. *J. Infrastruct. Syst.* **2013**, *19*, 99–107. [CrossRef]
8. Torres-Machi, C.; Pellicer, E.; Yepes, V.; Chamorro, A. Towards a sustainable optimization of pavement maintenance programs under budgetary restrictions. *J. Clean. Prod.* **2017**, *148*, 90–102. [CrossRef]
9. Herabat, P.; Tangphaisankun, A. Multi-Objective Optimization Model using Constraint-Based Genetic Algorithms for Thailand Pavement Management. *J. East. Asia Soc. Transp. Stud.* **2005**, *6*, 1137–1152.
10. T Chan, B.W.; Fwa, T.F.; Tan, C.Y. Road-maintenance planning using genetic algorithms. I: Formulation. *J. Transp. Eng.* **1994**, *120*, 693–709. [CrossRef]
11. Mishra, S.; Golias, M.M.; Sharma, S.; Boyles, S.D. Optimal funding allocation strategies for safety improvements on urban intersections. *Transp. Res. A* **2015**, *75*, 113–133. [CrossRef]
12. Fallah-Fini, S.; Triantis, K.; Rahmandad, H.; De La Garza, J.M. Optimizing highway maintenance operations: Dynamic considerations. *Syst. Dyn. Rev.* **2010**, *26*, 216–238. [CrossRef]
13. Fallah-Fini, S.; Triantis, K.; Rahmandad, H.; De La Garza, J.M. Measuring dynamic efficiency of highway maintenance operations. *Omega* **2014**, *50*, 18–28. [CrossRef]
14. Guevara, J.; Garvin, M.J.; Ghaffarzadegan, N. Capability Trap of the U.S. Highway System: Policy and Management Implications. *J. Manag. Eng.* **2017**, *33*, 04017004. [CrossRef]
15. Moazami, D.; Behbahani, H.; Muniandy, R. Pavement rehabilitation and maintenance prioritization of urban roads using fuzzy logic. *Expert Syst. Appl.* **2011**, *38*, 12869–12879. [CrossRef]

16. Organisation for Economic Co-operation and Development. *Road Maintenance and Rehabilitation: Funding and Allocation Strategies*; Organisation for Economic Co-operation and Development: Paris, France, 1994; ISBN 9789264142770.

17. Hicks, R.; Moulthrop, J.; Daleiden, J. Selecting a Preventive Maintenance Treatment for Flexible Pavements. In Proceedings of the Transportation Research Record, Washington, DC, USA, 10–14 January 1999; pp. 99–1025.

18. Butt, A.A. Life Cycle Assessment of Asphalt Pavements including the Feedstock Energy and Asphalt Additives. PH.D. Thesis, KTH Royal Institute of Technology, Stockholm, Sweden, 2012.

19. Harvey, J.; Kendall, A.; Saboori, A. *The Role of Life Cycle Assessment in Reducing Greenhouse Gas Emissions from Road Construction and Maintenance for Sustainable Transportation*; UC Davis, National Center for Sustainable Transportation: Berkeley, CA, USA, 2015.

20. Bujang, M.; Hainin, M.R.; Abd Majid, M.Z.; Idham Mohd Satar, M.K.; Azahar, W.N.A.W. Assessment framework for pavement material and technology elements in green highway index. *J. Clean. Prod.* **2018**, *174*, 1240–1246. [CrossRef]

21. Wang, Y.; Leng, Z.; Li, X.; Hu, C. Cold recycling of reclaimed asphalt pavement towards improved engineering performance. *J. Clean. Prod.* **2017**, *171*, 1031–1038. [CrossRef]

22. Egilmez, G.; Tatari, O. A dynamic modeling approach to highway sustainability: Strategies to reduce overall impact. *Transp. Res. A* **2012**, *46*, 1086–1096. [CrossRef]

23. Jullien, A.; Dauvergne, M.; Cerezo, V. Environmental assessment of road construction and maintenance policies using LCA. *Transp. Res. D* **2014**, *29*, 56–65. [CrossRef]

24. Saaty, T.L. How to make a decision: The Analytic Hierarchy Process. *Eur. J. Oper. Res.* **1994**, *48*, 9–26. [CrossRef]

25. Goebel, R.; Sanfelice, R.G.; Teel, A.R. *Hybrid Dynamical Systems: Modeling, Stability, and Robustness*; Princeton University Press: Princeton, NJ, USA, 2012; ISBN 9781400842636.

26. Forrester, J.W. *Industrial Dynamics*; MIT Press: Cambridge, MA, USA, 1961.

27. Schwab, K. The Global Competitiveness Report 2019. Available online: http://www3.weforum.org/docs/WEF_TheGlobalCompetitivenessReport2019.pdf (accessed on 8 January 2020).

28. Economic Commission for Latin America and the Caribbean. *The Economic Infrastructure Gap in Latin America and the Caribbean*; Economic Commission for Latin America and the Caribbean: Santiago, Chile, 2011.

29. Ruiz, A.; Guevara, J. Environmental and economic impacts of road infrastructure development: Dynamic considerations and policies. *J. Manag. Eng..* in press. [CrossRef]

30. Calderón, C.; Servén, L. *Trends in Infrastructure in Latin America*; The World Bank: Washington, DC, USA, 2004.

31. Vieria Gomes, S. Methodology for Benchmarking Road safety in LAtin America. Available online: https://trid.trb.org/view/1435842 (accessed on 8 January 2020).

32. Ministry of Public Works. *Proposiciones de Acciones de Mantenimiento (PAM) y Estado de la Calzada y Bermas para Caminos Pavimentados de la red Vial Nacional*; Ministry of Public Works: Santiago, Chile, 2017.

33. Torres-Machi, C.; Osorio-Lird, A.; Chamorro, A.; Videla, C.; Tighe, S.L.; Mourgues, C. Impact of environmental assessment and budgetary restrictions in pavement maintenance decisions: Application to an urban network. *Transp. Res. D Transp. Environ.* **2018**, *59*, 192–204. [CrossRef]

34. De Vialidad, D. *Proyecto Conservación Mayor Pista 17L/35R y Rodajes Asociados, Etapa 1*; Ministry of Public Works: Santiago, Chile.

35. De Vialidad, D. *Presupuesto Oficial. Conservación Periódica Ruta N°55, Sector km 36,00 al km 62,39, por Sectores. Comuna de Pinto, Provincia de Ñuble, Región del Biobío*; Ministry of Public Works: Santiago, Chile, 2014.

36. De Vialidad, D. *Conservación Periódica Ruta 55, Comuna de Pinto, Provincia de Ñuble, Región del Biobío*; Ministry of Public Works: Santiago, Chile, 2012.

37. Forrester, J.W. Counterintuitive behavior of social systems. *Technol. Forecast. Soc. Chang.* **1971**, *3*, 1–22. [CrossRef]

38. Sterman, J.D. *Business Dynamics. Systems Thinking and Modeling for a Complex World*; McGraw-Hill: New York, NY, USA, 2000; ISBN 007238915X.

39. Mallick, R.B.; Radzicki, M.J.; Zaumanis, M.; Frank, R. Use of system dynamics for proper conservation and recycling of aggregates for sustainable road construction. *Resour. Conserv. Recycl.* **2014**, *86*, 61–73. [CrossRef]

40. Leon, H.; Osman, H.; Georgy, M.; Elsaid, M. System Dynamics Approach for Forecasting Performance of Construction Projects. *J. Manag. Eng.* **2018**, *34*, 04017049. [CrossRef]

41. Ruiz, A.; Guevara, J. Assesment of $CO_2$ emissions in the road infrastructure sector through a system dynamics approach. In Proceedings of the 2019 TRB Annual Meeting, Washington, DC, USA, 13–17 January 2019.

42. Wang, P. *Greenhouse Gas Emissions Mitigation in Road Construction and Rehabilitation a Toolkit for Developing Countries Introduction to Greenhouse Gas Emissions in Road Construction and Rehabilitation*; World Bank Group: Washington, DC, USA, 2010.

43. Homer, J.B. Partial-model testing as a validation tool for system dynamics: EBSCOhost. In Proceedings of the International System Dynamics Conference, Massachusetts, MA, USA, 27–30 July 1983.

44. Triantaphyllou, E.; Mann, S.H. Using the analytic hierarchy process for decision making in engineering applications: Some challenges. *J. Ind. Eng. Appl. Pract.* **1995**, *2*, 35–44.

45. Thanki, S.; Govindan, K.; Thakkar, J. An investigation on lean-green implementation practices in Indian SMEs using analytical hierarchy process (AHP) approach. *J. Clean. Prod.* **2016**, *135*, 284–298. [CrossRef]

46. Rosa, L.V.; Haddad, A.N. Assessing the Sustainability of Existing Buildings Using the Analytic Hierarchy Process. *Am. J. Civ. Eng.* **2013**, *1*, 24–30. [CrossRef]

47. Ahsan, K.; Paul, S.K. Procurement Issues in Donor-Funded International Development Projects. *J. Manag. Eng.* **2018**, *34*, 04018041. [CrossRef]

48. Gerdsri, N.; Kocaoglu, D.F. Applying the Analytic Hierarchy Process (AHP) to build a strategic framework for technology roadmapping. *Math. Comput. Model.* **2007**, *46*, 1071–1080. [CrossRef]

49. Ramadhan, R.H.; Al-, H.I.; Wahhab, A.; Duffuaa, S.O. The use of an analytical hierarchy process in pavement maintenance priority ranking. *J. Qual. Maint. Eng.* **1999**, *5*, 25–39. [CrossRef]

50. Saaty, R.W. The analytic hierarchy process-what it is and how it is used. *Math. Model.* **1987**, *9*, 161–176. [CrossRef]

51. Saaty, T.L.; Vargas, L.G. *Models, Methods, Concepts & Applications of the Analytic Hierarchy Processs*, 2nd ed.; Springer: Berlin/Heidelberg, Germany, 2012.

52. Bhushan, N.; Rai, K. *Strategic Decision Making: Applying the Analytic Hierarchy Process*; Springer: Berlin/Heidelberg, Germany, 2004; ISBN 1852338644.

53. National Research Council. Pavement management system development transportation research board. In Proceedings of the Transportation Research Board, Washington, DC, USA, 14–18 January 1979.

54. Furuya, A.; Madanat, S.; Asce, M. Accounting for Network Effects in Railway Asset Management. *J. Transp. Eng.* **2013**, *139*, 92–100. [CrossRef]

55. Arif, F.; Bayraktar, M.E.; Chowdhury, A.G. Decision Support Framework for Infrastructure Maintenance Investment Decision Making. *J. Manag. Eng.* **2016**, *32*, 04015030. [CrossRef]

56. Sanchez-Silvia, M.; Klutke, G. *Reliability and Life—Cycle Analysis of Detoriorating Systems*; Springer: Berlin/Heidelberg, Germany, 2016; ISBN 9781447129653.

57. Saaty, T.L. Axiomatic Foundation of the Analytic Hierarchy Process. *Manage. Sci.* **1986**, *32*, 841–855. [CrossRef]

58. Saaty, T.L. *Theory and Applications of the Analytic Network Process*; RWS Publications: Pittsburgh, PA, USA, 2005; ISBN 1888603062.

59. Saaty, T.L. Relative Measurement and Its Generalization in Decision Making Why Pairwise Comparisons are Central in Mathematics for the Measurement of Intangible Factors the Analytic Hierarchy/Network Process. *RACSAM* **2008**, *102*, 251–318. [CrossRef]

60. Henning, T.F.P.; Alabaster, D.; Arnold, G.; Liu, W. Relationship between Traffic Loading and Environmental Factors and Low-Volume Road Deterioration. *Transp. Res. Rec. J. Transp. Res. Board* **2014**, *2433*, 100–107. [CrossRef]

61. Aaahajan, V.; Muller, E.; Srivastava, R.K. Determination of Adopter Categories by Using Innovation Diffusion Models. *J. Mark. Res.* **1990**, *27*, 37–50.

62. Roy, B. Paradigms and challenges. In *International Series in Operations Research and Management Science*; Springer New York LLC: New York, NY, USA, 2005; Volume 78, pp. 3–24.

63. Tiryaki, F.; Ahlatcioglu, B. Fuzzy portfolio selection using fuzzy analytic hierarchy process. *Inf. Sci.* **2009**, *179*, 53–69. [CrossRef]

64. Wong, J.K.W.; Li, H. Application of the analytic hierarchy process (AHP) in multi-criteria analysis of the selection of intelligent building systems. *Build. Environ.* **2008**, *43*, 108–125. [CrossRef]

65. Goepel, K.D. Implementation of an Online Software Tool for the Analytic Hierarchy Process-Challenges and Practical Experiences. *Working Pap.* **2017**, *10*, 1–20.

66. Dornan, D.L. Asset management: Remedy for addressing the fiscal challenges facing highway infrastructure. *Int. J. Transp. Manag.* **2002**, *1*, 41–54. [CrossRef]
67. Hicks, R.G.; Dunn, K.; Moulthrop, J.S. Framework for Selecting Effective Preventive Maintenance Treatments for Flexible Pavements. *Transp. Res. Rec. J. Transp. Res. Board* **1997**, *1597*, 1–10. [CrossRef]

*Article*

# Corporate Sustainability Strategies and Decision Support Methods: A Bibliometric Analysis


**Fotis Kitsios [1], Maria Kamariotou [1] and Michael A. Talias [2,*]**

[1] Department of Applied Informatics, University of Macedonia, 156 Egnatia st., 54636 Thessaloniki, Greece; kitsios@uom.gr (F.K.); mkamariotou@uom.edu.gr (M.K.)

[2] Healthcare Management Postgraduate Program, Open University Cyprus, P.O. Box 12794, Nicosia 2252, Cyprus

[*] Correspondence: michael.talias@ouc.ac.cy


Received: 21 November 2019; Accepted: 5 January 2020; Published: 10 January 2020


**Abstract:** Sustainability is becoming an increasing issue for decision-makers and scholars worldwide and many managers understand the significance of the strategic approach of corporate sustainability. However, they face difficulties in aligning sustainable development and strategic management as well as to implement it in practice. Thus, the purpose of this paper is to conduct a bibliometric analysis exploring the integration of strategic management, decision-making and corporate sustainability, providing a framework of interrelated issues according to the current literature in this area. 72 peer-reviewed papers were analyzed based on Webster's and Watson's (2002) methodology. The results of this review revealed that the number of publications in this domain has increased in the last decade, and there is a need to foster research (especially empirical) in this field because managers should find out ways to implement, in action, corporate sustainability strategies and integrate their action plans with their business strategy. This review concludes with a framework that includes the most commonly addressed issues of this topic and provides opportunities and challenges for further research.

**Keywords:** corporate sustainability; sustainable development; business strategy; decision-making; decision support systems


---

## 1. Introduction

Existing researchers [1–3] have noticed that sustainability is becoming an increasing issue for decision-makers and scholars worldwide because it is concerned with the sustainable development of an organization in terms of environmental, economic and social views. In organizations, managers formulate and implement sustainability strategies in order to respond to environmental and social issues [1]. Corporate sustainability presents the strategic position of a business with regard to sustainable development and provides many benefits for businesses that are becoming more sustainability-oriented, use resource-efficient technologies and offer products and services that are eco-friendlier. Unfortunately, the results of the existing literature show that many managers ignore the significance of the strategic approach of corporate sustainability and they face difficulties in aligning sustainable development and strategic management [4,5].

The formulation and implementation of a suitable corporate sustainability strategy is a challenging issue for businesses because each firm has specific characteristics (e.g., the industry sector, organizational structure and internal processes, capabilities, business policies, stakeholder interests, market changes, effects of external environment, etc.) [6]. The formulation and implementation of corporate sustainability strategies, as well as the alignment between corporate sustainability and business strategy, help businesses to deal with environmental and social challenges [7]. Thus, decision-makers should select the appropriate sustainability strategy in order to be aligned with the business strategy [6].

An important challenge for decision-makers who decide to formulate a corporate sustainability strategy is how to plan and implement it. Managers have already recognized the significance of formulating a corporate sustainability strategy but they face difficulties regarding the action plan [4,8,9]. How will they translate the corporate sustainability strategy into action in order to implement it? This translation should include appropriate plans, programs, management systems, decision support systems, organizational factors, policies, ethics and performance indicators [4,8,9]. Although decision-makers have paid attention to the formulation of corporate sustainability strategies, they do not focus on the practical implementation. Issues regarding sustainable development are not considered as a strategic decision by managers. Thus, there is a gap between the formulation and implementation of corporate sustainability strategies and there is a demand for more theoretical research by academics in this field [6,10].

In this view, limited structured literature reviews have been conducted in the area of strategic management and sustainability [7,11]. As the knowledge body in this area is growing, scholars have noticed that a literature review which will provide a better understanding about the state-of-the-art in this field is required. Thus, the purpose of this paper is to map studies regarding corporate sustainability, strategic management and decision-making. More specifically, this paper answers the following questions: How many papers are published each year? Which journals have published peer-reviewed publications and which have published the highest number of publications? Which are the most active institutions in the integration of the following fields: corporate sustainability, strategic management and decision-making based on number of publications? Who are the most active researchers in this domain by considering only journal publications? What is the frequency of keywords? What methods are used in empirical papers? What are the main concepts in this domain and how many papers are published in each one?

The contribution of this paper is twofold. First, the structured methodological framework which was used demonstrates how the academic interest in corporate sustainability strategy and decision-making has evolved over the years and it highlights areas that need further research. Secondly, this literature review can be useful for managers in order to increase the understanding of the complexity of areas regarding strategic management, corporate sustainability and decision-making.

The added value of this paper is the useful overview of the state of strategic management in sustainable development, which highlights issues among sustainability strategies research domains, while providing a complete overview of the literature from a different perspective, not implemented in existing analyses and, thus, it is a good starting point for further research. In contrast with previous literature reviews that are systematic mapping studies and provide narrative amount of knowledge in the area of corporate sustainability strategy and strategic management, this paper is a bibliometric study that provides a macro picture of a research field, its evolution and connections among studies, in order to be a starting point for future research. This literature review may be of interest to academics who are already studying corporate sustainability strategies and decision-making, or researchers who have been introduced to the field but are interested in examining more specific insights into where current research topics in this literature can be located, and how they may contribute to them.

The structure of the rest of the paper is as follows: Section 2 analyzes the methodology used for conducting the literature review. Then, the results of the analysis of the papers are provided in Section 3. Finally, the conclusion and suggestions for future research conclude the paper.

## 2. Materials and Methods

As it has already been stated, the aim of this paper is to examine the current state of corporate sustainability strategies and the decision-making process. Studies were identified using a three-phased literature review methodology, which was suggested by Webster and Watson (2002) [12], and has been previously used in the field of strategic management and Information Systems [13–22]. First, a search of the extant literature reviews was done to select the databases and keywords of the basic search. Then, the backward search was implemented to examine the references of the selected papers and finally, the forward search to examine the citations of the selected papers in order to increase their amount. After the selection of the papers, these were classified according to their content.

### 2.1. Previous Literature Reviews

The existing literature review papers from 2013 to 2016 are presented to place the current literature review alongside the existing knowledge about the field of strategic management and sustainable development and to examine the previous knowledge of this area, as well as to discuss the identified research questions based on the findings of previous studies. Additionally, previous literature reviews give an overview of the literature review methodologies used by researchers and highlight their importance and gaps in their implementation. Table 1 presents a summary of the existing literature reviews on this field.

**Table 1.** Previous literature reviews.

| Authors | Year | Methodology | Results |
|---|---|---|---|
| Crutzen and Herzig | 2013 | Searching for empirical papers in 2 databases using keywords regarding to sustainability, social, responsibility, environment, strategy, planning, decision-making, decision support, control, management system and accounting | 27 papers |
| Engert et al. | 2016 | Searching for peer-reviewed papers in 3 databases using keywords regarding to strategic management, corporate sustainability, responsibility, social, ethical, environment | 114 papers |

Crutzen and Herzig (2013) [11] conducted a literature review in order to examine the relationship between strategy, sustainability and management control. They concluded that many models and frameworks have emerged but companies have limited knowledge of the design or implementation of management control which will support corporate sustainability strategy. They analyzed papers in twelve peer-reviewed journals in the field of sustainability, and they categorized them based on the countries and the type of companies, the research methodology and the operationalization of management control and strategy. Except from this analysis, authors provided some avenues for further research about the use of management control in the process of sustainability strategy formulation and implementation.

Later, Engert et al. (2016) [7] conducted a content analysis in 114 peer-reviewed scientific journal papers in order to analyze the integration of strategic management into sustainability. They focused on organizational influences, internal and external drivers and supporting and hindering factors. Their analysis has been conducted on how companies integrate strategic management into corporate sustainability, the benefits for companies when they address stakeholders' requirements and the use of management tools and systems in this integration. Their literature review provides to academics a state-of-the-art in the fields of corporate sustainability and strategic management, and to managers, some guidelines about the process of integration. Authors support that future researchers should focus on whether or not companies need to integrate corporate sustainability into strategic management and how this process could be implemented.

### 2.2. Article Selection Process

The search was done in Scopus, Science Direct and Web of Science databases using combinations of the following keywords 'sustainability OR sustainable development', 'strategy OR strategic management OR strategic planning', and 'decision-making OR decision support' for papers published in peer-reviewed journals. These were selected without limiting them to a specific period. Books, book chapters, conference proceedings, technical reports and working papers were not included in the review. The admitted journals belonged to fields of strategic management, sustainability and decision-making. Articles included were only those with a focus on business management and contributed to the subject of corporate sustainability strategy, strategic management and sustainable development, decision-making and corporate sustainability. Articles with a focus on environmental issues, ethical issues or economic issues of sustainable development were excluded from the analysis. Finally, published papers were only in English.

During the data collection, a set of variables was extracted for each paper. The first variable refers to the list of authors of each paper. The second variable refers to the list of institutions of each paper. The third variable refers to the title of each paper. The fourth variable refers to the year of publication of each paper. The fifth variable refers to journals' names where papers have been published. The sixth variable refers to the h-index of each journal. The seventh variable refers to the number of citations of each paper in the Scopus database. The eighth variable presents the age of the paper in years and is an indicator because it was calculated by extracting the fourth variable from the current date. The ninth variable is also an indicator that presents the impact of each paper and it was calculated by dividing the number of citations in Scopus with the age of the paper in years [23,24].

Overall, 3067 articles were gathered using keywords in all databases. According to the limitations of language and the source of publication, the articles were reduced to 428. Duplicate articles were deleted and in scanning their titles, 107 articles were found relevant with the aim of this paper. Next, examining their abstract, 72 were accepted. A number of studies were rejected because their full text was not accessible. A prompt investigation was conducted to verify them. This second overview highlighted that all of them should be included. So, 57 articles were examined according to their full text. In these 57 articles, 6 were added from the backward search. Additionally, 9 more articles were added from the 'forward search' and thus, a total of 72 articles were revealed (Figure 1).

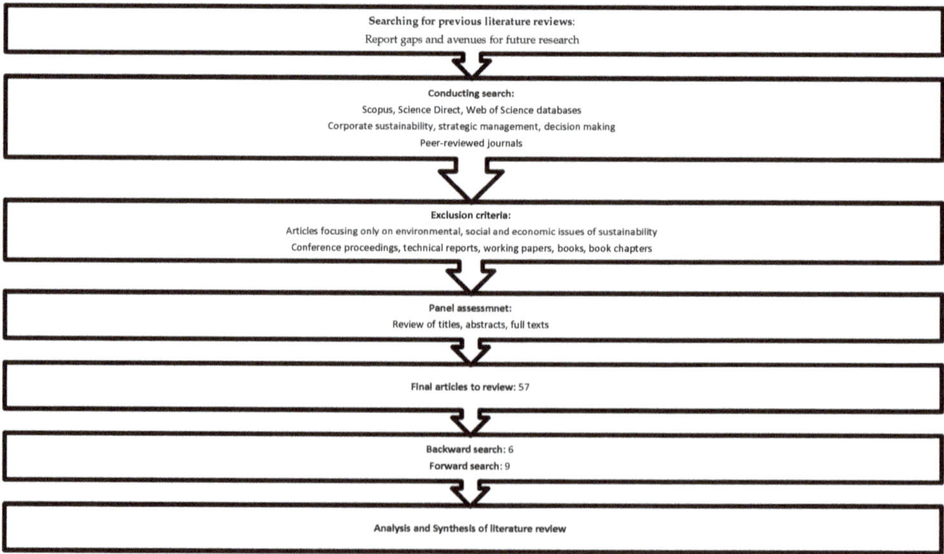


**Figure 1.** Article selection process.

The search was completed when it came to common articles from all databases and different combinations of keywords. Therefore, it was concluded that the critical mass of relevant literature sources had been collected (Webster and Watson, 2002) [12].

*2.3. Classification Framework for Analysis*

72 papers were analyzed based on a classification framework. These articles were analyzed on 13 broad dimensions which will provide a better understanding of the integration between corporate sustainability, strategic management and decision-making research, and will also help future researchers to expand the knowledge in this field. Papers were classified based on publication year, journals and publishers, universities and countries involved, authors, number of citations, keywords used, research method adopted, decision-making techniques, aspects of sustainability, drivers for sustainable development, sustainability strategies, business strategy planning and performance. Table A1 in Appendix A presents the main concepts of the analyzed papers.

## 3. Results

*3.1. Number of Published Articles per Year*

Although researchers in corporate sustainability and the strategic management area conducted studies two decades ago, the majority of the papers have only been published in the last seven years. Figure 2 presents the number of papers published each year. Especially, in the early 2000s, the awareness of strategic management into corporate sustainability was found to be very low as the majority of researchers focused only on the aspects of sustainable development, and they ignore the significance of the strategic aspect of sustainability. The strong practice of corporate sustainability strategy came into existence around 2013, when researchers realized the significance of the integration between strategic management and corporate sustainability and started examining drivers that affect sustainable performance combined with decision-making techniques. Such a finding highlights both the importance of the field and its continuous development. Figure 3 presents a clear increasing direction in the last five years.

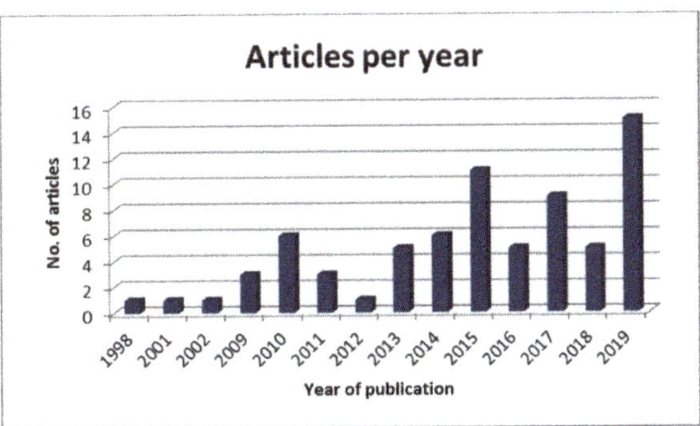


**Figure 2.** Number of papers per year.

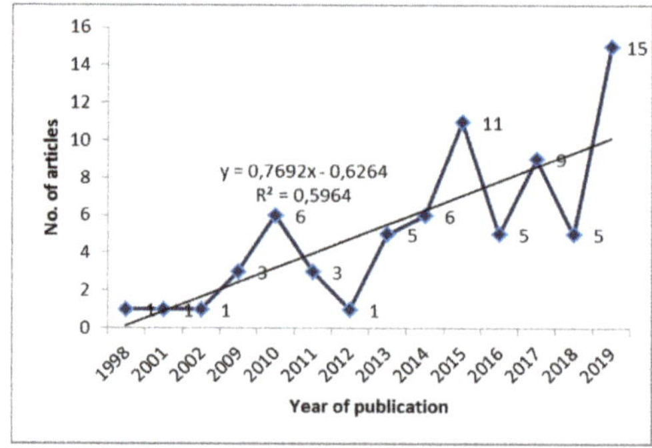


**Figure 3.** Papers based on the year of publication.

## 3.2. Number of Articles per Journal

Papers have been published in 35 peer-reviewed journals. Journal of Cleaner Production has published 20 papers, Business Strategy and the Environment and Corporate Social Responsibility and Environmental Management have published five papers each. Journal of Business Strategy and Technological Forecasting and Social Change have published three papers each. Table 2 presents the distribution of papers based on journals.

Regarding to publishers, the majority of papers were published in Elsevier journals (44.4%). Then, 12 papers were published in Emerald and Wiley peer-reviewed journals (16.7%). Springer contributed 6.95% followed by Taylor & Francis (5.56%) and Inderscience (4.17%). Other publishers have published less than three papers. Wiley and Taylor & Francis mainly have published papers based on sustainable development. On the other hand, Elsevier has published papers related to decision-making and corporate sustainability and Emerald peer-reviewed journals have focused on strategic management and sustainability. The classification of papers according to publishers is presented in Figure 4.

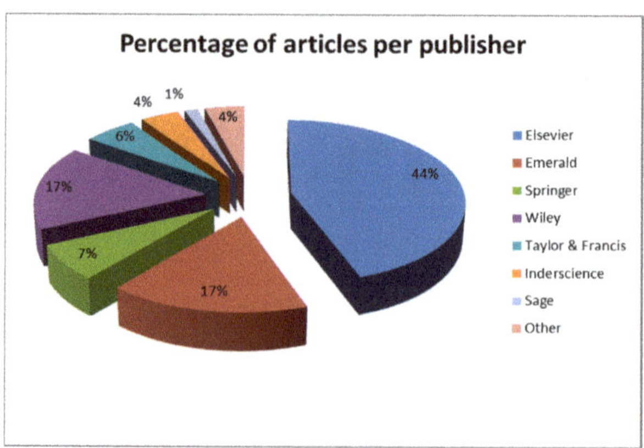


**Figure 4.** Papers per method.

**Table 2.** Distribution of papers based on journals.

| Journal Name | h-Index | Publisher | No. of Papers | % |
|---|---|---|---|---|
| Journal of Cleaner Production | 150 | Elsevier | 20 | 27.78 |
| Business Strategy and the Environment | 84 | Wiley | 5 | 6.94 |
| Corporate Social Responsibility and Environmental Management | 58 | Wiley | 5 | 6.94 |
| Journal of Business Strategy | 34 | Emerald | 3 | 4.17 |
| Technological Forecasting and Social Change | 93 | Elsevier | 3 | 4.17 |
| European Journal of Operational Research | 226 | Elsevier | 2 | 2.78 |
| Journal of Science and Technology Policy Management | 10 | Emerald | 2 | 2.78 |
| Long Range Planning | 89 | Elsevier | 2 | 2.78 |
| Sustainable Development | 51 | Wiley | 2 | 2.78 |
| Construction Management and Economics | 81 | Taylor & Francis | 1 | 1.39 |
| Engineering, Construction and Architectural Management | 49 | Emerald | 1 | 1.39 |
| EURO Journal on Decision Processes | - | Springer | 1 | 1.39 |
| European Business Review | 36 | Emerald | 1 | 1.39 |
| Industrial Management and Data Systems | 88 | Emerald | 1 | 1.39 |
| International Business Management | 14 | Medwell | 1 | 1.39 |
| International Journal of Business and Systems Research | 13 | Inderscience | 1 | 1.39 |
| International Journal of Business Performance Management | 18 | Inderscience | 1 | 1.39 |
| International Journal of Energy Sector Management | 17 | Emerald | 1 | 1.39 |
| International Journal of Logistics Systems and Management | 25 | Inderscience | 1 | 1.39 |
| International Journal of Production Economics | 155 | Elsevier | 1 | 1.39 |
| International Journal of Productivity and Performance Management | 48 | Emerald | 1 | 1.39 |
| International Journal of Project Management | 121 | Elsevier | 1 | 1.39 |
| International Journal of Public Sector Management | 48 | Emerald | 1 | 1.39 |
| Journal of Business Economics and Management | 30 | Taylor & Francis | 1 | 1.39 |
| Journal of Business Ethics | 147 | Springer | 1 | 1.39 |
| Journal of Change Management | 22 | Taylor & Francis | 1 | 1.39 |
| Journal of Engineering and Technology Management | 58 | Elsevier | 1 | 1.39 |
| Journal of Management and Governance | 44 | Springer | 1 | 1.39 |
| Journal of Small Business Strategy | 5 | Middle Tennessee State University | 1 | 1.39 |
| Management and Production Engineering Review | 9 | Polish Academy of Sciences | 1 | 1.39 |
| Management Decision | 82 | Emerald | 1 | 1.39 |
| Organization and Environment | 48 | Sage | 1 | 1.39 |
| Organization Management Journal | - | Taylor & Francis | 1 | 1.39 |
| Science and Engineering Ethics | 43 | Springer | 1 | 1.39 |
| Systemic Practice and Action Research | 31 | Springer | 1 | 1.39 |
| **Total** | | | **72** | **100** |

*3.3. Number of Articles per Country and Universities*

To develop the corporate sustainability and strategic management domain, a total of 101 universities across the world contributed through 72 papers. Table 3 presents the leading universities involved in

strengthening corporate sustainability and strategic management oriented research. The University of Graz is among the top universities involved in corporate sustainability and strategic management research followed by Åbo Akademi University, Blekinge Institute of Technology, Kedge Business School, Pontifical Catholic University of Parana, Rice University, Ryerson University, University of Kassel, University of Leoben and Utrecht University.

**Table 3.** Leading universities.

| Universities | Countries | No. of Articles |
| --- | --- | --- |
| University of Graz | Austria | 4 |
| Åbo Akademi University | Finland | 2 |
| Blekinge Institute of Technology | Sweden | 2 |
| Kedge Business School | France | 2 |
| Pontifical Catholic University of Parana | Brazil | 2 |
| Rice University | USA | 2 |
| Ryerson University | Canada | 2 |
| University of Kassel | Germany | 2 |
| University of Leoben | Austria | 2 |
| Utrecht University | The Netherlands | 2 |

*3.4. Authors Actively Involved in Publishing*

A total of 167 authors contributed to the 72 papers. Table 4 presents the main authors (three or more that two papers each) who have published articles on sustainability and strategic management. Baumgartner [2,4,6,25] appears to be a more prolific author in the field of corporate sustainability and strategic management and has published six papers, followed by Hahn with four published papers. Figge contributes to the research topic with three papers followed by four authors. Hallstedt, Lozano and Searcy published two papers. The results show that a vast majority of authors have contributed to just one article in the set of journals comprising the dataset.

**Table 4.** Main authors.

| Author | No. of Articles | h-Index (Retrieved from Scopus) |
| --- | --- | --- |
| Baumgartner R.J. | 6 | 16 |
| Hahn T. | 4 | 21 |
| Figge F. | 3 | 22 |
| Hallstedt S.I. | 2 | 13 |
| Lozano R. | 2 | 30 |
| Searcy C. | 2 | 25 |

Table 5 presents the most cited papers comprising this dataset in the field of corporate sustainability and strategic management. The number of citations were retrieved from Scopus for each one of these papers. Then, the age of each paper was calculated by extracting the year of publication from the current year (2019). Finally, the average annual number of citations was calculated by dividing the number of citations in Scopus with the age of the paper.

**Table 5.** Top cited papers.

| Title of Paper | No. of Citations (Retrieved from Scopus) | Age of the Paper (in Years) | Average Annual Number of Citations |
| --- | --- | --- | --- |
| The Sustainability Balanced Scorecard—Linking Sustainability Management to Business Strategy | 458 | 17 | 26.94 |
| Sustainability in Action: Identifying and Measuring the Key Performance Drivers | 261 | 18 | 14.5 |
| Corporate Sustainability Strategies: Sustainability Profiles and Maturity Levels | 234 | 9 | 26 |
| Strategy Development in Small and Medium- Sized Enterprises for Sustainability and Increased Value Creation | 169 | 10 | 16.9 |
| Tensions in Corporate Sustainability: Towards an Integrative Framework | 168 | 4 | 42 |
| A Holistic Perspective on Corporate Sustainability Drivers | 148 | 4 | 37 |
| Managing Corporate Sustainability and CSR: A Conceptual Framework Combining Values, Strategies and Instruments Contributing to Sustainable Development | 132 | 5 | 26.4 |
| Building Corporate Social Responsibility into Strategy | 116 | 10 | 11.6 |

## 3.5. Frequency of Keywords

Figure 5 shows the frequency of keywords that are used in each paper. The majority of keywords refer to sustainability, sustainable development, corporate sustainability and corporate social responsibility. Other keywords regarding to strategy, strategic planning and strategy management were also used. Finally, keywords regarding to decision-making or decision-making methods, such as Analytic Network Process were used by researchers.

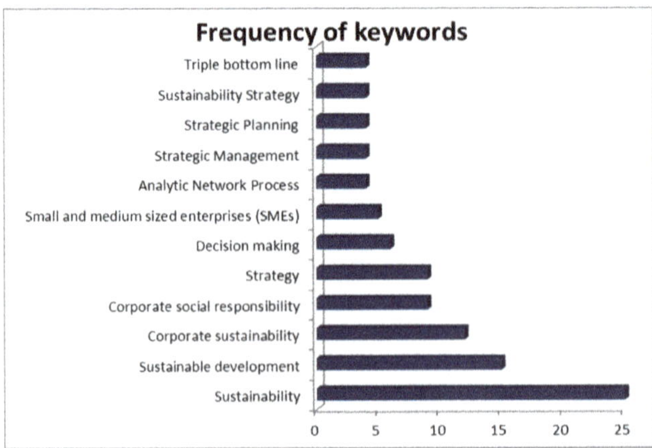


**Figure 5.** Frequency of keywords.

VOSviewer was used to indicate the most frequently used keywords of the 72 papers and the relationships among them. The network visualization, which is presented in Figure 6, shows the relationship among keywords and Figure 7 presents the heat map. The most frequent keywords are in the yellow area, which are "sustainability", "sustainable development", "integration", "competitiveness" and "corporate sustainability". Other keywords such as "strategies", "corporate social responsibility" and "decision-making" are not commonly used. This finding supports researchers who claim that studies ignore the strategic view of corporate sustainability [6,10].

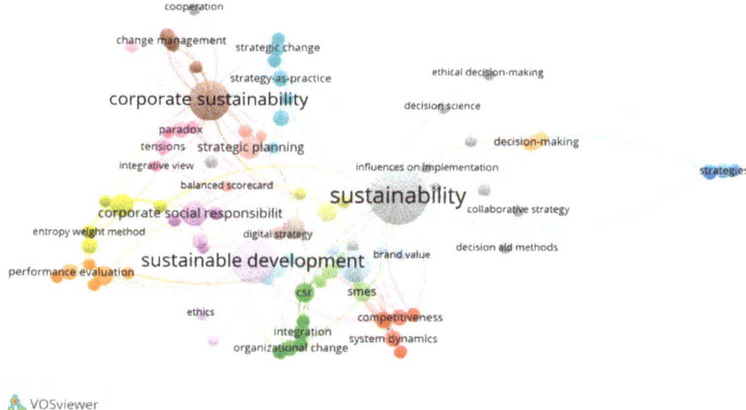

**Figure 6.** Network visualization of keywords.

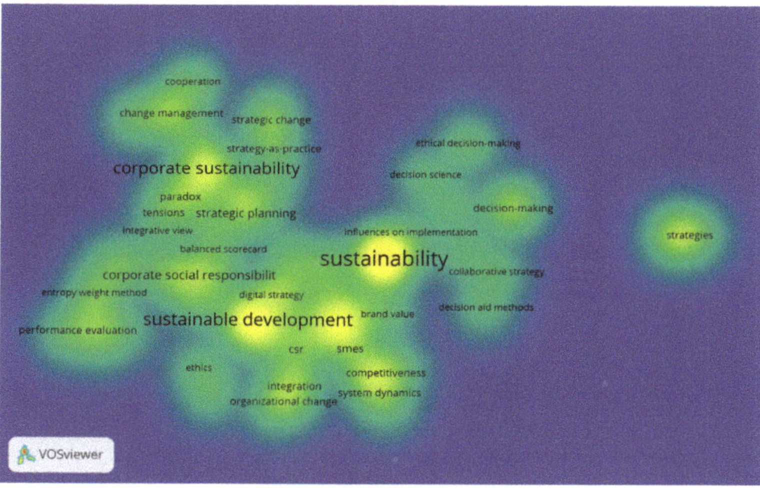

**Figure 7.** Heat map of keywords.

*3.6. Research Methods*

Figure 8 shows that 30.57% of the papers were categorized as conceptual, 26.39% of studies were case studies whereas limited studies were theoretical. 31.94% were classified to qualitative and quantitative surveys. The results confirm scholars' notifications that despite the importance of this field, limited empirical surveys have been implemented [1,26]. Empirical papers that have conducted quantitative surveys used Small-Medium Enterprises (SMEs) or large companies in their sample. Decision-making methods used were Fuzzy methods, Multi-Criteria Decision-making Methods (MCDM), Analytical Hierarchical Process (AHP), Technique for Order of Preference by Similarity to Ideal Solution (TOPSIS), Decision-making Trial and Evaluation Laboratory (DEMATEL), Graph Theory and Matrix Approach (GTMA) and Analytical Network Process (ANT).

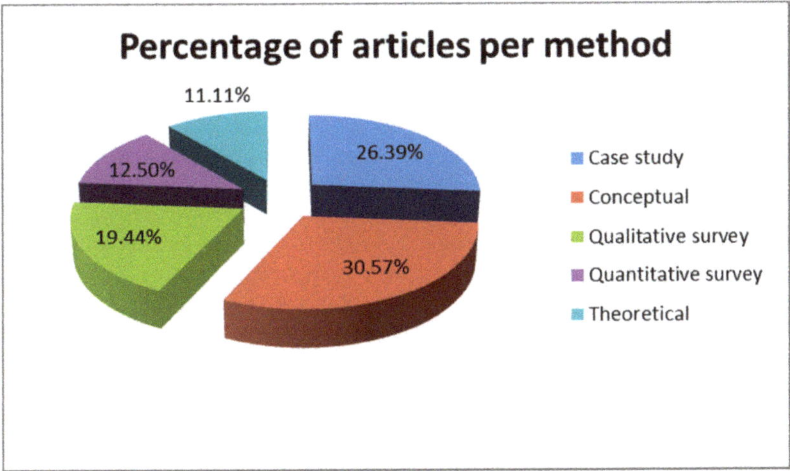



**Figure 8.** Papers per method.

*3.7. Number of Papers per Concept*

Based on the classification of papers that was presented in Table 2, Figure 9 shows the percentage of articles per concept. The majority of papers (36.11%) refer to sustainability strategies. 31.94% of papers are related to the integration of corporate sustainability and decision-making. 30.56% of papers combine the concept of sustainability with the concept of performance. Only 27.78% of papers combine the concepts of sustainability and business strategy, confirming researchers who claim that more research is required in order to examine how companies can formulate and implement sustainability strategies in practice and integrate them with their business strategy [10,26].

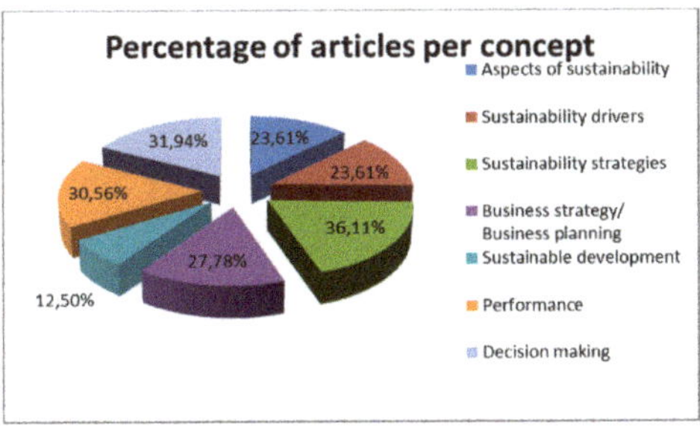



**Figure 9.** Papers per concept.

## 4. Concept Analysis

Sustainability is presented somewhat as a synonym of corporate sustainability and CSR, namely under the approach of the triple bottom line, i.e., considering the environmental, economic and social views [27]. In recent years, growing attention has been paid to sustainability as one of the most significant business goals because of organizations' concerns about human rights and the protection of the environment [28]. Researchers aim to explore how CSR can be integrated into strategic planning and how the three aspects of sustainability (environmental, economic and social views) can be aligned with the components of business strategy (mission, resources, market, customer needs, competitive advantage, stakeholder interests and value) [5,6,29].

A corporate sustainability strategy aligns social with environmental dimensions into the strategic management process, and highlights the company's strategic position with regard to sustainable development. Managers choose the sustainability approach in order to reduce the negative environmental and social impacts of business activities while increasing the financial performance of the organization [4]. Executives and academics have understood the significance of CSR for competitive success and they have used theoretical, conceptual and empirical perspectives to evaluate the relationship between CSR and performance [5,29,30]. In this view, [31] use a fuzzy AHP method to examine the selection of relevant sustainability issues and their integration into a company's strategic decision making process. The results show that the decisional criteria composing the fuzzy AHP hierarchies integrate the value chain activities and the dimensions of competitive advantage.

However, there is a gap between CSR and strategy [5,29,30]. Managers view CSR only in terms of social or legal responsibility. Many practitioners have connected the strategic view of sustainability with philanthropy or sponsorships for society in order to increase a firm's reputation. The business environment is complex and has many opportunities and threats for firms. Managers are obliged to make strategic decisions that do not ignore stakeholders' interests. It is important to find out how they can satisfy them in a strategic manner in order to increase a firm's competitiveness. Other researchers [8] and [32] proposed a framework, using Analytic Network Process, in order to examine the relationship between stakeholders' interests and sustainability strategy. The authors used 28 decision elements about stakeholders' interests. The survey was conducted in multinational manufacturing firms in the Philippines. The proposed decision model conveys a complex decision-making process and provides the content policies that should be considered in carrying out a sustainable manufacturing strategy. A decision- making model was proposed by [33] in order to examine the relationship between stakeholders' interests and financial performance. Results show that organizations need to evaluate diverse stakeholder interests in order to be aware of social and economic impacts and to be able to integrate this into decision-making. However, there is little guidance on the underlying process.

Thus, CSR can be fully integrated into the business strategy, developing a culture that is aligned with social factors that might affect a company. This alignment will help executives to inform shareholders, stakeholders and the society about their decisions, to fulfill firms' responsibilities with society and increase shareholders' value [5,29]. During decision-making, companies balance the importance and strength of stakeholder groups [34]. Furthermore, the alignment between sustainability strategy and business strategy does not only reflect a deep organizational commitment to a sustainable society but also establishes a perspective that an organization can increase economic success, profit and benefits for society [9]. The increase of economic success can support a vision that incorporates sustainability and includes economic, environmental and social elements. This vision can guide the decisions of managers and employees and helps an organization to earn profit by protecting the society and the environment. This vision also includes a strategic decision-making process that is based on decision-makers' commitment to sustainability. This strategy that refers to sustainability at all levels (corporate, business and functional), along with an organizational culture that promotes and supports sustainability efforts, positively contributes to business performance [35].

This gap requires the ability to translate sustainability strategy into action, developing plans, systems, goals and performance indicators. Findings show that executives have no common understanding concerning how sustainability is related to their daily business activities [7]. In this view, managers can combine, in the process of sustainability decision-making, external and internal drivers such as ethics, resources and cost savings, employees' shared values, leadership, reputation, market, laws, competition and customers' satisfaction into their strategic decision-making process, in order to make changes in their organizations and formulate a sustainability strategy that increases economic, ecological and social success [7,36–39]. For example, Fairfield et al. [40] examined how aspects of organizations, context, and decision-making processes can be aligned to influence the implementation and success of sustainability efforts. Drivers such as reputation, managers' values and attitudes, management support and organizational culture has a significant impact on sustainability strategy.

Other surveys indicated that leadership and organizational culture are fundamental drivers in order to promote the implementation of a sustainability strategy. Managers can motivate employees with their personal attitudes and values to understand the importance of sustainability [1,41,42]. Organizational structures that do not support collaboration and communication usually have a lack of trained employees, as well as the lack of clear vision of sustainability and policies about it, and are a significant obstacle for formulating and implementing a CSR strategy [43,44]. Size is also an important factor that affects an organization's willingness to formulate sustainability strategy. Large organizations have access to more resources and factors such as reputation, and stakeholder relationships play an important role in order to avoid environmental scandals and focus more on competitors' sustainability strategies [43,45,46].

Corporate sustainability strategy can affect the productivity and efficiency of processes, support the development of more sustainable products and services, reduce the risks associated with environmental and social impacts and improve the benefits for an organization. These benefits may reveal themselves in the form of an increase in economic performance or improved competitive success, such as reductions in costs and risks, and improvements in reputation [4]. The improvement of reputation can allow firms to access new markets, attract new customers, and retain good employees. Customers are expecting organizations to be responsible with a concern for environment and society. Managing these issues can allow firms to be sustainable and increase their economic performance [47]. These results are confirmed by Tseng et al. [48], who used decision-making methods in order to evaluate sustainability performance. A decision-making method was used by [49] in order to evaluate corporate sustainability performance based on the triple bottom-line concept. The results of this survey, conducted in 34 high-tech listed companies in Taiwan, can be used as an important basis for management decision-making, and can also serve as a reference for banks and investors when developing investment strategy. Another similar survey was conducted by Wicher et al. [50]. They evaluated sustainability performance of an industrial corporation using the TBL concept and a new generation of decision-support tools.

## 5. Discussion

The existing studies have given a solid ground and now, a conceptual framework can be developed based on the literature study. Using an open coding technique, in the content analysis of the 72 papers, with the purpose of dividing the categories to be used into the classification of the papers, gave readers a good indication of the issues of concern (Figure 10). These papers indicate that this field is still in its early stages and further research is required. Although, many papers have built a theoretical base for corporate sustainability and strategic management, only limited studies provided guidelines about the integration of decision-making, strategic management and corporate sustainability. This creates opportunities for future researchers to explore this gap and improve the sustainability performance through strategic management and decision-making processes.

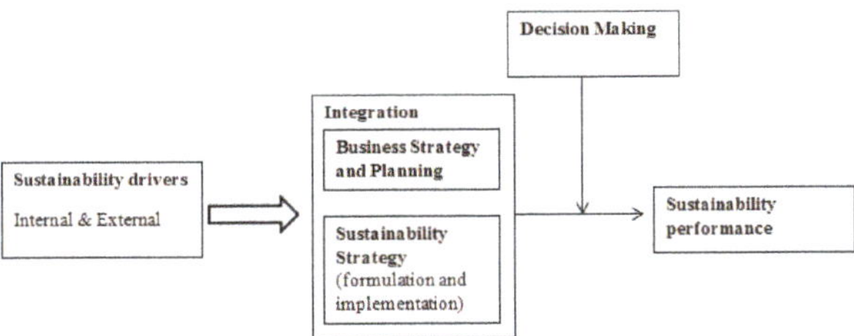


**Figure 10.** Conceptual Framework.

Furthermore, many papers conducted case studies in order to provide fruitful results, and researchers did not struggle through the deep penetration of corporate sustainability strategy because of unavailable generalized frameworks that provide guidelines about the formulation and implementation of action plans. Thus, there is a need for qualitative or quantitative research that provides conclusions about the effect of internal or external factors in the sustainability process, the integration between business strategy and corporate sustainability, the implementation of action plans, the challenges and problems of this integration, as well as the impact of this alignment on sustainability performance using decision-making methods.

The majority of existing studies explore the issue of integration from the sustainability perspective and not from the perspective of strategic management. Thus, researchers ignore a discussion about barriers and problems that they often face in practice during this integration. Such an analysis could lead to deeper insights because the alignment between business strategy and corporate sustainability is complex due to stakeholder demands, required competencies, external forces, market conditions, organizational culture, management structure, and it could also help managers to increase the quality of integration. Many questions have been raised, such as how can leaders motivate employees to participate in sustainability strategy? How can stakeholders be satisfied by the implementation of sustainability strategy? What factors affect the successful implementation of sustainability strategies? What should be the vision and the organizational structure in a company in order to improve the implementation of sustainability strategy? How can managers formulate and implement, in practice, sustainability strategies? How can decision-makers promote sustainability at all levels (corporate, business and functional)? How can the size of a firm affect the implementation of sustainability strategies?

## 6. Conclusions

The purpose of this paper was to map studies regarding corporate sustainability, strategic management and decision-making and provide a bibliometric analysis exploring the integration of strategic management, decision-making and corporate sustainability, providing a framework of interrelated issues according to the current literature in this area. The study is based upon an analysis of 72 papers, derived from databases and categorized according to the main issues of this emerging research area. The most significant findings of this paper are described as following. The first step for conducting this literature review was to study previous literature reviews. It was observed that previous literature reviews were systematic mapping studies and provided a narrative amount of knowledge in the area of corporate sustainability strategy and strategic management. This paper is a bibliometric study that provides a macro picture of a research field, and its evolution and connections among studies, in order to be a starting point for future research. This literature review helps academics who are already studying corporate sustainability strategies and decision-making, or researchers who have been introduced to the field but are interested in examining more specific insights into where current research topics in this literature can be located, and how they may contribute to them.

Regarding the descriptive analysis, Journal of Cleaner Production has published the majority of papers because its scope includes different topics such as corporate sustainability, environmental management systems or performance evaluation. There are also other journals such as Business Strategy and the Environment and Corporate Social Responsibility and Environmental Management that include similar topics and have published many papers. Austria, USA and Germany hold many of the total number of papers published in the field of sustainability and strategic management. This finding indicates that industry sectors in these countries are interested in those research issues. Furthermore, Baumgartner, Hahn and Figge have significantly contributed in this field. The majority of papers are conceptual models and case studies. Although many models have been proposed by researchers, they have not been tested using decision-making methods. Researchers who conducted case studies have presented successful adoption or implementation of a corporate sustainability strategy but there still exists a need for a generalized framework that can be used by different types of industries in order to adopt, formulate and successfully implement action plans of corporate sustainability strategies. Finally, many papers used MCDM techniques or decision-making methods (AHP, TOPSIS, DEMATEL, GTMA, ANT) but the usage of modern survey analysis techniques (e.g., SEM or PLS) is limited. AHP is a flexible method for decision-makers because it provides a good understanding of the problem and handling the uncertainty of human factor.

This paper has some limitations that are described below. First, criteria for searching articles in databases include keywords "sustainability or sustainable development", "strategy or strategic management or strategic planning" and "decision-making or decision support" in the title and abstract of the papers. There probably exists papers which lack these keywords in the title but still focus on the field of sustainability and strategic management. Furthermore, only peer-reviewed journals were included in the dataset, however, there are also related papers in conference proceedings or book chapters. Moreover, journals from Elsevier, Emerald, Wiley, Taylor & Francis, Springer and Inderscience were included in this paper but there are more journals that have published articles related to strategic management and sustainability. Another limitation is that only English papers were searched which may skip other publications in other languages. Thus, the use of different journals or papers from other sources could possibly lead to different findings regarding the most cited papers, or the most active researchers or institutions, and the percentage of papers per publisher.

Future researchers could expand on the existing models about the integration of strategic management into sustainability using decision-making methods in order to address the existing gaps. These models are expected to combine the aspects of sustainability, and the drivers that affect sustainability strategies with the business strategy planning. As this field is in the early stages and only limited studies provided guidelines about the integration of decision-making, strategic management and corporate sustainability, future researchers could explore this gap and conduct quantitative surveys in order to collect data from different industries and check the reliability of the theory developed, discussing the challenges and the problems of this integration. From a theoretical perspective, this paper is a bibliometric study that provides a macro picture of a research field, its evolution and connections among studies, in order to be a starting point for future research by highlighting issues among sustainability strategies research domains, while providing a complete overview of the literature from a different perspective, not implemented in existing analyses. Future researchers could expand this literature review and provide different bibliometric analyses such as co-author or co-citation.

**Author Contributions:** Conceptualization, F.K. and M.A.T.; methodology, F.K.; data collection, M.K.; writing—original draft preparation, F.K.; M.A.T. and M.K.; writing—review and editing, F.K.; M.A.T. and M.K.; supervision, F.K. All authors have read and agreed to the published version of the manuscript.

**Funding:** This research received no external funding.

**Conflicts of Interest:** The authors declare no conflict of interest.

# Appendix A

**Table A1.** Concept matrix table.

| No. | Authors | Year | Method | Aspects of Sustainability | Sustainability Drivers | Sustainability Strategies | Business Strategy/Business Planning | Sustainable Development | Performance | Decision-Making |
|---|---|---|---|---|---|---|---|---|---|---|
| | | | | | | | | | | |
| 1. | Engert and Baumgartner [1] | 2016 | Qualitative survey | | | x | | | | |
| 2. | Baumgartner and Korhonen [2] | 2010 | Conceptual | | | | x | x | | |
| 3. | Baumgartner and Rauter [4] | 2017 | Conceptual | | | | x | x | | |
| 4. | Galbreath [5] | 2009 | Conceptual | | | | x | x | | |
| 5. | Baumgartner [6] | 2014 | Conceptual | | | x | | | | |
| 6. | Epstein and Roy [8] | 2001 | Conceptual | | x | | x | | x | |
| 7. | Stead and Stead [9] | 2013 | Theoretical | | | x | x | | | |
| 8. | Kumar et al. [21] | 2018 | Quantitative survey | | x | | | | | |
| 9. | Baumgartner and Ebner [25] | 2010 | Theoretical | x | | x | | | | |
| 10. | Egels-Zandén and Rosén [26] | 2015 | Case study | | | x | | | | |
| 11. | Chang and Cheng [28] | 2019 | Quantitative survey | x | | | | x | | x |
| 12. | León-Soriano et al. [29] | 2009 | Conceptual | x | | | x | | | |
| 13. | Asif et al. [30] | 2013 | Case studies | x | | x | | | | |
| 14. | Calabrese et al. [31] | 2019 | Conceptual | | | x | | | | x |
| 15. | Ocampo [32] | 2016 | Qualitative survey | | | | | | x | x |
| 16. | Epstein and Widener [33] | 2011 | Qualitative survey | | | | | | x | x |
| 17. | Harangozó and Zilahy [34] | 2015 | Quantitative survey | | x | | | | | |
| 18. | Bonn and Fisher [35] | 2011 | Theoretical | | x | x | | | | |
| 19. | Cezarino et al. [36] | 2019 | Qualitative survey | | x | x | | | | |
| 20. | Lozano [37] | 2015 | Qualitative survey | | x | | | | | |
| 21. | Neugebauer et al. [38] | 2016 | Conceptual | | x | x | x | | | |
| 22. | Schrettle et al. [39] | 2014 | Conceptual | | x | | | | | x |
| 23. | Fairfield et al. [40] | 2011 | Quantitative survey | | x | | | | x | |
| 24. | Shields and Shelleman [41] | 2015 | Theoretical | | | x | x | | | |
| 25. | Thakhathi et al. [42] | 2019 | Case study | | x | | | | | |
| 26. | Kiesnere and Baumgartner [43] | 2019 | Qualitative survey | | | x | | | | |
| 27. | Lozano [44] | 2013 | Case studies | | x | x | | | | |
| 28. | Moore and Manring [45] | 2009 | Theoretical | | | | | x | | |
| 29. | Rosati and Faria [46] | 2018 | Quantitative survey | | x | | | | x | |
| 30. | McPhee [47] | 2014 | Conceptual | | x | | x | | | |
| 31. | Tseng et al. [48] | 2019 | Case study | x | | | | | x | |
| 32. | Ou [49] | 2016 | Case study | | | | | | x | |
| 33. | Wicher et al. [50] | 2019 | Conceptual | x | | | | | x | x |
| 34. | Beckmann et al. [51] | 2014 | Theoretical | x | | | x | | | |
| 35. | Aras and Crowther [52] | 2009 | Conceptual | | x | | | | x | |
| 36. | Bastons and Armengou [53] | 2017 | Conceptual | | | | | x | | x |
| 37. | Bianchi et al. [54] | 2015 | Case studies | | | | x | | x | x |

Table A1. *Cont.*

| No. | Authors | Year | Method | Aspects of Sustainability | Sustainability Drivers | Sustainability Strategies | Business Strategy/Business Planning | Sustainable Development | Performance | Decision-Making |
|---|---|---|---|---|---|---|---|---|---|---|
| | | | | | | | Concepts | | | |
| 38. | Brook and Pagnanelli [55] | 2014 | Qualitative survey | x | | | | | x | |
| 39. | Cagno et al. [56] | 2019 | Case studies | | | | | | x | |
| 40. | Chang et al. [57] | 2016 | Conceptual | x | | | | | x | |
| 41. | Christ et al. [58] | 2017 | Case study | | | | x | | x | |
| 42. | Daneshpour and Takala [59] | 2017 | Case study | | | x | | | | x |
| 43. | De Brucker et al. [60] | 2013 | Case study | | | | | x | | x |
| 44. | de Fátima Teles and de Sousa [61] | 2018 | Case studies | | | x | x | | | x |
| 45. | Eikelenboom and de Jong [62] | 2019 | Quantitative survey | | x | | | | x | |
| 46. | Figge et al. [63] | 2002 | Conceptual | x | | | x | | | |
| 47. | Garcia et al. [64] | 2016 | Case study | | | | | | x | x |
| 48. | Haffar and Searcy [65] | 2019 | Qualitative survey | | | x | | | x | |
| 49. | Hahn [66] | 2014 | Case study | x | | | | | | x |
| 50. | Hahn et al. [67] | 2015 | Conceptual | x | | x | | | | |
| 51. | Hallstedt et al. [68] | 2015 | Case study | x | | | | | | x |
| 52. | Hallstedt et al. [69] | 2010 | Qualitative survey | | | | | | | x |
| 53. | Hessami et al. [70] | 2019 | Case study | | | | | | x | |
| 54. | Ivory and Brooks [71] | 2018 | Conceptual | | | x | | | | |
| 55. | Janeiro and Patel [72] | 2015 | Theoretical | x | | | | | | x |
| 56. | Kelly [73] | 1998 | Conceptual | | | | | x | | x |
| 57. | Martin [74] | 2015 | Conceptual | | | | | | | x |
| 58. | Modrak and Dima [75] | 2010 | Conceptual | | | x | | | | x |
| 59. | Parisi [76] | 2013 | Quantitative survey | | x | | | | x | |
| 60. | Rodriguez et al. [77] | 2018 | Qualitative survey | | | x | | | | |
| 61. | Satyro et al. [78] | 2017 | Theoretical | | | x | x | | x | |
| 62. | Silvius et al. [79] | 2017 | Qualitative survey | | x | | | | | x |
| 63. | Singla et al. [80] | 2019 | Quantitative survey | | | | | x | | |
| 64. | Sreekumar and Rajmohan [81] | 2018 | Conceptual | x | | | | | | x |
| 65. | Sroufe [82] | 2017 | Qualitative survey | | | x | x | | | |
| 66. | Taghavi et al. [83] | 2014 | Qualitative survey | | | x | x | | | |
| 67. | Teh and Corbitt [84] | 2015 | Qualitative survey | x | | | x | | | |
| 68. | Teixeira and Junior [85] | 2019 | Case studies | | | x | x | | | |
| 69. | Ukko et al. [86] | 2019 | Quantitative survey | | | x | | | x | |
| 70. | Riccaboni and Leone [87] | 2010 | Case study | | | x | | | x | |
| 71. | Vazhayil and Balasubramanian [88] | 2012 | Case study | | | | x | | | x |
| 72. | Wollmann and Tortato [89] | 2019 | Conceptual | x | | | | | | x |

## References


1. Engert, S.; Baumgartner, R.J. Corporate sustainability strategy–bridging the gap between formulation and implementation. *J. Clean. Prod.* **2016**, *113*, 822–834. [CrossRef]
2. Baumgartner, R.J.; Korhonen, J. Strategic thinking for sustainable development. *Sustain. Dev.* **2010**, *18*, 71–75. [CrossRef]
3. Tsalis, A.T.; Nikolaou, E.I.; Grigoroudis, E.; Tsagarakis, P.K. A dynamic sustainability Balanced Scorecard methodology as a navigator for exploring the dynamics and complexity of corporate sustainability strategy. *Civ. Eng. Environ. Syst.* **2015**, *32*, 281–300. [CrossRef]
4. Baumgartner, R.J.; Rauter, R. Strategic perspectives of corporate sustainability management to develop a sustainable organization. *J. Clean. Prod.* **2017**, *140*, 81–92. [CrossRef]
5. Galbreath, J. Building corporate social responsibility into strategy. *Eur. Bus. Rev.* **2009**, *21*, 109–127. [CrossRef]
6. Baumgartner, R.J. Managing corporate sustainability and CSR: A conceptual framework combining values, strategies and instruments contributing to sustainable development. *Corp. Soc. Responsib. Environ. Manag.* **2014**, *21*, 258–271. [CrossRef]
7. Engert, S.; Rauter, R.; Baumgartner, R.J. Exploring the integration of corporate sustainability into strategic management: A literature review. *J. Clean. Prod.* **2016**, *112*, 2833–2850. [CrossRef]
8. Epstein, M.J.; Roy, M.J. Sustainability in action: Identifying and measuring the key performance drivers. *Long Range Plan.* **2001**, *34*, 585–604. [CrossRef]
9. Stead, J.G.; Stead, W.E. The coevolution of sustainable strategic management in the global marketplace. *Organ. Environ.* **2013**, *26*, 162–183. [CrossRef]
10. Klettner, A.; Clarke, T.; Boersma, M. The governance of corporate sustainability: Empirical insights into the development, leadership and implementation of responsible business strategy. *J. Bus. Ethics* **2014**, *122*, 145–165. [CrossRef]
11. Crutzen, N.; Herzig, C. A review of the empirical research in management control, strategy and sustainability. *Stud. Manag. Financ. Account.* **2013**, *26*, 165–195.
12. Webster, J.; Watson, R.T. Analyzing the past to prepare for the future: Writing a literature review. *MIS Q.* **2002**, *26*, 13–23.
13. Kitsios, F.; Kamariotou, M. Business strategy modelling based on enterprise architecture: A state of the art review. *Bus. Process Manag. J.* **2019**, *25*, 606–624. [CrossRef]
14. Kitsios, F.; Kamariotou, M. Mapping New Service Development: A Review and Synthesis of Literature. *Serv. Ind. J.* **2019**, 1–23. [CrossRef]
15. Kitsios, F.; Kamariotou, M. Service innovation process digitization: Areas for exploitation and exploration. *J. Hosp. Tour. Technol.* **2019**. [CrossRef]
16. Kitsios, F.; Kamariotou, M. Open data hackathons: An innovative strategy to enhance entrepreneurial intention. *Int. J. Innov. Sci.* **2018**, *10*, 519–538. [CrossRef]
17. Kitsios, F.; Papachristos, N.; Kamariotou, M. Business Models for Open Data Ecosystem: Challenges and Motivations for Entrepreneurship and Innovation. In Proceedings of the 19th IEEE International Conference on Business Informatics (CBI'17), Thessaloniki, Greece, 24–26 July 2017; pp. 398–408.
18. Kitsios, F.; Kamariotou, M. Critical success factors in service innovation strategies: An annotated bibliography on NSD. In Proceedings of the British Academy of Management (BAM) Conference 2016, Newcastle, UK, 6–8 September 2016; pp. 1–28.
19. Kitsios, F.; Kamariotou, M. Decision Support Systems and Business Strategy: A conceptual framework for Strategic Information Systems Planning. In Proceedings of the 6th IEEE International Conference on IT Convergence and Security (ICITCS2016), Prague, Czech Republic, 23–26 September 2016; pp. 149–153.
20. Kitsios, F.; Kamariotou, M. The impact of Information Technology and the alignment between business and service innovation strategy on service innovation performance. In Proceedings of the 3rd IEEE International Conference on Industrial Engineering, Management Science and Applications (ICIMSA 2016), Jeju Island, Korea, 23–26 May 2016; pp. 247–251.
21. Kumar, G.; Subramanian, N.; Arputham, R.M. Missing link between sustainability collaborative strategy and supply chain performance: Role of dynamic capability. *Int. J. Prod. Econ.* **2018**, *203*, 96–109. [CrossRef]
22. Metaxas, I.N.; Koulouriotis, D.E.; Spartalis, S.H. A multicriteria model on calculating the Sustainable Business Excellence Index of a firm with fuzzy AHP and TOPSIS. *Benchmarking Int. J.* **2016**, *23*, 1522–1557. [CrossRef]

23. Caputo, A.; Marzi, G.; Maley, J.; Silic, M. Ten years of conflict management research 2007–2017: An update on themes, concepts and relationships. *Int. J. Confl. Manag.* **2019**, *30*, 87–110. [CrossRef]

24. Karanatsiou, D.; Li, Y.; Arvanitou, E.M.; Misirlis, N.; Wong, W.E. A bibliometric assessment of software engineering scholars and institutions (2010–2017). *J. Syst. Softw.* **2019**, *147*, 246–261. [CrossRef]

25. Baumgartner, R.J.; Ebner, D. Corporate sustainability strategies: Sustainability profiles and maturity levels. *Sustain. Dev.* **2010**, *18*, 76–89. [CrossRef]

26. Egels-Zandén, N.; Rosén, M. Sustainable strategy formation at a Swedish industrial company: Bridging the strategy-as-practice and sustainability gap. *J. Clean. Prod.* **2015**, *96*, 139–147. [CrossRef]

27. Elkington, J. Towards the sustainable corporation: Win-win-win business strategies for sustainable development. *Calif. Manag. Rev.* **1994**, *36*, 90–100. [CrossRef]

28. Chang, A.Y.; Cheng, Y.T. Analysis model of the sustainability development of manufacturing small and medium-sized enterprises in Taiwan. *J. Clean. Prod.* **2019**, *207*, 458–473. [CrossRef]

29. León-Soriano, R.; Jesús Muñoz-Torres, M.; Chalmeta-Rosalen, R. Methodology for sustainability strategic planning and management. *Ind. Manag. Data Syst.* **2010**, *110*, 249–268. [CrossRef]

30. Asif, M.; Searcy, C.; Zutshi, A.; Fisscher, O.A. An integrated management systems approach to corporate social responsibility. *J. Clean. Prod.* **2013**, *56*, 7–17. [CrossRef]

31. Calabrese, A.; Costa, R.; Levialdi, N.; Menichini, T. Integrating sustainability into strategic decision-making: A fuzzy AHP method for the selection of relevant sustainability issues. *Technol. Forecast. Soc. Chang.* **2019**, *139*, 155–168. [CrossRef]

32. Ocampo, L.A.; Promentilla, M.A.B. Development of a sustainable manufacturing strategy using analytic network process. *Int. J. Bus. Syst. Res.* **2016**, *10*, 262–290. [CrossRef]

33. Epstein, M.J.; Widener, S.K. Facilitating sustainable development decisions: Measuring stakeholder reactions. *Bus. Strategy Environ.* **2011**, *20*, 107–123. [CrossRef]

34. Harangozó, G.; Zilahy, G. Cooperation between business and non-governmental organizations to promote sustainable development. *J. Clean. Prod.* **2015**, *89*, 18–31. [CrossRef]

35. Bonn, I.; Fisher, J. Sustainability: The missing ingredient in strategy. *J. Bus. Strategy* **2011**, *32*, 5–14. [CrossRef]

36. Cezarino, L.O.; Alves, M.F.R.; Caldana, A.C.F.; Liboni, L.B. Dynamic Capabilities for Sustainability: Revealing the Systemic Key Factors. *Syst. Pract. Action Res.* **2019**, *32*, 93–112. [CrossRef]

37. Lozano, R. A holistic perspective on corporate sustainability drivers. *Corp. Soc. Responsib. Environ. Manag.* **2015**, *22*, 32–44. [CrossRef]

38. Neugebauer, F.; Figge, F.; Hahn, T. Planned or emergent strategy making? Exploring the formation of corporate sustainability strategies. *Bus. Strategy Environ.* **2016**, *25*, 323–336. [CrossRef]

39. Schrettle, S.; Hinz, A.; Scherrer-Rathje, M.; Friedli, T. Turning sustainability into action: Explaining firms' sustainability efforts and their impact on firm performance. *Int. J. Prod. Econ.* **2014**, *147*, 73–84. [CrossRef]

40. Fairfield, K.D.; Harmon, J.; Behson, S.J. Influences on the organizational implementation of sustainability: An integrative model. *Organ. Manag. J.* **2011**, *8*, 4–20. [CrossRef]

41. Shields, J.; Shelleman, J.M. Integrating sustainability into SME strategy. *J. Small Bus. Strategy* **2015**, *25*, 59–78.

42. Thakhathi, A.; le Roux, C.; Davis, A. Sustainability Leaders' Influencing Strategies for Institutionalising Organisational Change towards Corporate Sustainability: A Strategy-as-Practice Perspective. *J. Chang. Manag.* **2019**, *19*, 246–265. [CrossRef]

43. Kiesnere, A.L.; Baumgartner, R.J. Sustainability management emergence and integration on different management levels in smaller large-sized companies in Austria. *Corp. Soc. Responsib. Environ. Manag.* **2019**. [CrossRef]

44. Lozano, R. Are companies planning their organisational changes for corporate sustainability? An analysis of three case studies on resistance to change and their strategies to overcome it. *Corp. Soc. Responsib. Environ. Manag.* **2013**, *20*, 275–295. [CrossRef]

45. Moore, S.B.; Manring, S.L. Strategy development in small and medium sized enterprises for sustainability and increased value creation. *J. Clean. Prod.* **2009**, *17*, 276–282. [CrossRef]

46. Rosati, F.; Faria, L.G.D. Business contribution to the Sustainable Development Agenda: Organizational factors related to early adoption of SDG reporting. *Corp. Soc. Responsib. Environ. Manag.* **2019**, *26*, 588–597. [CrossRef]

47. McPhee, W. A new sustainability model: Engaging the entire firm. *J. Bus. Strategy* **2014**, *35*, 4–12. [CrossRef]

48. Tseng, M.L.; Wu, K.J.; Ma, L.; Kuo, T.C.; Sai, F. A hierarchical framework for assessing corporate sustainability performance using a hybrid fuzzy synthetic method-DEMATEL. *Technol. Forecast. Soc. Chang.* **2019**, *144*, 524–533. [CrossRef]

49. Ou, Y.C. Using a hybrid decision-making model to evaluate the sustainable development performance of high-tech listed companies. *J. Bus. Econ. Manag.* **2016**, *17*, 331–346. [CrossRef]

50. Wicher, P.; Zapletal, F.; Lenort, R. Sustainability performance assessment of industrial corporation using Fuzzy Analytic Network Process. *J. Clean. Prod.* **2019**, *241*, 1–14. [CrossRef]

51. Beckmann, M.; Hielscher, S.; Pies, I. Commitment strategies for sustainability: How business firms can transform trade-offs into win–win outcomes. *Bus. Strategy Environ.* **2014**, *23*, 18–37. [CrossRef]

52. Aras, G.; Crowther, D. Making sustainable development sustainable. *Manag. Decis.* **2009**, *47*, 975–988. [CrossRef]

53. Bastons, M.; Armengou, J. Realism and Impartiality: Making Sustainability Effective in Decision-Making. *Sci. Eng. Ethics* **2017**, *23*, 969–987. [CrossRef]

54. Bianchi, C.; Cosenz, F.; Marinković, M. Designing dynamic performance management systems to foster SME competitiveness according to a sustainable development perspective: Empirical evidences from a case-study. *Int. J. Bus. Perform. Manag.* **2015**, *16*, 84–108. [CrossRef]

55. Brook, J.W.; Pagnanelli, F. Integrating sustainability into innovation project portfolio management—A strategic perspective. *J. Eng. Technol. Manag.* **2014**, *34*, 46–62. [CrossRef]

56. Cagno, E.; Neri, A.; Howard, M.; Brenna, G.; Trianni, A. Industrial sustainability performance measurement systems: A novel framework. *J. Clean. Prod.* **2019**, *230*, 1354–1375. [CrossRef]

57. Chang, R.D.; Zuo, J.; Soebarto, V.; Zhao, Z.Y.; Zillante, G. Dynamic interactions between sustainability and competitiveness in construction firms: A transition perspective. *Eng. Constr. Archit. Manag.* **2017**, *24*, 842–859. [CrossRef]

58. Christ, K.L.; Burritt, R.L.; Varsei, M. Coopetition as a potential strategy for corporate sustainability. *Bus. Strategy Environ.* **2017**, *26*, 1029–1040. [CrossRef]

59. Daneshpour, H.; Takala, J. Decision making towards integration of sustainability into project management; a multilevel theory building approach. *Manag. Prod. Eng. Rev.* **2017**, *8*, 13–21. [CrossRef]

60. De Brucker, K.; Macharis, C.; Verbeke, A. Multi-criteria analysis and the resolution of sustainable development dilemmas: A stakeholder management approach. *Eur. J. Oper. Res.* **2013**, *224*, 122–131. [CrossRef]

61. de Fátima Teles, M.; de Sousa, J.F. Linking fields with GMA: Sustainability, companies, people and Operational Research. *Technol. Forecast. Soc. Chang.* **2018**, *126*, 138–146. [CrossRef]

62. Eikelenboom, M.; de Jong, G. The impact of dynamic capabilities on the sustainability performance of SMEs. *J. Clean. Prod.* **2019**, *235*, 1360–1370. [CrossRef]

63. Figge, F.; Hahn, T.; Schaltegger, S.; Wagner, M. The sustainability balanced scorecard–linking sustainability management to business strategy. *Bus. Strategy Environ.* **2002**, *11*, 269–284. [CrossRef]

64. Garcia, S.; Cintra, Y.; Rita de Cássia, S.R.; Lima, F.G. Corporate sustainability management: A proposed multi-criteria model to support balanced decision-making. *J. Clean. Prod.* **2016**, *136*, 181–196. [CrossRef]

65. Haffar, M.; Searcy, C. How organizational logics shape trade-off decision-making in sustainability. *Long Range Plan.* **2019**, *52*, 101912. [CrossRef]

66. Hahn, W.J. Making decisions with multiple criteria: A case in energy sustainability planning. *EURO J. Decis. Process.* **2015**, *3*, 161–185. [CrossRef]

67. Hahn, T.; Pinkse, J.; Preuss, L.; Figge, F. Tensions in corporate sustainability: Towards an integrative framework. *J. Bus. Ethics* **2015**, *127*, 297–316. [CrossRef]

68. Hallstedt, S.I.; Bertoni, M.; Isaksson, O. Assessing sustainability and value of manufacturing processes: A case in the aerospace industry. *J. Clean. Prod.* **2015**, *108*, 169–182. [CrossRef]

69. Hallstedt, S.; Ny, H.; Robèrt, K.H.; Broman, G. An approach to assessing sustainability integration in strategic decision systems for product development. *J. Clean. Prod.* **2010**, *18*, 703–712. [CrossRef]

70. Hessami, A.R.; Faghihi, V.; Kim, A.; Ford, D.N. Evaluating planning strategies for prioritizing projects in sustainability improvement programs. *Constr. Manag. Econ.* **2019**, 1–13. [CrossRef]

71. Ivory, S.B.; Brooks, S.B. Managing corporate sustainability with a paradoxical lens: Lessons from strategic agility. *J. Bus. Ethics* **2018**, *148*, 347–361. [CrossRef]

72. Janeiro, L.; Patel, M.K. Choosing sustainable technologies. Implications of the underlying sustainability paradigm in the decision-making process. *J. Clean. Prod.* **2015**, *105*, 438–446. [CrossRef]

73. Kelly, K.L. A systems approach to identifying decisive information for sustainable development. *Eur. J. Oper. Res.* **1998**, *109*, 452–464. [CrossRef]

74. Martin, L. Incorporating values into sustainability decision-making. *J. Clean. Prod.* **2015**, *105*, 146–156. [CrossRef]

75. Modrak, V.; Dima, I.C. Conceptual framework for corporate sustainability planning. *Int. Bus. Manag.* **2010**, *4*, 139–144. [CrossRef]

76. Parisi, C. The impact of organisational alignment on the effectiveness of firms' sustainability strategic performance measurement systems: An empirical analysis. *J. Manag. Gov.* **2013**, *17*, 71–97. [CrossRef]

77. Rodriguez, R.; Svensson, G.; Eriksson, D. Organizational positioning and planning of sustainability initiatives: Logic and differentiators. *Int. J. Public Sect. Manag.* **2018**, *31*, 755–774. [CrossRef]

78. Satyro, W.C.; Sacomano, J.B.; Contador, J.C.; Almeida, C.M.; Giannetti, B.F. Process of strategy formulation for sustainable environmental development: Basic model. *J. Clean. Prod.* **2017**, *166*, 1295–1304. [CrossRef]

79. Silvius, A.G.; Kampinga, M.; Paniagua, S.; Mooi, H. Considering sustainability in project management decision making; An investigation using Q-methodology. *Int. J. Proj. Manag.* **2017**, *35*, 1133–1150. [CrossRef]

80. Singla, A.; Ahuja, I.S.; Sethi, A.S. An examination of effectiveness of technology push strategies for achieving sustainable development in manufacturing industries. *J. Sci. Technol. Policy Manag.* **2019**, *10*, 73–101. [CrossRef]

81. Sreekumar, V.; Rajmohan, M. Supply chain sustainability strategy selection using integrated multi-criteria decision-making method. *Int. J. Logist. Syst. Manag.* **2018**, *31*, 483–505. [CrossRef]

82. Sroufe, R. Integration and organizational change towards sustainability. *J. Clean. Prod.* **2017**, *162*, 315–329. [CrossRef]

83. Taghavi, M.; Bakhtiyari, K.; Taghavi, H.; Olyaee Attar, V.; Hussain, A. Planning for sustainable development in the emerging information societies. *J. Sci. Technol. Policy Manag.* **2014**, *5*, 178–211. [CrossRef]

84. Teh, D.; Corbitt, B. Building sustainability strategy in business. *J. Bus. Strategy* **2015**, *36*, 39–46. [CrossRef]

85. Teixeira, G.F.G.; Junior, O.C. How to make strategic planning for corporate sustainability? *J. Clean. Prod.* **2019**, *230*, 1421–1431. [CrossRef]

86. Ukko, J.; Nasiri, M.; Saunila, M.; Rantala, T. Sustainability strategy as a moderator in the relationship between digital business strategy and financial performance. *J. Clean. Prod.* **2019**, *236*, 1–9. [CrossRef]

87. Riccaboni, A.; Luisa Leone, E. Implementing strategies through management control systems: The case of sustainability. *Int. J. Prod. Perform. Manag.* **2010**, *59*, 130–144. [CrossRef]

88. Vazhayil, J.P.; Balasubramanian, R. Hierarchical multi-objective optimization of India's energy strategy portfolios for sustainable development. *Int. J. Energy Sect. Manag.* **2012**, *6*, 301–320. [CrossRef]

89. Wollmann, D.; Tortato, U. Proposal for a model to hierarchize strategic decisions according to criteria of value innovation, sustainability and budgetary constraint. *J. Clean. Prod.* **2019**, *231*, 278–289. [CrossRef]

 *sustainability*

*Article*

# The Use of Fuzzy Estimators for the Construction of a Prediction Model Concerning an Environmental Ecosystem

**Georgia Ellina** [1,*], **Garyfalos Papaschinopoulos** [2] and **Basil Papadopoulos** [1]

1   Department of Civil Engineering, Democritus University of Thrace, 67100 Xanthi, Greece;
    papadob@civil.duth.gr
2   Department of Environmental Engineering, Democritus University of Thrace, 67100 Xanthi, Greece;
    gpapas@env.duth.gr
*   Correspondence: geor.ellina@gmail.com

Received: 13 August 2019; Accepted: 12 September 2019; Published: 15 September 2019

**Abstract:** As a variable system, the Lake of Kastoria is a good example regarding the pattern of the Mediterranean shallow lakes. The focus of this study is on the investigation of this lake's eutrophication, analyzing the relation of the basic factors that affect this phenomenon using fuzzy logic. In the method we suggest, while there are many fuzzy implications that can be used since the proposition can take values in the close interval [0,1], we investigate the most appropriate implication for the studied water body. We propose a method evaluating fuzzy implications by constructing triangular non-asymptotic fuzzy numbers for each of the studied parameters coming from experimental data. This is achieved with the use of fuzzy estimators and fuzzy linear regression. In this way, we achieve a better understanding of the mechanisms and functions that regulate this ecosystem.

**Keywords:** implications; fuzzy estimators; lake eutrophication; prediction model

---

## 1. Introduction

Water resources are considered to be a valuable commodity of inestimable value. However, because of the effects of climate change (desertification—drought, water rise, floods, groundwater pollution, coastal erosion and erosion, wetland degradation; heat waves—fires, windstorms or siphons, reduction of biodiversity, etc.) and human intervention, the quality of surface and groundwater is continuously degrading and drinking water reserves are constantly decreasing [1–3]. Meanwhile, the need for increased use of water resources stems from population growth and ever-increasing demands on water and food, but also from the growth that generates activities and hence demands larger quantities of water. Unfortunately, urban water bodies usually have limited surface areas and poor mobility. As a result, they are susceptible to ecological deterioration with smaller water bodies and experience the phenomenon of eutrophication [4]. Strict measures should be taken by authorities in order to preserve and restore the quality and quantity of water basins [5,6], especially in areas that have been categorized as climate change hot spots, such as the Mediterranean Basin.

The extremely limited amount of freshwater on our planet, coupled with the imminent climate change, global warming due to greenhouse gas emissions, and rising water demand, put the availability of water in critical condition and make it imperative to develop control and management systems that aim to optimize the disposal of water resources, which is now called Water Resources Management. The Intergovernmental Panel on Climate Change reports that the climate warming in our century has regionally influenced the global water cycle [7].

One of the most important objects of investigation in Water Resources Management is lake ecosystems, which are environmental goods of particular importance. The lakes are associated with aquatic ecosystems with multiple significance and value, both for humans and for the natural environment. Their water is used in a variety of ways (e.g., water, irrigation, industry), but very often they become the final recipients of urban, industrial, and agricultural waste. This results in the pollution of their waters and disturbance of the ecological balance of the lake ecosystem. Many lakes, in Greece and around the world, face significant water quality problems, which are the subject of intense study by scientists [8,9]. Specifically for the lakes in the Mediterranean, researchers face more complex mechanisms because of the concentrations of phosphorus and nitrates [10].

In Western Macedonia and in the western part of Kastoria Prefecture, Lake Kastoria spreads around the homonymous city. As far as its biological status is concerned, it is a eutrophic lake with lakeside forests, rich in fish and birdlife. The city and the communities that are built on its banks and the catchment area have been heavily burdened by the waters of the lake. Since 1990 a sewage treatment plant has been operating in the city and the lakeside villages, resulting in gradual upgrading of water quality and wetland life. However, fertilizers and pesticides still used on a large scale in agriculture crops in the lakeside area pollute the soil and the underground aquifer, with the lake as the final recipient. This situation has the effect of enhancing the eutrophication of Lake Kastoria.

The lake concentrates the waters of a large catchment area with many watercourses, but through them travel fertilizers and various nutrients along with toxic substances that lead to the eutrophication phenomenon and consequently the degradation of water quality [11]. The amount of dissolved oxygen in water is an indicator of its pollution. The solution of re-cooling the water is not always able to restore the dissolved oxygen values to the levels necessary for the proper aquatic life. In order to study the quality of the Kastoria lake waters, the quantities of dissolved oxygen, as well as some factors affecting or affected by the above (e.g., temperature, pH) have been measured.

The interpretation and prediction of physical, chemical, and biological functions of lakes has so far been studied using widely available empirical and dynamic models along with multi-criteria analysis methods such as WASP5, EUTROMOD, PCLake, and CAEDYEM.

The theory of fuzzy logic is now being applied in several research studies relevant to the assessment of water quality and the trophic state of aquatic ecosystems [12,13]. The fact that fuzzy logic better approaches human logic makes this method more realistic for the description of complex systems like these when compared to classic logic. Multifactor systems such as lakes are governed by some rules. In classic logic, these rules/implications depend only on the question "is the statement true or not true?". All propositions take the values 0 or 1—holds or does not hold. In fuzzy logic, the true or false of a fuzzy proposition take values in the close interval [0,1]. Consequently, fuzzy implications generalize those of classical logic [13].

The key of the investigation of these complex systems is the use of the proper fuzzy implication, using fuzzy estimators for the construction of fuzzy numbers. Moreover, by applying fuzzy linear regression [14], we observe that by comparing the results from fuzzy estimators, we take the same implication which describes the best-studied ecosystem. In this study, we find the fuzzy implication that best describes the studied Lake Kastoria. The ability of selecting the most appropriate implication among others for each study case and calibrating the fuzzy inference systems is a useful tool for the construction of an accurate prediction model. By having such a model, every researcher can achieve a better understanding of the mechanisms that affect the biological and chemical functions in the ecosystem.

## 2. Materials and Methods

### 2.1. Study Area

Lake Kastoria is located in the northwestern part of Greece in the mainly mountainous prefecture of Kastoria, with limited cultivated land—most of which is lakeside, as shown in Figure 1. The lake is at an altitude of 630 m. Its area is about 28 km$^2$,with a maximum depth of about 9 m. Its geographical coordinates at the center are: latitude 40°31′N and longitude 21°18′E [15]. Its water volume is approximately 100,000,000 cubic meters and its coastline is 30.8 km.

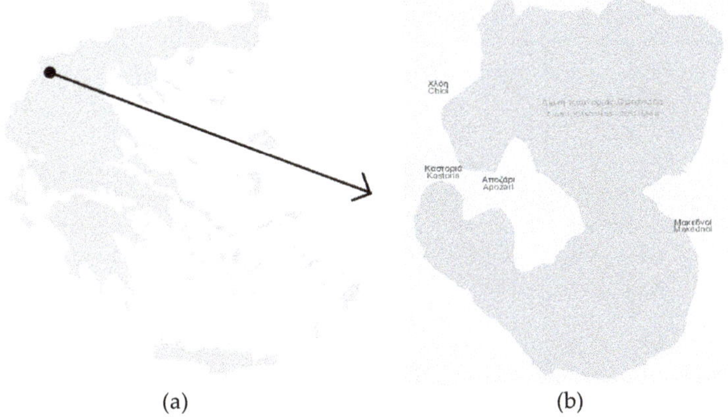

(a)        (b)

**Figure 1.** The location of the study case: (**a**) the map of Greece, (**b**) Lake Kastoria.

The lake is fed by many lakeside sources and by the rainwater that either falls directly to its surface or ends up in it with the surface runoff through the torrents located mainly in its northern and eastern parts. In the southern part, a canal (Guilli stream) connects the lake with the Aliakmonas River, where the excess water is drained using a gate that was recently modernized by the Municipality of Kastoria.

A very important natural ecosystem with diverse individual habitats supports a large biodiversity including rare and endangered species. The most typical ones are four species of herons, Dalmatian pelicans, mute swans, wild ducks, night herons, pygmy cormorants, cormorants, and many waterside birds. In addition, it is the only natural lake in Greece where there are still significant riparian forests of hydrophilic trees, which today are one of the rare habitats in the European area. In terms of fishing, Kastoria is the second most productive lake. The fish species inhabiting the lake are the following: perch, carp, chub, wells, muskellunge, roach, and butterfly ray. The existence of a great variety of plants and animals next to a highly developed city is truly remarkable. For the above reasons, the lake is protected by national and international institutional frameworks and since 1974 has been designated by the Ministry of Culture as a Monument of a Special Natural Beauty.

The fact that the lake has been registered with the Nature Network 2000 (GR 1320001) is an element that places the lake area as a priority in promoting the creation of such protection and management structures.

*2.2. Database and Model Application in Lake Kastoria*

In order to estimate the eutrophic status of this ecosystem, we investigated four of the most representative factors that influence the trophic status of a waterbody. For this research, we used the average measurements of four sampling stations. They were taken every fifteen days for a total duration of 24 months (2015–2017) with the following parameters as depicted in the Appendix A. For this research, the Department of Environment and Water Policies of Kastoria provided the measured data stated below:

- Water temperature: Temperature is directly related to the solubility of oxygen, the metabolism of aquatic organisms, and the process of decomposition of the organic substances it contains. The rise in temperature results in a reduction in dissolved oxygen, necessary for the survival of the aquatic organisms, and reduces the water density. The lightest water rises to the surface and stays there, creating a warmer layer with lower atmospheric oxygen dissolution capacity [16]. This may be fatal for organisms housed or preserved in water (e.g., plankton, fish, shells, amphibians, etc.). In addition, high temperatures promote the proliferation and growth of bacteria and other microorganisms.

- pH: Natural waters have pH values ranging between 4 and 9. Legislation sets the permitted limits for pH in the 6.5–8.5 range for drinking water. Natural fresh water has a slightly alkaline pH because of the presence of carbonate and bicarbonate ions. Because of agriculture drainage, pH affects the nitrogen and phosphorus release of the deposit sediment [17] and influences the eutrophication phenomenon in water bodies like rivers and lakes. This parameter has a great importance for water environments [18].

- DO: Dissolved oxygen plays a key role in aquatic ecosystems. Most life forms need oxygen to survive. The water receiver has the ability to maintain a maximum dissolved oxygen concentration, called the saturation concentration. Decreasing the concentration of dissolved oxygen to levels below the saturation value leads to degradation of the organic matter, while the aquatic lives are significantly affected or even killed.

- Chlorophyll: The knowledge of the chlorophyll concentration in an ecosystem provides useful information for assessing the phytoplankton biomass in a water area, and is an indicator of pollution from eutrophication [11].

### 2.3. Description of the Fuzzy Estimators and Fuzzy Implications

In classic logic, every proposition has two values: 0 or 1, true or false, holds or does not hold. The true or false of a fuzzy proposition depends on a degree—in contrast to classical logic, which takes values in the close interval [0,1] for the true or false [19,20]. Fuzzy implications use a more general and gradient model instead of classical logic. The main goal of this study was to investigate some of the eutrophication factors using fuzzy implications in the water body of Lake Kastoria. We aimed to select the most suitable implication that best expresses this lake.

More precisely, we used the membership functions of fuzzy estimators based on confidence intervals defined by extending the traditional confidence interval estimation of the mean of a normal distribution to the fuzzy domain; for readers' convenience, we state the following propositions without their proof [21]:

Proposition (non-asymptotic fuzzy mean of normal distribution-large samples). Let $x_1, \ldots, x_n$ be values assumed by our studied data sets, where $n = 48$. Let also $\gamma \in (0,1)$. The membership function of a fuzzy number is:

$$\mu_\gamma(x) = \begin{cases} \frac{2}{1-\gamma}\Phi\left(\frac{x-\bar{x}}{\sigma/\sqrt{n}}\right) - \frac{\gamma}{1-\gamma}, & \bar{x} - \frac{\sigma}{\sqrt{n}}\Phi^{-1}\left(1 - \frac{\gamma}{2}\right) \leq x \leq \bar{x} \\ \frac{2}{1-\gamma}\Phi\left(\frac{\bar{x}-x}{\sigma/\sqrt{n}}\right) - \frac{\gamma}{1-\gamma}, & \bar{x} \leq x \leq \bar{x} + \frac{\sigma}{\sqrt{n}}\Phi^{-1}\left(1 - \frac{\gamma}{2}\right) \\ 0, & \text{otherwise} \end{cases}, \tag{1}$$

$$^a\widetilde{\mu}_\gamma = \left[\bar{x} - z_{h(a)}\frac{\sigma}{\sqrt{n}}, \; \bar{x} + z_{h(a)}\frac{\sigma}{\sqrt{n}}\right], \; a \in (0,1], \tag{2}$$

$$Z_{h(a)} = \Phi^{-1}(1 - h(a)), \; h(a) = \left(\frac{1}{2} - \frac{\gamma}{2}\right)\alpha + \frac{\gamma}{2}, \tag{3}$$

where $\sigma$ is the standard deviation and $\bar{x}$ the mean value, the support of the membership function is exactly the $\gamma = 0.001$ confidence interval for $\mu$, and the $\alpha - cuts$ of this fuzzy number are the closed intervals and $\Phi$ denotes the cumulative distribution function of the standard normal distribution. We found the parameter values from the fuzzy numbers defined in (1). In this way, we estimated for every parameter all the truth values ($\alpha - cuts$) that these constructed fuzzy numbers can provide.

Our next step was to use these truth values for the estimation of a number of implications. The implications are a vital point for the prediction model's construction, since they are considered as the rules that express the relations among the abiotic and biotic parameters in ecosystems. The fuzzy symmetric and asymmetric implications [12,13,22] used in this paper are the following:

$$J_{Mamdani}(x, y) = \min\{x, y\}, \tag{4}$$

$$J_{Larsen}(x, y) = x \cdot y, \tag{5}$$

$$J_{Zadeh}(x, y) = \max(\min\{x, y\}, 1 - x), \tag{6}$$

$$J_{Reichenbach}(x, y) = 1 - x + x \cdot y, \tag{7}$$

$$J_{Luckasiewicz}(x, y) = \min(1 - x + y, 1), \tag{8}$$

$$J(x, y) = \frac{1 - x + y(1 + \lambda x) - y(1 - x)}{1 + \lambda x}, \lambda > -1. \tag{9}$$

For the last implication above, the t-norm (Probor) was used [22], while for this research we took $\lambda = -0.99$ [12].

## 3. Results

Our first step was the construction of the non-asymptotic fuzzy numbers. Figure 2 shows the form of the fuzzy numbers constructed based on confidence intervals defined by extending the traditional confidence interval estimation of the mean of a normal distribution to the fuzzy domain. At this point, it should be mentioned that the mean values of the experimental data correspond to the peak of the following fuzzy numbers. In this method, it is considered that the mean value is the "absolute right" value of every data set, and therefore this value corresponds to the unit. For this application, we also considered water temperature, pH, and dissolved oxygen as independent variables, while the value of chlorophyll was a dependent factor. Below we state the computed membership functions of the constructed fuzzy numbers along with their depiction in Figure 2.

a. $\mu_\gamma^{(temp)}(x) = \begin{cases} 2.002\Phi(x/1.097 - 13.7) - 10^{-3}, & 11.41 \le x \le 15.04 \\ 2.002\Phi(-x/1.097 + 13.7) - 10^{-3}, & 15.04 \le x \le 18.7 \\ 0, & otherwise \end{cases}$

b. $\mu_\gamma^{(DO)}(x) = \begin{cases} 2.002\Phi(x/0.3 - 25.5) - 10^{-3}, & 6.68 \le x \le 7.68 \\ 2.002\Phi(-x/0.3 + 25.5) - 10^{-3}, & 7.68 \le x \le 8.67 \\ 0, & otherwise \end{cases}$

c. $\mu_\gamma^{(pH)}(x) = \begin{cases} 2.002\Phi(x/0.031 - 268.71) - 10^{-3}, & 8.26 \le x \le 8.36 \\ 2.002\Phi(-x/0.031 + 268.71) - 10^{-3}, & 8.36 \le x \le 8.47 \\ 0, & otherwise \end{cases}$

d. $\mu_\gamma^{(Chl)}(x) = \begin{cases} 2.002\Phi(x/0.99 - 9.75) - 10^{-3}, & 6.4 \le x \le 9.67 \\ 2.002\Phi(-x/0.99 + 9.75) - 10^{-3}, & 9.67 \le x \le 12.95 \\ 0, & otherwise \end{cases}$

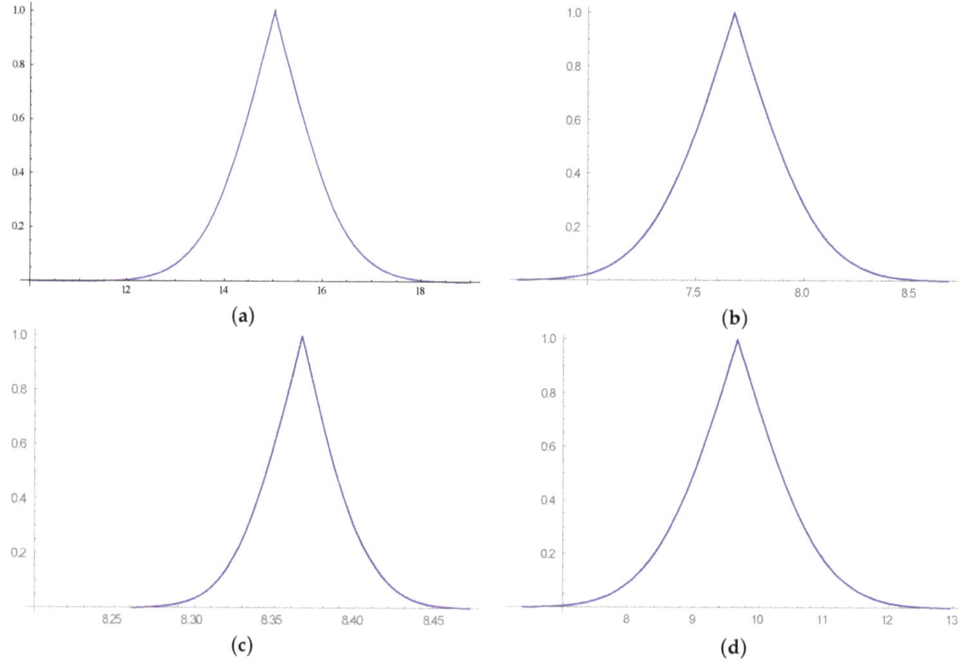

**Figure 2.** Non-asymptotic fuzzy numbers of the examined factors. (**a**) Water temperature (temp); (**b**) Dissolved oxygen (DO); (**c**) pH; (**d**) Chlorophyll (chl).

After computing all the fuzzy numbers representing the relation between the independents (water temperature, DO, and pH) and the dependent variable (chlorophyll), we estimated the deviations of all the studied implications given by the following relation [12]:

$$\sigma_{p,i} = \sqrt{\left(1 - \mu_{p,i}^{(1)}\right)^2 + \left(1 - \mu_{p,i}^{(2)}\right)^2 + \ldots + \left(1 - \mu_{p,i}^{(j_p)}\right)^2}, \tag{10}$$

where:

$\mu$ : the true values of the corresponding implications
$p = 1, 2,$ and 3: the examined independent parameters (water temperature, DO, pH);
$i = 1, 2, 3, 4, 5,$ and 6: the implications (Mamdani, Larsen, Zadeh, Lukasiewicz, Reichenbach, Probor);
$j_p$ : the number of true values.

For readers' convenience, below we state a numerical example:

Let us consider the relationship between the DO and chlorophyll. If the value of DO is 7.4 mg/L and chlorophyll is 9.615 mg/L, then from Figure 2 we take the following implication:

$$7.4 \Rightarrow 9.615,$$

which means that equals to the following relationship estimating the true values:

$$0.35 \Rightarrow 0.95.$$

At this point, we estimate the true value of this implication by the relation of Mamdani:

$$\mu_{2,1} = J_{\text{Mamdani}}(x, y) = \min\{x,\ y\} = \min\{0.35, 0.95\} = 0.35.$$

We followed the same procedure for the rest data sets for all of our studied couples of independent–dependent parameters to find the deviation of the implications ($\sigma_{p,i}$).

The smaller the deviation of the fuzzy parameter, the more accurate we are about the best implication which expresses our study case. Figure 3 presents the deviation of every independent parameter studied in relation to the dependent one. The method of Probor was the best applied implication. All of the studied parameters had the smallest deviation in the last implication. The case of Probor is special because each user can calibrate the λ variable so that they find the lowest possible deviation. In this way, minimizing the deviation, we also minimize the possible errors in predicting parameters that we are called to investigate. Apart from this method, the implications of Reichenbach, Zadeh, and Lucasiewicz are accurate tools for prediction since their deviations were small enough for almost all parameters.

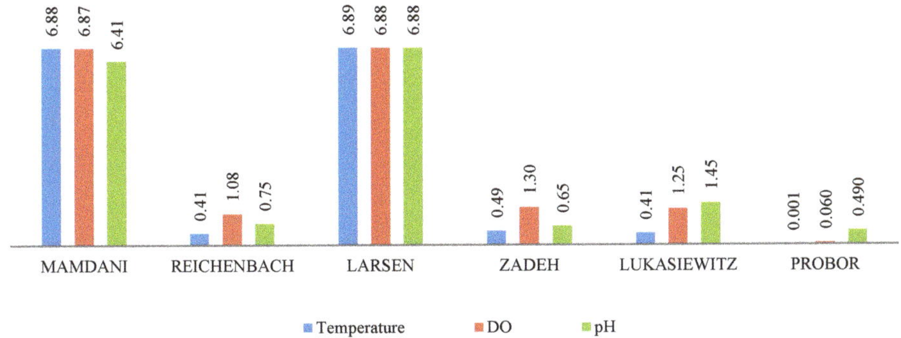


**Figure 3.** Deviation of the fuzzy parameters influenced by the dependent factor of chlorophyll.

## 4. Discussion

We obtained similar results to those we obtained in previous papers [12,22] investigating another Mediterranean lake—Lake Karla in Thessaly, which also has a hypertrophic character [16]. We used fuzzy linear regression for the construction of the parameters' fuzzy numbers [21–24] and compared the results with the method of fuzzy estimators. Similarly, we considered that the dependent variable was chlorophyll and all the others were independent. Using the fuzzy linear regression for all the couples of independent–dependent variables, we found all the fuzzy numbers of chlorophyll and consequently the truth values of chlorophyll. Then, we considered that the independent parameter of this ecosystem was chlorophyll. Thus, all the others were the dependent ones. We again applied the method above to estimate all the fuzzy numbers and the truth values that express the water temperature, dissolved oxygen, and pH.

Again applying the fuzzy implications mentioned in Section 2.3, similar results can be observed in Figure 4.

We calculated the true values by applying fuzzy linear regression, and then, using the same implications, we concluded that the implication of Probor was the best for our study case. The smallest deviation was observed in the implications of Probor and Reichenbach. It seems that this method has the advantage of calibrating one parameter ($\lambda$) in order to choose the smallest possible value for the studied ecosystem [12]. These fuzzy inference systems have the ability to select the most appropriate implication. Computing the deviation of each implication, it is easy to choose the implication with the smallest deviation. The smaller the deviation is, the more proper the implication for the examined study area.

Having this information about the examined ecosystem, it is easier to understand the mechanisms that influence the trophic state of this water body [25]. The method of fuzzy estimators is an alternative approach in the area of fuzzy logic, and replaces the use of fuzzy linear regression. Prediction models need accuracy in order to have the best solutions in dissemination and exploitation of the results. Monitoring components such as biotic and abiotic parameters is an ideal way to evaluate their effectiveness and their influence on the ecosystem's trophic state. The results are very useful for understanding and determining the correlation of the studied parameters [26]. Having realistic results, the effort for the ecosystem's restoration and protection is of vital importance. This method can be applied across a wide range of research and forecasting software [27,28]. With this mathematical model, we can adapt fuzzy inference systems for the construction of accurate prediction tools for the studied ecosystem. In this way, we obtain a better picture of the trophic state of the lake, having the proper software (i.e., Matlab) for future improvement by monitoring the relations among the parameters.

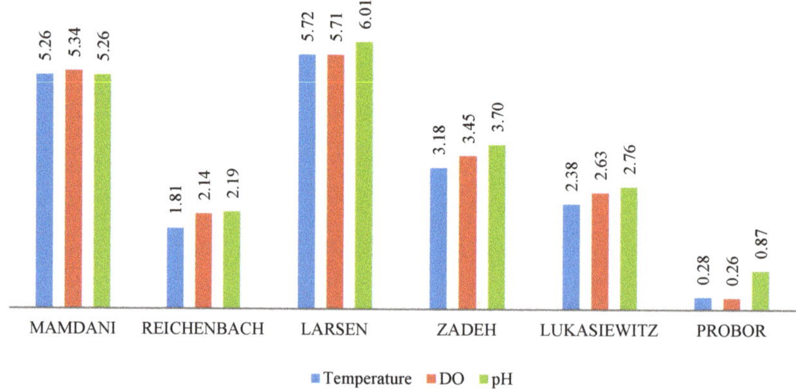


**Figure 4.** Deviation of the fuzzy parameters constructed by fuzzy linear regression.

## 5. Conclusions

The waters of Lake Kastoria receive substances due to human activities and natural processes. This results in significant quantities of fertilizers, insecticides, and pesticides being added to the soil each year, with increasing annual rates. Moreover, livestock farming contributes to the production and input of pollutants in the lake. The nutrients resulting from these substances end up in the waters of the lake. Of course, agricultural development and livestock farming cannot be limited, but they must be modernized to help protect the aquatic and terrestrial environment. The effect of water pooling on various nutrients and toxic substances is to create conditions of eutrophication, less water transparency, less dissolved oxygen, and adverse conditions for aquatic organisms—especially fish. Generally, water bodies cope with significant problems. The alternative method of fuzzy logic gives the opportunity to combine many approaches. The method proposed in this paper investigates and selects the most appropriate fuzzy implication using real water quality observations in Lake Kastoria. The use of fuzzy estimators for the construction of the fuzzy numbers is a useful method, and its use is suggested when there is a great deal of data to investigate. Comparing it with the method of fuzzy linear regression, we came to similar results. The implication of Probor was the best option when investigating this ecosystem. By already having the most accurate mathematical relations among parameters, the next step is to create the appropriate software in order to use it as a forecasting tool compared to already known general multi-criteria analysis software.

**Author Contributions:** B.P., G.P. and G.E. came up with the concept of this research. G.P. and G.E. analyzed the data and used the software. G.E. wrote the paper. B.P. and G.P. supervised the whole research.

**Funding:** This research received no external funding.

**Acknowledgments:** Authors would like to thank the reviewers for their helpful suggestions.

**Conflicts of Interest:** The authors declare no conflicts of interest.

## Appendix A

We state the following data, obtained by the department of environmental and water policies of Kastoria:

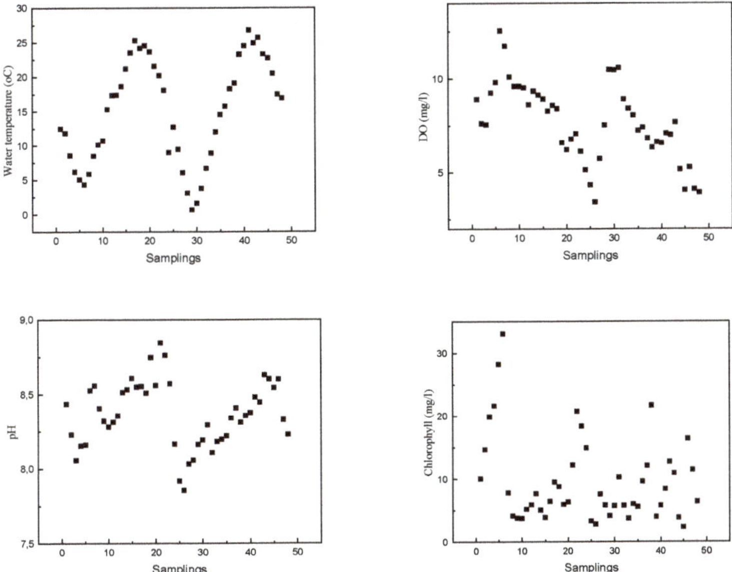

**Figure A1.** Measurements of the studied parameters of water temperature, dissolved oxygen (DO), pH and chlorophyll, for the years 2015 to 2017 taken every fifteen days.

## References


1. Stefanidis, K.; Dimitriou, E. Differentiation in Aquatic Metabolism between Littoral Habitats with Floating-Leaved and Submerged Macrophyte Growth Forms in a Shallow Eutrophic Lake. *Water* **2019**, *11*, 287. [CrossRef]
2. Demertzi, K.; Papadimos, D.; Aschonitis, V.; Papamichail, D. A Simplistic Approach for Assessing Hydroclimatic Vulnerability of Lakes and Reservoirs with Regulated Superficial Outflow. *Hydrology* **2019**, *6*, 61. [CrossRef]
3. Xu, C.; Wang, H.U.; Yu, Q.; Wang, H.Z.; Liang, X.M.; Liu, M.; Jeppesen, E. Effects of Artificial LED Light on the Growth of Three Submerged Macrophyte Species during the Low-Growth Winter Season: Implications for Macrophyte Restoration in Small Eutrophic Lakes. *Water* **2019**, *11*, 1512. [CrossRef]
4. Zheng, T.; Cao, H.; Liu, W.; Xu, J.; Yan, Y.; Lin, X.; Huang, J. Characteristics of Atmospheric Deposition during the Period of Algal Bloom Formation in Urban Water Bodies. *Sustainability* **2019**, *11*, 1703. [CrossRef]
5. Doulgeris, C.; Papadimos, D.; Kapsomenakis, J. Impacts of climate change on the hydrology of two Natura 2000 sites in Northern Greece. *Reg. Environ. Chang.* **2016**, *16*, 1941–1950. [CrossRef]
6. Wang, W.; Lee, X.; Xiao, W.; Liu, S.; Schultz, N.; Wang, Y.; Zhang, M.; Zhao, L. Global lake evaporation accelerated by changes in surface energy allocation in a warmer climate. *Nat. Geosci.* **2018**, *11*, 410–414. [CrossRef]
7. IPCC. Summary for policymakers. In *Climate Chang. 2013: The Physical Science Basis. Contribution of Working Group I to the Fifth Assessment Report of the Intergovernmental Panel on Climate Change*; Stocker, T.F., Qin, D., Plattner, G.K., Tignor, M., Allen, S.K., Boschung, J., Nauels, A., Xia, Y., Bex, V., Midgley, P.M., Eds.; Cambridge University Press: Cambridge, UK; New York, NY, USA, 2013.
8. Leon-Munoz, J.; Echeverria, C.; Marce, R.; Riss, W.; Sherman, B.; Iriarte, J.L. The combined impact of land use change and aquaculture on sediment and water environmental quality in oligotrophic Lake Rupanco. *J. Environ. Manag.* **2013**, *128*, 283–291. [CrossRef]

9.   Lin, B.; Chen, X.; Yao, H.; Chen, Y.; Liu, M.; Gao, L.; James, A. Analyses of land use change impacts on catchment runoff using different time indicators based on SWAT model. *Ecol. Indic.* **2015**, *58*, 55–63. [CrossRef]

10.  Matzafleri, N.; Psilovikos, A.; Neofytou, C.; Kagalou, I. Determination of the Trophic Status of Lake Kastoria, Western Macedonia, Greece. In Proceedings of the Small and decentralised water and wastewater treatment plants IV, Volos, Greece, 25–27 October 2013; ISBN 978-960-6865-72-5.

11.  Richardson, K.; Jørgensen, B.B. *Eutrophication in a Coastal Marine Ecosystem*; Coastal and estuarine studies; American Geophysical Union: Washington, DC, USA, 1996; Volume 1, pp. 1–9.

12.  Ellina, G.; Papaschinopoulos, G.; Papadopoulos, B.K. Fuzzy inference systems: Selection of the most appropriate fuzzy implication in terms of statistical data. *Environ. Process.* **2017**, *4*, 923–935. [CrossRef]

13.  Botzoris, G.; Papadopoulos, K.; Papadopoulos, B.K. A method for the evaluation and selection of an appropriate fuzzy implication by using statistical data. *Fuzzy Econ. Rev.* **2015**, *XX*, 19–29. [CrossRef]

14.  Ellina, G.; Papaschinopoulos, G.; Papadopoulos, B.K. Research of Fuzzy Implications via Fuzzy Linear Regression in Data Analysis for a Fuzzy Model. *J. Comput. Methods Sci. Eng.* **2019**. accepted for publication.

15.  Moustaka-Gouni, M.; Vardaka, E.; Michaloudi, E.; Kormas, K.A.; Tryfon, E.; Mihalatou, H.; Gkelis, S.; Lanaras, T. Plankton food web structure in a eutrophic polymictic lake with a history in toxic cyanobacterial blooms. *Limnol. Oceanogr.* **2006**, *51*, 715–727. [CrossRef]

16.  Ellina, G.; Kagalou, I. Selection of the most appropriate parameter for the chlorophyll-a estimation of an artificial lake via fuzzy linear regression. *Eur. Water* **2016**, *55*, 105–114.

17.  Tao, Y.; Wang, S.; Guan, X.; Xu, D.; Chen, H.; Ji, M. Study on Characteristics of Nitrogen and Phosphorus Loss under an Improved Subsurface Drainage. *Water* **2019**, *11*, 1467. [CrossRef]

18.  Fisher, L.H. *Effect of Water-Column pH on Sediment-Phosphorus Release Rates in Upper Klamath Lake, Oregon, 2001*; U.S. Geological Survey: Reston, VA, USA, 2004.

19.  Profillidis, V.A.; Papadopoulos, B.K.; Botzoris, G.N. Similarities in fuzzy regression models and application on transportation. *Fuzzy Econ. Rev.* **1999**, *4*, 83–98. [CrossRef]

20.  Papadopoulos, B.; Tsagarakis, K.P.; Yannopoulos, A. Cost and land functions for wastewater treatment projects: Typical simple linear regression versus fuzzy linear regression. *J. Environ. Eng.* **2007**, *133*, 581–586. [CrossRef]

21.  Sfiris, D.S.; Papadopoulos, B.K. Non-asymptotic fuzzy estimators based on confidence intervals. *Inf. Sci.* **2014**, *279*, 446–459. [CrossRef]

22.  Ellina, G.; Papaschinopoulos, G.; Papadopoulos, B.K. Research of Fuzzy Implications via Fuzzy Linear Regression in a Eutrophic Waterbody. In *AIP Conference Proceedings 1978*; AIP Publishing: New York, NY, USA, 2018; p. 290007. [CrossRef]

23.  Papadopoulos, B.K.; Sirpi, M.A. Similarities in fuzzy regression models. *J. Optim. Theory Appl.* **1999**, *102*, 373–383. [CrossRef]

24.  Chrysafis, K.A.; Papadopoulos, B.K. Cost–volume–profit analysis under uncertainty: A model with fuzzy estimators based on confidence intervals. *Int. J. Prod. Res.* **2008**, *47*, 5977–5999. [CrossRef]

25.  Brito, D.; Neves, R.; Branco, M.A.; Prazeres, A.; Rodrigues, S.; Maria, C.; Gonçalves, M.C.; Ramos, T.B. Assessing Water and Nutrient Long-Term Dynamics and Loads in the Enxoé Temporary River Basin (Southeast Portugal). *Water* **2019**, *11*, 354. [CrossRef]

26.  Buriboev, A.; Kang, H.K.; Ko, M.C.; Oh, R.; Abduvaitov, A.; Jeon, H.S. Application of Fuzzy Logic for Problems of Evaluating States of a Computing System. *Appl. Sci.* **2019**, *9*, 3021. [CrossRef]

27.  Pagouropoulos, P.; Tzimopoulos, C.; Papadopoulos, B. Selecting the most appropriate fuzzy implication based on statistical data. *Int. J. Fuzzy Syst. Adv. Appl.* **2016**, *3*, 32–42.

28.  Pagouropoulos, P.; Tzimopoulos, C.; Papadopoulos, B. A Method for the Detection of the Most Suitable Fuzzy Implication for Data Applications. In *Engineering Applications of Neural Networks, Communications in Computer and Information Science*; Springer: Heidelberg, Germany, 2017; Volume 744, pp. 242–255.

 *sustainability*

*Article*

# Stochastic Analysis of Embodied Carbon Dioxide Emissions Considering Variability of Construction Sites

**Dongyoun Lee [1], Goune Kang [2], Chulu Nam [3], Hunhee Cho [1,*] and Kyung-In Kang [1]**

[1] School of Civil, Environmental and Architectural Engineering, Korea University, 145, Anam-ro, Seongbuk-gu, Seoul 02841, Korea

[2] Korea Institute of Civil Engineering and Building Technology, 283 Goyang-daero, Ilsanseo-gu, Goyang, Gyeonggi 10223, Korea

[3] Deajeon-Sejong-Chungnam Headquarters, Korea Electric Power Corporation, 170, Songchonnam-ro 11beon-gil, Dong-gu, Daejeon 34537, Korea

* Correspondence: hhcho@korea.ac.kr; Tel.: +82-2-921-5920; Fax: +82-2-923-4229

Received: 30 June 2019; Accepted: 1 August 2019; Published: 4 August 2019

**Abstract:** The current method of estimating $CO_2$ emissions during the construction phase does not consider the variability that can occur in actual work. Therefore, this study aims at probabilistic $CO_2$ estimation dealing with the statistical characteristics in activity data of building construction work, focused on concrete pouring work and based on field data. The probabilistically estimated $CO_2$ emissions have some differences from $CO_2$ emissions measured by current deterministic methods. The results revealed that the minimum difference was 11.4%, and the maximum difference was 132.7%. This study also used Monte Carlo simulations to derive information on a probability model of $CO_2$ emissions. Results of the analysis revealed that there is a risk of underestimating emissions because the amount of emissions was estimated at a level that exceeds the 95% confidence interval of the simulation results. In addition, the probability that $CO_2$ emissions using the measured activities data were less than the estimated $CO_2$ emissions using the bill of quantity was 73.2% in the probability distribution model.

**Keywords:** $CO_2$ emissions; construction phase; stochastic analysis; Monte Carlo simulation

---

## 1. Introduction

The building sector consumes about 40% of the world's annual energy production, which accounts for approximately a quarter of the world's annual $CO_2$ emissions [1]. Continuous efforts have been made to reduce the amount of $CO_2$ emitted from buildings. Various technological developments and political measures related to eco-friendly building materials or renewable energy usage have been undertaken in the public and private sectors. However, these have tended to focus on specific phases in the whole life of the building, namely, material manufacturing and operational building phases [2,3]. Previous studies also have mentioned that it is important to deal with the embodied carbon emissions generated in the production, construction, maintenance, and disposal of building materials [4]. Among these phases, the construction phase accounts for approximately 10–30% of the entire building life cycle, even if it represents a short period. The aforementioned highlights the necessity for accurate and detailed estimation of $CO_2$ in order to devise means of reducing emissions from the construction phase [5].

However, estimating or predicting the amount of $CO_2$ generated during the construction phase is difficult because there is variability at the construction site. There is much variability in the input of resource types or quantities because construction activities are performed in outdoor environments [6].

Several researchers have utilized bills of quantity or standard productivity information prepared in the design stage, so values were almost fixed by design documents and cannot reflect the actual condition of the construction site.

Therefore, this study aims at probabilistic $CO_2$ estimation dealing with the statistical characteristic in activity data of building construction work, based on field data and considering variability. According to 2006 Intergovernmental Panel on Climate Change (IPCC) guidelines, variability is an inherent property of the system of nature and not of the analyst [7]. In this study, variability is defined as an inherent characteristic of the construction phase work. In addition, the result of the intrinsic characteristics is judged to be the working time. This study used actual activity data gathered during the construction phase and performed Monte Carlo simulation to determine the probabilistic interval of $CO_2$ emissions, focusing on concrete pouring work. Activity data were collected by measuring the operating time of concrete pump cars and concrete mixer trucks, which consume fossil fuels and directly emit $CO_2$. In this study, correlation analysis is performed to derive the correlation coefficients for the emission amount of each activity and to reflect them in the simulation. We executed the simulation test 10,000 times at a 95% confidence interval.

## 2. Literature Review

A deterministic $CO_2$ emission estimation obtained by multiplying the $CO_2$ emission coefficient with the amount of input material may not be accurate because these inputs do not take into account uncertainties at the construction site. Therefore, a previous study evaluated variability when estimating $CO_2$ emissions and has presented a prediction method that considers this level of variability [8–12].

This study also considered the most-consumed building materials to perform an analysis of statistical properties. Based on this, it probabilistically predicted the materials' construction stage emissions [8,9]. Other studies analyzed the variability of emission coefficients by evaluating the most-consumed building materials based on accumulated greenhouse gas emissions, as well as the energy emissions coefficient, in order to assess the level of uncertainty and variability when estimating the emissions of the apartment housing construction stage [10,11]. Another study suggested a method of calculating $CO_2$ emissions by calculating the amount of use of construction equipment on construction sites based on the schedule information of the building construction [12].

The only datum that can be used to quantify the environmental impact of equipment in the construction process is the daily report on the type and quantity of equipment that is deployed at the worksite. However, this data source lacks detailed information on equipment usage. To calculate $CO_2$ emissions using traditional life cycle impact assessment metrics and methods, historical performance data must be compiled with an accurate construction inventory (i.e., materials and equipment used for construction and maintenance work) [13,14]. Previous studies estimating $CO_2$ emissions have used simulation methods. The $CO_2$ of equipment used in road construction was calculated by measuring the engine load of the equipment. Moreover, the measured data were applied to the simulation considering the variability of the engine load [15–17]. Regarding equipment, simulation studies also exist. Similarly, other studies used CYCLONE simulation to predict emissions probabilistically while considering the variability of fuel consumption according to the activity of earthwork equipment [18,19].

However, the previous studies mentioned above have limitations, since they focus only on materials, or there is a lack of consideration of activity data, such as the operation time of construction equipment at the activity level. Previous studies have estimated only the amount of $CO_2$ emissions from construction equipment based on the amount of oil the equipment consumes. In addition, the variability in embodied carbon emissions of various building types has been explored [20,21]. However, there is still a lack of statistical information for the activity data for construction. In particular, the distribution type, parameters, and correlations required as input variables for Monte Carlo simulation have not been investigated yet [22].

Therefore, in this study, we gathered actual data about the operating time of construction equipment used at the activity level. Moreover, this study presents an estimation method using probabilistic $CO_2$ emissions for analyzing the variability during the construction phase.

## 3. Analyzing Variability in $CO_2$ Emissions

### 3.1. Overview of Activity Data Collection

In order to estimate $CO_2$ emissions, it is necessary to estimate the amount of oil consumption by measuring construction equipment operation time and equipment specification information. Therefore, data were collected on equipment operation time at the activity level (Table 1).

**Table 1.** Overview of the construction site with activity data collected.

|  | Content |
| --- | --- |
| Project | K Research Building (new construction) |
| Location | Seongbuk-gu, Seoul, Korea |
| Land area | 130,606 m$^2$ |
| Building area | 33,521 m$^2$ |
| Gross floor area | 22,910 m$^2$ |
| Structure type | Reinforced concrete |
| Floors | Six floors and two floors underground |
| Building use | Office building (education and research facility) |

This study collected data about concrete work for a typical floor slab. In this study, it was difficult to collect activity data for all tasks. Therefore, this study collected data focused on heavy equipment, which uses heavy oil in concrete pouring work; we installed a camcorder and a stopwatch to measure the operating time of each piece of construction equipment, namely, concrete pump cars and concrete mixer trucks.

Based on the specifications of the construction equipment, this study surveyed the equipment that meets the applicable standards of construction machinery expenditure as illustrated by Korean professional construction association information.

### 3.2. Calculating Emissions from Equipment Operation Times

In this study, one work cycle of concrete pouring work was defined from when a concrete mixer truck enters to when the truck leaves the construction site. A total of 37 work cycles were measured. However, interrupted working hours during the work cycle were excluded from the measurement.

The equipment operation time for each detailed task was measured, and 21 data sets were acquired in five categories and used in the analysis. The work of a concrete mixer truck was classified into four detailed tasks: Stopping after entering, waiting after stopping, pouring concrete, and leaving. The detailed work of the concrete pump cars was classified as a single task: Pumping concrete.

To estimate the working time of a concrete pump car for one concrete mixer truck (6 m$^3$), we referred to the average productivity value of the amount pumped per hour (87.5 m$^3$/h) provided in the Construction Equipment Cost Estimates Table (CECET). The CECET is an annual national standard released by the Construction Association of Korea (CAK) and the Ministry of Land, Infrastructure and Transport (MOLIT) and is calculated and published using data on the hourly hire of machines, miscellaneous materials, and drivers of construction machines listed in the standard production unit system. In this study, the concrete mixer trucks and concrete pump cars used at the construction site were collected through the CECET.

In this study, the $CO_2$ emission coefficients of materials and equipment were used according to the data presented in the 2006 IPCC guidelines [23]. The IPCC provides $CO_2$ emission coefficients through its own data surveys and reports submitted by each country. Recently, each country has used IPCC guidelines to calculate $CO_2$ emission coefficients that fit the situation for that country [17].

This study does not consider the working time of concrete mixer trucks at the work planning phase because the number of concrete mixer trucks is calculated at the time of work planning. The working time of the concrete pump car was estimated using the formula (Equation (1)) proposed in the standard for reinforced concrete construction estimation [24]:

$$T_p = t_p \times f_1 \times f_2 \times Q \tag{1}$$

where:

$T_p$: Concrete pump car working time;
$t_p$: Reference time, estimated time it takes for the concrete pump car to pour 1 m$^2$;
$f_1$: Facility type consideration coefficient;
$f_2$: Concrete mixer truck entry condition coefficient; and
$Q$: Workload.

In this study, a reference time ($t_p$) value was applied—1.25 min corresponding to 15 cm of the reinforced concrete slump—by referring to the information provided on the site where the activity data were collected [24]. The working time was calculated under the conditions that $f_1$ is normal (1.2) and $f_2$ is good (1.0) [24].

The average and the standard deviation of the measured equipment operation time are shown in Table 2. The planned equipment operation time is expressed as a single value. The variation of work time was the largest for waiting after stopping and was relatively low for stopping after entering and leaving. The average measured operation time of concrete pumping work was calculated to about 50 s longer than the value given by the machine cost calculation table. In addition, the concrete pumping work was more than twice the average measurement time in the case of the bills of quantity.

**Table 2.** Measured operation time and planned operation time (in seconds).

| Activity | Stopping after Entering | Waiting after Stopping | Pouring Concrete | Leaving | Pumping Concrete | Cycle |
|---|---|---|---|---|---|---|
| Average of the measured equipment operation time (standard deviation) | 36.3 (27.3) | 190.5 (243.2) | 343.8 (95.5) | 9.3 (3.5) | 299.6 (83.2) | 879.4 (279.35) |
| Planned equipment operation time | 36.3 | 190.5 | 343.8 | 9.3 | 246.9 | 826.8 |

### 3.3. Estimating CO$_2$ Emissions

In this study, CO$_2$ emissions from the oil consumption of construction equipment were estimated. The oil consumption was calculated using the measured working time at the construction site, the fuel efficiency of the construction equipment, and the engine load factor of construction equipment (Equation (2)):

$$F = T \times FE \times LF \tag{2}$$

where:

$F$: Oil Consumption by Work;
$T$: Working time;
$FE$: Fuel efficiency of construction equipment;
$LF$: Engine load factor of construction equipment.

The fuel consumption of the construction equipment is based on the equipment specification information (Table 3). The engine load factors are: Low (20–30% output, short distance travel); medium (30–40% output); waiting after stopping (when output is idling at 10%); and acceleration (100% output and continuous acceleration) [24]. When output is 40–50%, it indicates that the slope has suddenly changed, or there was a long distance of high driving resistance.

**Table 3.** Engine load factor by detailed operation.

|  | Stopping after Entering | Waiting after Stopping | Pouring Concrete | Leaving | Pumping Concrete |
|---|---|---|---|---|---|
| Engine load factor | Low (25%) | Waiting (10%) | Acceleration (100%) | Low (25%) | Acceleration (100%) |

Using the estimated amount of oil consumption, $CO_2$ emissions for detailed tasks were calculated using Equation (3):

$$CO_2 = F \times TE \times CF \times \frac{44}{12} \tag{3}$$

where:

$TE$: Petroleum conversion coefficient; and

$CF$: Carbon emission coefficient.

In this formula, the type of oil consumed is converted to equivalent oil consumed using an oil conversion factor, since the amount of $CO_2$ consumed differs depending on the type of oil. The converted value is multiplied by the $CO_2$ emission factor to estimate the $CO_2$ emission. The carbon molecular weight ratio to the carbon atom is multiplied by the constant 44/12 to calculate the final $CO_2$ emitted. In this study, the diesel oil conversion factor is 0.000845 toe/ton, and the emission factor is 0.837 ton-C/toe (where toe is ton of oil equivalent).

## 3.4. Comparative Analysis of Emissions

The average $CO_2$ emissions calculated based on the measured operating time and the planned operating time of the equipment were compared. Concrete mixer trucks do not have planned emissions, so they cannot be compared with actual $CO_2$ emissions. A descriptive analysis of the $CO_2$ emissions for the activities and the whole cycle was conducted to examine the variability of $CO_2$ emissions data (Table 4).

**Table 4.** Descriptive statistics of $CO_2$ emissions for measured operation time ($gCO_2$).

| Activity | | Stopping after Entering | Waiting after Stopping | Pouring Concrete | Leaving | Pumping Concrete | Cycle |
|---|---|---|---|---|---|---|---|
| Emissions for measured equipment operation time | Average | 85.1 | 178.4 | 3219.7 | 21.7 | 3819.7 | 7324.4 |
| | Median | 63.2 | 112.4 | 3118.4 | 18.7 | 3664.9 | 7220.7 |
| | Standard Deviation | 64.0 | 227.8 | 894.2 | 8.1 | 1061.4 | 1877.2 |
| | Coefficient of variation | 0.8 | 1.3 | 0.3 | 0.4 | 0.3 | 0.3 |
| | Kurtosis | 12.4 | 5.7 | 1.6 | 1.8 | −0.2 | 0.4 |
| | Skewness | 3.3 | 2.5 | 1.0 | 1.4 | 0.4 | 0.6 |
| | Min. Value | 44.5 | 9.4 | 1947.9 | 14.0 | 1879.6 | 3984.3 |
| | Max. Value | 334.8 | 889.6 | 5637.5 | 44.5 | 5903.4 | 11,333.1 |
| Emissions for planned equipment operation time | | 85.1 | 178.4 | 3219.7 | 21.7 | 3442.6 | 6947.5 |

The standard deviation, similar to the mean value, appeared to have a large value in the order of pumping concrete, pouring concrete, waiting after stopping, stopping after entering, and leaving. However, the coefficient of variation had the largest value of 1.3 for waiting after stopping, followed by stopping after entering with 0.8, leaving with 0.4, and pouring and pumping with 0.3. Therefore, the variation was relatively small in pouring concrete, pumping concrete, and leaving, while it was relatively large for waiting after stopping and stopping after entering. On the other hand, the coefficient of variation for the whole cycle was similar to that of pouring concrete and pumping concrete. Similarly, this is the reason why the emissions of both activities have a high emission rate.

Overall, the amount of $CO_2$ emissions during pouring works is influenced by the high emission ratios of pouring and pumping. Thus, the distribution of emissions and the total emissions from these two activities represent a relatively regular form of distribution. However, since the coefficient of variation exceeds 0.3 for both the detailed and the complete work, the analyzed deviation is relatively large. Moreover, several activities showed a relatively high coefficient of variation and a biased distribution in the emission data even though the emission rate was low. Thus, the estimation of $CO_2$ emissions using a single value may cause a significant error due to an insufficient reflection of the variability. In order to increase the reliability of $CO_2$ emission estimation, it is necessary to consider the variability of $CO_2$ emissions for each activity.

## 4. Predicting $CO_2$ Emissions Using a Monte Carlo Simulation

### 4.1. Monte Carlo Simulation

In this study, Monte Carlo simulation was adopted for probabilistic estimation of $CO_2$ emissions. Monte Carlo simulation was used because it is one of the essential techniques used for probabilistic analysis and it generates a stochastic model for uncertain variables and presents statistical results from the simulated experiment [25]. These statistical results enable effective decision-making under uncertain conditions.

In this study, Crystal Ball software was used to effectively perform Monte Carlo simulations. This software supports probabilistic model analysis and simulation testing of the data required for Monte Carlo simulations. For the application of Monte Carlo simulations in this study, $CO_2$ emissions of one work cycle were set as the result value. In addition, the variables that affect the resultant value were set at $CO_2$ emissions for each activity. Afterward, the Kolmogorov–Smirnov (K–S) test was performed on the activity data to derive the probability distribution model for $CO_2$ emissions from the activities. A correlation analysis was performed between the $CO_2$ emission data for each activity. The simulation was performed by generating random numbers using the correlation analysis results. As a result, the probability model of the resultant value was derived, and the $CO_2$ emissions were analyzed for each activity.

### 4.2. The Kolmogorov–Smirnov (K–S) Test

In this study, the K–S test was used to focus on a typical situation where the number of observations is higher than in certain situations where the equipment takes a long time to run. Comparing the K–S test with the Anderson–Darling test, the K–S test has the advantage that it is more sensitive in the middle distribution than in the tail [26]. In the K–S test, 14 types of probability distributions are available in the Crystal Ball software. The K–S test was used to test the goodness-of-fit of the emissions data from the measured operation time.

Since the test statistic of the critical value has a significance level of 0.05 and an N value of 21, it is confirmed to be 0.287 for the K–S test for goodness of fit [27]. The test results indicated that the null hypothesis of the K–S test method adopted in Table 5 for all the sub-tasks is lower than the critical value of 0.287. Therefore, the probability model of each variable derived from this section sufficiently reflects the distribution of actual data. Analysis showed that the distributions for stopping after entering, waiting after stopping, and leaving activities followed a log normal distribution, whereas pouring concrete followed a logistic distribution, and pumping concrete followed an extremum value distribution (Figure 1).

**Table 5.** Kolmogorov–Smirnov (K–S) test of $CO_2$ emissions by activities.

|  | Stopping after Entering | Waiting after Stopping | Pouring Concrete | Leaving | Pumping Concrete |
|---|---|---|---|---|---|
| Fit distribution | Log normal | Log normal | Logistics | Log normal | Extremum value |
| D | 0.108 | 0.108 | 0.137 | 0.158 | 0.076 |

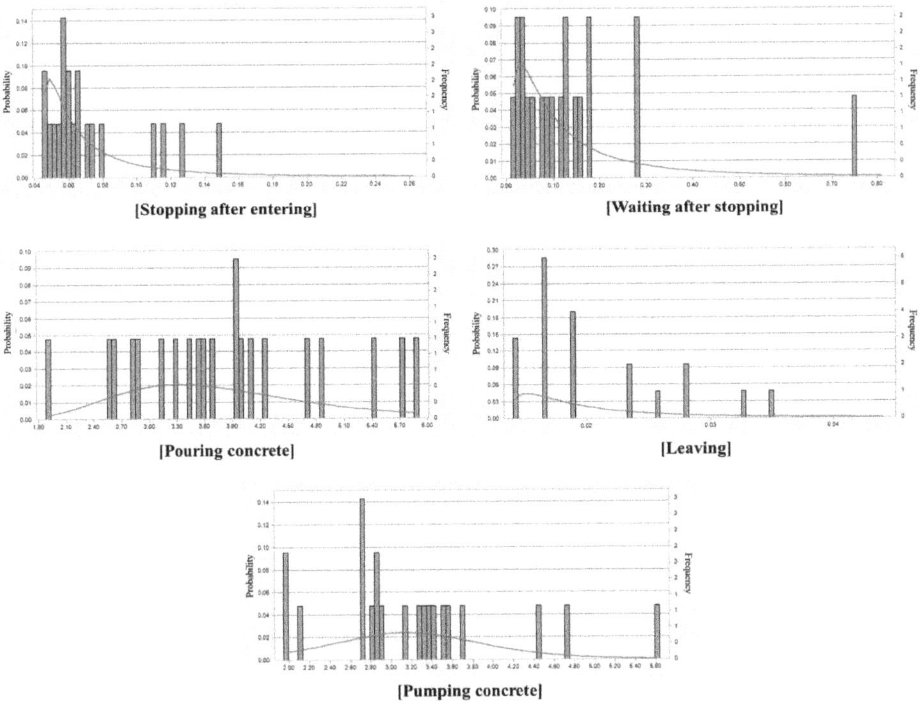

**Figure 1.** The plots of distribution by activities.

## 4.3. Correlation Analysis

In this section, correlation analysis is performed to derive the correlation coefficients for the emission amount of each activity and to reflect them in the simulation. The correlation analysis indicated the magnitude of the correlation among the activities through a correlation coefficient. The correlation coefficient range is from −1 to +1, and the signs of the coefficients indicated whether the relationship of the two variables is proportional or inversely proportional. The absolute value of the coefficient indicates the strength of the relationship. Thus, any value close to +1 or −1 indicates a strong correlation.

This study investigated the correlation between each variable, determined through the Spearman correlation coefficient. The Spearman correlation coefficient is a non-parametric measure of the degree of qualitative correlation of the increase or decrease when the distribution of two successive variables fits a normal distribution or provides rank-scale data. Therefore, the application of Spearman's correlation coefficient can be suitably considered a detailed work-specific fit distribution model that was verified in the previous section.

As a result of the correlation analysis with a significance level of 0.05, it was confirmed that the correlation between stopping after entering and pouring concrete, stopping after entering and leaving, pumping concrete and waiting after stopping, pumping concrete and pouring concrete was significant

(Table 6). The highest correlation was found between pumping concrete and pouring concrete with a correlation coefficient of 0.806 because the concrete pump car can pump concrete while concrete pouring work is underway. The correlations between pouring concrete and stopping after entering (0.524), and between stopping after entering and leaving (0.469) appeared to be strongly positive. The reason for the two positive correlations is that these activities are particularly affected by the proficiency of handling the equipment at a busy construction site and of the concrete mixer truck driver. The correlation between waiting after stopping and pumping was −0.469, indicating a strong negative correlation. This is because the waiting time is longer as the concrete pumping work is delayed.

**Table 6.** Results of correlation analysis between activities.

| Activity | Stopping after Entering | Waiting after Stopping | Pouring Concrete | Leaving | Pumping Concrete |
|---|---|---|---|---|---|
| Stopping after entering | 1 | | | | |
| Waiting after stopping | −0.040 | 1 | | | |
| Pouring concrete | 0.339 * | −0.079 | 1 | | |
| Leaving | 0.284 | 0.617 ** | 0.140 | 1 | |
| Pumping concrete | 0.387 * | −0.298 | 0.854 ** | 0.110 | 1 |

* Significant at 0.05 level of significance. ** Significant at 0.01 level of significance.

*4.4. Simulation Test Results and Analysis*

In this study, we executed the simulation test 10,000 times at a 95% confidence interval. To calculate the confidence interval, instead of a mathematical formula, we used an analytical bootstrapping method [28]. The simulation tests were performed by taking into account the probability distribution model obtained from the fit and the calculated correlation coefficient. The random values were generated considering the correlation to the probability distribution of a particular activity, and the results were obtained stochastically. The upper and lower limits of emissions were determined by adjusting the confidence interval.

The results of the simulation were similar to the corresponding figures of the measured operating time-based emissions data (Table 7). However, in the case of kurtosis and skewness, a significant difference was noticed in the corresponding figures in the $CO_2$ emission data based on the measured operating hours. The significant differences are because the tendency in the form of $CO_2$ emission distribution was strongly reflected by the performance of the repeated imitating experiment. The degree of deviation was significantly different from the corresponding values of the emission data based on the measured operating time of equipment.

As a result of the simulation, the $CO_2$ emission probability model of the pouring work follows the log normal distribution (Figure 2). At a 95% confidence interval, the lower limit of emissions is 4171.7 g $CO_2$, the upper limit is 11,705.4 g $CO_2$, and the average is 7329.7 g $CO_2$.

This study analyzed $CO_2$ emissions estimated by current deterministic methods as a single value using the derived probability distribution model. The results of the analysis of the estimated current $CO_2$ emissions were outside the 95% confidence interval of the probability distribution model or about 20% different from the mean value (Figure 3). In the probability distribution model, the probability of $CO_2$ emissions being less than those estimated by the CECET was found to be about 0.6% (i). Therefore, it is estimated that there is a risk of underestimating emissions because the level of emissions was estimated at a level that exceeds the 95% confidence interval of the simulation results. In addition, the probability of actual measured $CO_2$ emissions being less than the estimated $CO_2$ emissions using a bill of quantity was 73.2% in the probability distribution model (iii). The value of the result is within the 95% confidence interval of the probability distribution model. However, it is considered that the $CO_2$ emission calculation method using a bill of quantity may over-calculate compared to actual $CO_2$

emissions because the corresponding probability differs by more than 20% from the average value of the probability distribution model.

**Table 7.** Simulated test results (units are $gCO_2$).

| Activity | | Stopping after Entering | Waiting after Stopping | Pouring Concrete | Leaving | Pumping Concrete | Cycle |
|---|---|---|---|---|---|---|---|
| Simulated test result emissions | Average | 86.1 | 186.4 | 3164.6 | 22.4 | 3871.1 | 7329.7 |
| | Median | 64.4 | 98.4 | 3168.6 | 18.7 | 3.668.7 | 7165.1 |
| | Standard Deviation | 74.8 | 316.6 | 875.1 | 11.8 | 1186.3 | 186.9 |
| | Coefficient of variation | 0.9 | 1.7 | 0.3 | 0.5 | 0.3 | 0.3 |
| | Kurtosis | 107.7 | 226.5 | 4.2 | 47.4 | 6.0 | 3.5 |
| | Skewness | 7.58 | 10.5 | 0.0 | 4.8 | 1.2 | 0.5 |
| | Lower * | 45.0 | 11.0 | 1428.1 | 13.9 | 2152.3 | 4171.7 |
| | Upper * | 262.6 | 890.2 | 4941.6 | 53.5 | 6694.1 | 11,705.4 |
| Emissions based on planned equipment operation time | | 85.8 | 175.1 | 2718.2 | 21.6 | 3448.2 | 6831.4 |

* 95% confidence interval.

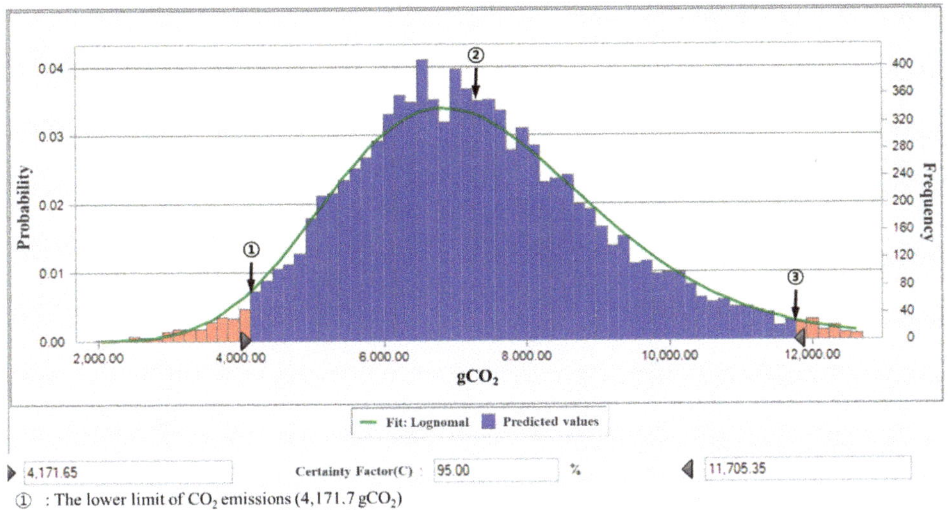


① : The lower limit of $CO_2$ emissions (4,171.7 $gCO_2$)
② : The average of $CO_2$ emissions (7,329.7 $gCO_2$)
③ : The upper limit of $CO_2$ emissions (11,705.4 $gCO_2$)
* Standard Deviation : 186.9 $gCO_2$

**Figure 2.** The probability model of $CO_2$ emissions in concrete pouring work according to simulation results.

Consequently, using actual measured data, probability-based estimation and analysis of $CO_2$ emissions will be more realistic and accurate for the construction planning phase as well as for the construction phase. The current methods of estimating $CO_2$ emissions are analyzed to have differences with actual $CO_2$ emissions, so further verification or improvement of the current methods of estimation is needed. It is also possible to assess quantitatively whether $CO_2$ emissions are likely to be over- or under-estimated by reviewing the probability distribution model for specific construction work.

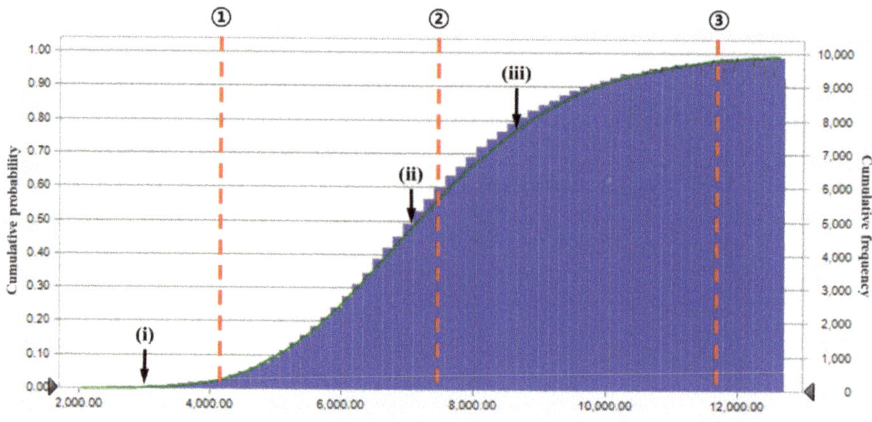



① : The lower limit of $CO_2$ emissions (4,171.7 $gCO_2$)
② : The average of $CO_2$ emissions (7,329.7 $gCO_2$)
③ : The upper limit of $CO_2$ emissions (11,705.4 $gCO_2$)
(i) : $CO_2$ emissions as calculated by Construction Equipment Expense Calculation Table (3,147.5 $gCO_2$, 0.6%)
(ii) : Amount of emissions equal to 50% of the probability model (7,134.8 $gCO_2$, 50.0%)
(iii) : $CO_2$ emissions as calculated by the bills of quantity (8,262.3 $gCO_2$, 73.2%)

**Figure 3.** Analysis of $CO_2$ emissions using the probability model.

## 5. Conclusions

This study gathered data on the operation time of construction equipment at the activity level and performed a probabilistic analysis using Monte Carlo simulations to determine the variability of $CO_2$ emissions during the construction phase.

This study used actual activity data to account for variability in the construction phase, and probabilistic estimation showed that the current deterministic methods are insufficient to predict $CO_2$ emissions at the planning and design phases. It is suggested that actual activity data collection and probabilistic analyses for activity data are necessary to present highly reliable $CO_2$ emission estimation. Risk information from the probabilistic emission calculation methods presented in this study may support decision makers to establish more realistic reduction goals in developing greenhouses gas and energy consumption reduction strategies.

This study dealt with the $CO_2$ emissions of the construction process stage within the life-cycle assessment (LCA) system boundaries. The $CO_2$ emissions from the construction process stage are known to account for less than those of the operational use stage in the building life cycle. Thus, the variability of the construction phase covered in this study may not have a significant impact on the $CO_2$ emissions of the whole life cycle. Nevertheless, the importance of embedded carbon, including $CO_2$ emissions from the current construction process stage, has been emphasized. In particular, according to the zero-energy building concept, which has recently become a reality, $CO_2$ emissions from the operational use stage will be close to zero in the future, and the calculation of carbon emissions from the construction process stage will be important.

However, this study was limited to the activity data gathered from concrete work only and considered variability at the activity level only in terms of construction equipment operation time. In future studies, we will gather further actual activity data during the construction process. In addition, we will derive the additional factors affecting $CO_2$ emissions during the construction phase and calculate the derived weights of the factors. The $CO_2$ emissions from the construction process will also be implemented as a probabilistic model with calculated weights.

**Author Contributions:** Data curation, C.N.; Formal analysis, D.L. and C.N.; Investigation, D.L. and C.N.; Methodology, D.L., G.K. and C.N.; Project administration, H.C and K.-I.K.; Supervision, K.-I.K.; Writing—original draft, D.L.; Writing—review & editing, D.L., G.K. and H.C.; funding acquisition, K.-I.K.

**Funding:** This research was funded by National Research Foundation of Korea grant number 2016R1A2B3015348 and Architecture & Urban Development Research Program funded by the Ministry of Land, Infrastructure and Transport of the Korean Government grant number 19AUDP-B121595-04.

**Acknowledgments:** This research was supported by National Research Foundation of Korea grant number 2016R1A2B3015348 and Architecture & Urban Development Research Program funded by the Ministry of Land, Infrastructure and Transport of the Korean Government grant number 19AUDP-B121595-04.

**Conflicts of Interest:** The authors declare no conflict of interest.

## References


1. United Nations Environment Programme. *Sustainable Buildings & Contruction Initiative: Information Note*; United Nations Environment Programme: Nairobi, Kenya, 2006.
2. Ibn-Mohammed, T.; Greenough, R.; Taylor, S.; Ozawa-Meida, L.; Acquaye, A. Operational vs. embodied emissions in buildings—A review of current trends. *Energy Build.* **2013**, *66*, 232–245. [CrossRef]
3. Zuo, J.; Pullen, S.; Rameezdeen, R. Green building evaluation from a life-cycle perspective in Australia: A critical review. *Renew. Sustain. Energy Rev.* **2017**, *70*, 358–368. [CrossRef]
4. Geng, S.; Wang, Y.; Zuo, J.; Zhou, Z.; Du, H.; Mao, G. Building life cycle assessment research: A review by bibliometric analysis. *Renew. Sustain. Energy Rev.* **2017**, *76*, 176–184. [CrossRef]
5. Daewoo Engineering & Construction Co., Ltd. *2014 Daewoo E&C Sustainability Report*; Daewoo Engineering & Construction Co., Ltd.: Seoul, Korea, 2014.
6. Gonzalez, V.; Echaveguren, T. Exploring the environmental modeling of road construction operations using discrete-event simulation. *Autom. Constr.* **2012**, *24*, 100–110. [CrossRef]
7. Intergovernmental Panel on Climate Change. 2006 IPCC Guidelines for National Greenhouse Gas Inventories Volume 1 General Guidance and Reporting. Available online: https://www.ipcc-nggip.iges.or.jp/public/2006gl/ (accessed on 22 July 2019).
8. Monahan, J.; Powell, J. An embodied carbon and energy analysis of modern methods of construction in housing: A case study using a lifecycle assessment framework. *Energy Build.* **2011**, *43*, 179–188. [CrossRef]
9. Roh, S.; Tae, S. An Analysis of the Life Cycle Embodied Carbon Emissions of Apartment Buildings Using Probabilistic Analysis Method. *J. Archit. Inst. Korea Struct. Constr.* **2017**, *33*, 65–72. [CrossRef]
10. Choi, D.; Jeon, H.; Shin, D.; Cho, K. A Study of Fuel Consumption and Carbon Dioxide Emission at the Construction Operation Phase of Apartment Buildings through Analysis on Construction Statements. *J. Archit. Inst. Korea Struct. Constr.* **2015**, *31*, 277–282.
11. Hong, J.; Shen, G.Q.; Feng, Y.; Lau, W.S.T.; Mao, C. Greenhouse gas emissions during the construction phase of a building: A case study in China. *J. Clean. Prod.* **2015**, *103*, 249–259. [CrossRef]
12. Jun, S.E.; Lim, T.K.; Lee, D.E. A System for Estimating $CO_2$ Emission using the Construction Project Schedule Information. *J. Archit. Inst. Korea Struct. Constr.* **2011**, *27*, 169–176.
13. Ahn, C.B.; Lee, S.H. Importance of Operational Efficiency to Achieve Energy Efficiency and Exhaust Emission Reduction of Construction Operations. *J. Constr. Eng. Manag.* **2012**, *139*, 404–413. [CrossRef]
14. Cass, D.; Mukherjee, A. Calculation of Greenhouse Gas Emissions for Highway Construction Operations by Using a Hybrid Life-Cycle Assessment Approach: Case Study for Pavement Operations. *J. Constr. Eng. Manag.* **2011**, *137*, 1015–1025. [CrossRef]
15. Zhang, H. Simulation-based estimation of fuel consumption and emissions of asphalt paving operations. *J. Comput. Civ. Eng.* **2014**, *29*, 04014039. [CrossRef]
16. Kim, B. Correlation Analysis on the Duration and CO Emission Following the Earth-Work Equipment Combination. *J. Korean Soc. Civ. Eng.* **2011**, *31*, 603–611.
17. Kim, B.; Lee, H.; Park, H.; Kim, H. Greenhouse gas emissions from onsite equipment usage in road construction. *J. Constr. Eng. Manag.* **2011**, *138*, 982–990. [CrossRef]
18. Yi, C.Y.; Gwak, H.S.; Lee, D.E. Stochastic carbon emission estimation method for construction operation. *J. Civ. Eng. Manag.* **2017**, *23*, 137–149. [CrossRef]
19. Yi, C.Y.; Gwak, H.S.; Lee, D.E. Methodology for Estimating Stochastic CO Emission for Construction Operation. *J. Archit. Inst. Korea Struct. Constr.* **2014**, *30*, 45–54. [CrossRef]

20. Nam, C.; Lee, D.; Kang, G.; Cho, H.; Kang, K.I. Activity based prediction of on-site $CO_2$ emissions containing uncertainty. In Proceedings of the 34th International Symposium on Automation and Robotics in Construction, Taipei, Taiwan, 28 June–1 July 2017; pp. 1085–1092.

21. Kang, G.; Kim, T.; Kim, Y.W.; Cho, H.; Kang, K.I. Statistical analysis of embodied carbon emission for building construction. *Energy Build.* **2015**, *105*, 326–333. [CrossRef]

22. Acquaye, A.; Duffy, A.; Basu, B. Stochastic hybrid embodied $CO_2$-eq analysis: An application to the Irish apartment building sector. *Energy Build.* **2011**, *43*, 1295–1303. [CrossRef]

23. Intergovernmental Panel on Climate Change. 2006 IPCC Guidelines for National Greenhouse Gas Inventories Volume 2 Energy. Available online: https://www.ipcc-nggip.iges.or.jp/public/2006gl/ (accessed on 22 July 2019).

24. Korea Institue of Construction Technology. *2018 Standard of Construction Estimate*; Korea Institue of Construction Technology: Goyang, Korea, 2018.

25. Touran, A.; Wiser, E.P. Monte Carlo technique with correlated random variables. *J. Constr. Eng. Manag.* **1992**, *118*, 258–272. [CrossRef]

26. Huynh, V.; Nakamori, Y.; Lawry, J.; Inuiguchi, M. *Integrated Uncertainty Management and Applications*; Springer Science & Business Media: Berlin, Germany, 2010.

27. Sachs, L. *Angewandte Statistik: Anwendung Statistischer Methoden*; Springer: Berlin/Heidelberg, Germany, 2013.

28. Oracle EPM Information Development Team. Oracle Crystal Ball Reference and Examples Guide, Release 11.1.2.4.850. 2017. Available online: https://docs.oracle.com/cd/E57185_01/CBREG/toc.htm (accessed on 22 July 2019).

*Article*

# Dynamic Lifecycle Assessment in Building Construction Projects: Focusing on Embodied Emissions

**Goune Kang [1],*, Hunhee Cho [2] and Dongyoun Lee [2]**

[1]  Korea Institute of Civil Engineering and Building Technology, 283 Goyang-daero, Ilsanseo-gu, Goyang, Gyeonggi 10223, Korea

[2]  School of Civil, Environmental and Architectural Engineering, Korea University, 145, Anam-ro, Seongbuk-gu, Seoul 02841, Korea

*  Correspondence: gounekang@kict.re.kr; Tel.: +82-31-910-0564

Received: 17 May 2019; Accepted: 5 July 2019; Published: 8 July 2019

**Abstract:** Understanding the structure and behavior of emissions in building systems is the first step toward improving the reliability of the environmental impact assessment of buildings. The shortcomings of current building lifecycle assessment (LCA) research is the lack of understanding of embodied emissions and static analysis. This study presents a methodology for the dynamic LCA of buildings, combined with the system dynamics technique. Dynamic factors related to recurrent embodied emissions are explored through a literature review. Applying the dynamic factors based on the review, a causal map and stock-flow diagram are invented. Collecting the field data and establishing the assumptions based on the literature, a case study is performed for the proposed model. As a result, through dynamic analysis, it was found that recurrent embodied emissions have a considerably different behavior from static ones during their whole life. Additionally, it was found that the environmental impacts changed by more than 10%, according to the variation of the users' required performance level in sensitivity analysis. This result thoroughly addressed the necessity and appropriateness of dynamic LCA. The dynamic LCA model developed in this study can contribute to the long-term behavioral understanding of the embodied environmental impacts of building LCA.

**Keywords:** lifecycle assessment; recurrent embodied carbon; system dynamics; buildings

---

## 1. Introduction

Buildings are key factors in energy consumption and global warming, consuming as much as 40% of the resources entering the global economy [1]. Life cycle assessment (LCA), which quantifies the environmental impacts during the whole life of a product, is a useful decision-making tool for green buildings [2]. It has been increasingly used by researchers to assist with decision-making for environment-related strategies and to reduce buildings' life cycle environmental impacts for the last 25 years [3]. Geng S. et al. [4] reviewed the literature related to building LCA using bibliometrics and showed that the number of publications related to building LCA grew steadily over the past 15 years and more rapidly since 2010.

Additionally, embodied carbon (EC) has recently become especially crucial for estimating the life-cycle carbon of buildings. EC refers to carbon dioxide emitted during the manufacture, transport, and construction of building materials, compared to operational carbon (OC), which means carbon dioxide emitted from the use of buildings, including heating, cooling, and lighting. In a recent review paper by Anand and Amor [5], numerous building LCA studies were explored, and it was shown that the areas of embodied energy had seen the maximum growth in the most recent years. Nevertheless, Pomponi and Moncaster [6] showed an extremely incomplete and short-sighted approach to life cycle studies through their meta-analysis of EC. According to the review, most studies only assess the manufacturing stages, often completely overlooking impacts occurring during the occupancy stage or at the end of life of the building. In other words, carbon emissions from construction and maintenance and repair (M&R) recurring during the building usage have been paid little attention so far. They emphasized that the LCA research community has the responsibility to address such shortcomings and work towards more complete and meaningful assessments. Considering the growing M&R proportion in the contemporary construction industry, embracing recurrent EC caused by M&R intervention activities in building lifecycle must not be overlooked.

Another limitation of LCA is that its analytical method is static. Traditional LCA methods are used to conduct building environmental impact assessment, with little consideration of influential factors that vary in time. Because the lifecycle of a building is quite long, such details have a significant influence on the accuracy of evaluation results [7]. Recently, dynamic LCA (DLCA) studies on buildings were conducted. Collinge [8–11] built DLCA research and identified significant time-related changes of variables, then developed a dynamic model based on the general LCA equation. Fouquet et al. [12] used a DLCA method to assess the global warming impact of three low-energy houses over time, considering the future electricity mix and innovation of materials for refurbishment. Su et al. [7] developed a dynamic assessment framework based on LCA principles and identified four dynamic building properties.

Previous building DLCA studies have accounted for economic and social progress (e.g., energy mix and Input–Output matrix) and characterization factors (e.g., global warming potential (GWP)). Su et al. [7] have proposed a DLCA framework that considers resident behavioral dynamics. They deal with the dynamic changes, focusing on the operational impacts, but embodied impacts from repair or replacement were not included. Fouquet et al. [12] added the materials used for refurbishment in the dynamic analysis, but they did not reflect the dynamic behavior in the usage stage, only dealing with the differences in the material types themselves. Overcoming the present limitations, this study developed the DLCA model, covering dynamic factors in recurrent embodied impacts in the usage stage and applying the system dynamics.

The objective of the research is to explore the impact of dynamic factors and their interaction with recurrent EC to obtain a better understanding of the recurrent EC in building LCA. This study utilizes system dynamics methodology for DLCA simulation. System dynamics involves the ability to represent and assess the dynamic complexity of the behavior that arises from the interaction of a factors in a system over time [13]. It is an advantageous modeling technique to reflect circular causality in simulation. Since a feedback loop is investigated while searching dynamic factors, system dynamics is considered a highly suitable method for DLCA simulations in this study. Meanwhile, LCA has defined a systematic set of procedures for compiling and examining the inputs and outputs of materials and energy and the associated environmental impacts, directly attributable to the functioning of a product or service system throughout its life cycle [14]. While this study focuses on the usage phase during the whole building life, the systematic process of the LCA principle is used for the equivalent calculation of embodied carbon dioxide in building systems.

The research procedure is as follows. First, dynamic factors related to recurrent EC are discussed in a literature review. According to condition-based management, intervention causing recurrent embodied emissions is determined by the performance. Several factors that impact the building performance are investigated. Second, the DLCA model for recurrent embodied impacts in the usage stage is developed using system dynamics. Applying the dynamic factors based on the review, a causal map and a stock-flow diagram are invented. The causal map is described for the relationship among the dynamic factors. The stock and flow diagram were invented for the DLCA simulation. Third, collecting the field data and establishing the assumptions based on the literature, a case study is performed to validate the proposed model. A base run is performed, with the optimization benchmarking the guideline data, substituted for real data. The base run simulation result, which displays the DLCA, is compared to the static analysis. Sensitivity analysis for embodied recurrent impacts is also performed, varying according to the change of the dynamic factors. This result can address the necessity and appropriateness of DLCA.

## 2. Literature Review

Most global policy has a tendency to focus on reducing OC since the majority of carbon emissions arise from the building usage [15]. However, recent studies have shown the growing significance of EC because much effort has already been invested into reducing OC [16]. Anand and Amor [5] showed that the research related to embodied energy has been significantly increased in recent years. Gavotsis and Moncaster [15] demonstrates that embodied carbon is also a significant proportion of the whole life impacts from buildings through a detailed case study of a low-energy school building. They also discussed about the uncertainties for post-construction stages. Brown et al. [17] pair attention to significant impacts arising from material production for buildings, and evaluated the importance of EC from refurbishment for operational energy reduction. This study displayed that EC for refurbishment actions take considerable share of the reduction of carbon dioxide emissions achieved by the refurbishment. Dixit [18] conducted a systematic survey of literature to identify parameters specifically affecting the recurrent embodied energy of buildings. It emphasized the need to standardize the parameters and quantify their uncertainties by developing appropriate models. Especially, service life, durability, aesthetics fashion, technology change, tenant change, and functional appropriateness factor, which are strongly related to supplied or required performance of building, were found as parameters affecting the recurrent embodied energy.

Several studies dealing with DLCAs have been recently conducted on the environmental impact of buildings. Negishi et al. [19] identified the time-dependent characteristics of a building system for performing DLCA. In this study, degradation of technical performances of building components, replacement and refurbishment with new technology factors were identified as a part of the key dynamic characteristics of building system. Su et al. [20] formalized four identified dynamic assessment elements in their recent study by examining the data transformation pathway in accordance with the standard LCA framework. Dynamics related to recurrent embodied consumption containing maintenance and demolition were included in the DLCA model.

Su et al. [7] organized dynamic factors in building LCA, classifying them into four categories: technological progress, variation in occupancy, dynamic characteristic factors, and dynamic weighting factors. Dynamic characteristic factors, dynamic weighting factors, and technological progress are related to calculation by the impact category. These factors reflect the variation of social change. The characteristic factors for impact categories could change according to the investigated relative impact for the substance. The weighting factors could vary according to the region, environmental issue, or social trend. Technological progress means the variability of the energy mix, which causes variation in the emission factor value. Variation in occupancy means the variability of the energy consumption behavior. Bringing the variation in the occupancy factor into the embodied sector, it can be expressed as a variation in the intervention behavior. Since this study deals only with the greenhouse effect among the impact categories, dynamic weighting factors are not considered in identifying the dynamic model variables. Moreover, this study primarily explores the factors related to a building itself.

## 3. Dynamic Factors in Recurrent EC

Since this study focuses on recurrent EC, it is concentrated on the variation in the intervention behavior to explore the dynamic factors. Intervention behavior is a dynamic factor, influenced by the performance of the building. That is, the performance of the building determines the time and amount of intervention. This relationship has been observed in the study of Tarefder & Rahman [21], which developed the lifecycle cost (LCC) model of airport pavement maintenance. In this study, they compared the performance improvement and LCC of the maintenance strategy using two condition indexes of airport pavement, with system dynamics. The initial condition, minimum acceptable condition, condition rise after maintenance, and deterioration were used as model parameters.

The relationship between performance and intervention also appears in existing asset management models. In the US federal facilities portfolios [22], the performance indicators that are used for maintenance decisions are the Facilities Condition Index (FCI) and the Facilities Rehabilitation Rate (FRR). In this model, if the FCI exceeds the acceptable condition level, intervention is applied. The FRR accounts for the required repairs and upgrades. The acceptable condition level will vary according to the mission, agency, organization and importance of specific facilities.

In addition, intervention behavior may differ from the occupants' characteristic, and likewise, the energy consumption is affected by the occupant behavior in terms of the operational impact. Observing that some home-owners invest heavily in repairs and improvements of their home, but some do not, Leather et al. [23] studied the reason why some occupants delay the maintenance of their home. In their report, different points of view in identifying repair needs, difficulties in finding trustworthy builders, financial problems and several other reasons are revealed as reasons for delayed maintenance. This report clearly showed the effect of occupants' characteristics on housing maintenance. Additionally, several studied the mentioned users' tendency to perform maintenance action in mechanical maintenance. Bitan and Meyer [24] examined users' tendencies to perform preventive maintenance actions. Shavartzon et al. [25] suggested a personalized alert agent for optimal user performance in computing, considering the users' preferences. It is possible to introduce the occupants' tendencies to perform the intervention in the housing maintenance field.

As a result of the literature review, dynamic factors to explain the relationship between intervention behavior and performance are identified: (Initial or current) performance, acceptable performance, deterioration, intervention rate, performance rise, and occupant's tendency. Applying values in a range for the application of several parameters in the existing literature shows the dynamic nature of these factors. This study utilizes these dynamic factors in DLCA modelling.

## 4. System Dynamics Model for DLCA

### 4.1. Causal Map and Feedback Loop

In the system dynamics methodology, a system may be represented as a causal map [26]. A causal map is a simple map of a system, with all its constituent components and their interactions. By capturing interactions and feedback loops, a causal map reveals the structure of a system. By understanding the structure of a system, it becomes possible to ascertain a system's behavior over a certain time period [27].

Since intervention, which causes recurrent embodied carbon during the building usage, strongly depends on the performance of a building, this study invented a causal map, containing the building performance. There is a simple causal relationship among the performance, intervention, and environmental impact. When the performance of a building deteriorates over time, intervention activities, such as repair or replacement, occur. These interventions generate environmental impacts and lead to a performance improvement. Thus, the performance of the building is recovered. If the building is in a sufficiently good condition, the intervention activity does not occur, and the environmental impact also does not occur. Without intervention for a while, the performance of the building is degraded again, and intervention activities occur again. This process will continue to be repeated during the life of buildings. This relationship is shown in the causal map below (Figure 1).

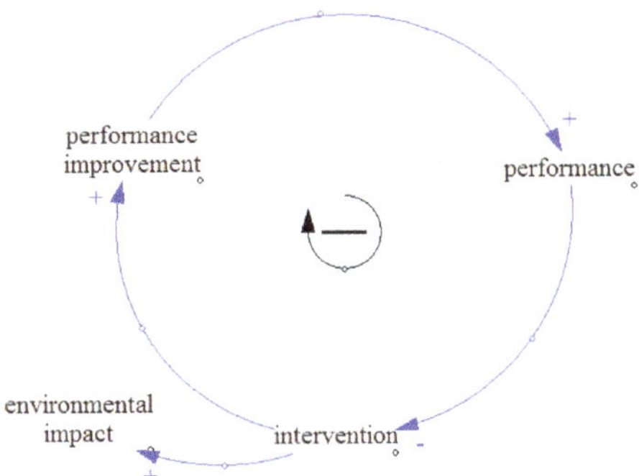


**Figure 1.** Causal map of building performance and environmental impact.

Especially, the causal map showed the feedback loop among the performance, performance improvement, and intervention. Combining the positive and negative relationships between the variables, the causal map has a negative feedback loop (balancing loop). This means that an increase of a parameter in the feedback loop consequently caused a decrease of itself. For example, a performance increase engenders zero or less intervention, and an intervention decrease causes a performance decrease. This balancing feedback loop suggests that the embodied environmental impact of the usage phase will converge, instead of diverting upward or downward.

Figure 2 displays the causal relationship in the initial construction and intervention. The typical environmental impact calculation process is also presented. The environmental impact is affected by the quantity of construction activity and emission factors, which are calculated by multiples of them. Additionally, recurring environmental impacts from intervention during the usage phase is influenced by the initial construction, because the quantity of materials and activities of intervention is likely to be determined based on the initial one.

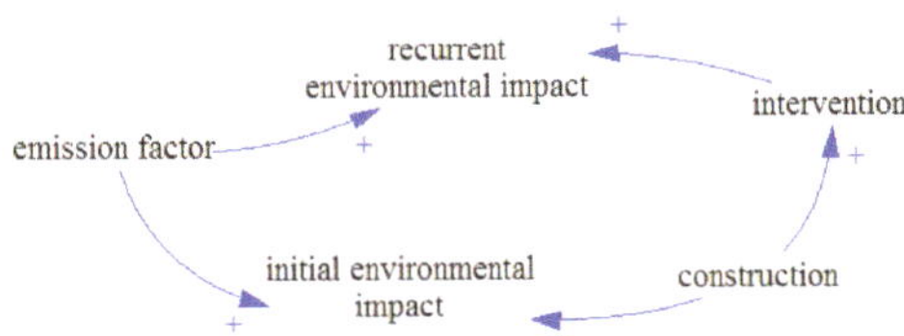


**Figure 2.** Causal map of construction activity and environmental impact.

### 4.2. Stock and Flow Diagram (SFD)

To perform a more detailed quantitative analysis, a causal map is transformed into an SFD. A stock and flow model aids in the study and analysis of the system in a quantitative way. A stock is the term for any entity that accumulates or depletes over time. A flow is the rate of change in a stock.

Considering the invented causal map, which contains the feedback loop, SFD was developed. In this process, additional parameters are required to answer several questions for the quantitative model: (1) When should intervention activities be applied? (2) How much intervention should be applied? (3) How will performance be improved by intervention activities?

(1) When should intervention activities be applied: ratio of supplied performance (SP) to required performance (RP), and maintenance of strategy strength.

Studies that estimate the existing M&R-related embodied environmental impacts are based on several guidelines that provide the durability of construction materials. The guidelines are based on 'InterNACHI's estimated life expectancy chart [28]' and the 'Study of life expectancy of home components' [29]. Previously, the material life expectancy was provided by the experimental data under the daily conditions, and the latter provides a value based on the data surveyed by the manufacturer. The purpose of these guidelines is mainly to provide a lifetime warranty of home components for residents. Many respondents noted that this lifetime is variable in terms of maintenance levels and emphasized that the lifetime of the component is changed before consumers are satisfied [30]. However, previous studies have deterministically decided the timing of interventions based on a guideline with a fixed lifecycle.

This study deals with the intervention times, as a variable that is determined by the performance. The intervention activity will occur when the supplied performance is lower than the required performance. Executing repair or replacement, the supplied performance can be improved to a level that satisfies the required performance. Therefore, this study uses the ratio of supplied performance to the required performance (SP/RP) as a parameter. If the value is greater than 1, the supply performance is good. If the value is less than 1, the supply performance does not satisfy the required performance.

In addition to the ratio of the supplied performance to the demand performance, this study introduces a parameter that can represent the maintenance strategy strength ($\lambda$). The occupants or building managers may want to prevent falling below the required performance through proactive maintenance, although the SP/RP is greater than 1. On the other hand, even if the SP/RP is less than 1, the building may not require immediate intervention to save maintenance costs or for other reasons, unless it is physically dangerous. $\lambda$ is a parameter representing this occupants' tendency. If $\lambda$ is greater than 1, a preemptive strategy is adopted. If $\lambda$ is less than or equal to 1, a cost-effective strategy is adopted. Intervention is performed when SP/RP becomes smaller than $\lambda$, as shown below.

$$\text{IF THEN ELSE (''SP/RP''} < \lambda, \text{intervention}, 0))$$

(2) How much intervention should be applied: repairing rate.

According to the Multi-Family Housing Management Act in Korea, the repairing rate is defined as a percentage of the cost of repairs for unpredictable partial damage or failure of a specific part of a building. The Enforcement Decree of the Multi-Family Housing Management Act in Korea proposed the repairing cycle and the repairing rate for each component in the 'Guidelines for a long-term repair program'. These guidelines propose what repairing cost is charged for fixed repairing periods for establishing long-term repairing plans.

The concept of this repairing rate can be used to estimate the extent of repair. The original rate is used to estimate the repairing cost, but in this study, it is used to estimate the quantity. In assessing environmental impacts, the volume of the material is the key factor that determines the amount of activity involved. The intervention quantity can be defined as the product of the initial construction quantity and the repairing rate, as shown below.

$$\text{intervention} = \text{initial construction} \times \text{repairing rate}$$

As the repair time varies according to the supplied performance and the required performance of the building, the repairing quantity will also be variably determined, depending on the performance of the period. In this study, the repairing rate is defined as the ratio of the required performance to the gap between the repairing performance and the supplied performance, as shown in the following. Combined with the definitions mentioned above, the final repairing time and quantity are defined as follows.

$$\text{repairing rate} = (\text{RP} - \text{SP})/\text{RP}$$
$$\text{IF THEN ELSE (''SP/RP''} < \lambda, \text{initial construction} \times \text{repairing rate}, 0))$$

(3) The extent to which the performance will be improved by intervention activities: ratio of intervention quantity to initial construction.

When intervention activities are performed, performance is improved. Performance improvement will increase the supplied performance of buildings and meet the required performance. The question is the extent to which the performance improvement is created by the intervention activity. As a result of investigating the asset management tools of the US federal agencies [22], indices are developed to accumulate and predict information about the supplied performance, the amount of intervention, and the timing of intervention. However, quantitative indicators of performance improvement by intervention are not found. This may be because the asset management or facility management deals with the building condition, focusing on the budget expenditure.

Quantitative indicators representing the degree of performance improvement are not utilized in the present asset management and facility management fields. It could be intuitively deduced that the performance improvement due to intervention is proportional to the amount of intervention. In this study, it is intended to quantify the performance improvement by introducing a coefficient ($\rho$) that can represent a proportional relationship according to this intuition.

$$\text{performance improvement} = (\text{intervention/initial construction}) \times \rho$$

The above definition is based on the assumption that the performance improvement is simply proportional to only the intervention ratio, compared to the initial construction quantity. $\rho$ is a coefficient indicating the relation between intervention quantity ratio and function improvement. Here, $\rho$ is a coefficient, with a yet unknown value. This value could be found through a model calibration in a further case study.

The SFD was created by adding the introduced variables, described above, and setting the stock and flow variable. The supplied performance is determined according to the stock variable, where the performance deterioration is the outflow variable, and the performance improvement is the inflow variable. In addition, the cumulative recurrent environmental impact is added as a stock variable to identify the total amount of recurrent environmental impacts during the building service life. The model includes the required service life parameter of the building, so that there is no intervention activity when the required service life is reached. The reference variable is a parameter generated for the calculation to refer to the initial construction value throughout the analysis period and does not affect the relationship between the variables in the environmental impact system of the usage stage. The SFD is shown in Figure 3. The bold arrows represent the feedback loop, analyzed in the causal map.

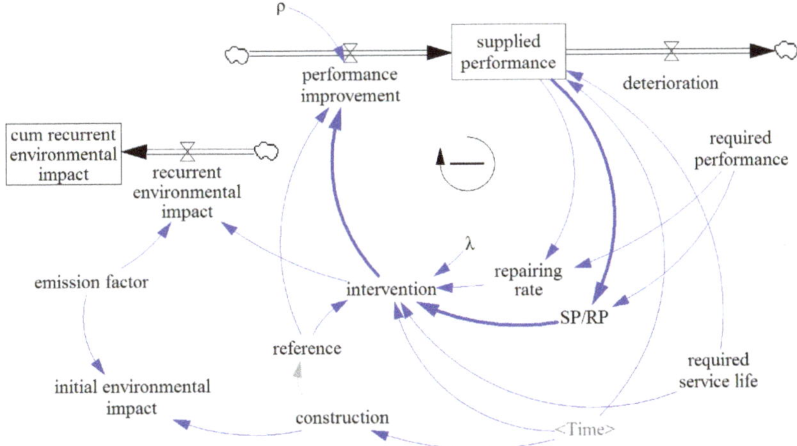


**Figure 3.** Stock and Flow Diagram of environmental impact in building usage stage.

## 5. Case Study

### 5.1. Data Information

The case building is a 30-story apartment building, with a reinforced concrete structure. This building consists of five types of floor plans, with a floor area of 146.48 m$^2$, 158.89 m$^2$, 161.25 m$^2$, 161.75 m$^2$, and 272.8 m$^2$, accommodating 117 generations. A summary of the building construction and the bill of quantity (BoQ) were collected for the building.

Referencing the existing research [31,32] that thoroughly analyzed the target materials of the Korean apartment building for LCA, this study determines the main materials for the case study: Rebar, ready-mixed concrete, concrete brick, cement, insulations, and paints are selected. However, existing studies tend to overlook the materials that are used in the finishing work, probably because of their lightweight and fewer emissions, comparing the structural components by cut-off rules. Since this study focuses on the intervention activities, some non-structural materials, which are repaired or replaced during the building usage, need to be considered. In this context, board, tile, granite, wallpaper, and flooring are also selected.

Using the National lifecycle inventory database by KEITI [33] and KICT [34], Greenhouse Gases (GHGs) emission factors are calculated. Three types of GHGs, carbon dioxide, methane, and nitrous oxide are considered among the lots of GHGs, considering the GWP and its importance, according to the IPCC report. Table 1 below shows the selected main materials and respective emission factors used in the case study.

**Table 1.** Main materials and emission factors.

| Item | Unit | Emission Factors (kgCO$_{2eq}$/Unit) | Reference |
|---|---|---|---|
| High tensile rebar | kg | $3.962 \times 10^{-1}$ | KICT |
| Ready-mixed concrete, 25-210-12 | m$^3$ | $4.001 \times 10^2$ | KEITI |
| Ready-mixed concrete, 25-240-15 | m$^3$ | $4.196 \times 10^2$ | KEITI |
| Concrete brick | kg | $1.222 \times 10^{-4}$ | KICT |
| Cement | ton | $1.049 \times 10^3$ | KEITI |
| EPS foam | kg | 2.024 | KICT |
| Tile | kg | $3.454 \times 10^{-1}$ | KICT |
| Granite | kg | 4.421 | KEITI |
| Gypsum board | kg | $1.383 \times 10^{-1}$ | KEITI |
| Mineral wool board | kg | 1.482 | KEITI |
| Paint, Water-soluble emulsion type | ton | $3.089 \times 10^2$ | KEITI |
| Paint, Amino-alkyd type | ton | $7.833 \times 10^2$ | KEITI |
| Paint, Epoxy type | ton | $3.158 \times 10^3$ | KEITI |
| MDF flooring | EA | 1.844 | KICT |
| PVC wallpaper | m$^2$ | 1.953 | KICT |

EA (each) = $1.2 \times 0.19 \times 0.01$ m$^3$ (size of one MDF flooring panel).

### 5.2. Assumptions in the Case Study

In EN 15978 [35], the system boundary for assessment is described, and the assessment modules in the building lifecycle are defined, as shown in Figure 4. Among the assessment modules, A1–A3 in the product stage, A4 and A5 in the construction process stage, B2–B5 in the usage stage, and C1–C4 in the end-of-life stage are relevant to the embodied impact (faded part in Figure 4). Since the DLCA model is focused on the recurrent EC, the case study is performed within modules B2–B5. The carbon emissions in modules A1–A5 were also calculated ahead of the simulation, because several variables in the DLCA model refer to initial construction information.

This study introduced several assumptions in the case study due to the availability of data. Table 2 shows the assumptions applied in the emissions calculation, with an LCA and Dynamic simulation.

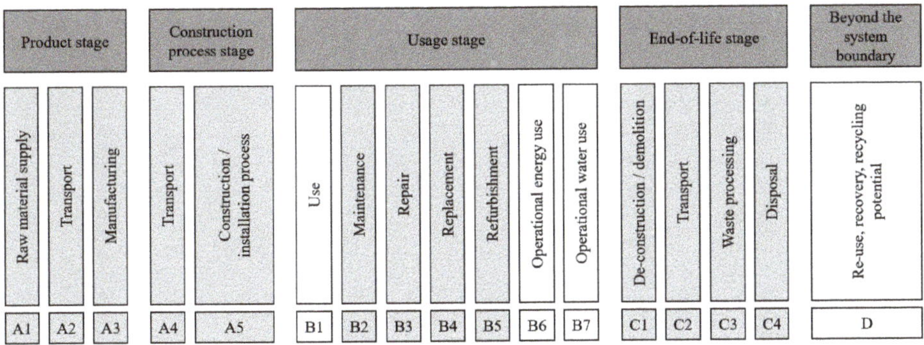

**Figure 4.** System boundary and assessment modules in building lifecycle assessment.

Building embodied emissions are caused by two factors, material use (the energy consumption in production) and construction activities (the energy consumption by transportation and execution). Material type and quantity data are quite well found in BoQ or the standard unit cost book. However, activity data, such as transportation or carriage-on-site, are not easy to obtain, since they are not well documented during construction. Moreover, M&R activity data, such as the type of repair machinery, work time, and productivity, are even harder to acquire, compared with new construction. Consequently, even if several activities cause carbon emissions, they are excluded from the calculation.

Meanwhile, several variables applied in SFD for dynamic simulation do not allow the real data to be acquired. M&R history data, such as the repair method, time, and quantity in relation to private buildings, are not yet recorded and organized in a database. Accordingly, the guidelines from Enforcement Decree of the Multi-Family Housing Management Act in Korea are assumed. Performance is quantified in a five-point scale, based on the condition grading system for building assets in the International Infrastructure Management Manual [36]. Additionally, there is no system for the performance assessment of private buildings, so deterioration (performance history) data are virtually assumed, referencing the existing study [37].

**Table 2.** Assumptions of the case study.

| Case Study | Variables | Assumption |
|---|---|---|
| LCA | Transportation | Delivery distance is assumed to be 20 km for every material. |
| | Lifting | Allocation rules with lifting weight are applied to common equipment (tower crane, construction lift). |
| | Carriage-on-site | While they are emission sources, they are not accounted for. |
| | M&R activities | While they are emission sources, they are not accounted for. |
| Simulation | Performance | Performance is quantified in a five-point scale. |
| | Deterioration | Deterioration curve refer to the existing literature [37]. |
| | M&R history | M&R history data are assumed from the guidelines of the Enforcement Decree of the Multi-Family Housing Management Act. |

Keshavarzrad et al. [37] presented deterioration curves using the National Asset Management Support data. They used the Markov model for the prediction of the deterioration trend. A Markov chain has been used in deterioration prediction in relation to bridges [38] and sewers [39]. Sharabah [40] introduced a weighting model for building assemblies using Markov process data, collected from Victorian city councils. Edirisinghe et al. [41] also applied the Markov model for building deterioration prediction.

The state transition probabilities and deterioration curves might be different for each building component. Components using materials with short service lives will have a high probability of transitioning to the next state (curve 2), and components using durable materials will have a significantly higher probability of staying current than transitioning to the next state (curve 1). Since a well-founded transition matrix that can be referred to in the literature is limited, two types of condition curves are virtually adopted. Figure 5 shows the two types of condition curves, resulting from the Markov model calculation. Stone work and interior finishing work, which recommend a repair time of more than 25 years in the guidelines, are applied to deterioration data form curve 1, assuming their performance is relatively slowly degraded. The plastering, tiling, painting and decoration works are applied data from curve 2.

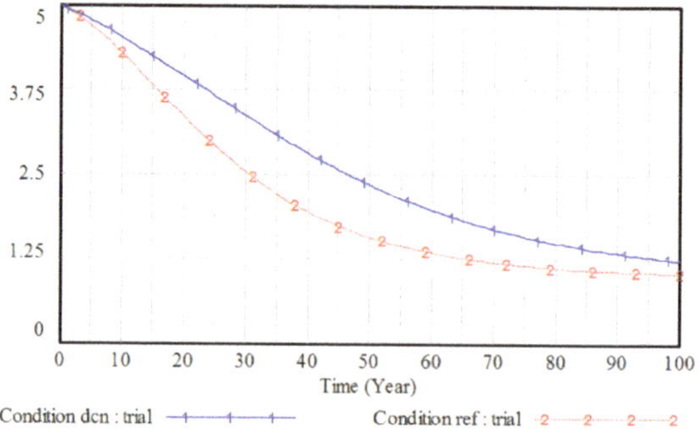


**Figure 5.** Condition curve of reference and discretion.

### 5.3. Simulation

Ideal data for the model calibration of this study would be the actual repair or replacement details, collected for the apartment building. In particular, information, such as the repair time, type of repair activity, and quantity of repair, is required for each object. However, there is still a lack of data accumulation on the repair content of buildings. Therefore, the guidelines in the Housing Act is applied, instead of the real-world value for the calibration of the model. Despite the fact that the guidelines of the Housing Act do not match the real world, calibration based on this reference means reflecting a long-term repair documentary plan in Korean apartment housing.

Comparing carbon emissions and cumulative carbon emissions during the building service life, based on the repairing time and rate from the Housing Act guidelines, $\lambda$, $\rho$, and the required performance parameters are optimized. Initial running with an initial $\lambda = 0.9$, the initial performance = 4 is simulated for the model variables for the 50-year study period. Optimized $\lambda$, $\rho$, and the required performance value from the calibration, with a payoff, is displayed in Table 3.

**Table 3.** Optimized $\lambda$, $\rho$, and the required performance from the calibration.

| Elements | $\lambda$ | $\rho$ | Required Performance | Simulations (Times) | Payoff |
|---|---|---|---|---|---|
| M-RO | 0.90 | 14.40 | 4.01 | 490 | $-1.70 \times 10^8$ |
| M-IF | 0.90 | 14.85 | 4.00 | 661 | $-2.67 \times 10^9$ |
| T-RO | 0.90 | 35.12 | 4.06 | 652 | $-2.52 \times 10^4$ |
| T-IW | 1.20 | 35.41 | 4.01 | 833 | $-6.80 \times 10^{10}$ |
| T-IF | 0.90 | 148.87 | 4.16 | 734 | $-1.19 \times 10^7$ |
| S-EW | 0.93 | 50.00 | 4.05 | 632 | $-1.20 \times 10^9$ |
| I-IC | 0.93 | 10.00 | 4.00 | 444 | $-3.30 \times 10^8$ |
| I-IW | 0.91 | 1.00 | 4.30 | 469 | $-2.95 \times 10^9$ |
| P-EW | 1.17 | 8.45 | 4.00 | 562 | $-4.80 \times 10^8$ |
| P-IC | 1.20 | 8.44 | 4.39 | 660 | $-1.14 \times 10^7$ |
| P-IW | 1.17 | 8.45 | 4.00 | 562 | $-8.97 \times 10^7$ |
| P-ST | 1.17 | 8.48 | 4.00 | 456 | $-1.25 \times 10^4$ |
| D-IF | 0.90 | 21.01 | 3.99 | 732 | $-1.40 \times 10^{12}$ |
| D-IW | 0.90 | 19.18 | 4.00 | 438 | $-3.38 \times 10^{10}$ |

M = plastering; T = tile; S = stone; I = interior finishing; P = painting; D = decoration; RO = roof; EW = exterior wall; IC = interior ceiling; IW = interior wall; IF = interior floor; ST = stair.

A base run with the parameters' input value from the calibration results is simulated. The total embodied GHG emissions are 1.97 million $kgCO_{2eq}$ in the dynamic model (Base run), whereas they are 1.528 million $kgCO_{2eq}$ in the static calculation from the Housing Act guidelines (Figure 6).

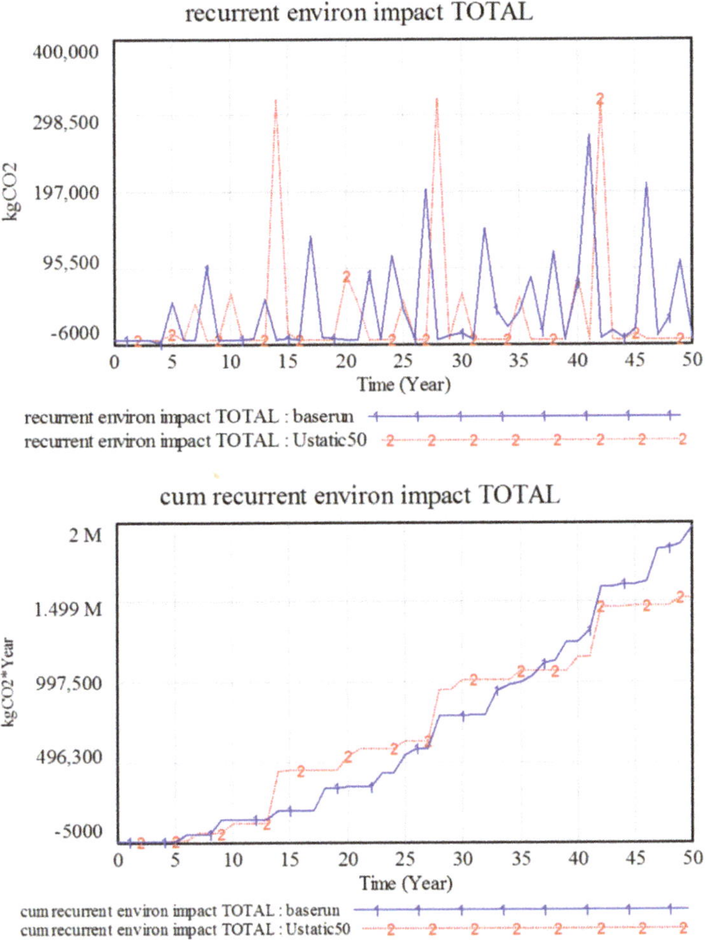


**Figure 6.** Comparison of the base run and Housing Act in the Greenhouse Gases of the usage stage.

Despite the optimization, the difference between the model and the guidelines is considerable. The dynamic model tends to have a lower intervention than the guideline value, at the beginning of the building service life, and this becomes more frequent as the building ages. Additionally, the intervention quantity is small at the beginning and grows larger at the end of the building's life in the dynamic model. While the static calculation shows a periodic value, ignoring the building condition, the dynamic model can describe the changing period and quantity of the intervention. Regardless of the benchmark data reality, this conceptual method presents the specific tool for dynamic LCA. It also shows the possibility of anticipating a realistic intervention and its environmental impact in advance.

### 5.4. Sensitivity Analysis

Using the developed dynamic model with a calibrated base run allows for an analysis of the variability involved in the environmental impact of construction activity in the usage stage. In this study, sensitivity analysis is performed on the variability of environmental impacts using two parameters. First, we analyzed the $\lambda$, which shows the intensity of the maintenance strategy. The $\lambda$ variability of the sensitivity analysis was set to $\pm10\%$ of the values derived from each trade and object in the calibration (Table 4). Sensitivity analysis applies the Monte Carlo simulation method, which generates random numbers, according to the probability distribution. Since we do not know about the specific probability distributions for $\lambda$, a simply uniform distribution was assumed. The analysis results are shown in Figure 7. With $\lambda$ changes, the total embodied emissions in the usage stage from the base run of 1970 $tCO_{2eq}$ could range from at least 1930 $tCO_{2eq}$ to 2140 $tCO_{2eq}$.

**Table 4.** Sensitivity analysis results concerning the maintenance strategy intensity ($\lambda$).

| | $\lambda$ | | GHGs Emission ($kgCO_{2eq}$) | | | |
|---|---|---|---|---|---|---|
| Component | Min | Max | Min | Max | Stdev | (Max-Min)/Avg |
| M-RO | 0.81 | 0.99 | 14,188 | 15,085 | 320 | 6.1% |
| M-IF | 0.81 | 0.99 | 55,720 | 60,821 | 1417 | 8.8% |
| T-RO | 0.81 | 0.99 | 168 | 201 | 11 | 17.6% |
| T-IW | 1.08 | 1.32 | 94,314 | 94,314 | 0 | 0.0% |
| T-IF | 0.81 | 0.99 | 4453 | 6881 | 991 | 44.4% |
| S-EW | 0.84 | 1.03 | 16,762 | 43,086 | 7296 | 101.8% |
| I-IC | 0.84 | 1.03 | 18,331 | 18,331 | 0 | 0.0% |
| I-IW | 0.82 | 1.00 | 44,641 | 44,641 | 0 | 0.0% |
| P-EW | 1.05 | 1.28 | 26,235 | 26,235 | 0 | 0.0% |
| P-IC | 1.08 | 1.32 | 4039 | 4039 | 0 | 0.0% |
| P-IW | 1.05 | 1.28 | 35,845 | 35,845 | 0 | 0.0% |
| P-ST | 1.05 | 1.29 | 134 | 134 | 0 | 0.0% |
| D-IF | 0.81 | 0.99 | 1,417,770 | 1,600,330 | 52,475 | 12.5% |
| D-IW | 0.81 | 0.99 | 195720 | 195,720 | 0 | 0.0% |
| Total | - | - | 1,930,130 | 2,140,150 | 53,182 | 10.6% |

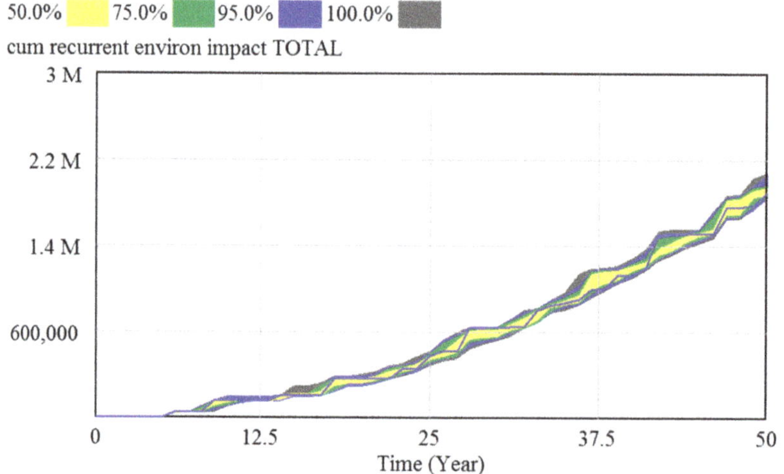



**Figure 7.** Sensitivity analysis results on maintenance strategy intensity ($\lambda$).

In other words, each $\lambda$ variation of 20% resulted in a total embodied emissions value, in the usage stage variation, of approximately 10.6%. The variation ratio of the embodied emissions for each trade and object in accordance with the $\lambda$ variation is described in Table 4. There were several components that did not change, regardless of the $\lambda$ variation, tile work in the interior wall, interior finishing work in the interior ceiling and wall, painting work in every object, and decoration work in the interior wall. Stone work is the most sensitive work affected by $\lambda$, even if it occupies a small percentage in the total emissions. Tile work in the interior floor, tile work in the roof, decoration work in the interior floor, plastering work in the interior floor, and plastering work in the roof were found to have a high sensitivity in sequence.

Next, the sensitivity is analyzed for the required performance of the building users. The required performance variation was ±10% of the values derived from each trade and object in the calibration in the sensitivity analysis (Table 5). The specific probability distribution of the required performance is also unknown, so a uniform distribution is assumed for the Monte Carlo simulation. The analysis results are shown in Figure 8. As the required performance changes, the total embodied emissions in the usage stage, from 1970 $tCO_{2eq}$ in the base run, can vary from at least 1848 $tCO_{2eq}$ to 2111 $tCO_{2eq}$.

**Table 5.** Sensitivity analysis results concerning the required performance.

| | Required Performance | | GHGs Emission ($kgCO_{2eq}$) | | | |
|---|---|---|---|---|---|---|
| Component | Min | Max | Min | Max | Stdev | (Max−Min)/Avg |
| M-RO | 3.61 | 4.42 | 13,733 | 15,344 | 425 | 11.2% |
| M-IF | 3.60 | 4.40 | 54,868 | 61,930 | 1644 | 12.3% |
| T-RO | 3.66 | 4.47 | 169 | 227 | 14 | 31.1% |
| T-IW | 3.61 | 4.41 | 75,569 | 139,221 | 14,571 | 64.7% |
| T-IF | 3.75 | 4.58 | 2929 | 7154 | 1074 | 97.6% |
| S-EW | 3.65 | 4.46 | 20,787 | 20,787 | 0 | 0.0% |
| I-IC | 3.60 | 4.40 | 18,331 | 18,331 | 0 | 0.0% |
| I-IW | 3.87 | 4.73 | 44,641 | 44,641 | 0 | 0.0% |
| P-EW | 3.60 | 4.40 | 26,235 | 26,235 | 0 | 0.0% |
| P-IC | 3.95 | 4.83 | 4039 | 4039 | 0 | 0.0% |
| P-IW | 3.60 | 4.40 | 35,845 | 35,845 | 0 | 0.0% |
| P-ST | 3.60 | 4.40 | 134 | 134 | 0 | 0.0% |
| D-IF | 3.59 | 4.39 | 1,357,820 | 1,622,000 | 62,445 | 18.1% |
| D-IW | 3.60 | 4.40 | 195,720 | 195,720 | 0 | 0.0% |
| Total | - | - | 1,848,360 | 2,111,350 | 61,997 | 13.4% |

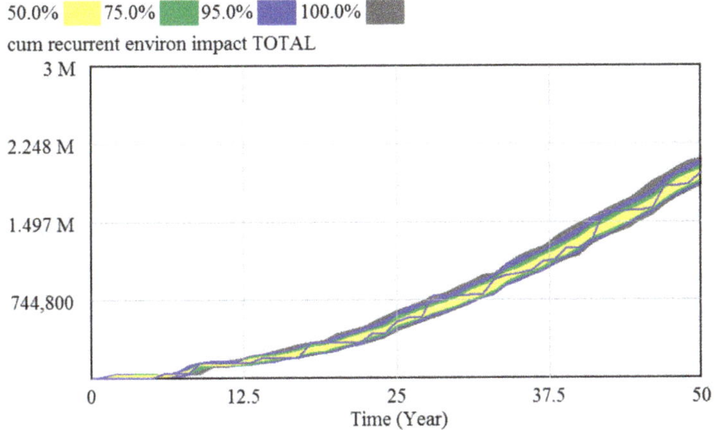


**Figure 8.** Sensitivity analysis results concerning the required performance.

Each required performance variation of 20% resulted in the embodied emission variation of approximately 13.4% during the usage stage. The variation ratio of the embodied emissions for each trade and object, in accordance with the required performance variation, is described in Table 5. There were several components that did not change, regardless of the required performance, stone work, interior finishing work in the interior ceiling and wall, painting work in every object, and decoration work in the interior wall. Tile work in the interior floor is the most sensitive work affected by the required performance, even if it occupies a small percentage of the total emissions. Tile work in the interior wall, tile work in the roof, decoration work in the interior floor, plastering work in the interior floor, and plastering work in the roof were found to have a high sensitivity in sequence.

Under the same conditions, the required performance was found to be more sensitive to changes in emissions for the components and total value than $\lambda$. The rate of change in tile work tends to have the largest ratio, and the painting work shows no change for both $\lambda$ and the required performance. Stone work was the most sensitive work in terms of $\lambda$ but did not show any variation in the required performance.

Adding a larger amount of data accumulation and elaborate assumptions, this analysis can be a useful tool for preliminarily verifying the variability of the repair scenarios for each trade and object. In addition, the evaluation of the life cycle environmental impacts, taking into account the variability of the intervention scenarios, could be useful for risk analysis in building environmental management.

## 6. Conclusions

Maintenance activities mainly depend on changes in building performance over time, but the static methodology of traditional LCA does not take this variability into account. Besides, current embodied environmental impact assessment has tended to focus on the structural materials, and decorating materials used in M&R are overlooked. This study combines system dynamics with LCA to assess the recurrent embodied carbon emissions. It visualizes the long-term behavior of the environmental impacts caused by the feedback structure between the building performance and intervention. Additionally, the variability of the environmental impacts, from the changes in users' required performance and maintenance strategy intensity, is analyzed.

The results of this study show the possibility of acquiring a great amount of important information that could not be captured by the traditional LCA methodology. It shows how the estimates of environmental impacts, assuming the application of fixed repair cycles and ratios, differ from the actual performance-based maintenance concept. This implies that it is possible to provide statistical information on the uncertainty of the forecast when estimating the emissions, and it can support the reliability of the environmental impact estimation of buildings.

This study has some limitations in terms of data collection and utilization. First, because the site of the case building was closed, it was impossible to collect activity data on the building. A simulation was performed using assumptions and data from the literature, and analysis based on the measured data should be performed in the next study. A lot of activity data, generated in M&R work during the usage stage, are not accessible and difficult to measure. Nevertheless, data acquisition for several sample works would be a meaningful starting place, before considering whole buildings.

Additionally, because of the lack of available information about performance degradation, it was replaced with data from the existing study. While the main interest of this study is the relationship between the building performance and the environmental impact, the accuracy of the deterioration data is important in the model, since performance is a major variable determining the timing of M&R. It is necessary to explore the performance deterioration, which is suitable for the LCA target building, through the accumulation of performance evaluation for representative buildings.

**Author Contributions:** Conceptualization, G.K.; methodology, G.K.; validation, G.K., H.C. and D.L.; formal analysis, G.K. and H.C.; investigation, G.K., H.C., and D.L.; data curation, G.K. and D.L.; writing—original draft preparation, G.K.; writing—review and editing, H.C. and D.L.; visualization, G.K. and D.L.; supervision, G.K.; project administration, H.C.; funding acquisition, H.C.

**Funding:** This research was funded by National Research Foundation of Korea grant number 2016R1A2B3015348 and Korea Institute of Civil Engineering and Building Technology grant number 20180543-001.

**Conflicts of Interest:** The authors declare no conflict of interest.

## References


1.  Ji, C.; Hong, T.; Jeong, J.; Kim, J.; Lee, M.; Jeong, K. Establishing environmental benchmarks to determine the environmental performance of elementary school buildings using LCA. *Energy Build.* **2016**, *127*, 818–829. [CrossRef]
2.  Zuo, J.; Pullen, S.; Rameezdeen, R.; Bennetts, H.; Wang, Y.; Mao, G.; Zhou, Z.; Du, H.; Duan, H. Green building evaluation from a life-cycle perspective in Australia: A critical review. *Renew. Sustain. Energy Rev.* **2017**, *70*, 358–368. [CrossRef]
3.  Atmaca, A. Life cycle assessment and cost analysis of residential buildings in south east of Turkey: Part 1—Review and methodology. *Int. J. Life Cycle Assess.* **2016**, *21*, 831–846. [CrossRef]
4.  Geng, S.; Wang, Y.; Zuo, J.; Zhou, Z.; Du, H.; Mao, G. Building life cycle assessment research: A review by bibliometric analysis. *Renew. Sustain. Energy Rev.* **2017**, *76*, 176–184. [CrossRef]
5.  Anand, C.K.; Amor, B. Recent developments, future challenges and new research directions in LCA of buildings: A critical review. *Renew. Sustain. Energy Rev.* **2017**, *67*, 408–416. [CrossRef]
6.  Pomponi, F.; Moncaster, A. Embodied carbon mitigation and reduction in the built environment—What does the evidence say? *J. Environ. Manag.* **2016**, *181*, 687–700. [CrossRef] [PubMed]
7.  Su, S.; Li, X.; Zhu, Y.; Lin, B. Dynamic LCA framework for environmental impact assessment of buildings. *Energy Build.* **2017**, *149*, 310–320. [CrossRef]
8.  Collinge, W.O.; Liao, L.; Xu, H. Enabling dynamic life cycle assessment of buildings with wireless sensor networks. In Proceedings of the 2011 IEEE International Symposium on Sustainable Systems and Technology, Chicago, IL, USA, 16–18 May 2011; pp. 1–6.
9.  Collinge, W.; Landis, A.E.; Jones, A.K.; Schaefer, L.A.; Bilec, M.M. Indoor environmental quality in a dynamic life cycle assessment framework for whole buildings: Focus on human health chemical impacts. *Build. Environ.* **2013**, *62*, 182–190. [CrossRef]
10. Collinge, W.O.; Landis, A.E.; Jones, A.K.; Schaefer, L.A.; Bilec, M.M. Dynamic life cycle assessment: Framework and application to an institutional building. *Int. J. Life Cycle Assess.* **2013**, *18*, 538–552. [CrossRef]
11. Collinge, W.O.; Landis, A.E.; Jones, A.K.; Schaefer, L.A.; Bilec, M.M. Productivity metrics in dynamic LCA for whole buildings: Using a post-occupancy evaluation of energy and indoor environmental quality tradeoffs. *Build. Environ.* **2014**, *83*, 339–348. [CrossRef]
12. Fouquet, M.; Levasseur, A.; Margni, M. Methodological challenges and developments in LCA of low energy buildings: Application to biogenic carbon and global warming assessment. *Build. Environ.* **2015**, *90*, 51–59. [CrossRef]
13. Rahim, F.H.A.; Hawari, N.N.; Abidin, N.Z. Supply and demand of rice in Malaysia: A system dynamics approach. *Int. J Sup. Chain. Mgt.* **2017**, *6*, 1–7.
14. ISO 14040:2006 Environmental Management-Life Cycle Assessment-Principles and Framework. Available online: https://www.iso.org/standard/37456.html (accessed on 8 July 2019).
15. Gavotsis, E.; Moncaster, A. Improved embodied energy and carbon accounting: recommendations for industry and policy. *Athens J. Tech. Eng.* **2015**, *2*, 9–23. [CrossRef]
16. Kang, G.; Kim, T.; Kim, Y.W.; Cho, H.; Kang, K.I. Statistical analysis of embodied carbon emission for building construction. *Energy Build.* **2015**, *105*, 326–333. [CrossRef]
17. Brown, N.W.O.; Olsson, S.; Malmqvist, T. Embodied greenhouse gas emissions from refurbishment of residential building stock to achieve a 50% operational energy reduction. *Build. Environ.* **2014**, *79*, 46–56. [CrossRef]
18. Dixit, M.K. Life cycle recurrent embodied energy calculation of buildings: A. review. *J. Clean. Prod.* **2019**, *209*, 731–754. [CrossRef]
19. Negishi, K.; Tiruta-Barna, L.; Schiopu, N.; Lebert, A.; Chevalier, J. An operational methodology for applying dynamic Life Cycle Assessment to buildings. *Build. Environ.* **2018**, *144*, 611–621. [CrossRef]
20. Su, S.; Li, X.; Zhu, Y. Dynamic assessment elements and their prospective solutions in dynamic life cycle assessment of buildings. *Build. Environ.* **2019**, *158*, 248–259. [CrossRef]

21. Tarefder, R.A.; Rahman, M.M. Development of system dynamic approaches to airport pavements maintenance. *J. Transp. Eng.* **2016**, *142*, 04016027. [CrossRef]
22. Davis, J.S.; Cable, J.H. *Key Performance Indicators for Federal Facilities Portfolios, Federal Facilities Council Technical Report Number 147*; The National Academies Press: Washington, DC, USA, 2005.
23. Leather, P.; Littlewood, M.; Munro, M. *Home-Owners and Housing Repair: Behavior and Attitudes*; Joseph Rowntree Foundation (JRF): York, UK, 1998.
24. Bitan, Y.; Meyer, J. Self-initiated and respondent actions in a simulated control task. *Ergonomics* **2007**, *50*, 763–788. [CrossRef]
25. Shavartzon, A.; Azaria, A.; Kraus, S.; Goldman, C.V.; Meyer, J.; Tsimhoni, O. Personalized alert agent for optimal user performance. In Proceedings of the Thirtieth AAAI Conference on Artificial Intelligence, Phoenix, AZ, USA, 12–17 February 2016.
26. Sterman, J.D. *Business Dynamics: Systems Thinking and Modeling for a Complex World*; McGraw: New York, NY, USA, 2000.
27. Meadows, D. *Thinking in Systems: A Primer*; Earthscan: London, UK, 2008.
28. International Association of Certified Home Inspectors (InterNACHI). InterNACHI's Standard Estimated Life Expectancy Chart for Home. Available online: https://www.nachi.org/life-expectancy.htm (accessed on 15 May 2019).
29. Seiders, D.; Ahluwalia, G.; Melman, S.; Quint, R.; Chaluvadi, A.; Liang, M.; Silverberg, A.; Bechler, C.; Jackson, J. *Study of Life Expectancy of Home Components*; National Association of Home Builders (NAHB), Bank of America (BoA) Home Equity: Washington, DC, USA, 2007.
30. Seiders, K.; Voss, G.B.; Godfrey, A.L.; Grewal, D. SERVCON: Development and validation of a multi-dimensional service convenience scale. *J. Acad. Mark. Sci.* **2007**, *35*, 144–156. [CrossRef]
31. Shin, S.; Tae, S.; Woo, J.; Roh, S. The development of environmental load evaluation system of a standard Korean apartment house. *Renew. Sustain. Energy Rev.* **2011**, *15*, 1239–1249. [CrossRef]
32. Choi, D.S.; Chun, H.C.; Ahn, J.Y. Prediction of Environmental Load Emissions from an Apartment House of Construction Phase through the Selection of Major Materials. *J. Archit. Inst. Korea Plan. Des.* **2012**, *28*, 237–246.
33. Korea Environmental Industry & Technology Institute (KEITI). Korea Life Cycle Inventory Database. 2004. Available online: http://www.epd.or.kr/lci/lciDb.do (accessed on 15 May 2019).
34. Korea Institute of Civil Engineering and Building Technology. *The Final Report of National DB on Environmental Information of Building Materials*; KICT: Goyang, Korea, 2008.
35. EN 15978:2011 Sustainability of construction works-Assessment of the environmental performance of buildings-Calculation methods. Available online: https://shop.bsigroup.com/ProductDetail?pid=000000000030256638 (accessed on 8th July 2019).
36. IPWEA. International Infrastructure Management Manual. Available online: https://www.ipwea.org/publications/ipweabookshop/iimm (accessed on 8th July 2019).
37. Keshavarzrad, P.; Setunge, S.; Zhang, G. Deterioration prediction of building components. In Proceedings of the 2014 Sustainability in Public Works Conference, Tweed Heads, Australia, 27–29 July 2014; pp. 1–9.
38. Madanat, S.; Mishanlani, R.; Ibrahim, W.H.W. Estimation of infrastructure transition probabilities from condition rating data. *J. Infrastruct. Syst.* **1995**, *19*, 120–125. [CrossRef]
39. Baik, H.S.; Jeong, H.S.; Abraham, D.M. Estimating Transition Probabilities in Markov Chain-Based Deterioration Models for Management of Wastewater Systems. *J. Water Resour. Plan. Manag.* **2006**, *132*, 15–24. [CrossRef]
40. Sharabah, A.; Setunge, S.; Zeephongsekul, P. A Reliability Based Approach for Service Life Modeling of Council Owned Infrastructure Assets. In Proceedings of the ICOMS Asset Management Conference, Melbourne, Australia, 22–25 May 2007.
41. Edirisinghe, R.; Setunge, S.; Zhang, G. Markov Model-Based Building Deterioration Prediction and ISO Factor Analysis for Building Management. *J. Manag. Eng.* **2015**, *31*, 04015009. [CrossRef]

*Article*

# Integration of a Balanced Scorecard, DEMATEL, and ANP for Measuring the Performance of a Sustainable Healthcare Supply Chain

**Eko Budi Leksono** [1,2,*], **Suparno Suparno** [2] **and Iwan Vanany** [2]

[1] Department of Industrial Engineering, Universitas Muhammadiyah Gresik, Gresik 61121, Indonesia
[2] Department of Industrial Engineering, Institut Teknologi Sepuluh Nopember (ITS),
Surabaya 60111, Indonesia
* Correspondence: eko_budileksono@umg.ac.id; Tel.: +62-81-2324-8573

Received: 29 May 2019; Accepted: 27 June 2019; Published: 2 July 2019

**Abstract:** The main purpose of this study is to develop a sustainable healthcare supply chain performance measurement (SHSCPM) model, which simultaneously considers intangible characteristics and sustainability aspects to ensure customer and/or stakeholder satisfaction. This model combines a balanced scorecard (BSC) with a decision-making trial and evaluation laboratory (DEMATEL) and analytical network process (ANP). After the arrangement and classification of perspectives and indicators from a literature review, the strategy map of the BSC is designed with DEMATEL. Furthermore, this study used a survey with in-depth interviews of seven expert respondents to make a pair-wise comparison between perspectives and indicators in order to determine the weights of indicators, perspectives, and sustainability aspects on ANP. The study finds the following. First, the innovation and learning perspective that reflects intangible assets has the most influence on the others but is not important, while the customer perspective has the most importance for SHSC performance. Second, the economic aspect has the greatest weight, followed by social and environmental aspects. Finally, indicators of the financial and customer perspectives as drivers of SHSC performance consist of profit, quality of service, revenue, customer satisfaction, and stakeholder satisfaction. Further, indicators of the economic aspect of sustainability have the most effect on SHSC performance, followed by social and environmental aspects. Furthermore, human resources, as an intangible asset and key factor in social aspects, are main factor in improving SHSC performance.

**Keywords:** sustainable healthcare supply chain; performance measurement; performance of perspectives and indicators; sustainability aspects; intangible characteristics

---

## 1. Introduction

In recent decades, the service sector's have made a significant contribution to gross domestic product (GDP) and affects the global economy [1]. This phenomenon has driven the development of the service supply chain (SSC) concept [2–4]. Methods of supply chain and performance measurement can be used to develop healthcare performance [5]. The healthcare supply chain (HSC) is an implementation of the SSC into healthcare businesses. HSC implementation stimulates healthcare service providers to collaborate with supply chain actors to ensure customer and/or stakeholder satisfaction [2] and cost reduction [6]. The HSC performance measurement (HSCPM) can be used to measure the success of the collaboration between healthcare service providers with other HSC actors.

In 2013, the service business contribution to Indonesia's GDP was 39.87% [7]. In the period from 2012 - 2015, the service business contribution to the Product Domestic Regional Brute (PDRB) of the East Java Province was an average of 71.11%. The service contributions show increasing trends in:

tourism, hotels and restaurants, professional services, banking and insurance, and healthcare [8]. On the other hand, pressure from globalization and stakeholders requires the HSC operation to give attention to the environment, community, economic, and intangible assets [9]. Implementation of sustainable service supply chain could minimize the negative effects of operation on the environment and society and maximize profits [10]. Based on the contribution to GDP and/or the PDRB and as well as globalization and stakeholder pressure, healthcare businesses need a sustainable healthcare supply chain (SHSC), especially in the Province of East Java, Indonesia. SHSC implementation is expected to be able to improve the performance of healthcare businesses. Besides, the development of SHSC is feasible because healthcare businesses are concerned with improving public health and wellbeing.

The SHSC performance requires a performance measurement system. Sustainable supply chain performance is defined as a company's capacity to reduce the use of materials, energy, or water and to find more eco-efficient solutions by improving the supply chain [11]. Performance measurement of the SHSC can be adapted from the sustainable service supply chain performance measurement models. The performance of sustainability aspects in the supply chain are often difficult to measure [12]. The scope of sustainable service supply chain performance measurement (SSSCPM) includes environmental management, social responsibility, management of health, safety and risk, and customer management [13]. SSSCPM is still dominated by economic and environmental aspects, and less by social aspects [14]. SHSC performance measurement (SHSCPM) is also still oriented toward economic and environmental (green) aspects, such as the environmental supply chain performance measurement [15], life cycle assessment, and life cost assessment [16]. The social aspect of SHSC is important and is becoming a key objective within the SHSC because healing patients is the primary outcome of the HSC, and the social aspect of the SHSC is concerned with human aspects [17]. Social aspects are similar to intangible assets and are the main characteristic of the services sector [18]. Thus, SHSCPM implementation must consider both intangible assets and sustainability aspects.

The concept of the SHSCPM has been developed and reported. There has been is littleness integration between economic, environmental, and social aspects simultaneously. Furthermore, SHSCPM research is suitable for development in East Java Province, Indonesia. This study aims to propose an SHSCPM by combining the balanced scorecard (BSC) with the decision-making trial and evaluation laboratory (DEMATEL) and the analytical network process (ANP), while with simultaneously considering to the intangibility characteristics and sustainability aspects. The BSC will be combined with DEMATEL to design a strategy map that represents the relationships between perspectives and indicators. The relationships are designed based on the level of importance and influence by DEMATEL. Finally, the BSC strategy map will be used as input in the design of the ANP structure model. The ANP structure model will determine the weights of the performance indicators, performance perspectives, and sustainability aspects.

Integration of the BSC with DEMATEL and ANP for SHSCPM can improve previous SHSCPM models, which SHSCPM has designed by integrating of the BSC and DEMATEL [9]. This study aims to add the ANP method into the SHSCPM model, along with BSC and DEMATEL, where the ANP output is the weight of the performance indicators based on the BSC strategy map. Finally, integration of the BSC with DEMATEL and ANP for SHSCPM can enrich the SHSCPM literature based on BSC.

This article is organized as follows. This study discusses the BSC and DEMATEL in the SHSCPM, including a strategy map framework from a previous study in Section 2. Section 3 presents the DEMATEL and ANP for performance measurement, especially SHSC. Then, the methodology that describes the research stages and the structural model are presented in Section 4. Section 5 discusses data from an expert preference questionnaire survey of professionals in the healthcare business; this questionnaire is used in the processes of the ANP, and in weighting the performance indicators, performance perspectives, and sustainability aspects. Next, the limitations of this study are discussed in Section 6. The last section presents conclusions that describe the phenomenon of SHSCPM and future research opportunities.

## 2. BSC and DEMATEL on SHSCPM

The actors in the healthcare supply chain consist of a producer, suppliers, healthcare service providers, and patients [19,20]. The orientation of HSCPM comprises eleven elements, i.e.: continuous improvement and customer satisfaction [21], demand, customer relationship, supplier relationship, capacity and resources, information technology [5], trust, knowledge exchange, IT integration between the supplier and service provider [22], and costs and benefit [23].

The BSC is a model of performance measurement that describes the relationship between the performance of perspectives and indicators as a strategy map. The strategy map is a business strategy that is related with financial and non-financial execution [24]. Besides that, BSC can be used simultaneously by several organizations that are collaborating together [25]. The BSC has been used to measure the performance of sustainable SCPM [26]. Finally, the BSC can be adapted to measure supply chain performance because a supply chain is a reflection of the collaboration between producers, suppliers, providers and customers.

The BSC is dynamic and innovative because it can integrate with other methods [27] for the design of sustainable or green SCPM practices. Examples include BSC and AHP (analytical hierarcy process) –PGP (pre-emptive goal programming) [28], BSC and game theory [29], BSC and a fuzzy analytical networking process (ANP) [30], BSC and data envelopment analysis (DEA) [31], and BSC and DEMATEL [9]. Based on this, the design of the SHSCPM can integrate the BSC with another method.

The DEMATEL is a multi-attribute decision-making (MADM) method that can be used as a tool to help in decision-making processes [32]. The DEMATEL has used to analyse the component structures from decision variables. It can analyse the direct or indirect relationships between variables [33]. It is used to determine the relationship between the impact of the performance of perspectives and indicators on the performance measurement, including on BSC [33]. Furthermore, the DEMATEL can describe both the level of importance and the level of influence of an attribute or a variable in a system and then use a matrix system to determine all the causal relationships between the attributes and variables [33]. Finally, the DEMATEL can be used to determine the level of importance, level of influence, and supplier selection on the sustainable supply chain [34,35].

Integration of the BSC and DEMATEL for SHSCPM has been published. This past research shows that the SHSCPM orientation has similarities with the perspectives used in services sector performance: finance, customers, operational, information, and innovation & growth [9]. The relationship between perspectives is shown in Figure 1.

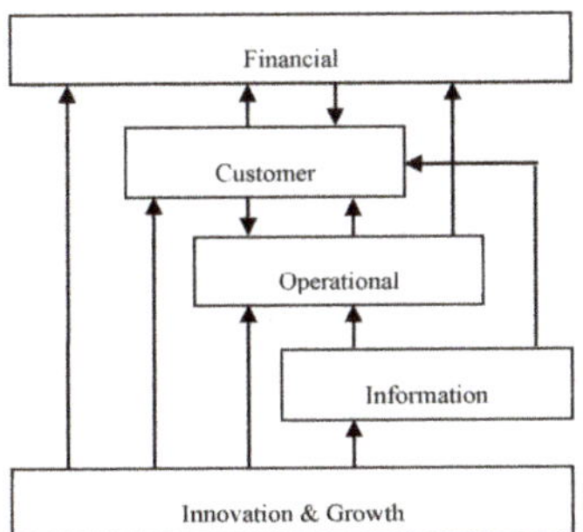


**Figure 1.** Strategy map of a sustainable healthcare supply chain performance measurement (SHSCPM).

**Table 1.** Relationship between performance indicators and perspectives performance based on sustainability aspects.

| Sustainability Aspects | Performance Perspectives | | | | | | | | | |
| --- | --- | --- | --- | --- | --- | --- | --- | --- | --- | --- |
| | Financial | | Customer | | Operational | | Information | | Innovation and Learning | |
| Economic | Demand (patient)<br>Effectiveness<br>Efficiency<br>Profit<br>Revenue | X1<br>X2<br>X3<br>X4<br>X5 | Quality of service<br>Delivery | X6<br>X7 | Inventory level<br>Standard of service<br>Flexibility<br>Supplier timeliness | X11<br>X12<br>X13<br>X14 | Integration of information system | X20 | Capacity and professionalism<br>Innovation<br>Training and education<br>Research and development | X24<br>X25<br>X26<br>X27 |
| Environmental | | | | | Green technology<br>Green material<br>Waste treatment<br>Work physic environment | X15<br>X16<br>X17<br>X18 | Environmental certification | X21 | | |
| Social | | | Customer satisfaction<br>Patient loyalty<br>Stakeholder satisfaction | X8<br>X9<br>X10 | Collab. with supplier | X19 | Medical information system<br>Sharing of information & knowledge | X22<br>X23 | Health and safety<br>Organization behavior | X28<br>X29 |

Once the strategy map has been determined, the performance indicators of the SHSCPM can be identified. The identification of the performance indicators was based on a literature review, and in this paper, twenty nine indicators are used [9]. Furthermore, the performance indicators have been classified into the sustainability aspects and performance perspectives that are show in Table 1.

Figure 1 and Table 1 still do not show the weights of the performance perspectives and indicators of the SHSCPM, which indicate their contribution to performance. More study is still needed to determine the performance weights of indicators and perspectives based on the relationships between them, and these aspects can be done by using the ANP.

## 3. DEMATEL and ANP for Performance Measurement

The ANP is a general theory of relative measurement used to derive a composite priority ratio from an individual ratio that represents the relative measurement of the influence of elements that interact with control criteria [36]. For the performance measurement, ANP is usually combined with the DEMATEL [32], where the DEMATEL is used to relate of indicators and perspectives, and then the relationships created are used to make the ANP model. Finally, the ANP model can determine the weights of perspectives and indicators based on inner and outer dependence.

The ANP can improve the limitation of AHP, especially for accommodating the relationships between criteria and alternative [36]. There are two relationships in ANP, namely inner and outer dependence. Inner dependence is the relationship between indicators within a cluster, and outer dependence is the relationship between indicators in different clusters.

The ANP provides a way to input judgment and measurements to derive the ratio of a priorities scale for the distribution of influence among the criteria and groups of criteria in the decision process [33]. ANP is a qualitative multi-attribute decision-making approach that provides structured communication to address the business model [30]. The weight of an indicator on performance measurement can be determined by the ANP. The ANP can help decision-makers to determine a strategy for improving performance based on the strategy map.

This section may be divided by subheadings. It should provide a concise and precise description of the experimental results, their interpretation, as well as the experimental conclusions that can be drawn. Figure 1 and Table 1 still do not show the weights of the performance perspectives and indicators of the SHSCPM, which can indicate their contribution to performance. More study is still needed to determine the performance weights of indicators and perspectives based on the relationships between them, and these relationships can be processed using the ANP.

## 4. Research Methodology

The healthcare business in the Province of East Java, Indonesia was used for the design of the SHSCPM. The healthcare business consists of producers, suppliers (distributors), health clinics, and hospitals. In this study, the BSC will be combined with DEMATEL to design a strategy map that represents relationships between perspectives and indicators. The relationships are designed based on the level of importance and influence by DEMATEL. Furthermore, the BSC strategy map will be used as an input for design of the ANP structure model. Finally, the ANP structure model can be determine the weights of the performance indicators, performance perspectives, and sustainability aspects as final output of this study.

### 4.1. Data Collection

The development of the SHSCPM model in this study was based on a survey with in-depth interviews with seven expert respondents for data collection. The respondents were representatives of the healthcare business as supply chain actors. The respondents consisted of: one pharmaceutical industry manager, one supplier owner, two private hospital professional managers, one public hospital professional manager, and two clinical professional managers. All of the respondents had more than 10 years of experience in their profession. The surveys were obtained for the period from March 2018 to December 2018. Respondents were asked about pair-wise comparisons between perspectives and indicators based on their perceptions.

*4.2. Stages of This Research*

The stages of development of the SHSCPM model by combining the balanced scorecard with DEMATEL and ANP, and using Super Decisions software version 2.8.0 for ANP data processing, were based on the respondents' perceptions. The research stages were as follows:

a.  Determine the BSC strategy map that describes the relationships between the performance of perspectives and indicators using DEMATEL.
b.  Development of ANP structure model based on the BSC strategy map.
c.  Survey to perform pair-wise comparisons between the performance of perspectives and indicators on a scale of 1–9. The survey results were processed according to the ANP steps [30] with Super Decisions software version 2.8.0. The validity of the pair-wise comparison considered the inconsistency value. If the value of inconsistency <1, the pair-wise comparison is valid [36].
d.  Running the process to determine the weights of performance indicators. These were classified into two types: performance indicator weights based on clusters (performance perspective) and indicator weights based on the system (partial indicator weight). Based on the performance indicator weights, the weights of the performance perspectives and sustainability aspects can be calculated.

*4.3. Determination of BSC Strategy Map by DEMATEL*

The DEMATEL used for design of strategy map. Survey result was processed for the DEMATEL steps. Survey to identify level of influence between perspectives and indicators with 0–4 scale (0 = no influence, 1 = low, 2 = normal, 3 = strong, 4 = very strong). The BSC strategy map was designed by using DEMATEL with following steps:

a.  Building direct relation matrix ($A$) based on average of influence value from $a_i$ to $a_j$ by survey:

$$A = [a_{ij}]_{nxn}$$

$$A = \begin{bmatrix} 0 & a_{12} & \cdots & a_{1n} \\ a_{21} & 0 & \cdots & a_{2n} \\ \vdots & \vdots & \cdots & \vdots \\ a_{n1} & a_{n2} & \cdots & 0 \end{bmatrix}.$$

b.  Building normalization matrix (X):

$$X = A \times s$$

$$s = \min \left[ \frac{1}{\max \sum\limits_{j=1}^{n} a_{ij}}, \frac{1}{\max \sum\limits_{i=1}^{n} a_{ij}} \right].$$

c.  Building total relationship matrix (T), $T = X(I - X)^{-1}$; I is a identity matrix. Calculating of importance level and influence level of the perspective and indicator. Element of $T = \left[t_{ij}\right]_{nxn}, i, j = 1, 2, \ldots, n$; where, $i$ = rows, and $j$ = columns; D and R represent direct and indirect relationships from rows and columns:

$$D = \left[\sum_{j=1}^{n} t_{ij}\right]_{nx1}, (i = 1, 2, \ldots, n)$$

$$R = \left[\sum_{i=1}^{n} t_{ij}\right]_{nx1}, (j = 1, 2, \ldots, n)$$

where, (D + R) indicates of importance level and (D − R) indicates of influence level.

d.  Building a significant matrix to describe the relationship between the perspective and the indicator. Steps to design of significant matrix:

-  Calculation of the average of T: $\overline{X} = \frac{\sum T}{\sum ij}$; where, $i$ = the sum of the row and $j$ = the sum of the column.
-  Reduction of all T with $\overline{X}$ or $(T - \overline{X})$, the value of the significant matrix indicates the level of the relationship

e.  The significant matrix describe of relationship between perspectives and indicators. Relationship between perspectives and indicators are made into a BSC strategy map, which is used for ANP structure model.

The direct relation matrix is shown in Table 2, and Table 3 shows the total relationships matrix. Then, the level of importance and influence is shown in Table 4. Finally, the significance matrix has illustrate of relationships between perspectives and indicators shown in Table 5.

Based on the level of importance (D + R), the important indicators on the SHSCPM are quality of service, stakeholder satisfaction, customer satisfaction, flexibility, collaboration with supplier, standard of service, innovations, and organization behavior. These are incorporated in the customer perspective and innovation and learning perspectives. After that, all of indicators incorporated into the innovation and learning perspective have the most influence compared to other indicators, with the values of all of the influence levels (D − R) being positive (+).

Finally, the indicators incorporated into the customer perspective and innovation and learning perspectives are closely related with human resources, so the existence of human resources is important in the SHSCPM.

**Table 2.** Direct relation matrix as initial matrix based on expert preferences.

| | X1 | X2 | X3 | X4 | X5 | X6 | X7 | X8 | X9 | X10 | X11 | X12 | X13 | X14 | X15 | X16 | X17 | X18 | X19 | X20 | X21 | X22 | X23 | X24 | X25 | X26 | X27 | X28 | X29 |
|---|---|---|---|---|---|---|---|---|---|---|---|---|---|---|---|---|---|---|---|---|---|---|---|---|---|---|---|---|---|
| X1 | | 4 | 2 | 4 | 4 | 3 | 3 | 4 | 4 | 4 | | | | | | | | | | | | | | | | | | | |
| X2 | 4 | | 4 | 3 | 0 | 4 | 3 | 4 | 4 | 3 | | | | | | | | | | | | | | | | | | | |
| X3 | 0 | 3 | | 4 | 4 | 3 | 0 | 0 | 0 | 4 | | | | | | | | | | | | | | | | | | | |
| X4 | 0 | 0 | 1 | | 0 | 3 | 0 | 0 | 0 | 4 | | | | | | | | | | | | | | | | | | | |
| X5 | 0 | 0 | 1 | 4 | | 4 | 0 | 0 | 0 | 4 | | | | | | | | | | | | | | | | | | | |
| X6 | 4 | 2 | 4 | 4 | 4 | | 4 | 4 | 4 | 4 | 3 | 2 | 3 | 0 | 3 | 1 | 3 | 3 | 4 | | | | | | | | | | |
| X7 | 4 | 4 | 3 | 2 | 4 | 4 | | 4 | 4 | 4 | 0 | 4 | 4 | 0 | 0 | 0 | 3 | 0 | 0 | | | | | | | | | | |
| X8 | 4 | 2 | 0 | 4 | 4 | 4 | 0 | | 4 | 4 | 0 | 4 | 4 | 0 | 0 | 0 | 0 | 0 | 4 | | | | | | | | | | |
| X9 | 4 | 0 | 0 | 4 | 4 | 4 | 0 | 4 | | 4 | 0 | 0 | 4 | 2 | 0 | 0 | 0 | 0 | 0 | | | | | | | | | | |
| X10 | 4 | 3 | 0 | 2 | 3 | 3 | 3 | 4 | 1 | | 0 | 0 | 3 | 3 | 3 | 0 | 3 | 3 | 4 | | | | | | | | | | |
| X11 | 4 | 4 | 4 | 4 | 3 | 3 | 4 | 3 | 4 | 1 | | 0 | 4 | 4 | 0 | 0 | 2 | 0 | 4 | | | | | | | | | | |
| X12 | 3 | 4 | 2 | 3 | 3 | 4 | 4 | 4 | 4 | 4 | 3 | | 4 | 3 | 3 | 0 | 4 | 4 | 0 | | | | | | | | | | |
| X13 | 4 | 4 | 4 | 4 | 4 | 4 | 4 | 4 | 4 | 4 | 4 | 4 | | 3 | 4 | 3 | 4 | 3 | 4 | | | | | | | | | | |
| X14 | 4 | 4 | 3 | 3 | 0 | 4 | 4 | 3 | 4 | 4 | 4 | 4 | 4 | | 0 | 0 | 0 | 0 | 4 | | | | | | | | | | |
| X15 | 4 | 4 | 4 | 4 | 3 | 4 | 3 | 4 | 3 | 4 | 0 | 3 | 4 | 0 | | 2 | 4 | 3 | 1 | | | | | | | | | | |
| X16 | 2 | 0 | 2 | 2 | 0 | 3 | 0 | 3 | 3 | 0 | 3 | 3 | 3 | 0 | 1 | | 4 | 0 | 3 | | | | | | | | | | |
| X17 | 3 | 3 | 4 | 3 | 4 | 4 | 0 | 2 | 4 | 4 | 3 | 0 | 0 | 0 | 4 | 0 | | 4 | 4 | | | | | | | | | | |
| X18 | 4 | 4 | 0 | 0 | 3 | 4 | 0 | 4 | 4 | 4 | 0 | 4 | 0 | 0 | 3 | 0 | 3 | | 1 | | | | | | | | | | |
| X19 | 4 | 3 | 3 | 4 | 3 | 4 | 3 | 3 | 3 | 4 | 4 | 0 | 3 | 4 | 4 | 2 | 4 | 0 | | | | | | | | | | | |
| X20 | | | | | | 4 | 4 | 3 | 3 | 4 | 4 | 0 | 4 | 4 | 2 | 0 | 2 | 0 | 3 | | 4 | 4 | 4 | | | | | | |
| X21 | | | | | | 4 | 0 | 3 | 3 | 3 | 0 | 2 | 2 | 0 | 4 | 3 | 4 | 3 | 3 | 4 | | 3 | 3 | | | | | | |
| X22 | | | | | | 4 | 4 | 4 | 4 | 4 | 4 | 0 | 4 | 4 | 3 | 2 | 2 | 3 | 4 | 4 | 2 | | 4 | | | | | | |
| X23 | | | | | | 4 | 4 | 4 | 4 | 4 | 4 | 3 | 4 | 4 | 4 | 2 | 4 | 0 | 4 | 4 | 0 | 4 | | | | | | | |
| X24 | 4 | 4 | 4 | 4 | 4 | 4 | 4 | 4 | 4 | 4 | 3 | 4 | 4 | 3 | 4 | 2 | 4 | 3 | 4 | 4 | 2 | 4 | 4 | | 4 | 3 | 4 | 4 | 4 |
| X25 | 4 | 4 | 4 | 4 | 4 | 4 | 4 | 4 | 4 | 4 | 4 | 4 | 4 | 4 | 4 | 4 | 4 | 4 | 4 | 4 | 4 | 4 | 4 | 4 | | 3 | 4 | 4 | 4 |
| X26 | 3 | 4 | 3 | 3 | 4 | 4 | 4 | 4 | 4 | 4 | 0 | 0 | 4 | 0 | 4 | 2 | 4 | 4 | 4 | 4 | 3 | 4 | 4 | 4 | 4 | | 4 | 4 | 4 |
| X27 | 4 | 4 | 4 | 4 | 4 | 4 | 3 | 4 | 3 | 4 | 3 | 4 | 4 | 3 | 4 | 4 | 4 | 3 | 4 | 4 | 3 | 3 | 4 | 4 | 4 | 4 | | 4 | 4 |
| X28 | 4 | 4 | 4 | 4 | 4 | 4 | 0 | 4 | 0 | 4 | 0 | 0 | 4 | 0 | 4 | 3 | 4 | 2 | 0 | 0 | 4 | 4 | 0 | 4 | 4 | 3 | 4 | | 4 |
| X29 | 4 | 4 | 4 | 4 | 4 | 4 | 4 | 4 | 4 | 4 | 3 | 0 | 4 | 3 | 4 | 4 | 4 | 4 | 4 | 4 | 4 | 4 | 4 | 4 | 4 | 4 | 4 | 4 | |

**Table 3.** Total relationships matrix.

| | X1 | X2 | X3 | X4 | X5 | X6 | X7 | X8 | X9 | X10 | X11 | X12 | X13 | X14 | X15 | X16 | X17 | X18 | X19 | X20 | X21 | X22 | X23 | X24 | X25 | X26 | X27 | X28 | X29 |
|---|---|---|---|---|---|---|---|---|---|---|---|---|---|---|---|---|---|---|---|---|---|---|---|---|---|---|---|---|---|
| X1 | | 0.11 | 0.09 | 0.12 | 0.11 | 0.11 | 0.09 | 0.11 | 0.1 | 0.11 | | | | | | | | | | | | | | | | | | | |
| X2 | 0.11 | | 0.09 | 0.11 | 0.08 | 0.11 | 0.1 | 0.11 | 0.1 | 0.09 | | | | | | | | | | | | | | | | | | | |
| X3 | 0.08 | 0.09 | | 0.11 | 0.1 | 0.09 | 0.07 | 0.09 | 0.07 | 0.1 | | | | | | | | | | | | | | | | | | | |
| X4 | 0.08 | 0.08 | 0.07 | | 0.08 | 0.09 | 0.07 | 0.08 | 0.07 | 0.09 | | | | | | | | | | | | | | | | | | | |
| X5 | 0.07 | 0.06 | 0.07 | 0.09 | | 0.09 | 0.06 | 0.08 | 0.07 | 0.09 | | | | | | | | | | | | | | | | | | | |
| X6 | 0.12 | 0.11 | 0.11 | 0.12 | 0.12 | | 0.11 | 0.12 | 0.11 | 0.13 | 0.1 | 0.11 | 0.11 | 0.07 | 0.1 | 0.08 | 0.11 | 0.09 | 0.12 | | | | | | | | | | |
| X7 | 0.1 | 0.09 | 0.08 | 0.09 | 0.09 | 0.1 | | 0.1 | 0.08 | 0.1 | 0.06 | 0.09 | 0.08 | 0.05 | 0.06 | 0.06 | 0.07 | 0.06 | 0.08 | | | | | | | | | | |
| X8 | 0.1 | 0.07 | 0.06 | 0.1 | 0.09 | 0.1 | 0.05 | | 0.09 | 0.1 | 0.06 | 0.06 | 0.08 | 0.04 | 0.06 | 0.05 | 0.06 | 0.05 | 0.09 | | | | | | | | | | |
| X9 | 0.09 | 0.06 | 0.05 | 0.09 | 0.07 | 0.08 | 0.05 | 0.08 | | 0.09 | 0.05 | 0.06 | 0.07 | 0.04 | 0.04 | 0.03 | 0.04 | 0.04 | 0.08 | | | | | | | | | | |
| X10 | 0.1 | 0.09 | 0.07 | 0.09 | 0.08 | 0.09 | 0.07 | 0.08 | 0.07 | | 0.06 | 0.09 | 0.08 | 0.06 | 0.08 | 0.05 | 0.08 | 0.08 | 0.1 | | | | | | | | | | |
| X11 | 0.1 | 0.1 | 0.09 | 0.11 | 0.09 | 0.1 | 0.09 | 0.1 | 0.08 | 0.08 | | 0.09 | 0.09 | 0.08 | 0.07 | 0.06 | 0.08 | 0.06 | 0.1 | | | | | | | | | | |
| X12 | 0.11 | 0.11 | 0.09 | 0.11 | 0.1 | 0.12 | 0.1 | 0.12 | 0.11 | 0.11 | 0.09 | | 0.11 | 0.07 | 0.1 | 0.07 | 0.1 | 0.09 | 0.09 | | | | | | | | | | |
| X13 | 0.12 | 0.11 | 0.1 | 0.12 | 0.11 | 0.12 | 0.1 | 0.12 | 0.11 | 0.12 | 0.1 | 0.11 | | 0.08 | 0.09 | 0.08 | 0.1 | 0.08 | 0.1 | | | | | | | | | | |
| X14 | 0.1 | 0.1 | 0.09 | 0.11 | 0.08 | 0.1 | 0.09 | 0.09 | 0.1 | 0.1 | 0.09 | 0.1 | 0.1 | | 0.06 | 0.05 | 0.07 | 0.06 | 0.1 | | | | | | | | | | |
| X15 | 0.12 | 0.1 | 0.1 | 0.12 | 0.1 | 0.11 | 0.08 | 0.11 | 0.1 | 0.11 | 0.07 | 0.1 | 0.1 | 0.07 | | 0.08 | 0.1 | 0.09 | 0.1 | | | | | | | | | | |
| X16 | 0.09 | 0.07 | 0.07 | 0.08 | 0.06 | 0.09 | 0.06 | 0.09 | 0.08 | 0.09 | 0.07 | 0.07 | 0.07 | 0.05 | 0.08 | | 0.09 | 0.07 | 0.08 | | | | | | | | | | |
| X17 | 0.1 | 0.09 | 0.09 | 0.1 | 0.09 | 0.1 | 0.07 | 0.09 | 0.09 | 0.1 | 0.07 | 0.09 | 0.08 | 0.06 | 0.08 | 0.07 | | 0.08 | 0.09 | | | | | | | | | | |
| X18 | 0.1 | 0.08 | 0.07 | 0.09 | 0.09 | 0.1 | 0.06 | 0.09 | 0.1 | 0.1 | 0.06 | 0.09 | 0.09 | 0.06 | 0.08 | 0.07 | 0.09 | | 0.08 | | | | | | | | | | |
| X19 | 0.12 | 0.1 | 0.1 | 0.12 | 0.1 | 0.1 | 0.09 | 0.11 | 0.09 | 0.12 | 0.1 | 0.09 | 0.09 | 0.09 | 0.1 | 0.08 | 0.1 | 0.08 | | | | | | | | | | | |
| X20 | | | | | | 0.11 | 0.1 | 0.1 | 0.1 | 0.11 | 0.09 | 0.11 | 0.1 | 0.07 | 0.08 | 0.07 | 0.09 | 0.07 | 0.09 | | 0.09 | 0.09 | 0.07 | | | | | | |
| X21 | | | | | | 0.08 | 0.06 | 0.09 | 0.08 | 0.09 | 0.07 | 0.08 | 0.07 | 0.05 | 0.08 | 0.07 | 0.09 | 0.08 | 0.09 | 0.07 | | 0.07 | 0.06 | | | | | | |
| X22 | | | | | | 0.11 | 0.09 | 0.1 | 0.09 | 0.1 | 0.08 | 0.1 | 0.09 | 0.07 | 0.07 | 0.07 | 0.08 | 0.08 | 0.09 | 0.09 | 0.07 | | 0.08 | | | | | | |
| X23 | | | | | | 0.1 | 0.08 | 0.1 | 0.09 | 0.1 | 0.09 | 0.09 | 0.09 | 0.07 | 0.07 | 0.06 | 0.08 | 0.05 | 0.09 | 0.09 | 0.07 | 0.09 | | | | | | | |
| X24 | 0.13 | 0.12 | 0.11 | 0.13 | 0.12 | 0.13 | 0.11 | 0.12 | 0.12 | 0.13 | 0.1 | 0.12 | 0.11 | 0.08 | 0.1 | 0.09 | 0.1 | 0.09 | 0.12 | 0.1 | 0.1 | 0.1 | 0.09 | | 0.1 | 0.1 | 0.08 | 0.1 | 0.12 |
| X25 | 0.13 | 0.13 | 0.12 | 0.14 | 0.13 | 0.14 | 0.11 | 0.13 | 0.12 | 0.14 | 0.11 | 0.13 | 0.12 | 0.1 | 0.1 | 0.1 | 0.12 | 0.11 | 0.13 | 0.12 | 0.11 | 0.12 | 0.13 | 0.1 | | 0.11 | 0.09 | 0.09 | 0.13 |
| X26 | 0.12 | 0.12 | 0.1 | 0.13 | 0.11 | 0.13 | 0.11 | 0.12 | 0.11 | 0.13 | 0.08 | 0.12 | 0.12 | 0.07 | 0.1 | 0.09 | 0.1 | 0.09 | 0.12 | 0.11 | 0.1 | 0.11 | 0.12 | 0.12 | 0.07 | | 0.09 | 0.1 | 0.12 |
| X27 | 0.11 | 0.11 | 0.1 | 0.12 | 0.12 | 0.11 | 0.09 | 0.12 | 0.1 | 0.12 | 0.09 | 0.1 | 0.1 | 0.07 | 0.1 | 0.09 | 0.11 | 0.1 | 0.11 | 0.1 | 0.1 | 0.09 | 0.12 | 0.11 | 0.1 | 0.07 | | 0.09 | 0.11 |
| X28 | 0.1 | 0.1 | 0.08 | 0.1 | 0.1 | 0.1 | 0.06 | 0.09 | 0.07 | 0.1 | 0.06 | 0.09 | 0.09 | 0.05 | 0.08 | 0.06 | 0.08 | 0.08 | 0.07 | 0.07 | 0.08 | 0.08 | 0.09 | 0.09 | 0.08 | 0.08 | 0.06 | | 0.09 |
| X29 | 0.13 | 0.13 | 0.12 | 0.13 | 0.12 | 0.13 | 0.11 | 0.13 | 0.12 | 0.13 | 0.1 | 0.12 | 0.12 | 0.09 | 0.1 | 0.09 | 0.11 | 0.1 | 0.12 | 0.12 | 0.1 | 0.12 | 0.13 | 0.13 | 0.11 | 0.11 | 0.1 | 0.09 | |

**Table 4.** Level of importance and influence.

| Performance Indicators | | D | R | D + R | D − R |
|---|---|---|---|---|---|
| Demand (patient) | X1 | 1.03 | 2.72 | 3.75 | −1.69 |
| Effectiveness | X2 | 0.97 | 2.53 | 3.5 | −1.55 |
| Efficiency | X3 | 0.85 | 2.29 | 3.13 | −1.44 |
| Profit | X4 | 0.79 | 2.83 | 3.62 | −2.05 |
| Revenue | X5 | 0.73 | 2.53 | 3.26 | −1.8 |
| Quality of service | X6 | 2.02 | 3.15 | 5.18 | −1.13 |
| Delivery | X7 | 1.5 | 2.49 | 3.99 | −0.99 |
| Customer satisfaction | X8 | 1.37 | 3.07 | 4.44 | −1.7 |
| Patient loyalty | X9 | 1.16 | 2.79 | 3.95 | −1.63 |
| Stakeholder satisfaction | X10 | 1.49 | 3.17 | 4.66 | −1.69 |
| Inventory level | X11 | 1.64 | 1.97 | 3.6 | −0.33 |
| Standard of service | X12 | 1.88 | 2.34 | 4.22 | −0.47 |
| Flexibility | X13 | 1.96 | 2.33 | 4.3 | −0.37 |
| Supplier timeliness | X14 | 1.63 | 1.63 | 3.26 | 0.001 |
| Green technology | X15 | 1.83 | 2.02 | 3.84 | −0.19 |
| Green material | X16 | 1.41 | 1.75 | 3.16 | −0.35 |
| Waste treatment | X17 | 1.59 | 2.19 | 3.78 | −0.6 |
| Work physic environment | X18 | 1.55 | 1.88 | 3.43 | −0.34 |
| Collaboration with supplier | X19 | 1.87 | 2.39 | 4.26 | −0.52 |
| Integration of information system | X20 | 1.61 | 0.99 | 2.59 | 0.62 |
| Environmental certification | X21 | 1.32 | 0.94 | 2.26 | 0.39 |
| Medical information system | X22 | 1.54 | 1 | 2.55 | 0.54 |
| Sharing of inform. and knowledge | X23 | 1.46 | 0.86 | 2.32 | 0.6 |
| Capacity and professionalism | X24 | 3.11 | 0.74 | 3.85 | 2.36 |
| Innovation | X25 | 3.41 | 0.77 | 4.18 | 2.65 |
| Training and education | X26 | 3.08 | 0.62 | 3.7 | 2.45 |
| Research and development | X27 | 2.97 | 0.65 | 3.62 | 2.32 |
| Health and safety | X28 | 2.36 | 0.58 | 2.93 | 1.78 |
| Organization behavior | X29 | 3.31 | 0.72 | 4.03 | 2.59 |

### 4.4. The ANP Structure Model

The ANP structure model was designed by BSC strategy map. The ANP is used to determine weight of perspectives and indicators performance. The ANP processes using survey of expert respondents. Survey has used to perform a pair-wise comparison between indicators with a scale of 1–9. Survey results were processed by ANP steps. The validity of the pair-wise comparison has considered the inconsistency value. If value of inconsistency <1, so the pair-wise comparison is valid. The ANP uses an initial matrix derived from average value of the survey result.

Weight calculation of perspectives and indicators on ANP can be process by inputting average values of pair-wise comparison into software of super decision version 2.8.0. The value of the consistency ratio can also be seen directly in the super decision software after values of pair-wise comparison inputted into the software system.

The structure model illustrates the inner dependence and outer dependence in this study shown in Figure 2. Inner and outer dependence were designed by relationships between perspectives and indicators as significant matrix. Inner dependence is relationship between indicators based on the same perspective. Outer dependence is relationship between indicators with other indicators based on the different perspectives.

**Table 5.** Relationship between perspectives and indicators.

| | Impact on | | | | | | | | | | | | | | | | | | | | | | | | | | | | |
|---|---|---|---|---|---|---|---|---|---|---|---|---|---|---|---|---|---|---|---|---|---|---|---|---|---|---|---|---|---|
| | X1 | X2 | X3 | X4 | X5 | X6 | X7 | X8 | X9 | X10 | X11 | X12 | X13 | X14 | X15 | X16 | X17 | X18 | X19 | X20 | X21 | X22 | X23 | X24 | X25 | X26 | X27 | X28 | X29 |
| X1 | — | √ | √ | √ | √ | √ | √ | √ | √ | √ | | | | | | | | | | | | | | | | | | | |
| X2 | √ | — | √ | √ | | √ | √ | √ | √ | √ | | | | | | | | | | | | | | | | | | | |
| X3 | | √ | — | √ | √ | √ | | √ | | √ | | | | | | | | | | | | | | | | | | | |
| X4 | | | | — | | √ | | | | √ | | | | | | | | | | | | | | | | | | | |
| X5 | | | | √ | — | √ | | | | √ | | | | | | | | | | | | | | | | | | | |
| X6 | √ | √ | √ | √ | √ | — | √ | √ | √ | √ | √ | √ | √ | | √ | | √ | √ | √ | | | | | | | | | | |
| X7 | √ | √ | | √ | √ | √ | — | √ | | √ | | √ | | | | | | | | | | | | | | | | | |
| X8 | √ | | | √ | √ | √ | | — | √ | √ | | | | | | | | √ | | | | | | | | | | | |
| X9 | √ | | | √ | | | | | — | √ | | | | | | | | | | | | | | | | | | | |
| X10 | √ | √ | | √ | | √ | | | | — | | √ | | | | | | | √ | | | | | | | | | | |
| X11 | √ | √ | √ | √ | √ | √ | √ | √ | | √ | — | √ | √ | | | | | | √ | | | | | | | | | | |
| X12 | √ | √ | √ | √ | √ | √ | √ | √ | √ | √ | √ | — | √ | | √ | | √ | √ | √ | | | | | | | | | | |
| X13 | √ | √ | √ | √ | √ | √ | √ | √ | √ | √ | √ | √ | — | | | | √ | | √ | | | | | | | | | | |
| X14 | √ | √ | √ | √ | | √ | √ | √ | √ | √ | √ | √ | √ | — | | | | | √ | | | | | | | | | | |
| X15 | √ | √ | √ | √ | √ | √ | | √ | √ | | √ | √ | √ | | — | | √ | √ | √ | | | | | | | | | | |
| X16 | √ | | | | | √ | | | | | | | | | | — | √ | | | | | | | | | | | | |
| X17 | √ | √ | √ | √ | √ | √ | | √ | | √ | | √ | | | | | — | | √ | | | | | | | | | | |
| X18 | √ | | | √ | √ | √ | | √ | √ | √ | | | | | | | | — | √ | | | | | | | | | | |
| X19 | √ | √ | √ | √ | √ | √ | √ | √ | √ | √ | √ | √ | √ | √ | √ | | √ | √ | — | | | | | | | | | | |
| X20 | | | | | | √ | √ | √ | √ | √ | √ | √ | √ | | | | √ | | √ | — | √ | | √ | √ | | | | | |
| X21 | | | | | | | √ | | | | | | | | | | √ | | √ | √ | — | | | | | | | | |
| X22 | | | | √ | √ | √ | √ | √ | √ | | | √ | √ | | | | | | √ | √ | √ | — | | | | | | | |
| X23 | | | | | | √ | √ | √ | √ | | √ | √ | √ | | | | | | √ | √ | √ | √ | — | | | | | | |
| X24 | √ | √ | √ | √ | √ | √ | √ | √ | √ | √ | √ | √ | √ | | √ | √ | √ | √ | √ | √ | √ | √ | √ | — | | √ | √ | √ | √ |
| X25 | √ | √ | √ | √ | √ | √ | √ | √ | √ | √ | √ | √ | | √ | √ | √ | √ | √ | √ | √ | √ | √ | √ | √ | — | √ | √ | √ | √ |
| X26 | √ | √ | √ | √ | √ | √ | √ | √ | √ | √ | √ | √ | √ | √ | √ | √ | √ | √ | √ | √ | √ | √ | √ | √ | √ | — | √ | √ | √ |
| X27 | √ | √ | √ | √ | √ | √ | √ | √ | √ | √ | √ | √ | √ | √ | √ | √ | √ | √ | √ | √ | √ | √ | √ | √ | √ | √ | — | √ | √ |
| X28 | √ | √ | √ | √ | √ | √ | | √ | √ | √ | √ | √ | √ | √ | √ | √ | √ | √ | √ | √ | √ | √ | √ | √ | √ | √ | √ | — | √ |
| X29 | √ | √ | √ | √ | √ | √ | √ | √ | √ | √ | √ | √ | √ | | √ | √ | √ | √ | √ | √ | √ | √ | √ | √ | √ | √ | √ | √ | — |

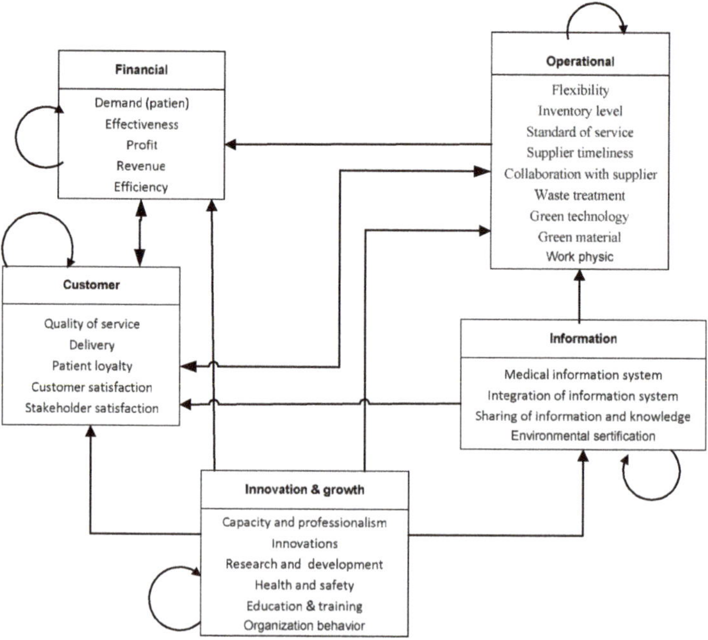

**Figure 2.** Inner and outer dependence.

Figure 2 makes two phenomena clear: the innovation and growth perspective is the most influential because it has an influence on all of the other perspectives, and the customer perspective is the most important perspective because it is influenced by all the other perspectives. Besides, performance indicators by inner and outer dependence can explain using examples: Demand indicator (X1) on the financial perspective has inner dependence with effectiveness (X2), efficiency (X3), profit (X4), and revenue (X5), and then, indicator of demand (X1) has outer dependence with indicators on the customer perspective (i.e. quality of service (X6), delivery (X7), customer satisfaction (X8), patient loyalty (X9), and stakeholder satisfaction (X10)). Furthermore, the organization behavior indicator (X29) on the innovation and learning perspective has inner dependence with capacity and professionalism (X24), innovation (X25), training and education (X26), research and development (X27), and health and safety (28). Besides, the organization behavior indicator (X29) on the innovation and learning perspective has outer dependence with all of indicators on the other perspectives.

## 5. Results and Discussion

This study uses Super Decisions software version 2.8.0 for data processing of the ANP. The data processing is the result of pair-wise comparison between performance perspectives and/or performance indicators derived from the inner dependence and outer dependence. Hence, the values of the pair-wise comparison are based on the preferences of the expert respondents.

### 5.1. Data Collection

Pair-wise comparison by the expert respondents' perceptions is used to collect the requirement data. Then, the data has been calculated to determine weights of the perspectives and indicators. Table 6 shows the results of pair-wise comparison for the performance perspectives.

**Table 6.** Normalized weight of pair-wise comparison between the performance perspectives.

| Perspectives | Normalized Weight |
| --- | --- |
| Financial | 0.245 |
| Customer | 0.448 |
| Operational | 0.091 |
| Information | 0.059 |
| Innovation and learning | 0.157 |
| Sum | 1 |

Validation of the pair-wise comparison result is based on the value of the inconsistency ratio. For pair-wise comparison of the performance perspectives, this value is 0.079, where $0.079 \leq 0.1$, which means that this perception by the respondents is valid.

### 5.2. Influence Analysis between Perspectives Based on Indicators Relationship

The influence between perspectives can be determined based on the relations between indicators. The Super Decisions software processed the values of influence between performance perspectives shown in Table 7.

**Table 7.** Influence between performance perspectives.

| Perspectives | Influence on | | | | |
| --- | --- | --- | --- | --- | --- |
| | **F** | **C** | **O** | **I** | **IL** |
| Financial (F) | 0.250 | 0.750 | | | |
| Customer (C) | 0.380 | 0.507 | 0.113 | | |
| Operational (O) | 0.258 | 0.637 | 0.105 | | |
| Information (I) | | 0.833 | | 0.167 | |
| Innovation and learning (IL) | 0.245 | 0.449 | 0.092 | 0.058 | 0.156 |

Table 7 shows that the performance indicators incorporated in the perspectives of innovation and learning had an influence on all of the perspectives, including a self-influence. The performance indicators in the information perspective, operational perspective, and financial perspective too have the greatest influence on the customer perspective. Besides, the customer perspective has the highest value of self-influence. From these phenomena, the customer perspective is seen to be the most important in the SHSCPM.

### 5.3. Weight of Performance Perspectives and Performance Indicators

Based on the Super Decisions processing, the indicator weights on the cluster (perspectives), indicator weights in the SHSCPM system, and the perspective weights can be calculated. Table 8 shows the calculation of the weights of the perspectives and indicators.

**Table 8.** Weights of the perspectives and indicators in SHSCPM.

| Performance Perspective | Performance Indicators | Indicator Weight on Cluster | Indicator Weight | Perspective Weight |
|---|---|---|---|---|
| Financial | Demand (patient) | 0.3190 | 0.0964 | 0.3021 |
|  | Effectiveness | 0.0808 | 0.0244 |  |
|  | Efficiency | 0.0576 | 0.0174 |  |
|  | Profit | 0.4773 | 0.1442 |  |
|  | Revenue | 0.0653 | 0.0197 |  |
| Customer | Quality of service | 0.3626 | 0.2296 | 0.6331 |
|  | Delivery | 0.0235 | 0.0149 |  |
|  | Customer satisfaction | 0.1915 | 0.1212 |  |
|  | Patient loyalty | 0.1949 | 0.1234 |  |
|  | Stakeholder satisfaction | 0.2274 | 0.1440 |  |
| Operational | Inventory level | 0.1579 | 0.0102 | 0.0648 |
|  | Standard of service | 0.1201 | 0.0080 |  |
|  | Flexibility | 0.1251 | 0.0081 |  |
|  | Supplier timeliness | 0.0137 | 0.0010 |  |
|  | Green technology | 0.0483 | 0.0030 |  |
|  | Green material | 0.0000 | 0.0000 |  |
|  | Waste treatment | 0.0389 | 0.0025 |  |
|  | Work physic environment | 0.0150 | 0.0010 |  |
|  | Collaboration with supplier | 0.4810 | 0.0310 |  |
| Information | Integration of information system | 0.0000 | 0.0000 | 0.0000 |
|  | Environmental certification | 0.0000 | 0.0000 |  |
|  | Medical information system | 0.0000 | 0.0000 |  |
|  | Sharing of inform. and knowledge | 0.0000 | 0.0000 |  |
| Innovation and learning | Capacity and professionalism | 0.0000 | 0.0000 | 0.0000 |
|  | Innovations | 0.0000 | 0.0000 |  |
|  | Training and education | 0.0000 | 0.0000 |  |
|  | Research and development | 0.0000 | 0.0000 |  |
|  | Health and safety | 0.0000 | 0.0000 |  |
|  | Organization behavior | 0.0000 | 0.0000 |  |
| Sum |  |  | 1.0000 | 1.0000 |

The indicator weights on the clusters in Table 8 indicate that profit and demand (patient) are very important for the financial perspective, while quality of service and stakeholder satisfaction are very important for the customer perspective. Then, collaboration with the supplier is very important for the operational perspective. Finally, the information perspective and innovation and learning perspective are not important indicators, because all of the indicators in this cluster have a value of zero (0).

Beside, big five dominant indicators in the SHSCPM are quality of service (0.2296), profit (0.1442), stakeholder satisfaction (0.1440), patient loyalty (0.1234), and customer satisfaction (0.1212). Furthermore, based on the performance perspectives, the customer perspective has the greatest weight compared with the other perspectives and is, therefore, the most influential on the performance of the SHSCPM, with a weight of 0.6331. Next, the financial perspective has a weight of 0.3021, followed by the operational perspective, with a weight of 0.0648. Finally, the perspectives of information and innovation and learning have no weight, so these perspectives have the least influence on the performance of the SHSCPM.

## 5.4. Weight of Sustainability Aspects

The weight of sustainability aspects can be processed by calculating the indicator weights based on the sustainability aspect classification. Table 9 shows the weights of the sustainability aspects based on the indicator weights.

**Table 9.** Weights of sustainability aspects.

| Sustainability Aspects | Performance Indicators | Indicator Weight | Weight of Aspects |
| --- | --- | --- | --- |
| Economic | Demand (patient) | 0.0964 | 0.5739 |
| | Capacity and professionalism | 0 | |
| | Effectiveness | 0.0244 | |
| | Inventory level | 0.0102 | |
| | Quality of service | 0.2296 | |
| | Standard of service | 0.0080 | |
| | Efficiency | 0.0174 | |
| | Profit | 0.1442 | |
| | Revenue | 0.0197 | |
| | Innovations | 0 | |
| | Flexibility | 0.0081 | |
| | Supplier timeliness | 0.0010 | |
| | Integration of information system | 0 | |
| | Delivery | 0.0149 | |
| | Training and education | 0 | |
| | Research and development | 0 | |
| Environmental | Green technology | 0.0030 | 0.0065 |
| | Green material | 0 | |
| | Waste treatment | 0.0025 | |
| | Environmental certification | 0 | |
| | Work physic environment | 0.0010 | |
| Social | Customer satisfaction | 0.1212 | 0.4196 |
| | Medical information system | 0 | |
| | Patient loyalty | 0.1234 | |
| | Collaboration with supplier | 0.0310 | |
| | Stakeholder satisfaction | 0.1440 | |
| | Health and safety | 0 | |
| | Sharing of inform. and knowledge | 0 | |
| | Organization behavior | 0 | |

Table 9 shows that the economic aspect has the greatest weight, and the environmental aspect has the lowest weight compared to the others. The social aspect was ranked second after the economic aspect, so this aspect was considered feasible in the SHSCPM. From the weight of the indicators, the performance indicators incorporated in the economic aspects still dominate compared to the others. The performance indicators incorporated in the environmental aspect have small weights, which means that environmental factors receive less attention from all the actors in the healthcare supply chain in Indonesia's Province of East Java.

The weight of the social aspect is 0.241, which means that social factors are highly regarded by actors in the healthcare supply chain in the Province of East Java. This is different from the opinion that social aspects were less explored in sustainable supply chain [14,37].

## 6. Conclusions

Integration of BSC with DEMATEL and ANP is a new model for measuring performance of sustainable SHSC. This model has more comprehensive with other models because all of the supply chain actors have involved to determining of performance indicators, strategy map, and weight of the performance indicators.

This study has been using five perspectives with twenty-nine indicators. The performance indicators included intangible characteristics and sustainability aspects. The performance indicators that reflect this intangibility are related to information and human resources, while performance indicators that reflect sustainability are related to economic, environmental, and social factors.

There are three major findings in this study. First, from the BSC and DEMATEL, the indicators incorporated in the customer perspective and innovation and learning perspectives were important, so the indicators incorporated into the innovation and learning perspective were the most influential on other indicators. Second, from DEMATEL and ANP, the innovation and learning perspective had the most influence on other perspectives, but on the other hand, this perspective was not important because it did not affect the performance value. Thus, the customer perspective is the most important because it has a major influence on the performance value. Third, based on the weights of the sustainability aspects, the economic aspect has the greatest weight, and the environmental aspect has the least weight compared to the others. The social aspect was ranked second after the economic aspect, so this aspect was considered feasible in the SHSC performance. Finally, the environmental aspect in the SHSC receives less attention in the healthcare business in East Java Province, Indonesia. Based on DEMATEL and the indicator weights from ANP, the performance indicators incorporated in the financial perspective and customer perspective are drivers of the SHSC's performance. The indicators driving the SHSC's performance consist of profit, quality of service, revenue, customer satisfaction, and stakeholder satisfaction. Besides, the performance indicators incorporated into the economic aspect of sustainability have the greatest effect on SHSC performance, followed by the social aspect and the environmental aspect. Furthermore, this study found a contradiction in the social aspect of sustainability, which received less attention than in some other studies, although this was an important aspect of sustainability after the financial aspect. Finally, human resources, as an intangible asset, are the main factor in the SHSC because they have a significant effect on the improvement of performance, especially from the customer perspective, innovation and learning perspective and the social aspect of sustainability.

Implementation of SHSCPM by integration of BSC with DEMATEL and ANP may help the management of the healthcare business to give more attention to human resources as one of intangible characteristics of a healthcare business, especially for innovation and organization behavior, because its exerts the most influence to other indicators. Beside that, the management of the healthcare business must be maintain customer satisfaction, patient loyalty, collaboration with suppliers, and stakeholder satisfaction as parts of social aspects for performance increasing. The SHSCPM model by combining BSC with DEMATEL and ANP can help management in the healthcare business to manage the company performance with simultaneously attention to intangible and sustainability aspects.

As with most empirical research, this article has limitations. First, the survey method with in-depth interviews was used with just seven expert respondents. So, the study could be improved by using more expert respondents. Second, the current study covered only East Java Province, so the findings cannot be generalized to other provinces in Indonesia.

Future research needs to explore the social and environmental aspects of sustainability in SHSCPM, where a contradiction was found. Furthermore, the SHSCPM system needs a new design with a simulation of system dynamics for predicting SHSC performance in the future based on the strategy map, weights of performance indicators and in the past performance values. The new design can help the healthcare business to prepare better its operations in order to achieve a high of the SHSC performance.

**Author Contributions:** Conceptualization, E.B.L.; Methodology, all authors, writing—original draft preparation, E.B.L.; writing—review and editing, all authors.

**Funding:** Partially of the research was funded by the grant scheme for doctoral dissertation research, Ministry of Research, Technology and Higher Education of the Republic of Indonesia, 2018.

**Acknowledgments:** This article is part of doctoral dissertation research at Department of Industrial Engineering, Institut Teknologi Sepuluh Nopember (ITS), Surabaya, Indonesia.

**Conflicts of Interest:** The authors declare no conflict of interest.

## References


1. Cho, D.W.; Lee, Y.H.; Ahn, S.H.; Hwang, M.K. A framework for measuring the performance of service supply chain management. *Comput. Ind. Eng.* **2012**, *62*, 801–818. [CrossRef]
2. Baltacioglu, T.; Ada, E.; Kaplan, M.D.; And, O.Y.; Kaplan, Y.C. A new framework for service supply chains. *Serv. Ind. J.* **2007**, *27*, 105–124. [CrossRef]
3. Giannakis, M. Conceptualizing and managing service supply chains. *Serv. Ind. J.* **2011**, *31*, 1809–1823. [CrossRef]
4. Drzymalski, J. Supply Chain Frameworks for the Service Industry: A Review of the Literature. *Eur. Int. J. Sci. Technol.* **2012**, *1*, 31–42.
5. Yap, L.L.; Tan, C.L. The effect of service supply chain management practices on the public healthcare organizational performance. *Int. J. Bus. Soc. Sci.* **2012**, *3*, 216–224.
6. Kumar, A.; Ozdamar, L.; Zhang, C.N. Supply chain redesign in the healthcare industry of Singapore. *Supply Chain Manag. Int. J.* **2008**, *13*, 95–103. [CrossRef]
7. The Central Bureau of Statistics. Pertumbuhan Ekonomi Indonesia Triwulan II-2014. *Statistics News of Indonesia*, 2014; 1–9.
8. BPS. *East Java in Figure 2015*; Statistics of Jawa Timur Province: Surabaya, Indonesia, 2015.
9. Leksono, E.B.; Suparno, S.; Vanany, I. Development of Performance Indicators Relationships on Sustainable Healthcare Supply Chain Performance Measurement Using Balanced Scorecard and DEMATEL. *Int. J. Adv. Sci. Eng. Inf. Technol.* **2018**, *8*, 115–122. [CrossRef]
10. Xu, X.; Gursoy, D. A conceptual framework of sustainable hospitality supply chain management. *J. Hosp. Mark. Manag.* **2015**, *24*, 229–259. [CrossRef]
11. Ortas, E.; Moneva, J.M.; Álvarez, I. Sustainable supply chain and company performance: A global examination. *Supply Chain Manag. Int. J.* **2014**, *19*, 332–350. [CrossRef]
12. Hassini, E.; Surti, C.; Searcy, C. A literature review and a case study of sustainable supply chains with a focus on metrics. *Int. J. Prod. Econ.* **2012**, *140*, 69–82. [CrossRef]
13. Hussain, M.; Khan, M.; Al-aomar, R. A framework for supply chain sustainability in service industry with Confirmatory Factor Analysis. *Renew. Sustain. Energy Rev.* **2016**, *55*, 1301–1312. [CrossRef]
14. Beske-janssen, P.; Johnson, M.P.; Schaltegger, S. 20 years of performance measurement in sustainable supply chain management—What has been achieved? *Supply Chain Manag. Int. J.* **2015**, *20*, 664–680. [CrossRef]
15. Chithambaranathan, P.; Subramanian, N.; Gunasekaran, A. Service supply chain environmental performance evaluation using grey based hybrid MCDM approach. *Intern. J. Prod. Econ.* **2015**, *166*, 163–176. [CrossRef]
16. Unger, S.; Landis, A. Assessing the environmental, human health, and economic impacts of reprocessed medical devices in a Phoenix hospital's supply chain. *J. Clean. Prod.* **2016**, *112*, 1995–2003. [CrossRef]
17. Hussain, M.; Ajmal, M.M.; Gunasekaran, A.; Khan, M. Exploration of Social Sustainability in Healthcare Supply Chain. *J. Clean. Prod.* **2018**, *203*, 977–989. [CrossRef]
18. Zigan, K.; Macfarlane, F.; Desombre, T. Intangible resources as performance drivers in European hospitals. *Int. J. Prod. Perform. Manag.* **2008**, *57*, 57–71. [CrossRef]
19. Smith, B.K.; Nachtmann, H.; Pohl, E.A. Quality measurement in the healthcare supply chain. *Qual. Manag. J.* **2011**, *18*, 50–60. [CrossRef]
20. Priyan, S.; Uthayakumar, R. Optimal inventory management strategies for pharmaceutical company and hospital supply chain in a fuzzy-stochastic environment. *Oper. Res. Health Care* **2014**, *3*, 177–190. [CrossRef]
21. Swinehart, K.; Smith, A.E. Internal supply chain performance measurement: A health care continuous improvement implementation. *Int. J. Health Care Qual. Assur.* **2005**, *18*, 533–542. [CrossRef]
22. Chen, D.Q.; Preston, D.S.; Xia, W. Enhancing hospital supply chain performance: A relational view and empirical test. *J. Oper. Manag.* **2013**, *31*, 391–408. [CrossRef]
23. Lega, F.; Marsilio, M.; Villa, S. An evaluation framework for measuring supply chain performance in the public healthcare sector: Evidence from the Italian NHS. *Prod. Plan. Control* **2013**, *24*, 931–947. [CrossRef]
24. Kaplan, R.S.; Norton, D.P. Linking the Balanced Scorecard to strategy. *Calif. Manag. Rev.* **1996**, *39*, 53–79. [CrossRef]
25. Kaplan, R.S.; Norton, D.P.; Rugelsjoen, B. Managing alliances with the Balanced Scorecard. *Harv. Bus. Rev.* **2010**, 114–121.

26. Reefke, H.; Trocchi, M. Balanced scorecard for sustainable supply chains: Design and development guidelines. *Int. J. Prod. Perform. Manag.* **2013**, *62*, 805–826. [CrossRef]
27. Voelpel, S.C.; Leibold, M.; Eckhoff, R.A. The tyranny of the Balanced Scorecard in the innovation economy. *J. Intellect. Cap.* **2006**, *7*, 421–428. [CrossRef]
28. Bhagwat, R.; Sharma, M.K. An application of the integrated AHP-PGP model for performance measurement of supply chain management. *Prod. Plan. Control* **2009**, *20*, 678–690. [CrossRef]
29. Naini, S.G.J.; Aliahmadi, A.R.; Jafari-Eskandari, M. Designing a mixed performance measurement system for environmental supply chain management using evolutionary game theory and balanced scorecard: A case study of an auto industry supply chain. *Resour. Conserv. Recycl.* **2011**, *55*, 593–603. [CrossRef]
30. Bhattacharya, A.; Mohapatra, P.; Kumar, V.; Dey, P.K. Green supply chain performance measurement using fuzzy ANP-based balanced scorecard: A collaborative decision-making approach. *Prod. Plan. Control* **2014**, *25*, 698–714. [CrossRef]
31. Haghighi, S.M.; Torabi, S.A.; Ghasemi, R. An integrated approach for performance evaluation in sustainable supply chain networks (with a case study). *J. Clean. Prod.* **2016**, *137*, 579–597. [CrossRef]
32. Gölcük, I.; Baykasoglu, A. An analysis of DEMATEL approaches for criteria interaction handling within ANP. *Expert Syst. Appl.* **2016**, *46*, 346–366. [CrossRef]
33. Chen, F.; Hsu, T.; Tzeng, G. A balanced scorecard approach to establish a performance evaluation and relationship model for hot spring hotels based on a hybrid MCDM model combining DEMATEL and ANP. *Int. J. Hosp. Manag.* **2011**, *30*, 908–932. [CrossRef]
34. Hsu, C.; Kuo, T.; Chen, S.; Hu, A.H. Using DEMATEL to develop a carbon management model of supplier selection in green supply chain management. *J. Clean. Prod.* **2013**, *56*, 164–172. [CrossRef]
35. Su, C.; Horng, D.; Tseng, M.; Chiu, A.S.F.; Wu, K.; Chen, H. Improving sustainable supply chain management using a novel hierarchical grey-DEMATEL approach. *J. Clean. Prod.* **2016**, *134*, 469–481. [CrossRef]
36. Saaty, T.L. Fundamentals of the analytic network process. In Proceedings of the ISAHP 1999, Kobe, Japan, 12–14 August 1999; pp. 12–14.
37. Varsei, M.; Soosay, C.; Fahimnia, B.; Sarkis, J. Framing sustainability performance of supply chains with multidimensional indicators. *Supply Chain Manag. Int. J.* **2014**, *19*, 242–257. [CrossRef]

 *sustainability*

*Article*

# An Integrated Approach for Modeling Ontology-Based Task Knowledge on an Incident Command System

**Kwoting Fang * and Shuoche Lin**

National Yunlin University of Science and Technology, Douliu City, Yunlin County 640, Taiwan; diablo79802@gmail.com
* Correspondence: fangkt@yuntech.edu.tw

Received: 15 April 2019; Accepted: 18 June 2019; Published: 25 June 2019

**Abstract:** This paper presents the TTIPP methodology, an integration of task analysis, task ontology, integration definition function modeling (IDEF0), Petri net, and Petri net mark language (PNML), to organize and model the task knowledge in the form of natural language expressions acquired during the knowledge-acquisition process. The goal of the methodology is to make the tasks more useful, accessible, and sharable through the web for a variety of stakeholders interested in solving a problem which is expressed mostly in linguistic form, and to shed light on the nature of problem-solving knowledge. This study provides a core epistemology for the knowledge engineer while developing the task ontology for a generic task. The proposed model overcomes the drawbacks of IDEF0, which are its static nature and Petri net which has no concept of hierarchy. A good number of countries lie on the typhoon and earthquake belts, which make them vulnerable to natural calamities. However, a practical incident command system (ICS) that provides a common framework to allow emergency responders of different backgrounds to work together effectively for standardized, on-the-scene, incident management has yet to be developed. There is a strong need to explicitly share, copy, and reuse the existing problem-solving knowledge in a complex ICS. As an example, the TTIPP model is applied to the task of emergency response for debris-flow during a typhoon as a part of an ICS.

**Keywords:** problem-solving; incident command system; task ontology; knowledge management

---

## 1. Introduction

In the past few decades, several large-scale earthquakes have occurred in various parts of the world, such as the 1994 Northridge earthquake in the U.S., the 1995 Hanshin-Awaji earthquake in Japan, the 1999 Chi-chi earthquake in Taiwan, and the 2001 Izmit earthquake in Istanbul, Turkey. In 2003, a series of earthquakes shook parts of the world, including Altay, Russia; Boumerdes, Algeria; Hokkaido, Japan; Bam, Iran. These were followed by the December 2004 earthquake in Indonesia, with its devastating large-scale tsunami. Recently, in 2019, an earthquake hit the Sulawest, Indonesia; off the East Coast of Honshu, Japan; South Sandwich Islands Region; and Banda Sea, respectively [1]. These natural calamities caused tragic loss of life, severe property damage, and a decline in regional prosperity.

Although a lot of effort has been spent on protecting people from natural calamities around the world, some countries which lie on the typhoon and earthquake belts, are still vulnerable to natural calamities. However, there is no practical incident command system (ICS) that provides a common framework allowing people to work together effectively for standardized, on-the-scene, incident management. There is a strong need to explicitly share, copy, and reuse prior or existing problem-solving knowledge in a complex ICS.

In general, there are four ICS stages: mitigation, preparedness, response, and post-disaster reconstruction, and each stage needs a complex and dynamic problem-solving method. ICS is a complex problem which requires collaboration and participation among many different stakeholders with conflicting interests and it also covers the complex aspects of environmental, economic, and social problems. With the vigorous development and rapid spread of the World Wide Web, a significant quantity of information is available to people. However, reusable and sharable problem-solving knowledge in an ICS has emerged much more slowly. A stream of existing knowledge representation techniques can be used, including semantic networks, frames, uncertain reasoning, ontology, and rules, etc. [2–6]. For a complex problem, such as an incident command system, simpler mechanisms, using declarative statements that are true or false, as semantic networks or frames are not appropriate. In a similar sense, uncertainty reasoning only provides solutions for situations when true or false conclusions cannot be reached, which may lead to conflicts with problem-solving knowledge in managerial practice [7].

To bridge the gap between real-world problem-solving methods and information found on the Internet and to enable people to communicate with computers by using accurate syntax and semantics, this research proposes a novel model not only for effectively capturing and representing real-world problem-solving knowledge, but also for overcoming the drawbacks of the existing methodologies for integration of heterogeneous information. The concept of task ontology is first adopted to build a three-level mediation representation for a task analysis. Second, an integrated methodology, integrating task analysis, task ontology, integration definition function modeling (IDEF0), Petri net, and Petri net mark language (PNML) (TTIPP), is proposed to systematically analyze the tasks and subtasks in terms of the inputs, outputs, mechanisms, and controls, using integration definition function modeling (IDEF0) and Petri net. Finally, Petri net mark language (PNML) is used as a standard interchange format to make the tasks of searching, displaying, integrating, and maintaining more assessable through the web. The practicality of the proposed model is demonstrated through an emergency response for debris-flow cases. It is hoped that this model lays the groundwork for understanding how to build reusable and sharable real-world problem-solving knowledge.

## 2. Related Work

### *2.1. Knowledge Management*

With the dramatic growth of globalization and the necessity to boost value creation, knowledge has continuously played the important role of being the major source of sustainable advantage [8]. Alonderienne et al. [9] defined knowledge as being the result of a process which combines ideas, rules, procedures, and information. Specifically, based on reasoning and understanding, made by the mind, through the posterior and the frontal hierarchy of cognitive networks, people can capture perceptual information and executive information through experience, learning, or introspection [10]. From an evolutionary economic standpoint, Erkut [11] pointed out that the generation of new knowledge, in turn of shaping markets, is associated with the perception of objectively available information in a system (called the nano dimension). Perception means that an individual classifies a certain experience according to his or her own categorization by means of a pattern recognition, through the interactions of hierarchical networks in the cerebral cortex, based on similar past experiences existing in an ever-evolving cerebral cortex [12]. Polanyi [13] divided knowledge into two categories: tacit (weak) and explicit (hard) knowledge. Explicit knowledge can easily be codified and transmitted through formal and systematic processes as published knowledge [14]. In contrast, tacit knowledge, with personal contextual expertise, has a cognitive component that intervenes in perception and learning [15].

In the past decade, perhaps the most dramatic evolution in business has been the dawn of the so-called "new economy" based on Internet and information technologies (IT), such as intelligent systems technologies. New economy is established based on an organization successfully shifting its economic value to intellectual assets, assets of information, product distribution, and affiliation.

In general, an analysis of the literature can identify a set of different approaches for data analysis, whether numerical or text. Essentially, they consign two dimensions of data analysis: linguistic techniques and statistical approaches. Roughly, they can be categorized into several groups based on differences and similarities in their features with each other, i.e., through similarity and nearest-neighbor methods, document similarity, decision rules, decision trees, and probabilities linear scoring methods [16].

Based on Twitter data from a large multinational telecommunication company and 200 managers, with three years of communication data, Fronzetti Colladon and Gloor [17] combined social network analysis (SNA) and text mining to investigate the effect of spammers' activity on the network structure. More recently, given process fragmentation and information exchange among port actors, Aloini et al. [18], by exploiting data from the Port Community Systems, integrated process mining (PM), SNA, and text mining to reconstruct, analyze, and evaluate the information exchange network of the freight export process.

Knowledge creation and knowledge management have been the new goals for organizations that want to increase their competitiveness. Unfortunately, due to a lack of absorptive capacity, many knowledge management projects are, in reality, information systems projects [19–21]. Gold et al. [22] mentioned that knowledge management becomes questionable when the projects only provide some consolidated data but lack innovation that is extended from prior knowledge or innovation which is unprecedented. Overall, to reach knowledge management from information management, in terms of exchange and combination, is a rather complex process that involves developing a knowledge structure that enables organizations to effectively generate knowledge [23–26].

Over the past few decades, there have been many methodologies born for building new knowledge bases, particular in ontologies, from scratch and from existing bases in a variety of settings. Combined with METHONTOLOGY and WebODE, in term of management and support activities, Corcho et al. [27] built a legal entity ontology in the context of the Spanish legal domain. From the viewpoint of knowledge workers on the day-to-day ontology life cycle, Kotis and Vouros [28] presented the human-centered ontology engineering methodology (HCOME) in living ontologies that can accentuate the role of knowledge workers in shaping their information by actively being involved in ontology engineering tasks. In 2011, Villazon-Terrazas et al. [29] developed a network of ontology networks, including local ontology networks and a reference ontology network, using the NeOn methodology to enable an exchange of curricula vitae and job offers in different languages in a semantic interoperability platform. Given existing ontology problems, Sofia Pinto, Tempich, and Staab [30] proposed the DILIGENT methodology that draws domain experts, users, knowledge engineers, and ontology engineers together to collaboratively build an ontology to solve the drawbacks of decentralization, partial autonomy, iteration, and non-expert builders.

There is burgeoning interest in the study of knowledge management which pertains to the interdisciplinary nature of research and practice in decision-making, with particular emphasis on ontology means and methods. From the standpoint of expert knowledge and complying with railway standards, Saa et al. [31] developed an ontology-driven decision support system for designing complex railway problems. Focused on integrating and restructuring methods in the repository, Ziemba et al. [32] adopted the algorithm to build a repository of knowledge about the methods for assessing the quality of a website. To satisfy particular accessibility needs in e-learning contexts, Elias, Lohmann, and Auer [33] presented rule-based queries, in terms of ontology, to retrieve relevant educational resources for learners with disabilities. Traditionally, water pollution accidents have been digitalized through the combination of monitoring sensors, management servers, and application software by adopting mechanistic water-quality models with achieved data. Meng et al. [34] provided the architecture of the ontology-underpinned emergency response system for water pollution accidents to make the water pollution information semantic and the referred applications intelligent. Due to a lack of knowledge systematization in the sustainability assessment domain, Konys [35] contributed knowledge-based mechanisms, with formal, practical, and technological guidance, to make the collected knowledge publicly available, reusable, and interoperable.

The growth of the Internet offers enormous potential for professionals and creating a significant body of virtual communities of practice can provide alternative channels for professionals to collaborate with their peers, manage information, and develop and spread knowledge. Research on social interaction falls into three broad categories [36]: (1) connectivity, (2) interactivity, and (3) language use. Through a seven-year longitudinal study with 14,000 members of 16 different healthcare virtual communities of practice, Antonacci et al. [37] pointed out that centralized structure, dynamic leaders, and complex language have driven the growth of the community. Moreover, by enriching the theoretical foundation or framework of knowledge creation and sharing, particularly in an online discussion forum, Barker [38] discovered that, from a continuous basis standpoint, an expert should play a proactive role to ensure new knowledge is created and shared by individuals.

Noy and Musen [39] point out that one of the major shortcomings of the current technology for knowledge-based building is the lack of both reusable and sharable knowledge. Because one must build knowledge bases based on "what one believes" and cannot take into consideration of "justified true beliefs" derived from the actual and potential resources, this increases the difficulty of building knowledge bases. Clearly, facilitating usable and useful knowledge should thus contribute to making it easier to build knowledge bases and fit them into the context within which they must be used. In order to achieve this, Therani [40] and Mizoguchi et al. [41] suggest expertise can be decomposed into a task-dependent but domain-independent portion in which applications can use common data for all domains but not for all tasks and a task-independent but domain-dependent portion in which applications can use common data for all tasks but not for all domains. The former is called "task knowledge", formalized knowledge for an independent problem-solving domain.

*2.2. Task Ontology*

Ontologies has been a field of study with growing importance in the academia from the late twentieth to the early twenty-first century. This phenomenon stems from both their conceptual use in organizing information and their practical use in communicating system characteristics [22,35,42].

In general, an ontology can be viewed as an information model that explicitly describes the various entities and abstractions that exist in a universe of discourse, along with their properties [43,44]. Furthermore, an ontology specifies a conceptual phrase partly to articulate knowledge-level theories of a certain field. From a system standpoint, ontologies provide an overarching framework and vocabulary with which to describe system components and relationships for communicating among architecture and domain areas [45]. Therefore, the more the essence of things is captured, the more possible it is for an ontology to be shared [46–49].

A number of different categorizations of ontologies have been proposed. Van Heijst et al. [50] classify ontologies according to the amount and type of structure of the conceptualization and the subject of the conceptualization, while Guarino [51] distinguishes the type of ontologies by their level of dependence on a particular task or point of view. Subsequently, Lassila and McGuinness [52] group ontologies from the perspective of the information the ontology needs to express and the richness of its internal structure.

Ontologies and problem-solving methods (PSMs), higher-order cognitive processes that require the modulation and control of more routine or fundamental skills, have been created to share and reuse knowledge and reasoning behavior across domains and tasks [20,46]. In general, ontologies are concerned with static domain knowledge, a given specific domain, while PSMs deal with modeling reasoning processes and describing the vocabulary related to a generic task or activity. Benjamins and Gomez-Perez [53] define PSMs as a way of achieving the goal of a task. PSMs have inputs and outputs, and many decompose a task into subtasks, and subtasks into methods. In addition, a PSM specifies the data flow between its subtasks. Guarino [51] defines task ontology as an ontology which formally specifies the terminology associated with a problem type, a high-level generic task which has characteristic generic classes of knowledge-based application. Chandrasekaren and Benjamins [54] also define task ontology as "a base of generic vocabulary that organizes the task knowledge for a generic task.". From a problem-solving viewpoint, Newell [55] illustrates that task ontology can be used to model the problem-solving behavior of a task, either at the knowledge level or the symbolic level. Thus, the advantage of task ontology is that it specifies not only a skeleton of the problem-solving process, but also the context in which domain concepts are used. In 2007, Mizoguchi et al. [56] developed an ontology-development tool known as Hozo which has the ability to deal with roles according to their context dependencies. Ikeda et al. [57] suggest that task ontology can be roughly interpreted in two ways: (1) task–subtask decomposition together with task categorization; and (2) an ontology for specifying problem-solving processes. They developed a conceptual level programming environment (CLEPE) based on task ontology in order to make problem-solving knowledge explicit and to exemplify its availability. Rajpathak et al. [58] formalized the task–method–domain–application knowledge modeling framework, which supports both constructing a generic scheduling task ontology to formalize the space of scheduling problems as well as constructing a generic problem-solving model of scheduling that generalizes from the variety of approaches to scheduling problem-solving.

With the volumes of information that continue to increase, the task of turning and integrating this resource dispersed across the Web into a coherent corpus of interrelated information has become a major problem [59]. The emergence of the Semantic Web, providing highly readable data without modifying any of the contents, has shown great promise for the next generation of more capable information technology solutions and mark another stage in the evolution of ontologies and PSMs [46]. Berners-Lee [60] who coined the term Semantic Web, comments that it is envisioned as an extension of the current Web, in which information is given well-defined meaning to better enable computers and people to work in corporations, effectively interweaving human understanding of symbols with machine processability [61]. The way to fulfillment of the corporation can be paved by sharing and re-using domain and task ontologies. The Semantic Web, with domain and task ontologies, can solve some problems much more simply than before and can make it possible to provide certain capabilities that have otherwise been very difficult to support [62–64].

## 3. Research Methodology

In this section we present the TTIPP methodology, which was an integration of task analysis, task ontology, integration definition function modeling (IDEF0), Petri net, and Petri net mark language (PNML), along with the framework shown in Figure 1, to organize and model the task knowledge acquired during the knowledge-acquisition process. The TTIPP methodology is aimed not only at reducing the brittle nature of traditional knowledge-based systems, but also at enhancing knowledge reusability and shareability over different real-world problem-solving applications. Moreover, the proposed model overcomes the drawbacks of IDEF0, namely its static nature and Petri net which has no concept of hierarchy. From the viewpoint of knowledge sharing, the TTIPP methodology can integrate heterogeneous information and distributed information sources to resolve the problems of information access on the Web by translating information into machine processable semantics to facilitate communication between machines and humans.

The TTIPP is composed of three layers and five phases (Figure 1). The top layer, the lexical level model, deals mainly with the syntactic aspect of the problem-solving description in terms of the task analysis phase and task ontology phase. The middle layer, called the conceptual level model, captures the conceptual level meaning of the activity description. The IDEF0 and Petri net models are used in this layer. The bottom layer, called the symbol level model, is the PNML corresponding to the executable program and specifying the computational semantics of problem solving.

This study provides a core epistemology for the knowledge engineer while developing the task ontology for a generic task. The five phases of the proposed integrated model are described in the following sub-sections.

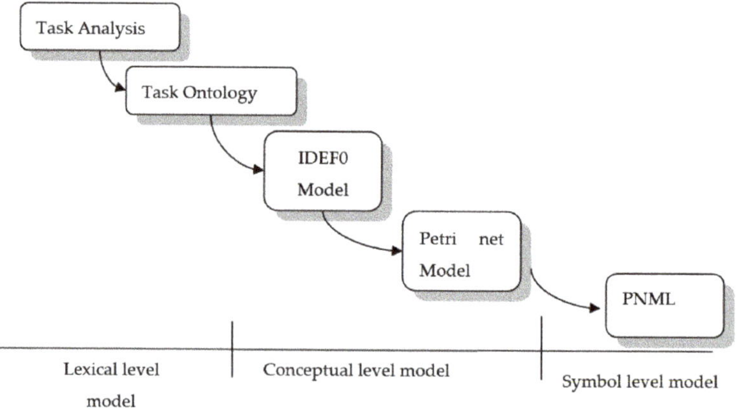


**Figure 1.** Research framework.

*3.1. Phase-I: Task Analysis*

During the first phase, the nature of a task needs to be thoroughly analyzed at a fine-grained level with diverse informational needs. Structured, semi-structured, or even unstructured knowledge is acquired and elicited from various sources, such as the available literature on the task, the test cases specific to the problem area, the actual interview of the domain experts, one's previous experience in the field, etc. Ikeda et al. [57] divide task analysis into two major steps: (1) rough identification and (2) detailed task analysis.

Based on various sources of knowledge, rough identification of task structure is a classification problem, while detailed task analysis is concerned with interaction with domain experts and then articulating how they perform their tasks. Once various knowledge sources, in terms of a variety of forms, such as document, fact, and records, are analyzed in detail, the important concepts from all of the different classes of application lead to a heightened awareness in such a way that this knowledge provides enough theoretical foundation for expressing the nature of the problem. According to this, the initial focus of the task analysis is to concentrate on the most important concepts around which the task ontology needs to be built

*3.2. Phase-II: Task Ontology*

Detailed categorization of concepts involved is indispensable for task knowledge description. This stage provides a fundamental understanding of the relationships among different concepts. Also, in accordance with the elicited concepts given in the previous phase, this stage provides the ontological engineer with an idea about the important axioms that need to be developed in order to decide on the competence of the task ontology. From the standpoint of granularity and generality, following Ikeda et al. [57], the lexical level task ontology consists of four concepts: (1) generic nouns representing objects reflecting their roles that appear in the problem-solving process; (2) generic verbs representing unit activities that appear in the problem-solving process; (3) generic adjectives and/or adverbs modifying the objects; and (4) generic constraints specific to the task. Figure 2 presents the hierarchy of the lexical level task ontology.

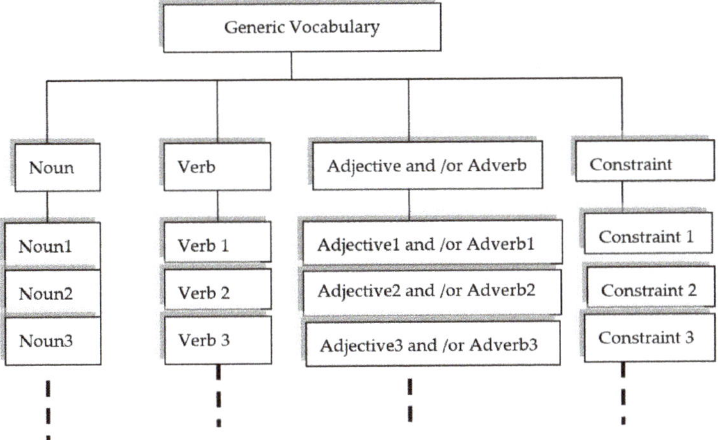

**Figure 2.** Lexical level of task ontology.

## 3.3. Phase-III: IDEF0 Model

During this phase, task ontology in the research framework can be operationalized by using a formal modeling language tool. IDEF0 is a widely-used, activity-oriented graphic notation and modeling approach for system specification and requirement analysis [65,66]. It transforms the concepts described at the natural language level into the formal knowledge modeling level in terms of structured graphical forms. A multi-level model with different classes and relations is created in order to decompose the complex problem into smaller and more detailed sub-problems until the purpose of the model building is reached. An IDEF0 diagram consists of an ordered set of boxes that represent activities performed by a given task. Each box or component in the diagram, representing a given activity, has a simple syntax, shown in Figure 3, with inputs of the activity entering from the left side and the results or outputs of the activity exiting from the right side. The mechanisms, indicated by arrows entering from the bottom of the box, represent resources such as machines, computers, operators, etc. The controls, shown by arrows entering from the top, represent control information, such as parameters and rules of the control systems. The boxes in an IDEF0 diagram, called ICOM for input–control–output–mechanism, are hierarchically decomposed in as many levels as necessary until there is sufficient detail on the basic activities to serve the tasks [21]. The mappings between elements of IDEF0 diagrams and generic vocabularies, from the lexical level to the conceptual level (see Figure 1), are shown in Table 1.

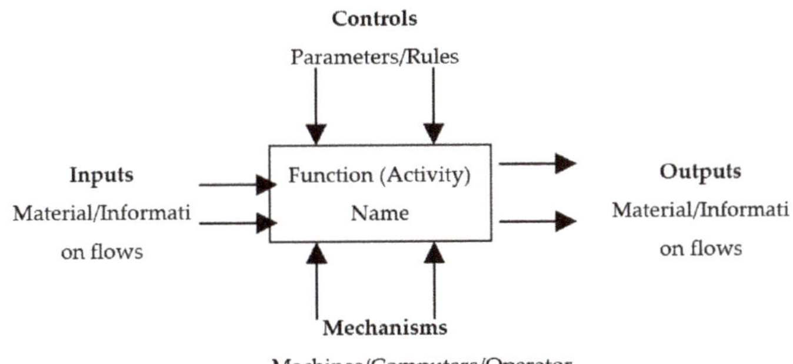


**Figure 3.** Component of the integration definition function modeling (IDEF0) model.

**Table 1.** The mapping between IDEF0 and vocabulary elements.

| IDEF0 | Vocabulary |
|---|---|
| Function | Verb and Noun |
| Input | Noun |
| Control | (1) Noun <br> (2) Adjective and Noun <br> (3) Adverb and Adjective and Noun <br> (4) Constraint and Noun |
| Output | Noun |
| Mechanism | (1) Noun <br> (2) Constraint and Noun |

### 3.4. Phase-IV: Petri Net Model

Broadly speaking, the IDEF0 model has a number of disadvantages such as its static nature and ambiguity in activity specification [62]. A Petri net consists of three entries: (1) the place, drawn as a circle; (2) the transition, drawn as a bar; and (3) the arcs, connecting places and transitions, as shown in Figure 4a [45]. Known as condition/event nets or place/transition nets, Petri nets are suitable for representation of the structure of hierarchical systems that exhibit concurrency, conflict, and synchronization [67]. To clearly visualize the information flow and control through transitions, a Petri net (PN) allows a place to hold none or a positive number of tokens pictured as small solid dots. Generally, the PN is defined as quintuple, PN = (P, T, I, O, m) [68], where:

$P = \{p_1, p_2, \dots, p_n\}$ finite set of places; where integer n > 0;

$T = \{t_1, t_2, \dots, t_s\}$ finite set of transitions, where integer s > 0, with $P \cup T \neq \emptyset$ and $P \cap T = \emptyset$;

I: $P \times T \rightarrow N$ input incidence function with n × s matrices containing the nonnegative integer that defines the set of directed arcs from P to T where $N = \{0,1,2,3 \dots \dots \}$;

O: $P \times T \rightarrow N$ output incidence function with n × s matrices containing the nonnegative integer that defines the set of directed arcs from T to P;

m: $P \rightarrow N$ marking vector whose $i^{th}$ component represents the number of tokens in the $i^{th}$ place. An initial marking is denoted by $m_0$.

The change of system states, called transition firing, will happen by an event when all the input places hold enough number of tokens. Cassandras and Lafortune [69] further explain that a transition is enabled when each input place P of T contains at least a number of tokens equal to the number of the directed arc connecting P to T. When an enabled transition $T_1$ fires as shown in Figure 4b, it removes the token from its input place and deposits it in its output place. Therefore, the execution rules of a PN include enabling and firing rules are shown as follows [68]:

(1) transition $t \varepsilon T$ is enabled if and only if $m(p) \geqq I(p,t)$, $\forall p \varepsilon P$

(2) enabled in a marking m, t fires and results in a new marking m′ following the rules:

$$m'(p) = m(p) - I(p, t) + O(p, t)$$
$$= m(p) + C(p,t), \forall p \varepsilon P$$

where $C = O - I$ and m′ is said to be immediately reachable from m.

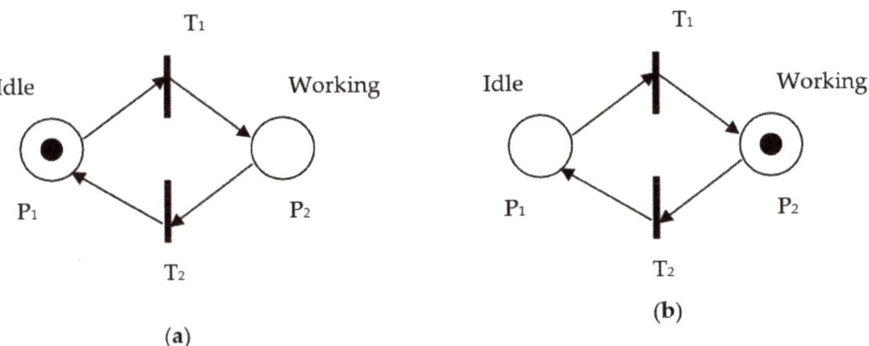

**Figure 4.** Component of a Petri net. (**a**) The token in its input place. (**b**) The token transfer into its output place.

Murata [70] presents invariant analysis methods, including P-invariant and T-invariant, to govern the dynamic behavior of concurrent systems modeled by Petri nets. He calls $C = O - I$ an incidence matrix, to prove that subsets of places over which the sum of the tokens may remain unchanged (P-invariant) and a transition firing sequence brings the marking back to the same one (T-invariant). In writing matrix equations, Murata [70] describes the execution rules as abiding by the following equation:

$$m_k = m_{k-1} + C_{uk} \qquad k = 1,2,3\ldots .. \tag{1}$$

where $m_k$ donates a marking immediately reachable from marking $m_{k-1}$. The $k^{th}$ firing vector or $u_k$ is an $s \times 1$ column vector and the $i^{th}$ column of C represents a change of a marking if transition $t_i$ fires at the $k^{th}$ firing, and then the $i^{th}$ position of $u_k$ is 1 and the other positions are filled with 0s. He further defines a P-invariant of the nonzero nonnegative integer solution x of $C^T x = 0$, then, the previously stated equation is rewritten as follows:

$$x^T m_k = x^T m_{k-1} + x^T C_{uk} \qquad k = 1,2,3\ldots .. \tag{2}$$

Since $C^T x = 0$, thus $x^T C = 0$, then

$$x^T m_k = x^T m_{k-1} \qquad k = 1,2,3\ldots .. \tag{3}$$

Therefore, $x^T m_k = x^T m_{k-1} = $ constant.

He also defines the nonzero nonnegative solution y of $Cy = 0$ as a T-invariant associated with firing a sequence of transitions, leading $m_o$ back to $m_0$, and the $i^{th}$ element of the aggregate firing vector u being the number of $t_i$s firing times in the sequence. Clearly, based on the equation $m_0 = m_0 + C_u$, $C_u = 0$ and u is a T-invariant.

To transform static models generated by the IDEF0 method into a dynamic PN model, Santarek and Buseif [21] developed the following transformation rules:

$Tr_1$: If activities exist, then transform them into a Petri net sequence: transition–place–transition.

$Tr_2$: If arrow outputs exist, then form PN place with tokens in them.

$Tr_3$: (1) If shared mechanisms exist, then form PN place with tokens in them.

(2) If a shared mechanism decomposed into a sub-mechanism exists, then there is no generation of PN place.

$Tr_4$: If all mechanisms used in the PN do not exist in any IDEF0 diagram, then transformation is completed.

The relationships, from a static perspective to dynamic viewpoint, between IDEF0 diagrams and Petri net components are presented in Table 2 [67].

**Table 2.** The relationship between Petri Net and IDEF0 components.

| IDEF0 | Petri Net | Petri Net Symbol |
|---|---|---|
| Activity (box) | PN place (condition) with two Petri net transitions (start and finished events) | |
| Input (arrow) | Place | |
| Output (arrow) | Place | |
| Control (arrow) | Place | |
| Mechanism (arrow) | Place | |

### 3.5. Phase-V: Petri Net Markup Language

The Petri net markup language (PNML) is an extensible markup language (XML) based on interchangeable formats for Petri nets. The PNML is designed to be a Petri net interchange format that is independent of specific tools and platforms. Moreover, the interchange format needs to support different dialects of Petri nets and must be extensible. Thus, PNML should necessarily include the following essential characteristics [71]: (1) Flexibility to represent any kind of Petri net with its specific extensions and features, (2) assurance to remove ambiguity for uniquely determining its PNML representation, and (3) compatibility to exchange as much information as possible between different types of Petri nets.

Even with a mature development of Petri net technology, it is difficult to know what is possible in the future. Certainly, PNML should shed light on the definition of Petri net types to support different versions of Petri nets and, in particular, future versions of Petri nets. Given the above-mentioned information, PNML is adopted as a starting point for a standard interchange format for Petri nets. For implementing purposes, XML was used for its platform, independent in terms of having many tools available for reading, writing, and validating. Table 3 presents the translation of the PNML meta model into XML elements, along with the attributes and their data types.

**Table 3.** Translation of the Petri net mark language (PNML) meta model into extensible markup language (XML) elements and attributes.

| Class | XML Elements | XML Attributes |
|---|---|---|
| PetriNetFile | <pnml> | |
| PetriNet | <net> | id: ID<br>type: anyURL |
| Place | <place> | id: ID |
| Transition | <transition> | id: ID |
| Arc | <arc> | id: ID<br>source: IDRef (Node)<br>traget: IDRef (Node) |
| Page | <page> | id: ID |
| RefPlace | <referencePlace> | id: ID<br>ref: IDRef (Place or RefPlace) |
| RefTrans | <referenceTransition> | id: ID<br>ref: IDRef (Transition or RefTrans) |
| ToolInfo | <toolspecific> | tool: string<br>version: string |
| Graphics | <graphics> | |

## 4. An Example Application

The development of an incident command system (ICS) for a natural disaster requires collaboration and participation at the national level, as well as the local community level. TTIPP methodology not only could help to integrate heterogeneous and distributed information sources with machine processable semantics, but also could help users browse the information based on semantics with common understanding to achieve the knowledge sharing purpose.

As an example application, the TTIPP model presented in this paper was used for management of the knowledge of debris-flow during typhoons in the past three decades in the area of Homchu in Nantou county located in Central Taiwan. Homchu village is located at an altitude of 550 to 750 m. It is connected to Mingde village in the north, Tongfu village in the south, and Chenyoulan River in the west (see Figure 5). There are over 70 families, with 451 males and 370 females, in the village and they plant grape orchards as their main source of income. The village was the site of significant damage resulting from the earthquake on September 21st, 1999. Such typhoons and the resulting debris flow originating from the mountainous region are an annual occurrence on the Chenyoulan River in the region. The Homchu community activity center, elementary school, and church are the main distribution centers when it comes to disaster prevention and refuge in the village. Prior to a disaster, they store refuge materials and prepare temporary shelters and medical emergency stations. The main task is to protect the public from debris flow, which includes water, rocks, soil, and tree trunks and it is divided into four subtasks: mitigation, preparedness, response, and post-disaster reconstruction. The subtasks correspond with the rough identification steps of Ikeda et al. [57].

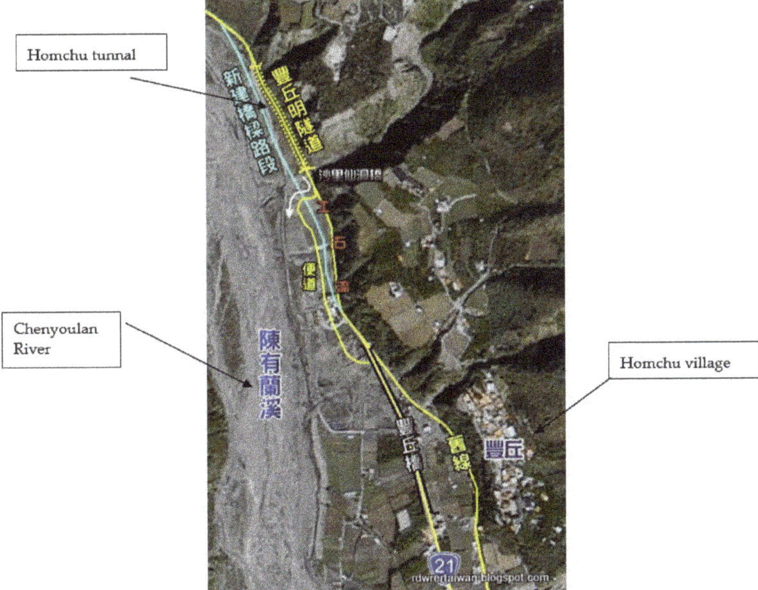


**Figure 5.** Study area.

### 4.1. Phase-I: Task Analysis

Based on the available literature on the task, an open-ended interview questionnaire, in terms of how, when, who, what, how many, and why, specific to the aforementioned subtasks was designed and the six stakeholders (different teams or domain experts) involved in the ICS operations were invited to share their experience and reconfirm the content of their interviews which were recorded and transferred verbatim for future analysis. Table 4 shows the profile of domain experts, along with their numbers of years of experience with the ICS for debris-flow management during typhoons.

**Table 4.** Profile of domain experts.

| Team | Responsibility | Number of Years of Experience with ICS |
|---|---|---|
| 1. Homchu Village Captain | Leader | 9 years |
| 2. Monitor and alert | Sub-Leader | 5 years |
| 3. Equipment and food supply management | Sub-Leader | 9 years |
| 4. Evacuation and accommodation | Sub-Leader | 2 years |
| 5. Urgent repair on structures | Sub-Leader | 9 years |
| 6. Emergency rescue, health, and sanitation | Sub-Leader | 9 years |

The tasks of the different teams are described briefly below:

1. Homchu village captain team: (1) Inform the sub-group teams to gather at the activity center. (2) Open a command center. (3) Command the execution of disaster relief tasks in each sub-group. (4) Request for assistance from outside. (5) Instruct the establishment of the shelter. (6) Command the rescue sub-group to prepare for the initial rescue and on-site rescue. (7) Report statistics on casualties. (8) Contact the ambulance staff and go to each shelter.

2. Monitor and alert team: (1) Examine, judge, and monitor the content and objects of the disaster. (2) Community security maintenance, disaster area control, and traffic guidance. (3) Monitor the disaster site. (4) Conduct evacuation guidance for sub-group members.

3. Equipment and food supply management team: (1) Supply of disaster relief food and water and warm clothing. (2) Provide rescue equipment. (3) Support and manage disaster relief materials and equipment from inside and outside of the community.

4. Evacuation and accommodation team: (1) Inform the responsible personnel of each district to be in place. (2) Monitor pre-set shelters to prepare for housing. (3) Inform everyone of temporary shelter and social assistance for residents. (4) Register and manage personnel in the shelter.

5. Urgent repair of structures team: (1) Collect rescue crew and equipment to go to the scene for initial rescue. (2) Deploy large equipment. (3) Rescue of victims. (4) Prevent the expansion of derivative disasters. (5) Report the rescue process to the command center.

6. Emergency rescue, health, and sanitation team: (1) Open a simple medical service station. (2) On-site emergency rescue work for the victims. (3) Send the victims to the hospital. (4) Register the casualties.

Interviews with domain experts constitute the detailed task analysis step mentioned earlier.

*4.2. Phase-II: Task Ontology*

The task knowledge of ICS operations from a variety of sources such as the available literature and the interview with the domain experts was analyzed for the lexical level task ontology. The four ICS stages were denoted as: mitigation (A1), preparedness (A2), response (A3), and post-disaster reconstruction (A4). Due to the complexity and the dynamic nature of problem-solving methods for ICS and space limitations, only the results for emergency response (A3) of debris flow were chosen to serve as an example for presenting the TTIPP framework. Focusing only on the important concepts of task analysis, Table 5 partially illustrates the generic vocabulary for describing the structure of ICS knowledge classified into nouns, verbs, adjectives, adverbs, and constraints.

**Table 5.** Vocabulary list of the incident command system (partial).

| Noun | People | Civil defense<br>refuge<br>victim<br>… … … | Noun | Place | church<br>village<br>… … … |
| | | | | | Diesel oil |
| | Information | Weather<br>information<br>… … … | | Material | Cellular phone<br>… … … |
| Verb | observe<br>patrol<br>… | | Verb | announce<br>inspect<br>… … … | |
| Adjective | hourly<br>compulsory<br>… … …. | | Adjective | small<br>weak<br>… … …. | |
| Adverb | completely<br>… … … | | Adverb | urgently<br>… … … | |
| Constraint | Hourly rainfall reaching 200 mm<br>(yellow alert)<br>… … … … | | Constraint | Hourly rainfall reaching 300 mm<br>(red alert)<br>… … … … | |

### 4.3. Phase-III: IDEF0 Model

At the modeling stage, an IDEF0 model was built for describing the function of the emergency response (A3). From a functional point of view, the present emergency response had six activities as shown in Figure 6. They were "Establish an advance command post"(A31), "Monitor any possible disaster locations"(A32), "Evacuate residents"(A33), "Urgent repair of structure"(A34), "Emergency rescue of people"(A35) and "Manage goods and materials" (A36). The ICOM of emergency response is presented in Table 6. For the purpose of making it simple to understand, the hierarchical decomposition of each activity into sub-activities was performed to reveal more detail at each level as shown in Figures 7–12.

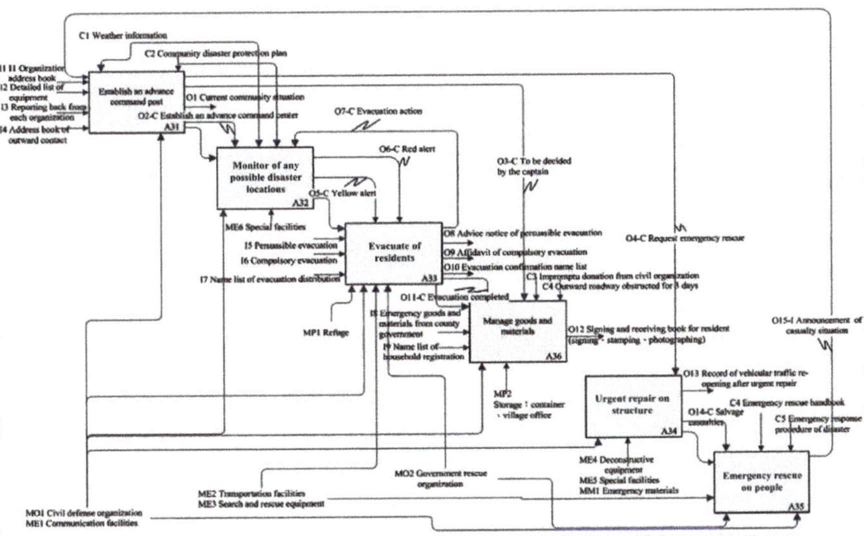


**Figure 6.** The IDFE0 model of the "emergency response" (A3).

**Table 6.** The input–control–output–mechanism (ICOM) of "Emergency response" (A3).

| Activity | Sub-Activity | Input | Control | Output | Mechanism |
|---|---|---|---|---|---|
| A3 Emergency response | A31 Establish an advance command post | I1 Organization address book<br>I2 Detailed list of equipment<br>I3 Reporting back from each organization<br>I4 Address book of outward contact<br>O15-I Announcement of casualty situation | C1 Weather information<br>1. Start announcement of disaster emergency response system<br>2. Start announcement of disaster emergency response system<br>3. Average hourly rainfall (200 mm)<br>4. C2 Community disaster protection plan | O1 Current community situation<br>O2-C Establish an advance command center<br>O3-C To be decided by the captain<br>O4-C Request emergency rescue | MO1 Civil defense organization<br>ME1 Communication facilities |
| | A32 Monitor of any possible disaster locations | | C1 Weather information<br>1. Start announcement of disaster emergency response system<br>2. Start announcement of debris flow monitoring system<br>3. Average hourly rainfall (200 mm)<br>C2 Community disaster protection plan<br>O2-C Establish an advance command center<br>O7-C Evacuation action completed | O5-C Yellow alert<br>O6-C Red alert | MO1 Civil defense organization<br>ME1 Communication facilities<br>ME6 Special facilities |
| | A33 Evacuate residents | I5 Persuasible evacuation<br>I6 Compulsory evacuation<br>I7 Name list of evacuation distribution | O5-C Yellow alert<br>O6-C Red alert | O7-C Evacuation action<br>O8 Advice notice of persuasible evacuation<br>O9 Affidavit of compulsory evacuation<br>O10 Evacuation confirmation name list<br>O11-C Evacuation completed | MO1 Civil defense organization<br>MO2 Government rescue organization<br>ME1 Communication facilities<br>ME2 Transportation facilities<br>ME3 Search and rescue equipment<br>MP1 Refuge |

**Table 6.** *Cont.*

| Activity | Sub-Activity | Input | Control | Output | Mechanism |
|---|---|---|---|---|---|
| | **A34 Urgent repair on structure** | | O4-C Request emergency repair | O13 Record of vehicular traffic re-opening after urgent repair<br>O14-C Salvage casualties | MO1 Civil defense organization<br>ME1 Communication facilities<br>ME4 Deconstructive equipment<br>ME5 Special facilities<br>MM1 Emergency materials |
| | **A35 Emergency rescue of people** | | O14-C Salvage casualties<br>C4 Emergency rescue handbook<br>C5 Emergency response procedure of disaster | O15-I Announcement of casualty situation | MO1 Civil defense organization<br>MO2 Governmental rescue organization<br>ME1 Communication facilities<br>ME2 Transportation facilities<br>ME3 Search and rescue equipment |
| | **A36 Manage goods and materials** | I8 Emergency goods and materials from county government<br>I9 Name list of household registration | O3-C To be decided by the captain<br>O11-C Evacuation completed<br>C3 Impromptu donation from civil organization<br>C4 Outward roadway obstructed for 3 days | O12 Signing and receiving book for resident (signing, stamping, photographing) | MO1 Civil defense organization<br>ME1 Communication facilities<br>MP2 Storage: container—village office |

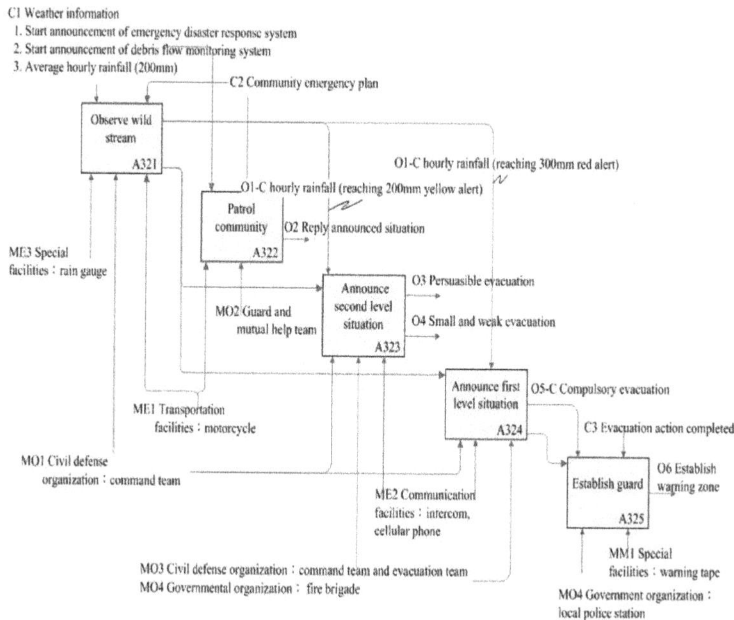

**Figure 7.** The IDFE0 model of the "Establish an advance command post" (A31).

**Figure 8.** The IDFE0 model of the ""Monitor any possible disaster locations" (A32).

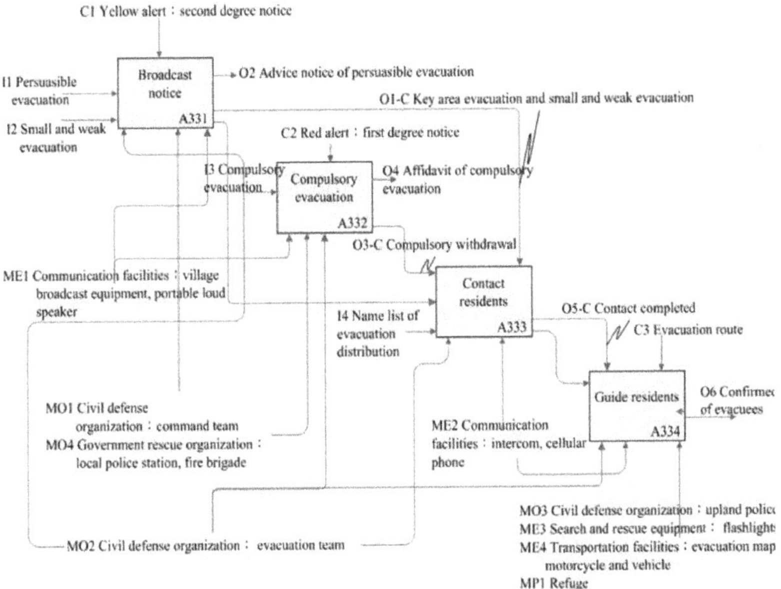

**Figure 9.** The IDFE0 model of the "Evacuate of residents" (A33).

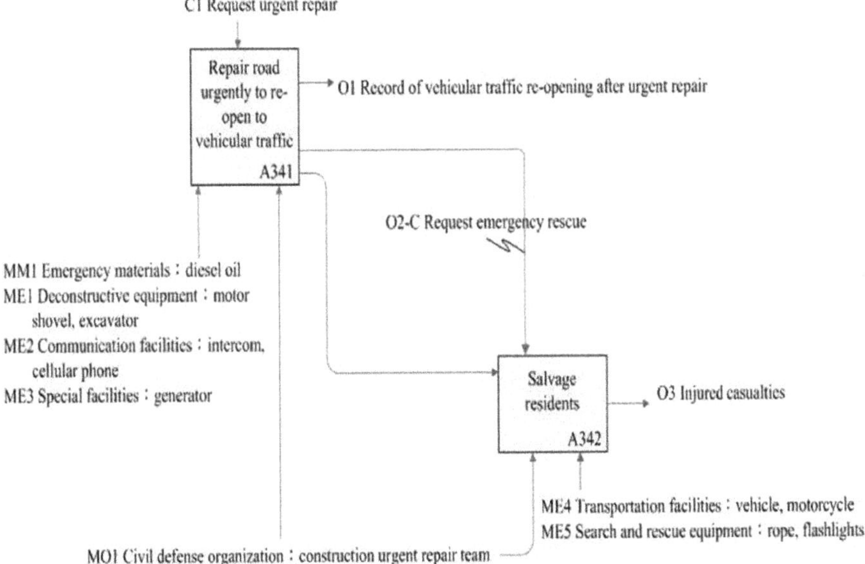

**Figure 10.** The IDFE0 model of the "Urgent repair of structure" (A34).

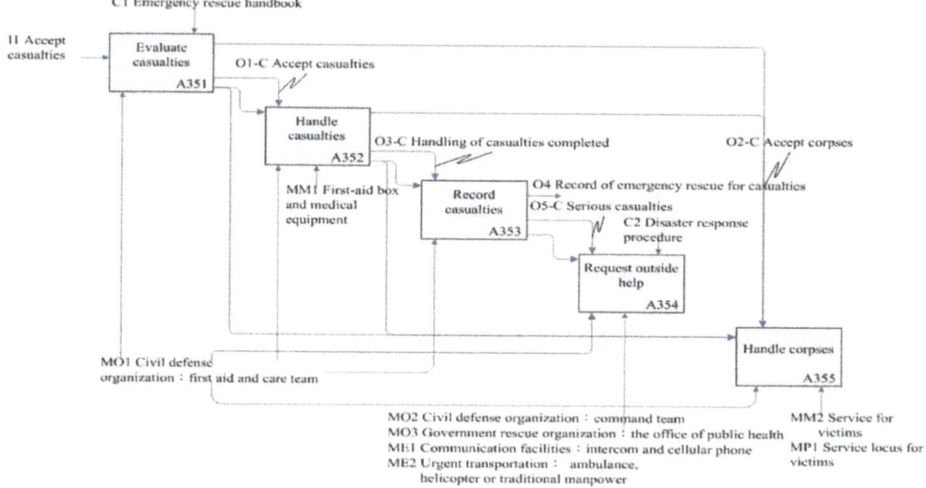

**Figure 11.** The IDFE0 model of the "Emergency rescue of people" (A35).

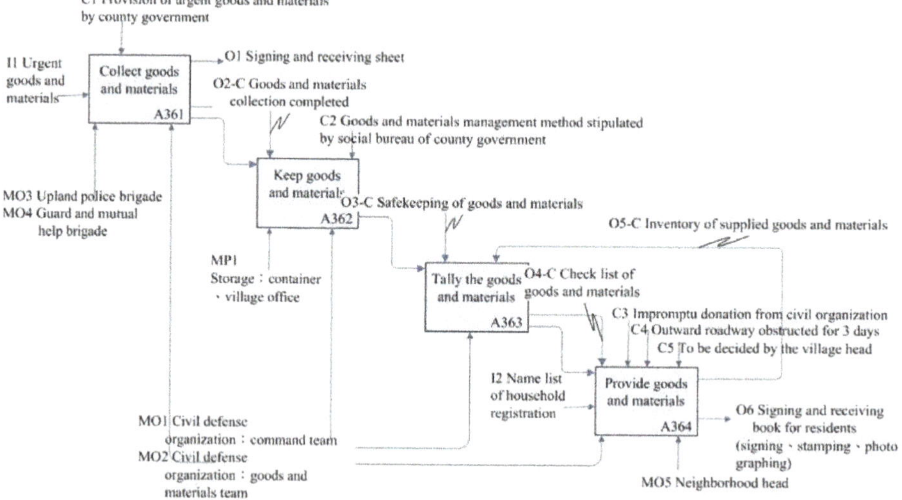

**Figure 12.** The IDFE0 model of the "Manage goods and materials" (A36).

"Monitor any possible disaster locations" (A32) was decomposed into five sub-activities, as shown in Figure 8: "Observe wild stream" (A321), "Patrol community" (A322), "Announce first level situation", (A323), "Announce second level situation" (A324), and "Establish guard" (A325), and its ICOM shown in Table 7. The different mechanisms support different sub-activities as shown in Figure 8. For example, the "Observe wild stream" (A321) and "Patrol community" (A322) were controlled by the "Weather information" (C1) and "Community emergency plan" (C2), while "Establish guard" (A325) was controlled by the "evacuation action completed" (C3). After obtaining the functional IDEF0 model, we then developed the Petri net model for behavior analysis at the next stage.

**Table 7.** The ICOM of "Monitor of any possible disaster location" (A32).

| Activity | Sub-Activity | Input | Control | Output | Mechanism |
|---|---|---|---|---|---|
| A32 Monitor of any possible disaster locations | A321 Observe wild stream | | C1 Weather information<br>1. Start announcement of emergency disaster response system<br>2. Start announcement of debris flow monitoring system<br>3. Average hourly rainfall (200 mm)<br>C2 Community emergency plan | O1-C hourly rainfall (reaching 200 mm yellow alert)<br>O1-C hourly rainfall (reaching 300 mm red alert) | MO1 Civil defense organization: command team<br>ME1 Transportation facilities: motorcycle<br>ME3 Special facilities: rain gauge |
| | A322 Patrol community | | C1 Weather information<br>1. Start announcement of emergency disaster response system<br>2. Start announcement of debris flow monitoring system<br>3. Average hourly rainfall (200 mm)<br>C2 Community emergency plan | O2 Reply announced situation | MO2 Guard and mutual help team<br>ME1 Transportation facilities: motorcycle |
| | A323 Announce second degree situation | | O1-C Hourly rainfall reaching 200 mm (yellow alert) | O3 Persuasible evacuation<br>O4 Small and weak evacuation | MO1 Civil defense organization: command team<br>MO3 Civil defense organization: command team and evacuation team<br>MO4 Governmental organization: fire brigade<br>ME2 Communication facilities: intercom, cellular phone |
| | A324 Announce first degree situation | | O2-C Hourly rainfall reaching 300 mm (red alert) | O5-C Compulsory evacuation | MO1 Civil defense organization: monitor team<br>MO3 Civil defense organization: command team and evacuation team<br>MO4 Government organization: fire brigade<br>ME2 Communication facilities: intercom, cellular phone |
| | A325 Establish guard | | C3 Evacuation action completed | O6 Establish warning zone | MO4 Government organization: local police station<br>MM1 Special facilities: warning tape |

*4.4. Phase-IV: Petri Net Model*

Based on the transformation rules, from static model into dynamic model as suggested by Santarek and Buseif [21], the PN model was constructed for A32's five activities (A321, A322, A323, A324, and A325) built in the IDEF0 model at the previous stage (Figure 8), as shown in Figure 13. Following the $Tr_1$ and $Tr_2$ transformation rules, the PN model generated the following places: P321, P322, P323, P324, P325, P32-C1, P32-C2, P32-C3, P32-O1-C, P32-O5-C and transitions: T321-1, T321-2, T322-1, T322-2, T323-1, T323-2, T324-1, T324-2, T325-1, and T325-2 described in Table 8. The transformation rule, $Tr_3$ (1) and (2), was used for the mechanisms to produce places P32-MO1, P32-MO3, P32-MO4, and P32-ME2 (Table 8). The PN model for "Monitor any possible disaster locations" (A32) consisted of 14 places and 10 transitions (Table 8).

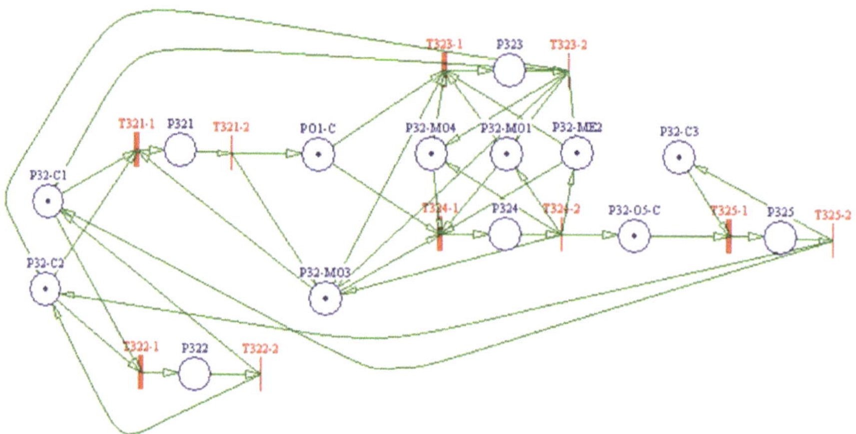

**Figure 13.** The Petri net (PN) model of the "Monitor any possible disaster locations" (A32).

**Table 8.** The notation of place and transition for "Monitor any possible disaster location" (A32).

| Place | Name | IDEF0 | Transition | Name | IDEF0 |
|---|---|---|---|---|---|
| P1 | P321 | A321 Observe wild stream | T1 | T321-1 | A321 Start "Observe wild stream" |
| P2 | P322 | A322 Patrol community | T2 | T321-2 | A321 End "Observe wild stream" |
| P3 | P323 | A323 Announce second level situation | T3 | T322-1 | A322 Start "Patrol community" |
| P4 | P324 | A324 Announce first level situation | T4 | T322-2 | A322 End "Patrol community" |
| P5 | P325 | P325 Establish guard | T5 | T323-1 | A323 Start "Announce second level situation" |
| P6 | P32-C1 | C1 Weather information | T6 | T323-2 | A323 End "Announce second level situation" |
| P7 | P32-C2 | C2 Community disaster protection plan | T7 | T324-1 | A324 Start "Announce first level situation" |
| P8 | P32-O1-C | O1-C Hourly rainfall reaching 200 mm (yellow alert) | T8 | T324-2 | A324 End "Announce first level situation" |
| P9 | P32-O5-C | O5-C Compulsory evacuation | T9 | T325-1 | A325 Start "Establish guard" |
| P10 | P32-C3 | C3 Evacuation action completed | T10 | T325-2 | A325 End "Establish guard" |
| P11 | P32-MO1 | MO1 Civil defense organization: monitor team | | | |
| P12 | P32-MO3 | MO3 Civil defense organization: command team and evacuation team | | | |
| P13 | P32-MO4 | MO4 Government organization: fire brigade | | | |
| P14 | P32-ME2 | ME2 Communication facilities: intercom, cellular phone | | | |

After constructing the PN model, a DaNAMiCS software package was used to verify the behavioral properties of the developed Petri net model. The incidence matrix (D), P-invariants, and T-invariants are shown in Figure 14. The nonzero nonnegative integer entries in a P-invariant, such as [1,1,1,1,1,1,1,0,0,0,0,0], [0,0,0,0,1,0,0,0,1,0,0,0,0], [0,0,1,1,0,0,0,0,0,1,0,0,0], [1,0,1,1,0,0,0,0,0,0,1,0,0], [0,0,1,1,0,0,0,0,0,0,0,1,0], [0,0,1,1,0,0,0,0,0,0,0,0,1], revealed that the weights associated with the corresponding place as such that the weighted sum of tokens on these places was constant for all marking reachable from an initial marking. Similarly, for the purpose of avoiding deadlock, the nonzero nonnegative integer entries in a T-invariant, such as [0,0,1,1,0,0,0,0,0,0], [1,1,0,0,1,1,0,0,0,0], [1,1,0,0,0,0,1,1,1,1], represented not only the firing counts of the corresponding transitions that belonged to a firing sequence transforming a marking m0 back to m0, but also the number of times these transitions appeared in this sequence.

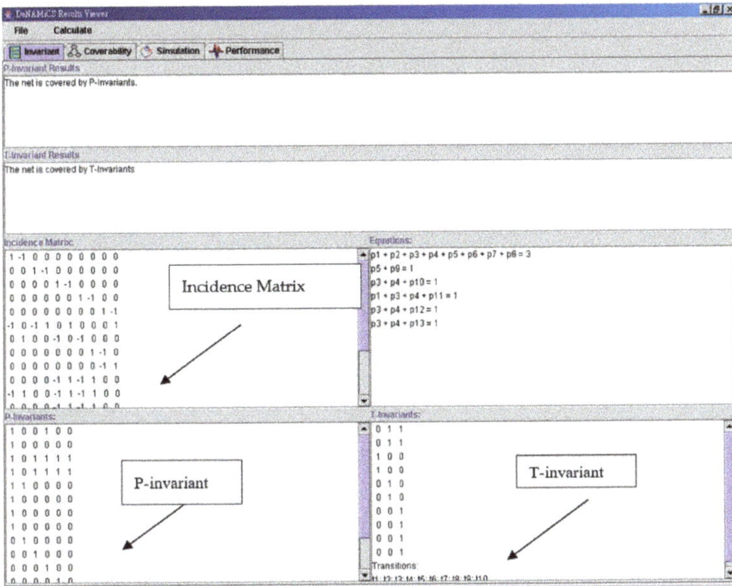


**Figure 14.** The places (P)-invariant and transition(T)-invariant values of A32.

### 4.5. Phase-IV: Petri Net Mark Language (PNML)

At the symbol level model (the bottom layer in Figure 1), in order to interweave understandable human symbols with a machine-readable format effectively, PNML was adopted as a starting point for a standard interchange format for Petri nets. The XML-based interchange format for the A32 Petri net model is shown in Figure 15. The relationships between the places and transitions and attributes of transitions are noted in this figure.

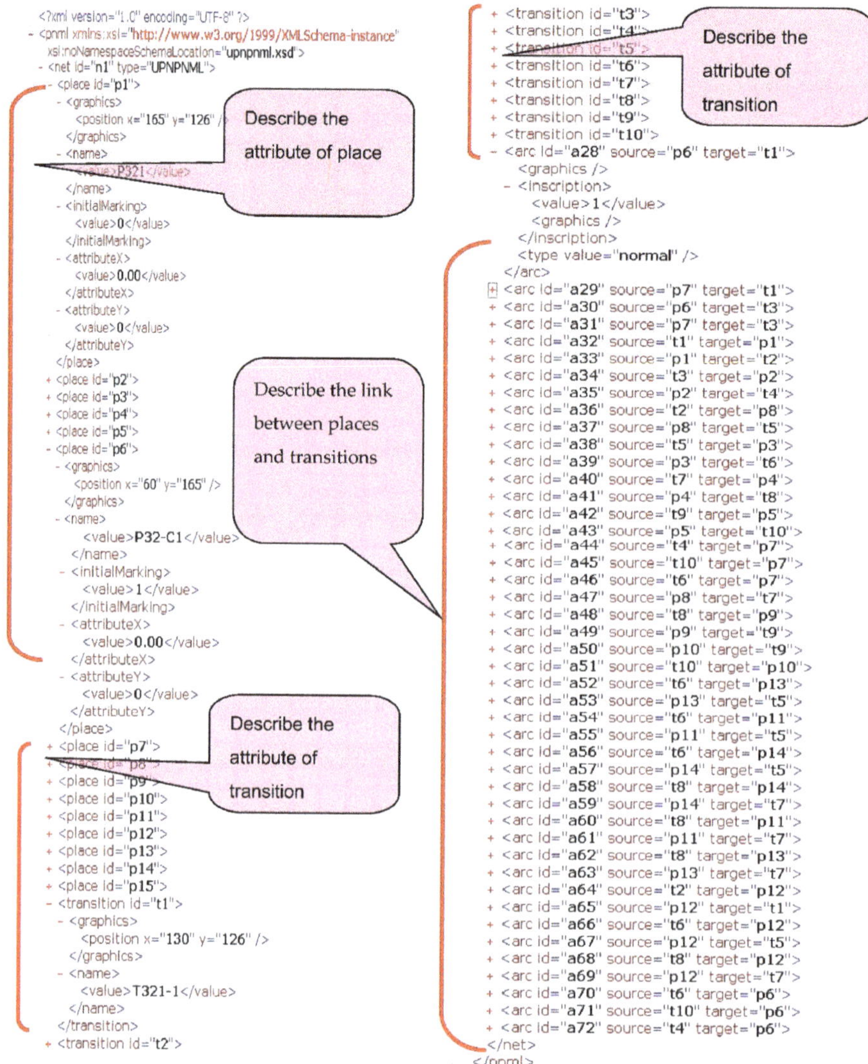


**Figure 15.** The XML-based interchange format for the Petri net model.

## 5. Conclusions and Future Work

Following the advent of Web service technology associated with the continuing rapid growth in knowledge management, problem-solving knowledge has increasingly grown dependent on the Internet, particularly in carrying out ICS operations. To bridge the gap in understanding and facilitate communications between computers and human beings, we presented the TTIPP framework and its related methodologies. The framework we developed consisted of three layers, including lexical, conceptual, and symbolic, and five phases: task analysis, task ontology, IDEF0 model, Petri net model, and PNML. The IDEF0 model was used to capture the requirements corresponding to the system specification at the stage of functional analysis. Subsequently, at the stage of behavior analysis, the Petri net model was constructed according to the IDEF0 model. Finally, at the implementation stage, the model could be realized by using PNML.

The TTIPP methodology is general and can be used to solve any linguistic problem. It provides a sound ontological foundation for different problem-solving approaches and can be used to support a great variety of task modeling, independent of the target shell or computational method. The TTIPP model overcomes the drawbacks of IDEF0 in terms of its static nature and Petri net with no hierarchy for concepts. Moreover, it not only enables better access to information and promotes shared understanding of real-world problem-solving knowledge for humans in an explicit and reusable manner, but also facilitates comprehension of information and better processing by computers.

Protecting people from natural calamities and maintaining the quality of the natural environment are complex problems. It requires the development of an effective incident command system which requires collaboration and participation among government agencies, academic institutes, private industries, non-governmental organizations, and local communities. As an example, the TTIPP model was applied to the task of emergency response for debris-flow during a typhoon as a part of an ICS.

Within the field of knowledge management, future research should focus on developing reusable and sharable real-world problem-solving knowledge models. We plan to use the TTIPP methodology as a major building block for developing generic semantic web problem-solvers.

**Author Contributions:** K.F. developed the concept of the TTIPP methodology, analyzed the task, developed the task ontology, built the IDEF0 and Petri Net model, transferred the PNML. S.L. interviewed the stakeholders and transferred verbatim. K.F. wrote the paper.

**Funding:** This research was supported in part by the following National Science Council (NSC) grants: NSC 93-2625-Z-224-002.

**Conflicts of Interest:** The authors declare no conflict of interest.

## References


1. Latest Earthquakes in the World. Available online: https://www.emsc-csem.org/Earthquake/world/ (accessed on 1 April 2019).
2. Dutta, A. Integrating ai and optimization for decision support: A survey. *Decis. Support Syst.* **1996**, *18*, 217–226. [CrossRef]
3. Geamsakul, W.; Yoshida, T.; Ohara, K.; Motoda, H.; Yokoi, H.; Takabayashi, K. Constructing a decision tree for graph-structured data and its applications. *Fundam. Inf.* **2005**, *66*, 131–160.
4. Ghenniwa, H.; Huhns, M.; Shen, W. Emarketplaces for enterprise and cross enterprise integration. *Data Knowl. Eng.* **2005**, *52*, 33–59. [CrossRef]
5. Nora, M.; El-Gohary, A.; El-Diraby, T. Dynamic knowledge-based process integration portal for collaborative construction. *J. Constr. Eng. Manag.* **2010**, *136*, 316–328.
6. Schreiber, G.H.; Akkermans, H.; Anjewierden, A.; de Hoog, R. *Knowledge Engineering and Management: The Common KADS Methodology*; MIT Press: Boston, MA, USA, 2000.
7. Kendal, S.; Creen, M. *An Introduction to Knowledge Engineering*; Springer: London, UK, 2007.
8. Nonaka, I. The knowledge-creating company. *Harv. Bus. Rev.* **1991**, *69*, 96–104.
9. Alonderienne, R.; Pundziene, A.; Krisciunas, K. Tacit knowledge acquisition and transfer in the process of informal learning. *Probl. Perspect. Manag.* **2006**, *4*, 134–145.
10. Fuster, J.M. Hayek in today's cognitive neuroscience. In *Advance in Austrian Economic*; Emerald Publishing Limited: Bingley, UK, January 2011.
11. Erkut, B. Product innovation and market shaping: Bridging the gap with cognitive evolutionary economics. *IJM* **2016**, *4*, 3–24.
12. Kaya, T.; Erkut, B. Tacit knowledge capacity: A comparison of university lecturers in Germany and North Cyprus. *Electr. J. Knowl. Manag.* **2018**, *16*, 131–142.
13. Polanyi, M. *The Tacit Dimension*; Garden City: New York, NY, USA, 1966.
14. Rahe, M. Subjectivity and cognition in knowledge management. *J. Knowl. Manag.* **2009**, *13*, 102–117. [CrossRef]
15. Nonaka, I.; Takeuchi, H. *The Knowledge Creating Company*; Oxford University Press: Oxford, UK, 1995.
16. Weiss, S.M.; Indurkhya, N.; Zhang, T. *Fundamentals of Predictive Text Mining*; Springer: Berlin, Germany, 2015.

17. Fronzetti Colladon, A.; Gloor, P.A. Measuring the impact of spammers on email and twitter networks. *Int. J. Inf. Manag.* **2018**. [CrossRef]
18. Aloini, D.; Benevento, E.; Stefanini, A.; Zerbino, P. Process fragmentation and port performance: Merging SNA and text mining. *Int. J. Inf. Manag.* **2019**, in press. [CrossRef]
19. Labuschagnea, C.; Brenta, A.C.; Ron, P.G.; Van Ercka, P.G. Assessing the sustainability performances of industries. *J. Clean. Prod.* **2005**, *13*, 373–385. [CrossRef]
20. Miemczyk, J.; Johnsen, T.E.; Macquet, M. Sustainable purchasing and supply management: A structured literature review of definitions and measures at the dyad, chain and network levels. *Supply Chain Manag.* **2012**, *17*, 478–496. [CrossRef]
21. Santarek, K.; Buseif, I.M. Modelling and design of flexible manufacturing systems using SADT and Petri nets tools. *J. Mater. Process. Technol.* **1998**, *76*, 212–218. [CrossRef]
22. Gold, A.H.; Malhotra, A.; Segars, A.H. Knowledge management: An organizational capabilities perspective. *J. Manag. Inf. Syst.* **2001**, *18*, 185–214. [CrossRef]
23. Adeli, H. *Knowledge Engineering—Volume One—Fundamentals*; McGraw-Hill Book Company: New York, NY, USA, 1990.
24. Adeli, H. *Knowledge Engineering—Volume Two—Applications*; McGraw-Hill Book Company: New York, NY, USA, 1990.
25. Moreno, V., Jr.; Cavazotteb, F. Using information systems to leverage knowledge management processes: The role of work context, job characteristics and task-technology Fit. *Proc. Comput. Sci.* **2015**, *55*, 360–369. [CrossRef]
26. Yang, W.; McKinnon, M.C.; Turner, W.R. Quantifying human well-being for sustainability research and policy. *Ecosyst. Health Sustain.* **2015**, *1*, 1–13. [CrossRef]
27. Corcho, O.; Fernandez-Lopez, M.; Gomez-Perez, A.; Lopez-Cima, A. Building Legal Ontologies with METHONTOLOGY and WebODE. *Lect. Notes Comput. Sci.* **2005**, *3369*, 142–157. [CrossRef]
28. Kotis, K.; Vouros, G. Human-centered ontology engineering: The HCOME methodology. *Knowl. Inf. Syst.* **2006**, *10*, 109–131. [CrossRef]
29. Villazon-Terrazas, B.; Ramirez, J.; Suarez-Figueroa, M.C.; Gomez-Perez, A. A network of ontology networks for building e-employment advanced systems. *Expert Syst. Appl.* **2011**, *38*, 13612–13624. [CrossRef]
30. Pinto, H.S.; Tempich, C.; Staab, S. Ontology engineering and evolution in a distributed world using DILIGENT. In *Handbook on Ontologies—International Handbooks on Information Systems*, 2nd ed.; Staab, S., Studer, R., Eds.; Springer: Berlin, Germany, 2009; pp. 153–176.
31. Saa, R.; Garcia, A.; Gomez, C.; Carretero, J.; Garcia-Carballeira, F. An ontology-driven decision support system for high-performance and cost-optimized design of complex railway portal frames. *Expert Syst. Appl.* **2012**, *39*, 8784–8792. [CrossRef]
32. Ziemba, P.; Wątróbski, J.; Jankowski, J.; Wolski, W. Construction and restructuring of the knowledge repository of website evaluation methods. In *Information Technology for Management–Lecture Notes in Business Information Processing*; Ziemba, E., Ed.; Springer Nature: Basel, Switzerland, 2016; Volume 243, pp. 29–52.
33. Elias, M.; Lohmann, S.; Auer, S. Ontology-based representation for accessible opencourseware systems. *Information* **2018**, *9*, 302. [CrossRef]
34. Meng, X.; Xu, C.; Liu, X.; Bai, J.; Zheng, W.; Chang, H.; Chen, Z. An ontology-underpinned emergency response system for water pollution accidents. *Sustainability* **2018**, *10*, 546. [CrossRef]
35. Konys, A. An ontology-based knowledge modelling for a sustainability assessment domain. *Sustainability* **2018**, *10*, 300. [CrossRef]
36. Gloor, P.A.; Fronzetti Colladon, A.; Grippa, F.; Giacomelli, G. Forecasting managerial turnover through e-mail based social network analysis. *Comput. Hum. Behav.* **2017**, *71*, 343–352. [CrossRef]
37. Antonacci, G.; Fronzetti Colladon, A.; Stefanini, A.; Gloor, P. It is rotating leaders who build the swarm: Social network determinants of growth for healthcare virtual communities of practice. *J.Knowl. Manag.* **2017**, *21*, 1218–1239. [CrossRef]
38. Barker, R. Management of knowledge creation and sharing to create virtual knowledge-sharing communities: A tracking study. *J. Knowl. Manag.* **2015**, *19*, 334–350. [CrossRef]
39. Noy, N.F.; Musen, M.A. Ontology versioning in an ontology management framework. *IEEE Intell. Syst.* **2004**, *19*, 6–13. [CrossRef]

40. Therani, M. Ontology development for designing and managing dynamic business process networks. *IEEE Trans. Ind. Inf.* **2007**, *3*, 173–185. [CrossRef]

41. Mizoguchi, R.; Tijerino, Y.; Ikeda, M. Task analysis interview based on task ontology. *Expert Syst. Appl.* **1995**, *9*, 15–25. [CrossRef]

42. Mohammad, A.; Zakaria, N.; Hidayati, N.; Darshana, S. Ontology-based knowledge management for enterprise systems. *Int. J. Enterp. Inf. Syst.* **2011**, *7*, 64–90.

43. Gruber, T.R.A. A translation approach to portable ontology specification. *Knowl. Acquis.* **1993**, *5*, 199–221. [CrossRef]

44. Musen, M.A. The Protégé project: A look back and a look forward. *AI Matters* **2015**, *1*, 4–12. [CrossRef] [PubMed]

45. David, R.; Alla, H. Petri Nets for modeling of dynamics systems—A survey. *Automatica* **1994**, *30*, 175–202. [CrossRef]

46. Gomez-Perez, A.; Fernandez-Lopez, M.; Corcho, O. *Ontology Engineering*; Springer: Berlin/Heidelberg, Germany, 2004.

47. Shih, D.H.; Chiang, H.S.; Lin, B. A generalized associative Petri Net for reasoning. *IEEE Trans. Knowl. Data Eng.* **2007**, *19*, 1241–1251. [CrossRef]

48. Bialas, A. Computer-aided sensor development focused on security issues. *Sensors* **2016**, *16*, 759. [CrossRef] [PubMed]

49. Park, H. Task model and task ontology based on mobile users' generic activities for task-oriented tourist information service. *Int. J. Smart Home* **2013**, *7*, 33–44.

50. Van Heijst, G.; Schreiber, A.T.; Wielinga, B.J. Using explicit ontologies in KBS development. *Int. J. Hum. Comput. Stud.* **1997**, *46*, 183–292. [CrossRef]

51. Guarino, N. Formal Ontology and Information System. 1998. Available online: http://www.loa-cnr.it/Papers/FOIS98.pdf (accessed on 11 June 2015).

52. Lassila, A.; McGuinness, D. *The Role of Frame-Based Representation on the Semantic Web (Ksl-01-02)*; Knowledge System Laboratory, Stanford University: Stanford, CA, USA, 2001.

53. Benjamins, V.R.; Gomez-Perez, A. Knowledge-system technology: Ontologies and problem-solving. 2004. Available online: http://hcs.science.uva.nl/usr/richard/pdf/kais.pdf (accessed on 6 October 2004).

54. Chandrasekaran, B.; Benjamins, V.R. What are ontologies, and why do we need them. *IEEE Intell. Syst.* **1999**, *14*, 20–26. [CrossRef]

55. Newell, A. The knowledge-level. *Artif. Intell.* **1982**, *18*, 87–127. [CrossRef]

56. Mizoguchi, R.; Sunagawa, E.; Kozaki, K.; Kitamura, Y. A model of roles within an ontology development tool: Hozo. *J. Appl. Ontol.* **2007**, *2*, 159–179.

57. Ikeda, M.; Seta, K.; Kakusho, O.; Mizoguchi, R. Task ontology: Ontology for building conceptual problem solving models. In Proceedings of the ECAI98 Workshop on Applications of Ontologies and Problem-Solving Model, Osaka, Japan, 23–28 August 1998; pp. 126–133.

58. Rajpathak, G.D.; Motta, E.; Zdráhal, Z.; Roy, R. A generic library of problem solving methods for scheduling applications. *IEEE Trans. Knowl. Data Eng.* **2006**, *18*, 815–828. [CrossRef]

59. Tho, Q.T.; Hui, S.C.; Fong, A.C.M.; Cao, T.H. Automatic fuzzy ontology generation for semantic Web. *IEEE Trans. Knowl. Data Eng.* **2006**, *18*, 842–856. [CrossRef]

60. Berner-Lee, T. *Weaving the Web: The Original Design and Ultimate Destiny of the World Wide Web by its Inventor*; HarperCollins Publishers: New York, NY, USA, 1999.

61. Corby, O.; Dieng-Kuntz, R.; Gandon, F.; Faron-Zucker, C. Searching the semantic web: Approximate query processing based on ontologies. *IEEE Intell. Syst.* **2006**, *21*, 20–27. [CrossRef]

62. Colguhoun, G.J.; Baines, R.W.; Crossley, R. A composite behavior modeling approach for manufacturing enterprise. *Int. J. Comput. Integr. Manuf.* **1996**, *9*, 463–475. [CrossRef]

63. Pan, J.Z. A flexible ontology reasoning architecture for the semantic web. *IEEE Trans. Knowl. Data Eng.* **2007**, *19*, 246–260. [CrossRef]

64. Farida, B.-D.; Catherine, C.; Malik, S.-M.; Pierre-Jean, C. Ontology based teaching domain knowledge management for E-Learning by doing systems. *Electr. J. Knowl. Manag.* **2015**, *13*, 156–171.

65. Ross, D.T. Application and extensions of SADT. *Computer IEEE* **1985**, *18*, 25–34. [CrossRef]

66. Feldmann, C.G. *The Practical Guide to Business Process Reengineering Using IDEF0*; Dorset House Publishing: New York, NY, USA, 1998.

67. Lee, J.S.; Hsu, P.L. An IDEF0/Petri Net approach to the system integration in semiconductor manufacturing systems. *IEEE Int. Conf. Syst. Man Cybern.* **2003**, *5*, 4910–4915.
68. Zhou, M.C.; Jeng, M.D. Modeling, analysis, simulation, scheduling, and control of semiconductor manufacturing systems: A Petri Net approach. *IEEE Trans. Semicond. Manuf.* **1998**, *11*, 333–357. [CrossRef]
69. Cassandras, C.G.; Lafortune, S. *Introduction to Discrete Systems*; Kluwer: Boston, MA, USA, 1999.
70. Murata, T. Petri Nets: Properties, analysis and application. *Proc. IEEE* **1989**, *77*, 541–580. [CrossRef]
71. Billington, J.; Christensen, S.; van Hee, K.; Kinder, E.; Kummer, O.; Petrucci, L. The Petri Net Markup Language: Concepts, Technology, and Tools. 2004. Available online: http://www.informtik.hu-berlin.de/top/PNX (accessed on 25 October 2008).

*Article*

# Inter-University Sustainability Benchmarking for Canadian Higher Education Institutions: Water, Energy, and Carbon Flows for Technical-Level Decision-Making

**Abdulaziz Alghamdi [1],*, Husnain Haider [2], Kasun Hewage [1] and Rehan Sadiq [1]**

[1]   School of Engineering, University of British Columbia (Okanagan), 3333 University Way, Kelowna, BC V1V 1V7, Canada; kasun.hewage@ubc.ca (K.H.); rehan.sadiq@ubc.ca (R.S.)
[2]   Civil Engineering Department, College of Engineering, Qassim University, Buraydah, Qassim 52571, Saudi Arabia; husnain@qec.edu.sa
*   Correspondence: aziz.ghamdi@alumni.ubc.ca; Tel.: +1-250-899-4073

Received: 16 April 2019; Accepted: 2 May 2019; Published: 6 May 2019

**Abstract:** The education sector is one of the major contributors to the total greenhouse gas (GHG) emissions in Canada, i.e., 16% of total emissions among 11 sectors. Canadian higher education institutions (HEIs) consume around 60% of the total energy fed to the educational sector. Existing tools holistically cover a wide array of functions to assess the sustainability of HEIs. The infrastructure (engineered) systems are the pivotal units responsible for the majority of energy and water consumption and may have been built, retrofitted, or replaced at different times using different materials and technologies. Consequently, infrastructures have varying efficiency, designs, building envelopes, and environmental impacts. For technical-level decision making for improving the engineered systems, HEIs need to be benchmarked on the basis of their water, energy, and carbon flows. A methodology is developed for sustainability assessment of 34 Canadian HEIs that are classified into small, medium, and large sizes based on their number of full-time equivalent students (FTE). Energy, water consumption, number of students, and floor area is measured in different units and are, thus, normalized. The study revealed that the energy source was the primary factor affecting the sustainability performance of an institution. The analysis also revealed that small-sized institutions outperformed medium-to-large-sized institutions.

**Keywords:** higher education institutions; sustainability assessment; benchmarking; energy flow; carbon flow; sustainable development

---

## 1. Introduction

Higher educational institutions (HEIs), such as universities and colleges, are mandated to educate citizens, engage communities, lead innovation, and serve as hubs for science and research [1]. Owing to the specific role of universities in fostering talent and educating future generations, it can be argued that universities should have higher positive impact on a community than any other public sector [2–4]. Universities also have large socio-economic impacts on their communities. For instance, 96 Canadian HEIs with a cumulative budget of $35 billion/year employ around 250,000 people [5]. This amount is approximately 2.79% of the nation's gross domestic production [6]. Appreciating their worth, HEIs have been promoting and implementing sustainable development (SD) principles in their growth plans over the past few decades [7].

In the 21st century, SD has been recognized as an important concept in the perspective of socio-economic and environmental entangled policies [8]. HEIs are assessing their overall sustainability in the wider context by situating the concepts of sustainability in all their functions, including academics (e.g., curriculum and research activities), engagement (e.g., employee orientation and community participation), planning and administration (e.g., participatory governance, diversity, and equity coordination), and operations (e.g., climate, buildings, energy, transportation, and water) [9]. For example, Bauer et al. [10] applied an empirical approach to improving the sustainability governance at universities in Germany which covers politics, profession, organization, knowledge, and visibility. Universities are complex physical infrastructure systems housing roads, water, and wastewater facilities, and different types of buildings that vary in scope, operation, characteristics, and area. Universities may be regarded as a small city [11]. The significance or magnitude of infrastructures in Canadian HEIs has yet to be adequately realized.

The National Science Foundation defined an engineered system as "a combination of components that work in synergy to collectively perform a useful function," such as manufacturing processes, technologies to improve quality of service, or infrastructure systems [12]. In HEIs, the operation of engineered (infrastructure) systems and other activities (i.e., educational services, recreation and sports, and other extra-curricular activities) utilize immense natural resources to meet water and energy requirements and, consequently, produce carbon emissions. The existing approaches discussed in the following section aggregate the performance of all the functions to rank HEIs based on their overall sustainability. The final outcomes mask the actual facts about the utilization of natural resources and environmental impacts in an HEI. Hence, a sustainability assessment of these engineered systems in terms of water, energy, and carbon (WEC) flows is essential for technical-level decision making about improving their performance on a continuous and long-term basis (e.g., greener energy source, efficient plumbing fixtures and lighting systems, wastewater and waste to energy recycling, and building retrofits).

Globally, HEIs have been identified as one of the major sources of greenhouse gas (GHG) emissions in an urban setting. A study in the UK revealed that energy consumption increased by 2.7% and the related $CO_2$ emissions rose by 4.3% between 2005 and 2006. Overall, GHG emissions generated through HEIs in the UK soared from 1.78 to 2.05 $tCO_2e$ (metric tons of carbon equivalent) during the period from 1990 to 2005 [13]. Similarly, in China, universities are the largest public sector contributing to the total GHG emissions by consuming 40% of the energy devoted to the public sector [14]. Emissions per student at the Norwegian University of Technology and Science were significantly higher than the national average emission reported per citizen. Consequently, HEIs have been investing more in terms of investments to reduce their carbon emissions [15].

Educational services have been subgrouped under Canada's largest activity category as the commercial and institutional (C&I) sector. Emissions from the Canadian C&I sector are predicted to increase by 23% by 2020 from the 2011 reported values [16]. The C&I sector consumes 10% of the total energy and emits 9% of the nation's GHG production as of 2013 [17]. It is estimated that educational services have consumed 16% of the total energy supplied to the C&I sector at a staggering value of 134.7 PJ (Petajoule) ~$1.34 \times 10^8$ GJ (Gigajoules) in 2009. If this trend continues, these services are likely to become the country's largest emission producing subgroup in the C&I category. Furthermore, HEIs alone represent 60% of the total energy consumption in the educational sector (i.e., both primary and higher education), which is equivalent to approximately 430,000 households in Canada [18,19]. Canadian HEIs primarily rely on the national electricity grid that is often produced by fossil fuels and emits high GHG emissions. As HEIs consume water and energy (i.e., input flows) and produce GHG emissions (i.e., output flows), the term WEC flows is used for convenience.

The main objectives of this study were are to: (i) comprehensively review the current state of sustainability in HEIs including international declarations, reporting systems, and size-based classification, (ii) establish sized-based classification criteria for universities, (iii) develop a WEC flow weighting criteria and assessment methodology to be used for sustainability assessment of engineered systems in HEIs, and (iv) benchmark the Canadian HEIs based on their WEC flows for effective technical-level decision making for continuous performance improvement and to achieve long-term sustainability.

The present study encompasses WEC flows for 34 HEIs where 617,325 full-time equivalent (FTE) students used 12.7 million $m^2$ of building area. The combined energy usage in these HEIs was approximately 20 GJ. Statistical analysis was performed to fit a relationship between the response variable (GHG emissions) and other parameters, such as climate, energy, water, area, and the number of occupants (FTEs). Finally, a mathematical model was developed to rank the universities based on their WEC flows. The findings provided a platform to identify the underperforming HEIs and compare them with the excelling HEIs to highlight the most important factors affecting the ranking score.

## 2. Background

### 2.1. Sustainable Development in HEIs

The idea of sustainability in HEIs was introduced in the Stockholm declaration held in 1972 [20]. The historical background of declarations on the sustainability of HEIs developed by different organizations around the globe is given in Appendix A. For several years, the emphasis of the declarations was on introducing, integrating, and implementing the concepts of sustainability in HEI curricula development, enhancing teaching techniques, securing research funding [21,22], promoting awareness in the general public [23], and developing regional and international networks to support the concept of sustainability in HEIs [21–24]. Until the start of the 21st century, although the declarations prompted HEIs to comply with the stated objectives, the declarations were vague and did not measure the sustainability performance of HEIs at an operational level [25].

Velazquez et al. [26] formally included the impact of resource utilization in the sustainability of a university campus. They defined a sustainable university as "a whole or as a part, that addresses, involves and promotes, on a global level, the minimization of negative environmental, economic, societal and health effects generated in the use of their resources in order to fulfill its functions of teaching, research, outreach and partnership, and stewardship in ways to help society make the transition to a sustainable lifestyle." A review of the declarations revealed that the G8 University Summit Sapporo Sustainability Declaration (No 19, Table A1, Appendix A) held in 2008 in Sapporo, Hokkaido (Japan), specified in the principles that universities should develop sustainable (green) campuses as experimental models (G8, 2008) [27]. The idea was appreciated in almost all subsequent declarations. As a result, universities, including Canadian HEIs, started to appraise the sustainability of their infrastructures in terms of natural resource utilization and associated environmental impacts. In this section, we briefly outline the state of operational-level sustainability assessment in Canadian universities.

The United Nations (in 2030 Agenda for Sustainable Development) released 17 new agenda goals and 169 ambitious targets aimed to engage and define sustainable development. The report called all parties to reduce the accelerating GHG emission and recognized the disappearances among the committed targets and results of holding global temperature below two degree Celsius or 1.5 degree Celsius above pre-industrial levels. The fourth goal proclaims the vital role of education (pedagogical), training and public engagement and access to information are emphasized in this agreement aiming to bridge the technological innovation across the areas of concern, namely energy and medicine [28]. There is an instrumental argument shaped by the notion that HEIs have the ability to shape the future leaders of the world, who morally have an obligation to engage sustainable development in their communities [29].

A university campus consists of different infrastructure systems that consume vast quantities of water and energy throughout their life cycles. Figure 1 shows that most of the energy over the entire lifespan of a building is consumed during the operational phase [30]. During the operational phase, buildings are believed to emit 50% to 80% of emissions in their entire lifespan [30–32]. Managing the environmental impacts and optimizing water and energy consumption are critical steps in the pursuit of objectives, such as zero-energy, green buildings, and sustainable campuses [33]. Top 10 GHG emitting HEIs in Canada are listed in Table 1 [34]. The average value for the HEIs listed in Table 1 is 4.41 tCO$_2$e per FTE student [34]. In the United States, average annual GHG emissions from all institutional classifications (colleges and universities) were reported as 7.67 tCO$_2$e per FTE student for the year 2010 [35]. With an increase in efforts towards sustainability, significant reductions have been noticed by the HEIs in the country during the present decade. A 13% reduction in just four years (2010 and 2014) from around 11 $tCO_2e$ per FTE student to 9.55 $tCO_2e$ per FTE student was reported by University of Nebraska-Lincoln [36]. Another example is the Massachusetts Institute of Technology where the total net emissions fell 20 % below 2014 baseline, primarily due to shifting to renewable energy sources. It is worth to mention that 3% increase in total emissions was observed due to an increase in the area of buildings [37]. Hence, continuous efforts would be needed for sustainable buildings' operations in future all over North America.

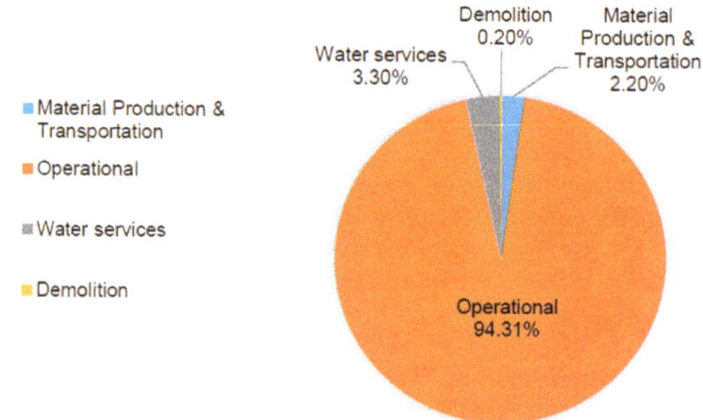


**Figure 1.** Primary energy usage in buildings.

**Table 1.** Top 10 Canadian greenhouse gas (GHG) per full-time equivalent students (FTEs) emitting higher education institutions (HEIs) [34].

| No | HEI | Fiscal Year | Number of FTEs | GHG Emissions ($tCO_2e$/ FTE) |
| --- | --- | --- | --- | --- |
| 1 | University of Saskatchewan | 2015/2016 | 18,082 | 8.43 |
| 2 | University of Alberta | 2015/2016 | 34,693 | 8.24 |
| 3 | University of Calgary | 2017/2018 | 28,860 | 6.31 |
| 4 | Dalhousie University | 2016/2017 | 17,610 | 4.94 |
| 5 | University of New Brunswick | 2015/2016 | 6748 | 4.59 |
| 6 | Northern Alberta Institute of Technology | 2013/2014 | 12,479 | 4.34 |
| 7 | MacEwan University | 2016/2017 | 12,623 | 2.22 |
| 8 | Nova Scotia Community College | 2015/2016 | 10,420 | 1.94 |
| 9 | Western University | 2016 | 32,529 | 1.71 |
| 10 | McGill University | 2014 | 31,755 | 1.36 |

HEIs have set targets and goals on individual bases through declarations. These declarations have not been effectively implemented due to lack of incentive structures [2,38–41]. Grindsted [38] studied 31 HEIs declarations and concluded that declarations aid the scope and intensity between HEIs and governments by: setting the role and function of universities (definition of SD in HEIs), declarations have set the base for a broader legislative measure on the national level, and finally increase competition between HEIs which, in turn, led to broader assessment tools. Lozano et al. [39] highlighted the importance of the declarations in three fronts, raise awareness of environmental degradation, communicate progress within universities and their stakeholders, and encourage universities to perform through benchmarking. However, many of these goals have not met their intended targets to date [23]. Studies on the effect of declarations on HEIs' sustainability are ambiguous, it is true that declarations played a positive role in setting up a broader definition of sustainable assessment tools in

HEIs [29,42]. Although these tools raised the awareness of the sustainability in HEIs, the real outcomes through their practical implementation could not be attained, so far [29,43].

The impacts of buildings in Canadian HEIs have not been clearly identified so far because of the difficulties in setting emission targets for their diverse classification, e.g., residential, administrative, and operational [13]. Buildings have been built or retrofitted at different times using a range of materials and technologies and exhibit varying process efficiency, characteristics, building envelopes, and environmental impacts during their lifetime. Energy and water metabolic flows have significantly increased because of rapid population growth, climate change, enhanced lifestyles, and technological advances. Canada's emission trend report estimated that the per capita emissions during 2011 were around 20 $tCO_2e$ [16]. Diverse trends in HEIs were observed. For instance, the Vancouver campus of the University of British Columbia (UBC) emitted nearly 53,000 $tCO_2e$ (~1.22 $tCO_2e$ /FTE) [44], and the University of Alberta, operating in extreme cold weather conditions, emitted nearly 300,000 $tCO_2e$ (~8.24 $tCO_2e$ /FTE) [34]. Therefore, the type of energy sources and heating and cooling requirements play an important role in the sustainability assessment process.

In Canada, regional goals have been set by local authorities to achieve specific targets for HEIs. For example, based on the 2007 baseline emission levels, UBC set goals for 33% reduction by 2015, 67% by 2020, and 100% by 2050 [44]. UBC was able to achieve a 28% reduction in its GHG emission by 2015 [45,46]. These results show the need for more aggressive efforts toward sustainability even in motivated HEIs like the UBC.

### 2.2. Sustainability Reporting Systems

Performance indicators were suggested as a broader tool to assess sustainability performance [47]. Wackernagel and Rees [48] developed an ecological footprint tool to evaluate the performance per unit area of land or number of students. Although the tool lacks standardization and has limited flexibility to include cost-related parameters and insufficient information for decision making [25], it was useful for communicating performance in terms of environmental impact on the basis of unit area or per student [49]. Such a basis provided the opportunity for benchmarking systems with dissimilar sources of water and energy.

HEIs report their SD status for cross comparison with their peers using three prominent systems: the sustainability tracking assessment and rating system (STARS), the campus sustainability assessment framework (CSAF), and the sustainability assessment questionnaire (SAQ). Alghamdi et al. [50] listed 12 assessment tools employed in HEIs based on their relevance, usability, accessibility, and holistic coverage. These tools aim to push the boundaries beyond the conventional declarations listed in Table A1. In addition to STARS and SAQ, they included benchmarking indicators questions (BIQ-AUI), graphic assessment of sustainability in university (GASU), The Green Plan (Green Plan), adaptable model for assessing sustainability in higher education (AMAS), sustainable university model (SUM), green metric (GM), university environmental management system (UEMS), sustainable campus assessment system (SCAS), assessment instrument for sustainability in higher education (AISHE), and unit-based sustainability assessment tool (USAT). These assessment tools are considered to be the most important progress toward the SD of HEIs [51]

STARS is a voluntary and self-reporting system that evaluates the performance of HEIs and ranks them based on five main categories: engagement, planning and administration, academics, innovation, and operations. Each of these categories has subcategories consisting of several indicators. Details on categories, subcategories, indicators, and the assessment procedure are given in Appendix B [9]. Figure 2a–c show sized-based performance results reported by STARS. In Figure 2a, the level of participation is similar across all sizes of HEIs. Figure 2b shows that none of the large-sized HEIs were ranked less than Silver. Conversely, in Figure 2c, 6% of the small- and medium-sized HEIs obtained a Bronze ranking. We used data from the STARS database to compare the overall sustainability performance scores with the WEC scores. Figure 3 shows that the normalized scores were higher than the individual WEC normalized average scores for all sizes of HEIs (i.e., 12 small-sized, 10 medium-sized, and 12 large-sized). None of the HEIs was able to secure the Platinum rating (refer to Appendix B).

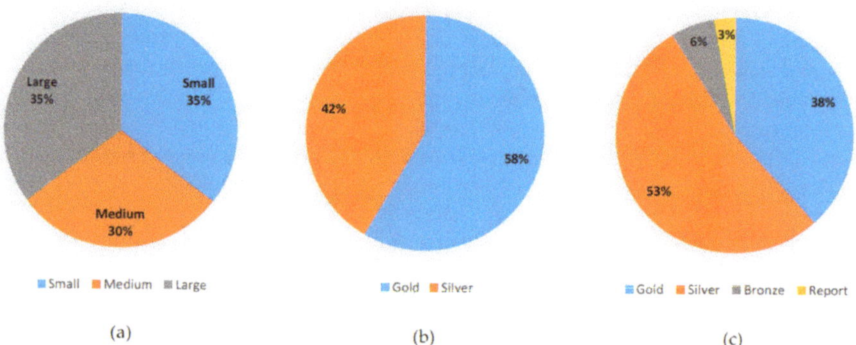


**Figure 2.** Summary of the sustainability tracking assessment and rating system (STARS) size-based evaluation: (**a**) distribution of reporting universities, (**b**) performance of large-sized universities, (**c**) performance of small- and medium-sized universities.

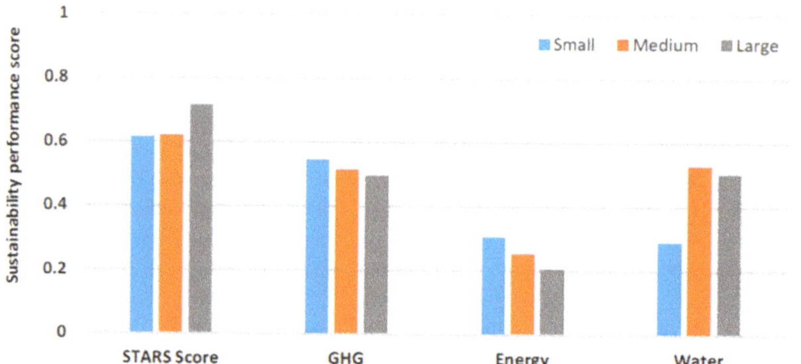


**Figure 3.** STARS scores vs. water, energy, and carbon (WEC) scores averaged for 12 small-sized, 10 medium-sized, and 12 large-sized HEIs.

Recently, Alghamdi et al. [50] stated: "Although there are many assessment tools and there has been a noticeable progress, this progress is not clearly measurable." Some of the systems require a large amount of data for analyses. For example, Tenley et al. [25] used the CSA framework for sustainability assessment with 175 indicators subgrouped into environmental, social, equity, and economics categories. They stated that different units of indicators also cause difficulties in prioritizing improvement actions.

As mentioned above, STARS uses a more comprehensive approach by covering five performance categories using 73 indicators. In the operations category, 11 indicators (22.7% of total weight) under five subcategories were related to WEC flows in HEIs, i.e., air and climate: two indicators, buildings: two indicators, energy: two indicators, waste: three indicators, and water: two indicators [9]. Here, indicators under waste were considered to appreciate the ideas of 'waste-to-energy' and 'waste minimization' for SD. Aggregating all the indicators and categorizing the HEIs as Gold, Silver, and Bronze are very useful for overall sustainability ranking.

Alba-Hidalgo et al. [52] made an interesting argument that the most efficient approach to practically implementing the policies towards sustainability is self-assessment while other approaches (participating in ranking systems) primarily enhance the competitiveness without confirming whether the improvement actions were actually planned to improve the sustainability at the university or global level or not.

Nevertheless, existing tools with their holistic approach cannot be directly used for sustainability evaluation of engineered systems in HEIs. Hence, technical-level decision making requires: (i) evaluation of WEC flows for engineered systems at the university level for inter-university benchmarking (at regional or national levels) and the setting of future targets for continuous sustainability performance improvement and (ii) classification of HEIs' infrastructure systems, particularly building, based on type and size to evaluate their WEC flows for implementing improvement actions at an intra-university level. The present work addresses the first step toward the sustainability of engineered systems in HEIs.

## 3. Methodology

To pursue SD in HEIs, both the top-down and bottom-up management approaches have been reported in the literature. Lozano [53,54] emphasized the importance of leadership and vision from higher management to the lower hierarchical levels in achieving sustainability targets, e.g., emissions reductions. Others have argued that a bottom-up approach is necessary for raising awareness and addressing the behavioral aspects of resource consumption from the users upwards [23,40,53,55–57]. As experts establish the evaluation criteria of the top-down approach [58], it seems more rational for the technical-level sustainability assessment of HEIs.

The methodology adopted to assess the performance of the WEC flows in this research is shown in Figure 4. Data for WEC flows were obtained from selected universities, government reports, and the STARS database. Based on the available data for a given climatic region, a regression analysis was conducted to interpret the missing data for three unreported universities: Selkirk College (S3), Camosun College (S12), and Concordia University (L3). GHG emissions in HEIs is a function of different factors (utilities) including energy, water consumption, number of FTE students, and floor area, which are measured in different units, and need to be normalized. The utility of overall GHG emissions for all the HEIs was estimated as a function of these different utilities. Normalized sustainability scores for WEC flows using STARS data were also estimated for comparison with the study results. All the HEIs were ranked based on their sustainability performance. Details of the steps of the methodology presented in Figure 4 are described in the following subsections.

### 3.1. Size-Based Classification of HEIs

There is no universally accepted process to classify universities on the basis of the number of FTE students. For example, Wann et al. [59] classified universities as small, medium, and large with numbers of FTE students less than 6589, 6590–18,631, and greater than 18,631. Herremans and Allwright [60] carried out a survey of North American HEIs and classified them as small with FTEs less than 10,000, medium with FTEs ranging between 10,000 and 20,000, and large with FTEs greater than 20,000. The later classification seems more rational with FTEs in the thousands and is used in the present study whenever reference is made to the size of a university.

Evidently, large-sized HEIs consume more water and energy, and generate higher GHG emissions as compared with small- and medium-sized HEIs. However, they provide larger economies-of-scale than their smaller counterparts when it comes to GHG emissions/number of FTE students. For example, large-sized HEIs can adopt a wide variety of sustainable solutions and technologies that small campuses cannot afford, e.g., biofuel generation as an alternate energy source can generate a huge amount of waste. Small-sized HEIs face challenges such as the affordability of hiring skilled staff, whereas, it is relatively easier to make decisions and implement strategies in smaller organizations [61].

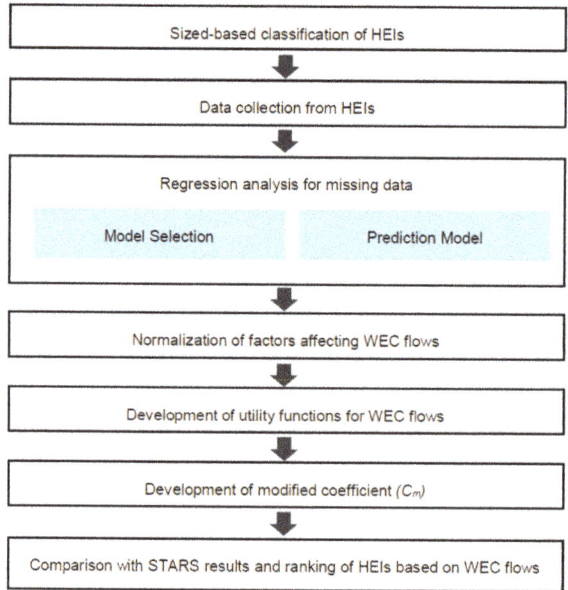


**Figure 4.** Methodology for inter-university sustainability benchmarking based on WEC flows.

### 3.2. Data Collection

The majority of the data for this study was collected from the STARS self-reporting database. Additional data was obtained from the sustainability reporting cards published by each university and the MOECC's report on the GHG emissions of large facilities in Canada, i.e., GHG scope 1 and 2 [62]. The World Resource Institute/World Business Council for Sustainable Development (WRI/WBCSD) classified GHG emissions into three scope groups. Scope 1 represents direct emissions resulting from onsite or within an organization, e.g., boilers, combustion, and mobile source combustion. Scope 2 consists of indirect sources from outside an organization, e.g., purchased electricity. Scope 3 includes other indirect sources, e.g., employee and student commuting and business travel [35].

Table 2 shows a statistical summary of the data collected. The baseline data collected for this research include climate (temperature), heating degree day (HDD), cooling degree day (CDD), the number of FTE students, building area in $m^2$, GHG emissions in $tCO_2e$, energy in GJ, and water consumption in gallons (1 $m^3$ = 264.172 US gallons). A thirty-year average was taken to assess the climate in °C. HDD and CDD indicate the cooling and heating requirements of a building. HDD is a measure of the number of degrees and days that the outside air temperature was lower than a specific base temperature that necessitated the operation of the heating system. If the temperature rose above a base temperature, the air conditioning system would start. Many of these parameters are widely used in energy performance assessment and have been extensively covered in the literature [63] in addition to factors like the behavioral aspects of consumption [64].

STARS, a self-reporting system, is a voluntary reporting system in which universities can correct a mistakenly incorrect input value. For example, the Sheridan Institute of Technology reported its mistakes related to GHG emission inventory parameters [65]. There is no provision for penalizing institutions for reporting incorrect data, so some abnormalities in the data have been noticed. For instance, the Dalhousie University reported a value of 84.91 $tCO_2e$ for the indicator of GHG (i.e., OP1 in Appendix B) [66] that was found to be significantly lower than that of all universities, i.e., 39,162 $tCO_2e$. On rechecking this value with the reported field in the university's sustainability report, it was found to be 93,907 $tCO_2e$ [67].

**Table 2.** Summary statistics of data collected.

| Input | Units | N | Min | Max | Mean | SD | Q1 | Q3 |
|---|---|---|---|---|---|---|---|---|
| Climate | °C | 33 | 2.975 | 10.4 | 7.014 | 2.492 | 4.188 | 9.342 |
| HDD | | 33 | 2136 | 11,296 | 4700 | 2179 | 3312 | 5666 |
| CDD | | 33 | 2 | 632 | 260 | 168 | 129.8 | 350.3 |
| FTE | | 33 | 971 | 56,123 | 17,860 | 13,363 | 8224 | 28,413 |
| Area | $m^2$ | 33 | 39,912 | 1,434,513 | 369,772 | 367,429 | 113,478 | 550,147 |
| GHG | $tCO_2e$ | 30 | 1021 | 299,127 | 42,953 | 67,409 | 4200 | 52,795 |
| Energy | GJ | 33 | 0.0368 | 2.4031 | 0.5815 | 0.6994 | 0.1162 | 0.9647 |
| Water | Gallons[1] | 31 | 6,326,921 | 806,266,185 | 132,466,219 | 183,924,619 | 21,774,381 | 197,980,185 |

Note: [1]($1\ m^3 = 264.172$ US Gallons).

### 3.3. Regression Analysis for Missing Data

When collecting the data, it was observed that the units and scope of the reported data used the same parameter in different reports (i.e., GHG values from different reporting systems were to cover only scopes 1 and 2). For example, the HDD and CDD were not available for some HEIs that participated in the STARS ranking, such as S1, S7, and M6, and these values could not be accessed in local climate databases. To overcome this limitation, the averages of each HEI serving in a given province (climatic region) were used to determine the missing factors (e.g., the average of all HEIs in Ontario was used to find missing information on universities in Ontario). Similarly, other missing factors, e.g., water usage and GHG emissions, were interpolated through as regression analysis.

Chung et al. [64] applied multiple regression analysis to develop a relationship between energy usage intensity (EUI) and various explanatory factors for 30 commercial buildings in Hong Kong. Sun et al. [68] used regression analysis on 95 buildings to benchmark their energy performance. In the past, Birtles and Grigg [69] studied the energy efficiency of buildings and benchmarked them via regression modeling.

In this research, the following linear regression model was developed using the pooled data obtained during data collection to assess unreported or missing data:

$$MU = a + b_1 x_1 + \cdots + b_n x_n + \varepsilon \tag{1}$$

where MU is the metabolic usage for GHG (GU), a is the intercept, $b_1$ is the coefficient for average climate, $b_2$ is the coefficient of HDD, $b_3$ is the coefficient of CDD, $b_4$ is the coefficient of FTE, $b_5$ is the coefficient of floor area of buildings, and $x_1$ to $x_n$ are the corresponding values for each component of water use. Equation (1) is a function of climate, HDD, CDD, FTE, and building area.

The data summarized in Table 2 does not follow a normal distribution, and a lognormal distribution was applied to all explanatory and response variables. Seven regressions were calculated and the best fit for the models was selected for the GHG function of energy consumption. The reason for the lack of statistical significance of other variables was their high multicollinearity. Energy consumption is highly correlated to the area of buildings, the number of FTE students, HDD, CDD, and water consumption. The results shown in Table 3 represent the best fits of R-squared (R2) coefficients found, i.e., 82.87 % and an adjusted R2 of 82.24 %. Similarly, the variance inflation factor and probability were found to be in reasonable ranges.

**Table 3.** Linear regression coefficients.

| Water Use Coefficents | | |
|---|---|---|
| Intercept | $X_1$ | −3.93 |
| Log Energy | $X_2$ | 1.0789 |

For the missing GHG values, a linear model was adjusted through curve fitting between the energy consumption and the amount of GHG emissions used per campus. The values of the two missing parameters were calculated using the curve fitting equation shown in Figure 5. For the two non-reported HEIs, i.e., GHG emissions based on the regression model were found to be 10,188.49 tCO$_2$e for Lakehead University and 22,081.03 tCO$_2$e for University of Regina.

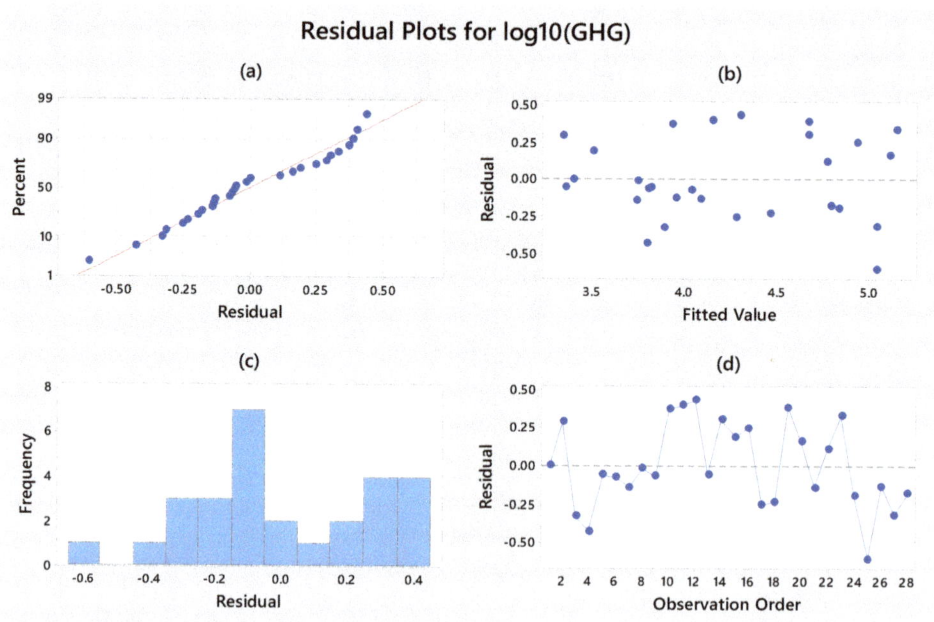


**Figure 5.** Statistical analysis results for GHG and energy consumption in Canadian HEIs: (**a**) normal residual probability plot, (**b**) residual versus fitted value, (**c**) histogram, and (**d**) residual versus observation order.

### 3.4. Normalization of Factors Affecting WEC Flows

The main assumption is that GHG emissions are a function of several factors, such as energy, water consumption, number of FTE students, and floor area, and these factors are considered as utilities in the following sections. The inputs are measured in different units, and are replaced with their utilities. Utility is defined as a parameter's energy usage, number of FTE students, floor area, and water consumption. For sustainable operations, these inputs need to be reduced to minimize GHG emissions. In this research, different universities were compared on the basis of the total GHG emitted from these input factors, while the total GHG emissions were assessed in terms of the total utility of a building. Thus, universities are labeled as $U_i$, where i = (1, ... N), and the total GHG emitted is denoted by $\widetilde{G}[U]_i$. The symbols used to define the utilities are derived from the Taylor series variables previously used by Faust and Baranzini [70] for studying the economic performance of 141 water utilities. Van Calker et al. [71] used utility functions to assess the economic, social, and ecological sustainability of dairy farms.

To obtain a rational functional relationship of GHG on such quantities, every quantity needs to be normalized for a rational comparison. The normalized GHG emitted by a university $U_i$ as $\widetilde{G}[U]_i$ can be defined as follows:

$$\widetilde{G}[U]_i = \frac{G[U]_i}{\overline{G}[U]}, \quad \overline{G}[U] = \sum_{i=1}^{N} G[U]_i \tag{2}$$

If $E[U]_i = Q[U]_{1i}$ denotes the total energy consumed, then $[U]_i = Q[U]_{2i}$. The total water consumed, $A[U]_i = Q[U]_{3i}$, corresponds to the total area heated, and $S[U]_i = Q[U]_{1i}$ represents the number of occupants in a building. These quantities depend on other factors such as climatic conditions, building envelope, and age of infrastructure. All these factors can be normalized to obtain a rational functional dependence of emitted GHG on these factors as follows:

$$\widetilde{E}[U]_i = \frac{E[U]_i}{\overline{E}[U]}, \quad \overline{E}[U] = \sum_{i=1}^{N} E[U]_i \tag{3}$$

$$\widetilde{W}[U]_i = \frac{W[U]_i}{\overline{W}[U]}, \quad \overline{W}[U] = \sum_{i=1}^{N} W[U]_i$$

$$\widetilde{A}[U]_i = \frac{A[U]_i}{\overline{A}[U]}, \quad \overline{A}[U] = \sum_{i=1}^{N} A[U]_i$$

$$\widetilde{S}[U]_i = \frac{S[U]_i}{\overline{S}[U]}, \quad \overline{S}[U] = \sum_{i=1}^{N} S[U]_i$$

where, $\widetilde{E}[U]_i$ is the normalized energy consumed, $\widetilde{w}[U]_i$ is the normalized water consumption, $\widetilde{A}[U]_i$ represents the normalized area heated, and the normalized number of people in the building can be defined as $\widetilde{s}[U]_i$. The overall GHG emissions can be considered a function of all these different quantities and can be calculated as follows:

$$\widetilde{G}(U)_i = f[\widetilde{E}(U)_i, \widetilde{W}(U)_i, \widetilde{A}(U)_i, \widetilde{S}(U)_i] \tag{4}$$

In an HEI, the total energy consumed during a year is a function of both the heating and cooling degree days corresponding to the winter and summer seasons. In energy calculations, consideration must be given to both seasons. In this regard, $\widetilde{Q}_{1i}$ needs to be estimated as the factor obtained from the heating degree day $HDDi$, and $\widetilde{Q}_{2i}$ as the factor obtained from the cooling degree days $CDDi$. These factors can be defined as follows:

$$\widetilde{Q}_{1i} = \frac{HDDi}{\overline{HDDi}}, \quad \overline{HDDi} = \sum_{i=1}^{N} HDDi \tag{5}$$

$$\widetilde{Q}_{2i} = \frac{CDDi}{\overline{CDDi}}, \quad \overline{CDDi} = \sum_{i=1}^{N} CDDi$$

These factors are included to normalize the effect of temperature variations because heating and cooling can increase energy consumption in cold and hot climates. High-order corrections are neglected. Only the linear functional dependence of these quantities is considered as a first approximation as follows:

$$\widetilde{G}(U)_i = C_i \left[ \widetilde{E}(U)_i + \widetilde{W}(U)_i + \widetilde{A}(U)_i + \widetilde{S}(U)_i \right] \tag{6}$$

Equation (6) can also be written as follows:

$$\widetilde{G}[U]_i = C_i \left[ \widetilde{E}[U]_i / \widetilde{Q}_{1i}\widetilde{Q}_{2i} + |\widetilde{W}[U]_i| + |\widetilde{A}[U]_i| + |\widetilde{S}[U]_i| \right] \tag{7}$$

$C_i$ is the coefficient used to measure the relative importance of energy when compared with the number of FTE students, floor area, and water usage. For example, water and energy cannot be compared because of the different units used in a single HEI. The comparison of the utility for each factor is more appropriate because all the quantities need to be converted into a consistent unit to incorporate different quantities into the benchmarking process.

### 3.5. Ranking of HEIs Based on WEC Flows

The coefficients $C_i$ with $i = (1, \ldots N)$ measure the relative utility of these quantities (i.e., WEC) as compared with normalized energy consumption. These quantities vary for different regions. For example, it is more difficult to obtain water in the Middle East (i.e., arid environment) compared with Canada, and energy is easy to generate in a region with extensive oil resources. A simple form of the model can be written as follows:

$$C_i = \frac{\widetilde{G}[U]_i}{\left|\frac{\widetilde{E}[U]_i}{\widetilde{Q}_{1i}\widetilde{Q}_{2i}}\right| + |\widetilde{w}[U]_i| + |\widetilde{A}[U]_i| + |\widetilde{s}[U]_i|} = \frac{\widetilde{G}[U]_i}{\overline{p}[U]i} \tag{8}$$

Now, if $C_1 > C_2$ and $\overline{p}[U]_1 = \overline{p}[U]_2$, then $\widetilde{G}[U]_1 > \widetilde{G}[U]_2$, and university $U_1$ is producing more GHG emissions for the same average utility as compared with $U_2$. This may be attributed to the use of cleaner or less emission-intensive energy sources. As a lower value of Ci indicates better environmental performance of a university, the final sustainability rank (SR) is calculated as follows:

$$SR = 1 - C_i \tag{9}$$

## 4. Results

Based on the relative utility coefficient 'Ci', 34 Canadian HEIs were benchmarked (ranked) using the methodology described above. As an example, the reported data of the UBC (Vancouver Campus) is presented in Table 4. The first step normalized all the values by the mean as shown in Equations (2) and (3). The normalized UBC values are listed in Table 4. The new utility (i.e., normalized energy) calculated with Equation (5) was 10.08. The total utility value calculated using Equations (6) and (7) was 22.35. The value of the coefficient calculated by using Equation (8) was 0.0607. Sustainability ranks were established for all the HEIs using Equation (9). The final sustainability ranks for the 34 HEIs are presented in Figure 6. The figure shows that the sustainability ranks for all the small-sized HEIs and 67% of the large-sized HEI were higher than the average. While 60% of the medium-sized HEIs underperformed with sustainability ranks lower than the average.

In Figure 6, the model data were normalized for simplicity with equal values across all the parameters for each HEI assuming they all had equal effects. The variability of the reported data may have an impact on the performance of an HEI. For example, the data was not collected in a cut-off year and the most recent reports were taken into consideration instead. Some of the latest reports published by HEIs were two to five years old. As per the average values of all the reporting HEIs shown in Figure 6, the ten lowest performing HEIs were L4, M3, L6, M4, M2, M7, L7, M5, L2, and M9.

The size-based classification of HEIs better justifies the relationship between the flow of energy and GHG. A strong correlation (i.e., $R^2 = 0.94$) between energy consumed and associated GHG emissions was observed for the small- and medium-sized HEIs in Figure 7. These results are contrary to the combined values of all HEIs, where the R2 value for all the HEIs regardless of size was 0.6442.

**Table 4.** Data and results of University of British Columbia (UBC, Vancouver).

| HEI | Climate(°C) | HDD | CDD | FTE | Area (m²) | GHG (tCO₂e) | Energy (GJ) | Water (Gallons)[1] |
|---|---|---|---|---|---|---|---|---|
| UBC (Vancouver) | 10.4 | 5093 | 95.8 | 43,509 | 1,434,513 | 52,350 | 1.2472 | 806,266,185 |
| HEI mean | - | 4710 | 262 | 18,157 | 374,902 | 35,453 | 0.5809 | 126,627,438 |
| Normalized UBC | - | 1.08 | 0.37 | 2.39 | 0.3558 | 1.34 | $3.4711 \times 10^{-6}$ | 6.05 |

Note: [1]($1\text{ m}^3 = 264.172$ US Gallons).

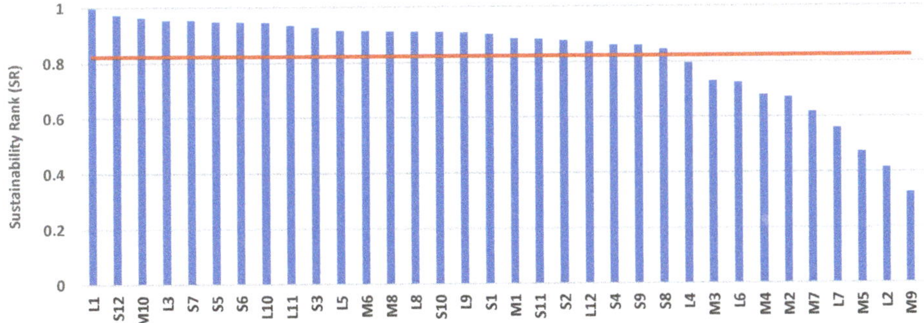

**Figure 6.** Sustainability ranking of all HEIs in Canada.

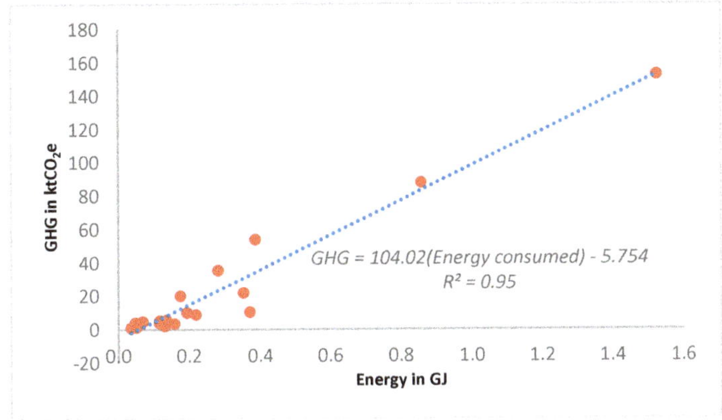



**Figure 7.** Relationship between energy consumed and GHG emissions for small- and medium-sized HEIs in Canada.

The HEIs with relatively poor performance and their sources of fuels are presented in Table 5. The performance of HEIs in terms of energy conservation varied depending on the mode of energy conservation, efficiency of different processes, and use of GHG intensive fuels, e.g., coal and fossil fuel. For instance, the University of Regina (M3) produced 100% of its electricity from coal. The Nova Scotia Community College (M2) met 53% of its electricity requirements from coal, and the MacEwan University (M4) produced 40.6% of its electricity from coal. Other underperforming HEIs primarily relying on fossil fuels include the University of Calgary (L4), University of Alberta (L7), and University of Saskatchewan (M9). The latest energy report indicates that Alberta's electricity generation is 51% fueled from coal [72]. However, precise information could not be obtained about these universities' inner-grid combinations (e.g., gas vs. electricity). The province of Saskatchewan generates approximately 34% of its electricity from coal [73].

The HEIs identified in Table 5 performed poorly because of inefficient energy usage in older buildings (e.g., in terms of building envelope and thermal characteristics). For example, the Nova Scotia Community College (M2) performed better than Dalhousie University (M7) despite both being in the same city and relying on the same electricity grid. The overall performance enhancement in the Nova Scotia Community College was the result of its improved energy efficiency or improved inner-grid distributions (more reliance on renewable energy sources, e.g., hydro or wind).

The data for the 34 HEIs primarily obtained from the STARS database shown in Figure 8a–c presents the aggregated sustainability scores for WEC flows. The values were normalized by dividing the score of each institution in the WEC category by the total grade. Figure 8a illustrates that, regardless of size, the performance of all HEIs was similar in terms of GHG emissions while some of the large-sized HEIs clearly underperformed, i.e., L2, L3, L7, and L12. In the case of energy consumption, small-sized HEIs outperformed large- and medium-sized HEIs as shown in Figure 8b. Figure 8c shows that the water-consumption performance of medium- and large-sized HEIs was similar.

**Table 5.** Ten worst-performing HEIs and their fuel sources.

| No. | HEI | Fuel Source Scope 1 | Fuel Source Scope 2 |
|---|---|---|---|
| L4 | University of Calgary | 84% natural gas, 0.22% propane | 73.8% coal, 20.61% natural gas, 5.39% renewables, 0.18% other |
| M3 | University of Regina | 100% natural gas | 100% coal |
| L6 | Western University | 98% natural gas, 2% electricity | 40.85% hydro, 55.67% nuclear, 3.49% natural gas, biomass, and coal |
| M4 | MacEwan University | 100% natural gas | 3% biomass, 40.65% coal, 6.4% hydro, 41.3% natural gas, 7.9% wind, 0.8% other sources |
| M2 | Nova Scotia Community College | 70% fuel oil, 8% other sources, 22% natural gas | 2% biomass, 53% coal, 14% hydro, 14% natural gas, 12% wind, 4% other sources |
| M7 | Dalhousie University | 8% biomass, 84% natural gas, 3% electricity, 5% fuel oil | 2% biomass, 63% coal, 7% hydro, 12% natural gas, 9% wind, 7% other sources |
| L7 | University of Alberta | 54% natural gas, 45.4% purchased electricity | 53% coal, 38% natural gas, 3% hydro, 1% wood biomass, 5% renewables |
| M5 | Northern Alberta Institute of Technology | - | - |
| L2 | University of Manitoba | 99.9% natural gas, 0.05% fuel oil | 95% hydro, 5% natural gas |
| M9 | University of Saskatchewan | 99.96% natural gas, 0.04% fuel oil | 37% coal, 20% hydro, 37% natural gas, 0.02% solar, 5% wind, 0.08% other sources |

The possible reasons for these results are outlined here. Small universities manage a smaller number of students, have smaller infrastructures, and relatively less-complex processes. The emissions of large-sized HEIs are normalized over a large number of students (i.e., $tCO_2$/FTE) and may result in a higher sustainability ranking. Large universities accommodate more students per water consuming activity, such as laboratories, washrooms, and treatment plants, with larger economies-of-scale. GHG generation per square foot in large-sized HEIs is less than that of medium- and small-size HEIs. Energy consumption is directly related to the size of a building. Therefore, medium-sized HEIs are more critical because buildings are constructed for longer lifecycles.

Regression analysis was performed between the floor area of buildings in the HEIs and their corresponding water consumption, energy consumption, and GHG emissions. The analysis revealed that the 'power function' best fits the relationship between the variables with high R2 values. Empirical constants for all three relationships are presented in Table 6. The results presented in Table 6 show a strong correlation between the floor area of HEI buildings and WEC flows.

**Table 6.** Empirical constants for power functions between floor area of HEI buildings and WEC flows.

| Floor Area of Building (thousand ft$^2$) [Y = a X$^b$] | | | | | | | | |
|---|---|---|---|---|---|---|---|---|
| X = Water Consumption (Million Gallons)[1] | | | X = Energy Consumption (GJ) | | | X = GHG Emissions (Million-ton CO$_2$ Equivalent) | | |
| a | b | R$^2$ | a | b | R$^2$ | a | b | R$^2$ |
| 0.0019 | 1.309 | 0.8790 | 11.908 | 1.273 | 0.942 | 0.2234 | 1.396 | 0.797 |

Note: [1](1 m$^3$ = 264.172 US Gallons).

A comparison between the STARS ranking and the ranks established in the present study for all sizes of HEIs are presented in Figure 9. The figure shows that of the 10 worst-performing HEIs in Figure 6, the sustainability ranks of four HEIs (M3, M4, M7, and L4) were approximately equal or higher than the STARS ratings. These HEIs need to: (i) implement the SD declarations by embedding sustainability in the fabric of the organizational structure and by planning and setting realistic goals and (ii) use renewable fuel sources. The sustainability ranks presented in Figure 9 were calculated based on comparative assessment. Hence, a higher SR does not ensure that there is no need for further improvement. To attain long-term sustainability, HEIs should adopt a continuous performance improvement approach.

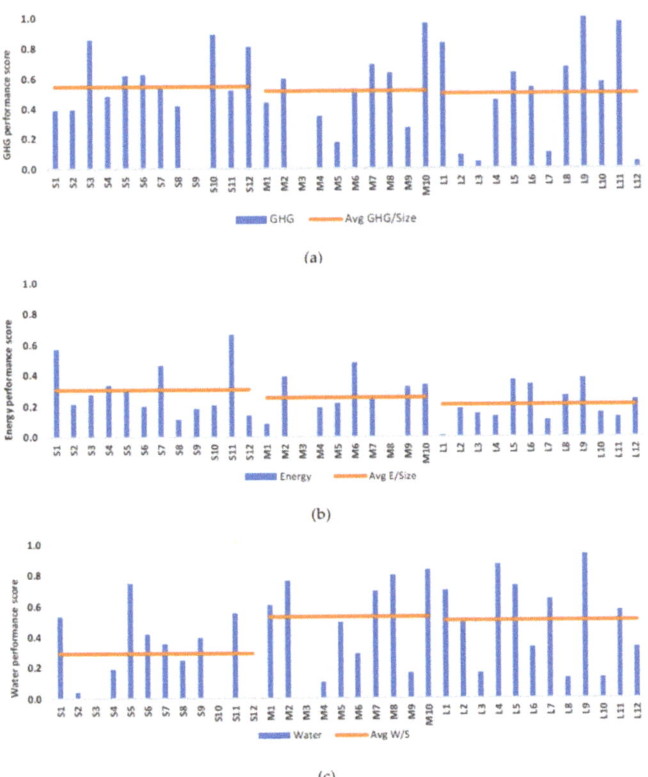

Figure 8. STARS ranking of Canadian universities by size: (**a**) GHG emissions performance, (**b**) energy performance, and (**c**) water performance.

**Figure 9.** Comparison between STARS overall ranking and SRs based on WEC flows estimated in present study.

## 5. Discussion

HEIs have sought to implement rating systems for their buildings as an attempt to reduce energy consumption and subsequently reduce GHG emissions. Research has shown that many of these building rating systems could not achieve their desired goals [74,75]. Tilton and ELAsmar [75] compared LEED certified buildings with non-certified buildings based on their functionality (e.g., dormitory, research buildings, and offices) and found that non-certified research buildings consumed less energy than their certified counterparts. According to the results presented in Figure 2, none of the Canadian HEIs attained Platinum rank from STARS (refer to Appendix B for details) and should adopt a continuous performance improvement approach for long-term sustainability.

Leadership in Energy and Environmental Design (LEED) certifies a green building by verifying that the design and construction aimed at improving the sustainability performance, including energy and water efficiency, reduction of carbon emissions, etc. [76]. Agdas et al. [74] studied the EUI of 10 LEED-certified buildings and compared them with non-LEED-certified buildings. It was found that the EUI of LEED-certified buildings was higher than that of the non-LEED-certified buildings. For example, they evaluated 24 campus buildings, 10 of which were LEED-certified and the remaining non-LEED-certified buildings. They could not find a statistically significant difference between both sets of the buildings, in fact, the EUI for the LEED buildings was found to be higher than those of non-certified buildings. Existing systems do not identify the causes of poor performance and the actions required to improve the sustainability performance of universities. This is due to the diverse nature of the parameters being measured (social, environmental, economic) [77]. Moreover, communicating the sustainability assessment of HEIs goals and scope with the stakeholders is not clearly defined [29]. There is a need for an integrated system that can lead HEIs toward continuous improvement in terms of sustainability while maintaining cost effectiveness and simplicity. There is also a need to shift the focus from corrective controls to proactive action [78].

Energy baseline targets (reported values for 2007) were self-imposed by universities in an attempt to conserve or reduce energy consumption. Figure 10 presents a comparison between the performance of HEIs and the baseline energy targets using STARS data. Approximately 58% of the large-sized universities succeeded in reducing their energy consumption below the baseline, while approximately 70% of medium-sized and 75% of small-sized HEIs performed better than the baseline value. The reasons for non-compliance can be unrealistic targeting, an increase in student enrollment, or the absence of any interventions. Overall, approximately 43% of Canadian HEIs fell behind their baseline values [9]. This is aligned with the data reported from the UK, where educational institutions account for more than 5% of the non-domestic buildings and are responsible for approximately 14% of the entire public-sector emissions. Public-funded HEIs had an average EUI of 293 kWh/m$^2$ in FY 2001, which gradually decreased to 287 kWh/m$^2$ by the FY 2006. This EUI level is still substantially below the best practice benchmark of 162 kWh/m$^2$ [79]. The GHG emissions followed similar patterns [80].

HEIs still rely heavily on local electricity providers, which requires that they strive for a net-zero stage. This requires an understanding of the inner components of buildings and how performance can be improved. Imposing taxes on HEIs' intensive fuel usage may result in a positive impact in terms of GHG reduction. To attain high sustainability standards, specific targets are required to be met. The assessment of the state of engineered systems in Canadian HEIs in Figure 10 requires more information for technical-level decision making, e.g., solar coefficient, split of energy used and purchased by sources, age of buildings to determine efficiency, and solar-power street lighting.

Similar to any other built infrastructure, a considerable amount of energy is required in all stages of water and wastewater operations, including water abstraction, water treatment, distribution, wastewater collection, wastewater treatment, and disposal and/or reuse [81]. Energy use can be optimized using efficient water appliances, adopting water conservation strategies, and effective conservation practices. These strategies can be combined with renewable energy use including wastewater energy recovery and onsite solar and wind energy generation for achieving net-zero energy and carbon emissions [82].

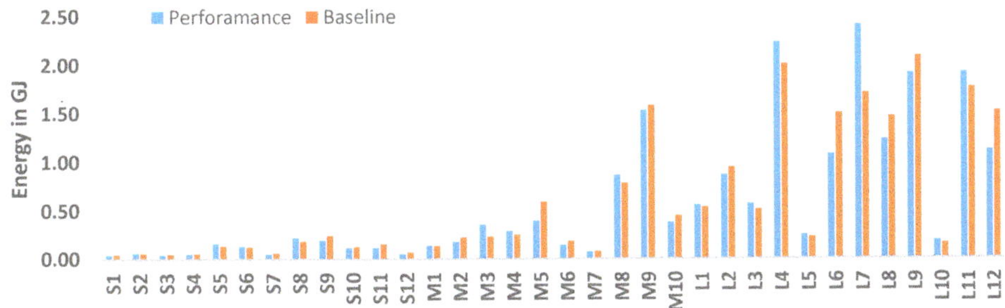

**Figure 10.** Energy baseline targets and performance of Canadian HEIs based on STARS reporting data.

HEIs require proper planning approaches to reduce both the short-term and long-term environmental impacts during their operations. Comprehensive data are needed for the effective assessment and prudent decision making for selecting rational interventions, i.e., day-to-day maintenance and major long-term improvements. These interventions have their own socioeconomic and environmental implications over the life of a building. An integrated technology–human effort is required for energy and water conservation, GHG emission reductions, and ecosystem protection in a cost-effective manner.

HEIs must address mounting issues owing to an increase in the number of FTE students and associated energy demand, costs associated to increasing fuel prices, aging of buildings and associated reductions in efficiency, uncertainties owing to emerging regulations aimed at curbing carbon emissions, and climate change and its impact on seasonal variations. These factors coupled with the limited available resources to improve these assets will have a significant impact on an institution's ability to forecast future trends, goals, and performance. Therefore, HEIs need to consistently address these changing factors and their impact to attain their sustainability goals.

In 2010, Universitas Indonesia (UI) developed a platform, known as the UI Green Metric World University Ranking [83], for sharing the state of sustainability in HEIs around the globe. The approach of this ranking system consists of six categories, including: (i) setting and infrastructure, (ii) energy and climate change, (iii) waste management, (iv) water usage, (v) transportation, and (vi) environmental education. In addition to improving the overall environmental conditions and promoting sustainability in HEIs, these green campus activities can improve the perception of stakeholders about HEIs and can increase student enrolment.

Continuous performance improvement (CPI) is a top-down management approach intended to manage the quality performance of processes and systems by continuously improving the performance of an organization [84]. CPI aims at reducing wastes, increasing competitiveness, and improving overall performance [85]. The conceptual CPI benchmarking process presented in Figure 11 can be implemented for both inter- and intra-university sustainability assessments. Details of the CPI application in benchmarking processes are provided in Bereskie et al. [86].

A benchmark can be established using a representative sample of similar-sized universities having comparable characteristics, e.g., number of students, and types and age of buildings. The benchmark can be applicable to a specific region to accommodate temporal and geographical factors, such as seasonal and climatic variations and differences in provincial/state regulations. For CPI, future benchmarks can also be established based on improving WEC flow projections from interventions, changing regulations, technological advances, and students' expectations. The goal of the CPI process is not only to improve upon the original benchmark for a factor or group of factors but also to cluster individual HEIs (or buildings in case of inter-university application) closer to the benchmark value.

The resent research provides an insight into the existing state of sustainability in Canadian HEIs. Establishing SRs based on WEC flows will be useful for technical-level decision making. Instead of relying on existing rating systems that cover a wider spectrum of sustainability, technical-level decisions can practically improve the existing global and regional GHG scenarios and be a step toward water and energy conservation.

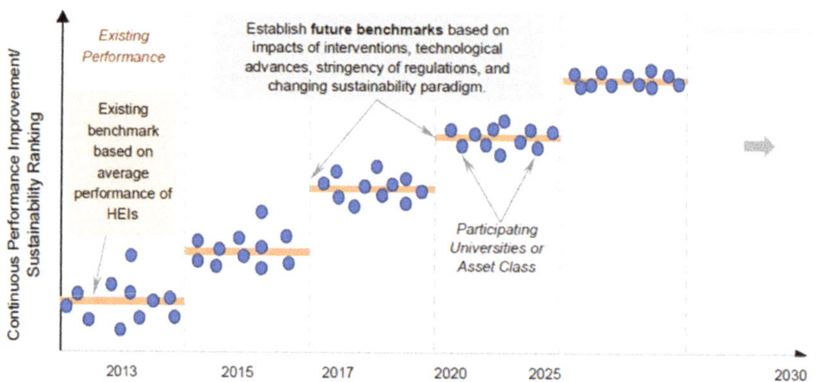



**Figure 11.** Continuous performance improvement (after Bereskie et al. [86]).

## 6. Conclusions

In Canada, HEIs consume 60% of the total energy allocated to the educational sector, which is equivalent to approximately 430,000 households. Accordingly, such institutions have been recognized as one of the major contributors to Canada's total GHG emissions. A review of the historical background of declarations revealed that the sustainability of engineered systems (infrastructures) in HEIs was highlighted over the last decade.

Engineered systems in HEIs have been constructed in different eras as a result of increasing enrolments and the establishment of new disciplines. Consequently, they have varying process efficiency, features, and environmental impacts. Existing sustainability ranking systems include a wide range of sustainability indicators and do not measure actual performance in terms of natural-resource consumption and generation of GHG emissions. Technical-level decision makers need to know how the infrastructures, particularly buildings, and processes are utilizing these resources. This can be accomplished by comparing the HEIs based on their energy utilization, water consumption, and production of associated GHG emissions. Instead of relying on existing global sustainability performance, decision makers can select and practically implement actions for long-term sustainability of HEIs, e.g., development of resource conservation strategies and building retrofits.

The inter-university benchmarking methodology developed in this research for sustainability assessment of Canadian HEIs was based on their WEC flows. The methodology incorporates the impact of economies-of-scale by taking the sizes of the HEIs into account for the rational benchmarking of 34 HEIs.

The proposed methodology addresses the difficulties encountered during benchmarking owing to the use of different HEI reporting systems. A regression analysis was used to interpret the missing data for a given climatic region. The GHG emissions of HEIs is a function of different factors (utilities), including energy, water consumption, number of FTE students, and floor area. These factors are measured in different units and were normalized in this research. It is recommended that all universities follow uniform parameters for a rational comparison in the future.

Regression analysis established power function, with strong correlations (i.e., R2 values higher than 0.8), between the floor area of buildings in the HEIs and their corresponding water consumption, energy consumption, and GHG emissions.

Canadian HEIs are ranked based on the sustainability performance of their engineered systems while considering the differences in climatic regions, size of campus, number of occupants, and WEC flows. The study results revealed that small-sized HEIs outperformed the large-sized HEIs, whereas, the performance of 60% of the medium-sized HEIs was found to be less than the average. The reasons for such results could be: (i) small HEIs deal with a smaller number of students, smaller infrastructure, and relatively less-complex physical processes, (ii) emissions in large-sized HEIs are normalized over a large number of students (i.e., $tCO_2e$/FTE), (iii) larger economies-of-scale, i.e., each water consuming facility serves a large number of students in large-sized HEIs, (iv) EUI is less in large-sized HEIs, and (v) energy consumption is directly related to the size of a building, and medium-sized HEIs serving a smaller number of students construct buildings for longer periods of use.

The study results suggest the need for more in-depth investigations to improve the sustainability of different infrastructures' operations through effective operational-level decision making (e.g., efficient retrofits, particularly for older institutions) for all sizes of HEIs. It is expected that future research will focus on the assessment of detailed environmental impacts during the entire life of HEIs, including all construction and operations phases.

**Author Contributions:** Abdulaziz Alghamdi (A.A.) is the lead author on this publication, having completed the initial writing of the paper, data collection, formal analysis and the methodology. Husnain Haider (H.H.) contributed to conceptualization of the paper, redefining the objectives, narrowing the focus, reviewing the paper and managing the overall research. Kasun Hewage (H.W.) contributed in the literature review and definition of the scope of the work. Rehan Sadiq (R.S.) contributed to the conceptualization of the paper, methodology refinement, and the statistical analysis.

**Funding:** This research was funded by the Saudi Cultural Bureau in Ottawa for funding this research under the King Abdullah Bin Abdulaziz Scholarship Program.

**Acknowledgments:** The author gratefully acknowledges the financial support of the Saudi Cultural Bureau in Ottawa for funding this research under the King Abdullah Bin Abdulaziz Scholarship Program. Also, the support of the STARS database and the labs and facilities at the University of British Columbia are highly appreciated. Authors also acknowledge the insights and help of Mir Faizal for his grateful contributions.

**Conflicts of Interest:** The authors declare no conflict of interest.

## Appendix A  Historical Background of Declarations on Sustainability of HEIs

The historical progress of declarations on the sustainability of HEIs is summarized in Table A1. The concept of sustainability of HEIs was first introduced in the Stockholm declaration in 1972 [19]. In 1990, more than 300 universities signed the Talloires declaration with an action plan to introduce sustainability and environmental literacy in HEIs. The UBC signed the Talloires declaration in 1990 and built its first green building in 1996. In 1997, the UBC was ranked first in Canada to adopt an SD policy and opened its sustainability office in 1998. It was also the first to launch the ECOTREK (energy and water intervention program) in Canada and to publish a comprehensive campus-wide strategy in 2006. It has undertaken many initiatives in its pursuit to achieve a sustainable campus [38]. In 1992, the UNCED conference established the correlation between buildings and the capacity for decision making for SD primarily in undeveloped nations [7]. The Earth Charter initiative in 2000 involved discussions between all levels of educational institutions, and not explicitly universities. This charter inspired a new sense of global interdependence by identifying 16 broad-based principles that were later applied to many organizations, cities, NGOs, and the education sector. The aim was to promote interdependence and shared responsibility for the well-being of the whole human family and the greater community of life [79]. In 2012, the Rio+20 declaration was signed by 78 universities, where the overwhelming majority was from Europe. This declaration covered five sustainability aspects including the greening of campuses, teaching SD concepts, supporting sustainability efforts, fostering and engaging in international collaboration, and encouraging research on SD issues [49]. Recently, the UNESCO launched a Global Action Plan in 2014 to generate and scale up actions at all levels and areas of education. This action plan identified five priorities: whole-institution approach, policy support, youth, local communities, and educators. Overall, 272 universities signed the initiative and shared their information to address climate change, 13% of the participating universities were North American [78]. More than 680 universities around the world have signed the American College and University

Presidents' Climate Commitment (AUPCC) agreement, which requires participating institutions to reduce GHG emissions [36].

**Table A1.** Declarations addressing sustainability in HEIs [2,35,80].

| No | Year | Declaration |
|---|---|---|
| 1 | 1972 | Stockholm Declaration on the Human Environment UNEP |
| 2 | 1977 | Tbilisi Declaration UNESCO |
| 3 | 1988 | Manga Charta of European Universities Association (EUA) |
| 4 | 1990 | University Presidents for a Sustainable Future: the Talloires Declaration ULSF |
| 5 | 1991 | Halifax Declaration (International Institute for Sustainable Development) |
| 6 | 1992 | Agenda 21 Report of the United Nations Conference on Environment and Development (UNCED) |
| 7 | 1993 | Ninth International Association of Universities Round Table: the Kyoto Declaration |
| 8 | 1993 | Association of Commonwealth Universities "Fifteenth Quinquennial Conference: Swansea Declaration |
| 9 | 1994 | CRE Copernicus Charter |
| 10 | 1997 | International Conference on Environment and Society: Education and Public Awareness for Sustainability UNESCO |
| 11 | 1998 | World Declaration on Higher Education for the Twenty-first Century: Vision and action UNESCO |
| 12 | 2000 | Earth Charter |
| 13 | 2001 | Luneburg Declaration UNESCO |
| 14 | 2002 | Ubuntu Declaration UN |
| 15 | 2002 | Johannesburg Plan of Implementation World Summit on Sustainable Development (WSSD) |
| 16 | 2005-2014 | The UN Decade Education for Sustainable Development UNESCO |
| 17 | 2005 | Graz Declaration on Committing Universities to Sustainable Development |
| 18 | 2006 | Declaration on the Responsibility of Higher Education for a Democratic Culture Citizenship, Human Rights and Sustainability |
| 19 | 2008-2017 | G8 University Summit Sapporo Sustainability Declaration |
| 20 | 2009 | Abuja Declaration on Sustainable Development in Africa |
| 21 | 2009 | Tokyo Declaration of HOPE ASSU |
| 22 | 2009 | Turin Declaration on Education and Research for Sustainable and Responsible Development, Italy |
| 23 | 2009 | World Conference on Higher Education UNESCO |
| 24 | 2010 | G8 University Summit |
| 25 | 2011 | Copernicus Charta 2.0 |
| 26 | 2012 | Peoples Sustainability Treaty on Higher Education |
| 27 | 2012 | UN Higher Education Sustainability Initiative with Rio+20 |
| 28 | 2015 | Higher Education Sustainability Initiative: Climate change action for sustainable development |

## Appendix B  Assessment Procedure of Sustainability Tracking Assessment and Reporting System (STARS)

The weights of these subcategories and indicators were assigned based on the recommendations of a panel consisting of members of the STARS's steering committee and staff of the Association for the Advancement of Sustainability in Higher Education (AASHE). The performance points were measured on the overall impact of reporting fields. Owing to the diverse nature of HEIs, STARS insists that the credits should be given in a flexible manner because the reporting fields do not apply to all HEIs. The aggregation of these points culminates at an overall rank, such as Platinum, Gold, Silver, or Bronze. An HEI with an overall score of 85 or more obtains a Platinum rating, Gold ranking starts at 65, Silver at 45, and the low-performing institutions with a score of 25 or less obtain a Bronze rating. If a university receives a cumulative score lower than 25, it is shown as incomplete reporting by the HEI. The holistic approach adopted by the STARS ranking in Appendix B may give a higher rank to an HEI that may not be performing the same in terms of its WEC flows.

**Table A2.** STARS groups/subgroups and points awarded [9].

| Category/Group | Subgroups | Indicators | Points |
|---|---|---|---|
| Academics (AC) | Curriculum (40 points) | AC1 Academic Courses | 14 |
| | | AC2 Learning Outcomes | 8 |
| | | AC3 Undergraduate Program | 3 |
| | | AC4 Graduate Program | 3 |
| | | AC5 Immersive Experience | 2 |
| | | AC6 Sustainability Literacy Assessment | 4 |
| | | AC7 Incentives for Developing Courses | 2 |
| | | AC8 Campus as a living Laboratory | 4 |
| | Research (18 points) | AC9 Academic Research | 12 |
| | | AC10 Support for Research | 4 |
| | | AC11 Access to Research | 2 |
| Engagement (EN) | Campus Engagement (21 points) | EN1 Student Educators Program | 4 |
| | | EN2 Student Orientation | 2 |
| | | EN3 Student Life | 2 |
| | | EN4 Outreach Materials and Publications | 2 |
| | | EN5 Outreach Campaign | 4 |
| | | EN6 Assessing Sustainability culture | 1 |
| | | EN7 Employee Educators Program | 3 |
| | | EN8 Employee Orientation | 1 |
| | | EN9 Staff Professional Development | 2 |
| | Public Engagement (20 points) | EN10 Community Partnerships | 3 |
| | | EN11 Inter-Campus Collaboration | 3 |
| | | EN12 Continuing Education | 5 |
| | | EN13 Community Service | 5 |
| | | EN14 Participation in Public Policy | 2 |
| | | EN15 Trademark Licensing | 2 |

**Table A2.** *Cont.*

| Category/Group | Subgroups | Indicators | Points |
| --- | --- | --- | --- |
| Operations (OP) | Air & Climate (11 points) | OP1 Greenhouse Gas Emissions<br>OP2 Outdoor Air Quality | 10<br>1 |
| | Buildings (8 points) | OP3 Building Operations and Maintenance<br>OP4 Building Design and Construction | 5<br>3 |
| | Energy (10 points) | OP5 Building Energy Consumption<br>OP6 Clean and Renewable Energy | 6<br>4 |
| | Food & Dining (8 points) | OP7 Food and Beverage Purchasing<br>OP8 Sustainable Dining | 6<br>2 |
| | Grounds (3-4 points) | OP9 Landscape Management<br>OP10 Biodiversity | 2<br>1-2 |
| | Purchasing (6 points) | OP11 Sustainable Procurement<br>OP12 Electronics Purchasing<br>OP13 Cleaning and Janitorial Purchasing<br>OP14 Office Paper Purchasing | 3<br>1<br>1<br>1 |
| | Transportation (7 points) | OP15 Campus Fleet<br>OP16 Student Commute Modal Split<br>OP17 Employee Commute Modal Split<br>OP18 Support for Sustainable Transportation | 1<br>2<br>2<br>2 |
| | Waste (10 points) | OP19 Waste Minimization and Diversion<br>OP20 Construction and Demolition Waste Diversion<br>OP21 Hazardous Waste Management | 8<br>1<br>1 |
| | Water (6-8 points) | OP22 Water Use<br>OP23 Rainwater Management | 4-6<br>2 |
| Planning & Administration (PA) | Coordination & Planning (8 points) | PA1 Sustainability Coordination<br>PA2 Sustainability Planning<br>PA3 Participatory Governance | 1<br>4<br>3 |
| | Diversity & Affordability (10 points) | PA4 Diversity and Equity Coordination<br>PA5 Assessing Diversity and Equity<br>PA6 Support for Underrepresented Groups<br>PA7 Affordability and Access | 2<br>1<br>3<br>4 |
| | Investment & Finance (7 points) | PA8 Committee on Investor Responsibility<br>PA9 Sustainable Investment<br>PA10 Investment Disclosure | 2<br>4<br>1 |
| | Well-being & Work (7 points) | PA11 Employee Compensation<br>PA12 Assessing Employee Satisfaction<br>PA13 Wellness Program<br>PA14 Workplace Health and Safety | 3<br>1<br>1<br>2 |
| Innovation & Leadership (IN) | (4 bonus points available) | Exemplary Practice (Catalog of credits available)<br>Innovation (4 credits available) | 0.5 each<br>1 each |

## References


1. World Health Organization (WHO). *Health Promoting Universities Concepts, Experience and Framework for Action*; Tsouros, A.D., Dowding, G., Thompson, J., Dooris, M., Eds.; World Health Organization: Copenhagen, Denmark, 1998.

2. Caeiro, S.; Filho, W.L.; Jabbour, C.; Azeiteiro, U.M. *Sustainability Assessment Tools in Higher Education Institutions*; Springer International Publishing: Gewerbestrasse, Switzerland, 2013; p. 432.

3. Cortese, A.D. The Critical Role of Higher Education in Creating a Sustainable Future. *Plan. High. Educ.* **2003**, *13*, 15–22.

4. Orr, D.W. Earth in Mind. In *The Heart of Higher Education: A Call to Renewal*, 10th ed.; Island Press: San Francisco, CA, USA, 2004.

5. Univcan. Facts and Stats. Available online: http://www.webcitation.org/782U49h9N (accessed on 25 April 2019).

6. Organisation for Economic Co-Operation and Development (OCED). *Education at a Glance: OCED Indicators*; OCED Publishing: Paris, France, 2014.

7. United Nations (UN). Agenda 21, Ch. 36: Promoting Education and Public Awarness and Training. Available online: http://www.webcitation.org/782UFY9pI (accessed on 24 April 2019).

8. Raj, B. Directed and Systematic Approaches Towards Sustainability in the Twenty-first Century, Chapter in The Mind of an Engineer. In *The Mind of an Engineer*; Ghosh, P., Raj, B., Eds.; Springer: Singapore, 2016.

9. AASHE. Technical Manual Version 2.1. Available online: http://www.webcitation.org/782VI1ulG (accessed on 4 October 2017).

10. Bauer, M.; Bormann, I.; Kummer, B.; Niedlich, S.; Rieckmann, M. Sustainability Governance at Universities: Using a Governance Equalizer as a Research Heuristic. *High. Educ. Policy* **2018**, *31*, 491–511. [CrossRef]

11. Alshuwaikhat, H.M.; Abubakar, I. An integrated approach to achieving campus sustainability: Assessment of the current campus environmental management practices. *J. Clean. Prod.* **2008**, *16*, 1777–1785. [CrossRef]

12. NSF. What's an Engineered System? Available online: erc-assoc.org/content/what%E2%80%99s-engineered-system (accessed on 28 November 2018).

13. Ward, I.; Ogbonna, A.; Altan, H. Sector review of UK higher education energy consumption. *Energy Policy* **2008**, *36*, 2939–2949. [CrossRef]

14. Geng, Y.; Liu, K.; Zue, B.; Fujita, T. Creating a "green university" in China: A case of Shenyang university. *J. Clean. Prod.* **2013**, *61*, 13–19. [CrossRef]

15. Larsen, H.N.; Pettersen, J.; Solli, C.; Hertwich, E.G. Investigating the Carbon Footprint of a University—The case of NTNU. *J. Clean. Prod.* **2013**, *48*, 39–47. [CrossRef]

16. Canada's Emission Trends. Environment Canada. 2013. Available online: https://www.ec.gc.ca/ges-ghg/985F05FB-4744-4269-8C1A-D443F8A86814/1001Canada\T1\textquoterights%20Emissions%20Trends%202013_e.pdf (accessed on 20 April 2015).

17. NRC. Energy Efficiency Trends in Canada 1990–2013. Available online: https://www.nrcan.gc.ca/sites/www.nrcan.gc.ca/files/energy/pdf/trends2013.pdf (accessed on 7 June 2015).

18. NRC. Consumption of Energy Survey for Universities, Colleges and Hospitals 2003. 2004. Available online: http://oee.nrcan.gc.ca/corporate/statistics/neud/dpa/data_e/consumption03/universities.cfm?attr=0 (accessed on 24 April 2017).

19. NRC. Comprehensive Energy Use Database. Available online: https://oee.nrcan.gc.ca/corporate/statistics/neud/dpa/showTable.cfm?type=CP§or=com&juris=ca&rn=3&page=0 (accessed on 8 December 2017).

20. United Nations (UN). *Stockholm Declaration on the Human Environment*; United Nations: Stockholm, Sweden, 1972; Volume 1650, p. 383.

21. Dlouhá, J.; Barton, A.; Janoušková, S.; Dlouhý, J. Social learning indicators in sustainability-oriented regional learning networks. *J. Clean. Prod.* **2013**, *49*, 64–73. [CrossRef]

22. Klein-Banai, C.; Theis, T.L. Quantitative analysis of factors affecting greenhouse gas emissions at institutions of higher education. *J. Clean. Prod.* **2013**, *48*, 29–38. [CrossRef]

23. Anand, C.K.; Bisaillon, V.; Webster, A.; Amor, B. Integration of sustainable development in higher education—A regional initiative in Quebec (Canada). *J. Clean. Prod.* **2015**, *108*, 916–923. [CrossRef]

24. Lozano, F.J.; Lozano, R. Developing the curriculum for a new Bachelor's degree in Engineering for Sustainable Development. *J. Clean. Prod.* **2014**, *64*, 136–146. [CrossRef]

25. Tenley, M.C.; Chelsea, D.; Jennifer, L.; Laura, B. Developing ecological footprint scenarios on university campuses: A case study of the University of Toronto at Mississauga. *Int. J. Sustain. High. Educ.* **2008**, *9*, 4–20. [CrossRef]

26. Velazquez, L.; Munguia, N.; Platt, A.; Taddei, J. Sustainable university: What can be the matter? *J. Clean. Prod.* **2006**, *14*, 810–819. [CrossRef]

27. Komiyama, H. G8 University Summit: Sapporo Sustainability Declaration (SSD). Available online: https://www.global.hokudai.ac.jp/wp-content/uploads/2017/05/SSD.pdf (accessed on 25 April 2019).

28. United Nations (UN). Transforming Our World: The 2030 Agenda for Sustainable Development. Available online: https://sustainabledevelopment.un.org/content/documents/21252030%20Agenda%20for%20Sustainable%20Development%20web.pdf (accessed on 26 April 2019).

29. Fischer, D.; Jenssen, S.; Tappeser, V. Getting an empirical hold of the sustainable university: A comparative analysis of evaluation frameworks across 12 contemporary sustainability assessment tools. *Assess. Eval. High. Educ.* **2015**, *40*, 785–800. [CrossRef]

30. Scheuer, C.; Keoleian, G.A.; Reppe, P. Life cycle energy and environmental performance of a new university building: Modeling challenges and design implications. *Energy Build.* **2003**, *35*, 1049–1064. [CrossRef]

31. Junnila, S.; Horvath, A.; Guggemos, A. Life-Cycle Assessment of Office Buildings in Europe and the United States. *J. Infrastruct. Syst.* **2006**, *12*, 10–17. [CrossRef]

32. Collinge, W.; Landis, A.; Jones, A.; Schaefer, L.; Bilec, M. Dynamic life cycle assessment: Framework and application to an institutional building. *Int. J. Life Cycle Assess.* **2013**, *18*, 538–552. [CrossRef]

33. Omer, A.M. Energy, environment and sustainable development. *Renew. Sustain. Energy Rev.* **2008**, *12*, 2265–2300. [CrossRef]

34. AASHE. The Sustainability Tracking, Assessment & Rating System. Available online: https://reports.aashe.org/institutions/participants-and-reports/ (accessed on 27 April 2019).

35. Sinha, P.; Schew, W.A.; Sawant, A.; Kolwaite, K.J.; Strode, S.A. Greenhouse Gas Emissions from U.S. Institutions of Higher Education. *J. Air Waste Manag. Assoc.* **2010**, *60*, 568–573. [CrossRef] [PubMed]

36. University of Nebraska-Lincoln (UNL). Greenhouse Gas Emissions. Available online: https://sustainability.unl.edu/greenhouse-gas-emissions (accessed on 27 April 2019).

37. Massachusetts Institute of Technology (MIT). Climate Action Plan Update 2018: Reducing MIT's Emissions. Available online: http://sustainability.mit.edu/sites/default/files/resources/2018-11/2018_ghg_summary_report_handout.pdf (accessed on 21 April 2019).

38. Grindsted, T.S. Sustainable universities—From declarations on sustainability in higher education to national law. *Environ. Econ.* **2011**, *2*, 29–36. [CrossRef]

39. Lozano, R.; Lukman, R.; Lozano, F.J.; Huisingh, D.; Lambrechts, W. Declarations for sustainability in higher education: Becoming better leaders, through addressing the university system. *J. Clean. Prod.* **2013**, *48*, 10–19. [CrossRef]

40. Wright, T. The Evolution of Sustainability Declarations in Higher Education. In *Higher Education and the Challenge of Sustainability: Problematics, Promise, and Practice*; Corcoran, P.B., Wals, A.E.J., Eds.; Springer: Dordrecht, The Netherlands, 2004; pp. 7–19.

41. Wright, T.S.A. Definitions and frameworks for environmental sustainability in higher education. *Int. J. Sustain. High. Educ.* **2002**, *3*, 203–220. [CrossRef]

42. Findler, F.; Schönherr, N.; Lozano, R.; Stacherl, B. Assessing the Impacts of Higher Education Institutions on Sustainable Development—An Analysis of Tools and Indicators. *Sustainability* **2018**, *11*, 59. [CrossRef]

43. Clarkson, R.E.; Samson, K.; Bekessy, S.A. The failure of non-binding declarations to achieve university sustainability: A need for accountability. *Int. J. Sustain. High. Educ.* **2007**, *8*, 301–316. [CrossRef]

44. UBC. Annual Sustainability Report 2013–2014. Available online: https://sustain.ubc.ca/sites/sustain.ubc.ca/files/uploads/CampusSustainability/CS_PDFs/PlansReports/Reports/2013-2014-Annual-Sustainability-Report.pdf (accessed on 10 September 2016).

45. UBC. UBC Vancouver Campus—2007 GHG Emissions Inventory. Available online: https://sustain.ubc.ca/sites/sustain.ubc.ca/files/uploads/CampusSustainability/CS_PDFs/ClimateEnergy/UBCGHGInventory_2007.pdf (accessed on 10 September 2016).

46. UBC. UBC Vancouver Campus—2015 Greenhouse Gas Emissions Inventory. Available online: https://sustain.ubc.ca/sites/sustain.ubc.ca/files/UBCGHGInventory_2015.pdf (accessed on 10 September 2016).

47. Martin, S.; Jucker, R. Educating Earth-literate Leaders. *J. Geogr. High. Educ.* **2005**, *29*, 19–29. [CrossRef]

48. Rees, W.; Wackernagel, M. Urban ecological footprints: Why cities cannot be sustainable—And why they are a key to sustainability. *Environ. Impact Assess. Rev.* **1996**, *16*, 223–248. [CrossRef]

49. Kate, F. Institutional ecological footprint analysis—A case study of the University of Newcastle, Australia. *Int. J. Sustain. High. Educ.* **2001**, *2*, 48–62. [CrossRef]

50. Alghamdi, N.; den Heijer, A.; de Jonge, H. Assessment tools' indicators for sustainability in universities: An analytical overview. *Int. J. Sustain. High. Educ.* **2017**, *18*, 84–115. [CrossRef]

51. Shriberg, M. Institutional assessment tools for sustainability in higher education: Strengths, weaknesses, and implications for practice and theory. *Int. J. Sustain. High. Educ.* **2002**, *3*, 254–270. [CrossRef]

52. Alba-Hidalgo, D.; Benayas del Álamo, J.; Gutiérrez-Pérez, J. Towards a Definition of Environmental Sustainability Evaluation in Higher Education. *High. Educ. Policy* **2018**, *31*, 447–470. [CrossRef]

53. Lozano, R. Incorporation and institutionalization of SD into universities: Breaking through barriers to change. *J. Clean. Prod.* **2006**, *14*, 787–796. [CrossRef]

54. Lozano, R. A tool for a Graphical Assessment of Sustainability in Universities (GASU). *J. Clean. Prod.* **2006**, *14*, 963–972. [CrossRef]

55. Lozano García, F.J.; Kevany, K.; Huisingh, D. Sustainability in higher education: What is happening? *J. Clean. Prod.* **2006**, *14*, 757–760. [CrossRef]

56. Disterheft, A.; Ferreira da Silva Caeiro, S.S.; Ramos, M.R.; de Miranda Azeiteiro, U.M. Environmental Management Systems (EMS) implementation processes and practices in European higher education institutions—Top-down versus participatory approaches. *J. Clean. Prod.* **2012**, *31*, 80–90. [CrossRef]

57. Yuan, X.; Zuo, J.; Huisingh, D. Green Universities in China—What matters? *J. Clean. Prod.* **2013**, *61*, 36–45. [CrossRef]

58. Haider, H.; Sadiq, R.; Tesfamariam, S. Intra-utility performance management model (In-UPM) for the sustainability of small to medium sized water utilities: Conceptualization to development. *J. Clean. Prod.* **2016**, *133*, 777–794. [CrossRef]

59. Wann, D.L.; Schrader, M.P.; Allison, J.A.; McGeorge, K.K. The Inequitable Newpaper Ceverage of Men's and Womens Athletics at Small, Medium, and Large Universities. *J. Sport Soc. Issues* **1998**, *22*, 79–87. [CrossRef]

60. Herremans, I.; Allwright, D.E. *Research: Environmental Management Systems at North American Universities: What Drives Good Performance?* ULSF: Washington, DC, USA, 2000.

61. Haider, H.; Sadiq, R.; Tesfamariam, S. Selecting performance indicators for small and medium sized water utilities: Multi-criteria analysis using ELECTRE method. *Urban Water J.* **2014**. [CrossRef]

62. ECCC. Greenhouse Gas Emissions from Large Facilities Canada. 2014. Available online: https://www.canada.ca/en/environment-climate-change/services/environmental-indicators.html (accessed on 15 February 2018).

63. Piper, J.E. *Operations and Maintenance Manual for Energy Management*; Routledge: Abingdon, UK, 1998.

64. Chung, W.; Hui, Y.V.; Lam, Y.M. Benchmarking the energy efficiency of commercial buildings. *Appl. Energy* **2006**, *83*, 1–14. [CrossRef]

65. AASHE. Dalhousie Univeristy Greenhouse Gas Emissions. Available online: https://stars.aashe.org/institutions/dalhousie-university-ns/report/2015-01-07/OP/air-climate/OP-1/ (accessed on 4 November 2017).

66. AASHE. Sheridan Institute of Technology and Advanced Learning OP-1: Greenhouse Gas Emissions. Available online: https://stars.aashe.org/institutions/sheridan-institute-of-technology-and-advanced-learning-on/report/2015-04-20/OP/air-climate/OP-1/ (accessed on 4 November 2017).

67. Dalhousie_University. Greenhouse Gas (GAS) Inventory Report 2014–2015. Available online: https://www.dal.ca/content/dam/dalhousie/pdf/dept/sustainability/Dalhousie%20GHG%20Inventory%202014-15%20Final.pdf (accessed on 4 November 2017).

68. Sun, H.; Lee, S.E.; Priyadarsini, R.M.T.; Wu, X.; Chia, Y.; Majid, S. Building energy performance benchmaking and simulation under tropical climatic conditions. In Proceedings of the International Workshop on Energy Perofrmance and Environmental Quality of Buildings, Milos Island, Greece, 6–7 July 2006.

69. Birtles, A.B.; Grigg, P. Energy efficiency of buildings: Simple appraisal method. *Build. Serv. Eng. Res. Technol.* **1997**, *18*, 109–114. [CrossRef]

70. Faust, A.-K.; Baranzini, A. The economic performance of Swiss drinking water utilities. *J. Product. Anal.* **2014**, *41*, 383–397. [CrossRef]

71. van Calker, K.J.; Berentsen, P.B.M.; Romero, C.; Giesen, G.W.J.; Huirne, R.B.M. Development and application of a multi-attribute sustainability function for Dutch dairy farming systems. *Ecol. Econ.* **2006**, *57*, 640–658. [CrossRef]

72. Alberta, E. Alberta's Electricity Generation. 2015. Available online: http://www.energy.alberta.ca/Electricity/682.asp (accessed on 4 October 2017).

73. SaskPower. A Powerful Connection. Available online: http://www.saskpower.com/wp-content/uploads/2015-16_SaskPower_annual_report.pdf (accessed on 4 October 2017).

74. Agdas, D.; Srinivasan, R.S.; Frost, K.; Masters, F.J. Energy use assessment of educational buildings: Toward a campus-wide sustainable energy policy. *Sustain. Cities Soc.* **2015**, *17*, 15–21. [CrossRef]

75. Tilton, C.; El Asmar, M. Assessing LEED versus Non-LEED Energy Consumption for a University Campus in North America: A Preliminary Study. In Proceedings of the ICSI 2014, Long Beach, CA, USA, 6–8 November 2014; pp. 1071–1076.

76. BU. What Is LEED. Available online: http://www.bu.edu/sustainability/what-were-doing/green-buildings/leed/%3E%20cited%20on%20November%202018 (accessed on 20 November 2018).

77. Waheed, B.; Khan, F.I.; Veitch, B. Developing a quantitative tool for sustainability assessment of HEIs. *Int. J. Sustain. High. Educ.* **2011**, *12*, 355–368. [CrossRef]

78. Boons, F.A.A.; Baas, L.W. Types of industrial ecology: The problem of coordination. *J. Clean. Prod.* **1997**, *5*, 79–86. [CrossRef]

79. HEEPI. Results of the HEEPI HE Building Energy Benchmarking Initative. 2004. Available online: http://heepi.org.uk/ (accessed on 8 December 2015).

80. Altan, H. Energy efficiency interventions in UK higher education institutions. *Energy Policy* **2010**, *38*, 7722–7731. [CrossRef]

81. Tuladhar, A.; Cheng, P.; Xiao, C.; Siemens, E. An Investigation into the Potential for PVC Reduction in Building Floorings. Available online: https://open.library.ubc.ca/collections/18861/items/1.0108802 (accessed on 20 April 2019).

82. Novotny, V. Water and energy link in the cities of the future—Achieving net zero carbon and pollution emissions footprint. *Water Sci. Technol.* **2011**, *63*, 184–190. [CrossRef]

83. Tiyarattanachai, R.; Hollmann, N.M. Green campus initiative and its impacts on quality of life of stakeholders in green and non-green campus universities. *SpringerPlus* **2016**, *5*, 84. [CrossRef] [PubMed]

84. Bhuiyan, N.; Baghel, A. An overview of continuous improvement: From the past to the present. *Manag. Decis.* **2005**, *43*, 761–771. [CrossRef]

85. Elmuti, D.; Kathawala, Y. An overview of benchmarking process: A tool for continuous improvement and competitive advantage. *Benchmark. Qual. Manag. Technol.* **1997**, *4*, 229–243. [CrossRef]

86. Bereskie, T.; Haider, H.; Rodriguez, M.J.; Sadiq, R. Framework for continuous performance improvement in small drinking water systems. *Sci. Total Environ.* **2017**, *574*, 1405–1414. [CrossRef]

MDPI

St. Alban-Anlage 66

4052 Basel

Switzerland

Tel. +41 61 683 77 34

Fax +41 61 302 89 18

www.mdpi.com

*Sustainability* Editorial Office

E-mail: sustainability@mdpi.com

www.mdpi.com/journal/sustainability